:hbuch!

Hofheinz, W.
VDE-Schriftenreihe Band 114
Schutztechnik mit Isolationsüberwachung
Grundlagen und Anwendungen ungeerdeter IT-Systeme in medizinisch genutzten Räumen, in der Industrie, auf Schiffen, in Elektro- und Schienenfahrzeugen und im Bergbau, DIN EN 61140 (VDE 0140 Teil 1), DIN VDE 0100-410 (VDE 0100 Teil 410), DIN VDE 0100-710 (VDE 0100 Teil 710), DIN VDE 0118-1 (VDE 0118 Teil 1) und andere mit Isolations-Überwachungsgeräten nach DIN EN 61557-8 (VDE 0413 Teil 8)
2003, 277 S., DIN A5, kart.
ISBN 3-8007-2750-1
20,– € / 33,80 sFr*

Hofheinz, W.
VDE-Schriftenreihe Band 117
Elektrische Sicherheit in medizinisch genutzten Bereichen
Normgerechte Stromversorgung und fachgerechte Überprüfung medizinischer elektrischer Geräte
DIN VDE 0100-410 (VDE 0100 Teil 410), DIN VDE 0100-610 (VDE 0100 Teil 610), DIN VDE 0100-710 (VDE 0100 Teil 710), DIN EN 60601-1 (VDE 0750 Teil 1), DIN VDE 0751-1 (VDE 0751 Teil 1)
2005, 172 S., DIN A5, kart.
ISBN 3-8007-2831-1
25,– € / 43,80 sFr*

* = Persönliche VDE-Mitglieder erhalten beim Kauf von Fachbüchern des VDE VERLAGs unter Angabe der Mitgliedsnummer 10 % Rabatt.

Bestellungen über den Buchhandel bzw. direkt beim Verlag. Preisänderung und Irrtum vorbehalten. Es gelten die Liefer- und Zahlungsbedingungen des VDE VERLAGs.

*Weitere Informationen zu unserem Buchprogramm finden Sie unter: **www.vde-verlag.de***

2. Remagener Physiktage 2004

Tagungsband der 2. Remagener Physiktage 2004

Aktuelle Methoden der
Laser- und Medizinphysik

RheinAhrCampus Remagen, 29. September bis 1. Oktober 2004

Herausgegeben von

U. Hartmann
RheinAhrCampus Remagen

M. Kohl-Bareis
RheinAhrCampus Remagen

P. Hering
Stiftung caesar Bonn, Uni Düsseldorf

G. Lonsdale
NEC Research Lab, St. Augustin

J. Bongartz
RheinAhrCampus Remagen

T. M. Buzug
RheinAhrCampus Remagen

VDE VERLAG GMBH • Berlin • Offenbach

Bibliografische Information Der Deutschen Bibliothek
Die Deutsche Bibliothek verzeichnet diese Publikation in der Deutschen Nationalbibliografie; detaillierte bibliografische Daten sind im Internet über http://dnb.ddb.de abrufbar

ISBN 3-8007-2838-9

© 2005   VDE VERLAG GMBH, Berlin und Offenbach
         Bismarckstraße 33, D-10625 Berlin

Alle Rechte vorbehalten (all rights reserved)

Druck und Bindung: „Thomas Müntzer" GmbH, Bad Langensalza   2005-01

Vorwort

Nach einem erfolgreichen Auftakt im Jahre 2002 wurde mit den *Zweiten Remagener Physiktagen*, die vom 29. September bis zum 1. Oktober 2004 am RheinAhrCampus Remagen stattfanden, wieder ein interdisziplinäres Forum für Wissenschaftler aus Hochschulen, Forschungszentren und Industrielaboren geschaffen. Um eine Vielzahl von Repräsentanten aus diesen Institutionen zu erreichen, wurde die Konferenz gemeinsam mit der Stiftung caesar in Bonn und dem Forschungslabor des japanischen Computerunternehmens NEC in St. Augustin organisiert. Auf diese Weise ist es gelungen, mehr als 100 Teilnehmer aus dem gesamten Bundesgebiet und dem benachbarten Ausland zu gewinnen.

Thematische Schwerpunkte der Tagung waren – entsprechend der Fokussierung der Forschungsaktivitäten am RheinAhrCampus – die Medizintechnik und die Lasertechnik. Aktuelle Fragestellungen und Problemlösungen aus diesen Bereichen wurden in Form von Vorträgen und Postern vorgestellt. Besondere Höhepunkte waren die eingeladenen Hauptvorträge von

- Dr. Guy Lonsdale
  (NEC Research Lab)
  zum Thema High Performance Computing in der Medizin,
- PD Dr. Gabriele Lohmann
  (Max-Planck-Institut Leipzig)
  zum Thema Computeranalyse von fMRI-Daten,
- Prof. Dr. Dr. Hans-Florian Zeilhofer
  (Universitätsspital Basel)
  zum Thema Laseranwendung in der Mund-, Kiefer- und Gesichtschirurgie,
- Prof. Dr. Georg Pretzler
  (Universität Düsseldorf)
  zum Thema Teilchenbeschleunigung mit Hochleistungslasern.

Auch für Studenten und Doktoranden gab es Gelegenheit zur Darstellung eigener Arbeiten. Der vorliegende Tagungsband dokumentiert die präsentierten Ergebnisse.

Der Erfahrungs- und Ideenaustausch zwischen Wissenschaftlern aus unterschiedlichen Institutionen und Fachgebieten wurde allgemein als fruchtbar und bereichernd empfunden. Aus den Diskussionen ergaben sich konkrete Projektdefinitionen, so dass die Remagener Physiktage auch zur Netzwerkbildung beitragen. Im Jahre 2006 werden die *Dritten Remagener Physiktage* wieder am RheinAhrCampus stattfinden.

Remagener Physiktage 2004

Organisationskomitee
Prof. Dr. Ulrich Hartmann (Vorsitz, RAC)
Prof. Dr. Jens Bongartz (CAESAR)
Prof. Dr. Thorsten M. Buzug (RAC)
Dr. Jochen Fingberg (NEC)
Prof. Dr. Peter Hering (CAESAR, Uni Düsseldorf)
Dr. Anke Hülster (RAC)
Prof. Dr. Matthias Kohl-Bareis (RAC)
Dr. Guy Lonsdale (NEC)
Prof. Dr. Georg Schmitz (RAC)
Prof. Dr. Thomas Wilhein (RAC)

Tagungssekretariat
Frau Gisela Niedzwetzki
RheinAhrCampus Remagen
Südallee 2
53424 Remagen
Tel.: 02642/932-300
Fax: 02642/932-301
E-Mail: physiktage@rheinahrcampus.de

Danksagung
Als Vorsitzender der Zweiten Remagener Physiktage möchte ich allen, die zum Gelingen der Tagung beigetragen haben, meinen herzlichsten Dank aussprechen. Neben dem Organisationskomitee haben vor allem die folgenden Personen zum Erfolg der Konferenz beigetragen (in alphabetischer Reihenfolge):

Tobias Bildhauer (RAC)
Dr. Michael Böttcher (RAC)
Günter Jungjohann (RAC)
Birgit Lentz (RAC)
Dr. Kerstin Lüdtke-Buzug (RAC)
Volker Luy (RAC)
Daniela Möhren (RAC)
Gisela Niedzwetzki (RAC)
Dr. Jens Uwe Schmidt (NEC)
Dr. Nils Schramm (FZ Jülich)
Dirk Thomsen (RAC)

Prof. Dr. Ulrich Hartmann   Remagen, Dezember 2004

Inhaltsverzeichnis

Bildgebung / Medical Imaging — 1

Multi-Pinhole SPECT in Small Animal Research — 2
N. U. Schramm, C. Lackas, J. Hoppin and H. Halling

T-SPECT: Initial Results of a Novel Method for Small Animal Imaging — 4
C. Lackas, N. U. Schramm, J. W. Hoppin and H. Halling

Pinhole SPECT Calibration — 8
J. Hoppin, C. Lackas, N. U. Schramm and H. Halling

Tomographische Bildrekonstruktion für die hochauflösende Positronen-Emissions-Tomographie (PET) — 12
P. Musmann, N. U. Schramm und S. Weber

Implementation of a Fourier-Rebinning Algorithm for a High-Resolution PET-Scanner — 16
M. Oehler, S. Weber, P. Musmann and T. M. Buzug

Non-Standard Medical Imaging Modalities — 22
T. M. Buzug, D. Holz, M. Kohl-Bareis and G. Schmitz

Thermography-Based Detection of Skin Cancer — 28
S. Schumann, L. Pfaffmann, U. Reinhold, M. Marklewitz, J. Ruhlmann and T. M. Buzug

Simulation von Detektorsystemen für die medizinische Physik — 34
D. Krücker, M. Khodaverdi, J. Perez, H. Herzog und U. Pietrzyk

Multimodale Abbildung der Haut mit hochfrequentem Ultraschall und optischer Kohärenztomographie — 38
M. Vogt, A. Knütel, T. Grünendick, K. Hoffmann, P. Altmeyer und H. Ermert

3D-Reconstruction of Microscopic Translucent Silicate-Based Marine and Freshwater-Organisms — 43
D. Schmitz, R. Herpers, D. Seibt and W. Heiden

Development of a High Resolution Positron Emission Tomograph — 49
S. Weber

Bildverarbeitung / Image Processing — 53

Immersion Square - A Mobile Platform for Immersive Visualization — 54
R. Herpers, F. Hetmann, A. Hau and W. Heiden

An Open-Source System for Radiation Therapy — 60
O. Grünwald, M. Oehler, S. Ostrowitzki, U. Altenburger, C. Haller, J. Ruhlmann and T. M. Buzug

Hardware Accelerated 2D/3D Image Segmentation for Real-Time Medical and Industrial Applications — 66
P. H. Dillinger, J.-F. Vogelbruch, J. Leinen, S. Suslov, M. Ramm, R. Patzak, K. Zwoll, H. Winkler, K. Schwan and H. Halling

Segmentierung individueller Hirnatlanten — 72
G. Wagenknecht, H. Belitz, L. Wischnewski, J. Castellanos, H. Herzog, H.-J. Kaiser, U. Büll und O. Sabri

Brain Tumour Segmentation with Quantitative MRI — 78
S. Wildenburg, H. Neeb and N. J. Shah

Ein semiautomatisches Verfahren zur Segmentierung von Hirnregionen in MRT-Daten — 84
L. Wischnewski, G. Wagenknecht and R. Dillmann

Modellbasierte Segmentierung von MRT-Daten des menschlichen Gehirns — 90
H. Belitz, G. Wagenknecht and H. Müller

Advances of a 3D Noise Suppression Method for MRI Data   96
J. Castellanos, G. Wagenknecht and T. Tolxdorff

Hemodynamic Modeling of Abdominal Aorta Aneurysms – From Imaging to Visualization   102
U. Kose, M. Breeuwer, K. Visser, R. Hoogeveen, F. Laffargue, J.-M. Rouet,
B. M. Wolters, S. de Putter, H. v. d. Bosch and J. Buth

**Dreidimensionale Rekonstruktion zahnmedizinischer Präparate
aus 2D-Schichtaufnahmen und μCT-Scans**   106
C. Bourauel, A. Rahimi, L. Keilig, S. Reimann, A. Jäger und T. M. Buzug

Approaches to Visualization in 3D Coronary MRA   112
U. Blume, M. Stuber, M. Solaiyappan and D. Holz

Robotik / Robotics   119

Robotic-Guided Dental Laser Interventions   120
J. Bongartz, M. Ivanenko, P. Hering and T. M. Buzug

Safety Analyses for Robotics: An Example from Space Technology   126
H. Schäbe

Medical Navigation Based on Various Visualization Methods   131
R. R. Evbatyrov, G. G. Gubaidullin, M. Kunkel and T. E. Reichert

Euler Angles and Quaternions in Robotics   137
G. G. Gubaidullin

Robot, Robotics, Medical Robotics   144
G. G. Gubaidullin

Biomechanik / Biomechanics   149

**Biomechanische Untersuchung der Lage des Widerstandzentrums
des oberen Frontzahnblocks**   150
S. Reimann, C. Bourauel, L. Keilig und A. Jäger

Anwendung der Biomechanik bei der ergonomischen Arbeitsgestaltung   156
R. Ellegast

Zur Biomechanik des Fahrradfahrens   159
T. Bildhauer, O. Schulyzk und U. Hartmann

Ein Verfahren zur Vermessung der Pronation beim Laufen   163
S. Richarz, S. Preissler und U. Hartmann

Neue Messverfahren in der Golfschwunganalyse   165
J. Bongartz, G. Laschinski und U. Hartmann

Biosignalverarbeitung / Bio-Signal Processing   169

**Messung der Komplexität in Herzfrequenz- und
Blutdruckvariabilitätszeitreihen mittels Kompressionsentropie**   170
A. Voss und M. Baumert

**Die Erholungsfunktion des Hörnerven und ihr Einfluss
auf die Anpassung von Cochlear-Implants**   175
A. Morsnowski, G. Pfister und J. Müller-Deile

Sound Investigation for Electronic Artificial Larynx   179
T. M. Buzug and M. Strothjohann

Novel Modulating Deep Brain Stimulation Techniques
based on Real Time Model in the Loop Concepts 185
C. Silex, M. Schiek, N. Hermes, H. Rongen, U. B. Barnikol, C. Hauptmann,
H. J. Freund, V. Sturm and P. A. Tass

Zeitaufgelöste diffuse Nahinfrarot-Reflektometrie zur Bestimmung
der Hirndurchblutung von Schlaganfallpatienten am Krankenbett 187
M. Möller, A. Liebert, H. Wabnitz, J. Steinbrink, R. Macdonald und H. Obrig

Muscle Oxygenation and Blood Content of Muscle at Rest and during
Exercise Assessed by Near-Infrared Spectroscopy 193
O. Rohm, C. Andre, R. Gürtler, P. Neary, M. Kohl-Bareis

System for the Measurement of Blood Flow and Oxygenation in
Tissue Applied to Neurovascular Coupling in Brain 198
M. Kohl-Bareis, R. Gürtler, U. Lindauer, C. Leithner, H. Sellien, G. Royl and U. Dirnagl

Forensik / Forensics 205

Forensic Facial Reconstruction 206
T. M. Buzug, D. Thomsen and J. Müller

3D Warping for Forensic Soft-Tissue Reconstruction 212
J. Müller, A. Mang and T. M. Buzug

Gradient Vector Flow Based Active Contours for Facial Reconstruction 218
A. Mang, J. Müller and T. M. Buzug

**Forensische Schussrekonstruktion mit 3-dimensionalen individuell angepassten
Opfer- und Tätermodellen im CAD-generierten 3D Tatort** 224
J. Subke

Zerstörungsfreie Materialprüfung / Non-Destructive Testing 231

Micro-CT in der Archäologie – Untersuchung eines Neandertalerzahnes 232
T. M. Buzug und A. von Berg

Development of a C-Arm-Based Meso-CT for NDT and Educational Purposes 238
S. Schneider, B. Bruckschen and T. M. Buzug

X-Ray Based NDT of Accumulator Membranes 244
D. Thomsen, T. Kurz, R. Kaesler and T. M. Buzug

Lasermesstechnik / Laser-Measurement Technology 247

**Bestimmung der optischen Eigenschaften und Farbwirkung
von zahnfarbenen Füllungsmaterialien** 248
K. Weniger und G. Müller

**Ultraschallübertragung über Quarzglasfasern:
Endoskopische Gewebeentfernung in der Neurochirurgie** 254
K. Liebold

Konzept eines neuartigen Miniatur-Scanners für kleinste Hohlräume 256
K. Stock, M. Müller und R. Hibst

No-Motion OCT-Verfahren 260
E. Koch, A. Popp, P. Koch and D. Boller

**Videofluoreszenz-Mikroskopie für die Beobachtung von
Oberflächenkoronargefäßen am isoliert schlagenden Herzen** 266
A. Popp, S. Stehr, A. Deußen und E. Koch

Ultraschnelle holografische Gesichtsprofilvermessung mit vollautomatischer Hologrammentwicklung N. Ladrière, S. Frey, A. Thelen, S. Hirsch, J. Bongartz, D. Giel und P. Hering	272
Laserosteotomie mit gepulsten CO_2-Lasern M. Werner, S. Afilal, M. Ivanenko, M. Klasing und P. Hering	275
Acoustic Monitoring of Bone Ablation Using Pulsed CO_2 Lasers A. Rätzer-Scheibe, M. Klasing, M. Werner, M. Ivanenko and P. Hering,	281

Lasertechnik / Laser Technology   287

Prozessüberwachung beim Laserreinigen und Laserschichtabtrag durch Anwendung der Plasmaanalyse mittels "low resolution" Spektrometer M. Lentjes, K. Dickmann und J. Meijer	288
Untersuchung und Manipulierbarkeit der Photophysik einzelner Fluorophore mit Hilfe der Zwei-Farben Einzelmolekülspektroskopie R. Kasper, M. Heilemann, P. Tinnefeld und M. Sauer	294
Druckwellen in Gewebe bei Ablation mit ultrakurzen Laserpulsen – Modellmessungen in Wasser J. Eichler, K. Beop-Min, L. Dünkel, C. Schneeweiss und L. Cibik	299
New Spatial Resolving Detector for Infrared Laser Radiation P. Kohns, M. Oehler, T. M. Buzug, G. Kokodi and V. Kuzmihov	300
Realisierung eines monochromatischen Laser-Doppler Profilsensors unter Verwendung von Frequenzmultiplexing P. Pfeiffer, T. Pfister, L. Büttner, K. Shirai und J. Czarske	306

Autorenverzeichnis / Author Index   311

Bildgebung

Medical Imaging

Multi-Pinhole SPECT in Small Animal Research

N. U. Schramm, C. Lackas, J. Hoppin, H. Halling
ZEL, Research Center Jülich, Leo-Brandt-Straße, 52425 Jülich, Germany

Abstract

In small-animal SPECT high resolution is typically achieved using pinhole collimation. In order to improve the sensitivity of single-pinhole systems we employ a novel collimation approach called multi-pinhole imaging. In this contribution we report on the performance of a dual-headed multi-pinhole SPECT. The system is characterized by a series of phantom measurements and tested on numerous animal studies. We will show that the system yields excellent image quality with a reconstructed resolution of 1.2mm and a sensitivity of up to 1600cps/MBq. In addition to regular semi-quantitative single-isotope studies, we will present data on dual-isotope imaging, absolute tracer quantification and the fusion of the SPECT images with MR data of the same animal.

1 Introduction

Pinhole SPECT employing conventional gamma cameras is known to provide high resolution and excellent image quality when the object is small compared to the available detector area. Thus, it can be very useful in pharmaceutical and pre-clinical research where new radio-pharmaceuticals have to be tested in small animal studies. Pinhole SPECT has also become a valuable tool in the emerging field of molecular imaging where small animal imaging experiments are carried out to study biochemical pathways and biological mechanisms or disorders on a more fundamental, i.e. molecular level. However, due to the poor geometric efficiency of the high-resolution collimator the sensitivity in conventional single-pinhole tomography is limited. In order to improve the system sensitivity without degrading spatial resolution we have developed and tested a multi-pinhole collimation technique for high-resolution and high-sensitivity imaging of mice and rats.

This novel collimation technique is an extension of single-pinhole tomography to more than one pinhole per collimator. The pinholes are typically arranged in a way that each pinhole views only a certain part of the object with all pinholes together covering the whole field of view. To accomplish this the pinholes are tilted in the axial and transaxial direction. Another important feature of multi-pinhole imaging is that projections through different pinholes may overlap partially on the detector resulting in a more efficient coverage of the detector area and thus, leading to a considerable increase in system sensitivity. Compared to single-pinhole tomography the multi-pinhole system exhibits a more homogeneous sampling of object space and is less susceptible to inconsistencies arising from incomplete sampling of projection space and the violation of the data sufficiency condition.

In this contribution we report on the performance of two multi-pinhole imaging systems: a dual-headed Siemens ECAM and a triple-headed Trionix TRIAD. The big-headed ECAM being upgraded with two 10-pinhole collimators, while the medium-sized detectors of the TRIAD were equipped with three 7-pinhole apertures. Both systems were characterized by a series of phantom measurements and tested on a variety of animal studies.

2 Materials & Methods

The multi-pinhole collimator designs are based on conventional collimator frames and consist of a 12mm pyramidal lead shielding and an interchangeable multi-pinhole aperture made of 10mm tungsten alloy (HPM1850). The collimator depth, i.e. the distance from the image plane to the aperture plane, totals 140mm. The conical pinholes have an inner diameter of 1 to 2mm and an acceptance angle between 40 to 60° depending on the application. For mouse studies both systems are operated at a radius of rotation of 30 to 35mm with corresponding magnifications between 4.7 and 4. A rat is typically measured at a ROR of 40 to 45mm yielding magnification values between 3.5 and 3.1. The overlap fraction in the multi-pinhole projections typically ranges from 40 to 60%.

High-resolution pinhole imaging requires the exact knowledge of the geometrical imaging parameters, i.e. the collimator depth, the radius of rotation, the center of rotation shift, and the aperture offset in the axial and the transaxial direction. In order to determine these parameters we have developed a calibration method based on the SPECT acquisition of a three-point source phantom. This calibration estimates the geometric parameters to within ±0.2mm and is typically repeated every 6 to 12 month.

The tomographical image reconstruction is carried out using a dedicated multi-pinhole algorithm. This iterative reconstruction is based on the maximum likelihood approach that solves the set of linear equations connecting the unknown activity distribution with the measured projections. The entries of the system matrix are pre-calculated numerically using a dedicated ray-tracing technique, thus, physical effects such as aperture penetration at the pinhole edges or varying photon transmission and absorption at different angles of incidence are incorporated in the system model. To speed up the reconstruction we have implemented the method of ordered subsets. The reconstruction is controlled by a robust and easy-to-use graphical interface and is integrated into the clinical network environment via DICOM and Interfile data exchange.

3   Results & Discussion

The imaging capabilities of the two multi-pinhole imagers were tested thoroughly on measured phantom and animal data. It was found that both systems yield excellent image quality with a reconstructed resolution of 1.2mm (**see Fig. 1**) an average sensitivity of up to 1600cps/MBq. Depending on the tracer used in the mouse studies small anatomical details such as 2mm tumors, cortex versus medulla of the kidneys, the mouse striatum or small bone structures were clearly resolved. Even at very low doses (< 50µCi) and poor counting statistics (< 5kcts/view) the multi-pinhole systems provided remarkably good image quality. The acquisition time for a complete SPECT study ranges from 5 to 15min. while the reconstruction times vary from 5 to 10min. depending on the number of pinholes per collimator.

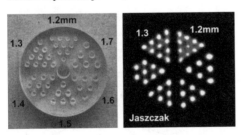

Figure 1 **Multi-pinhole SPECT of a micro Jaszczak phantom providing 1.2 mm resolution.**

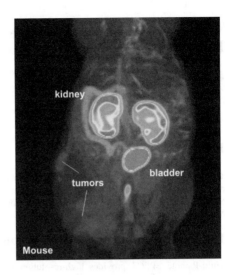

Figure 2 **3D image fusion of a high-resolution SPECT study with MR data of the same animal**

We have also investigated the possibility of performing dual-isotope imaging with Tc-99m and In-111. Both isotopes are imaged simultaneously and their corresponding activity distributions are separated using energy discrimination. Further, as many imaging applications require the knowledge of the absolute tracer concentrations (µCi/ml) we have studied quantification capabilities of our multi-pinhole imagers. It was found that absolute activity quantification in small animals is feasible and that the error typically amounts to less than 3%. Finally, we have investigated the fusing of high-resolution SPECT data with MR images of the same animal. The MR data was acquired on a clinical 1 Tesla MR system using a miniature receive-only RF coil and a double-echo imaging sequence. Image fusion was carried out by a semi-automatic procedure employing a mutual information algorithm (**see Fig. 2**).

T-SPECT: Initial Results Of A Novel Method For Small Animal Imaging

Christian Lackas, Nils U. Schramm, John W. Hoppin, Horst Halling,
Forschungszentrum Jülich, Jülich, Deutschland

Kurzfassung

T-SPECT is a novel imaging technique based on a simple mechanical setup using stationary detectors, while the object is translatory moved in their field of view. Thus, the method is called Translatory SPECT or short T-SPECT. It provides high resolution and high sensitivity by using multi-pinhole collimators and a dedicated MLEM based reconstruction algorithm.

1 Background and Introduction

Multi-Pinhole SPECT provides high-resolution and high-sensitivity imaging for small animal research. While each pinhole views only a portion of the object, the set of pinholes covers the entire field of view (FOV). The axes of the pinholes (see figure 1(b)) are tilted in the axial and transaxial directions to focus on different parts of the object. For reasonable radii of rotation (~50 mm) and FOV (~40x40x90 mm) we are able to achieve sensitivities of up to 350 cps/MBq using a 7-pinhole aperture with diameters of 1.5 mm. The increase in sensitivity provides us with better statistics and allows for shorter measurement periods and/or smaller amounts of radioactivity. In figure 1(a) we display a schematic diagram of our collimator. Due to magnification in our pinhole system, we are able to achieve reconstructed spatial resolutions as low as 1.5 mm.

For typical imaging setups using a multi-pinhole aperture, individual projections corresponding to each pinhole overlap on the detector. This overlapping or multiplexing (see figure 3) allows for more efficient use of the detector surface. Too much overlap, however, introduces a loss of uniqueness which in turn results in a deterioration of the underlying system matrix and the resultant reconstructions. Typically, the total overlap of a 7-pinhole aperture ranges between 15 and 60 %.

Conventional SPECT systems require the rotation of (often heavy) detectors around the examined object. Such a rotation requires a complex mechanical setup in order to position the detectors with the necessary precision. One alternative is to rotate the animal rather than the detector, an approach which can lead to tissue displacement and physiological stress. We propose a novel high-resolution and high-sensitivity imaging technique using two stationary detectors arranged perpendicularly to one another (see figure 1 (c)). Rather than rotate the object or the detector(s), we simply translate the object through the field of view. We call our new approach Translatory SPECT or T-SPECT. In this poster we describe T-SPECT and present preliminary results. We also compare the performance of T-SPECT with rotation SPECT (R-SPECT).

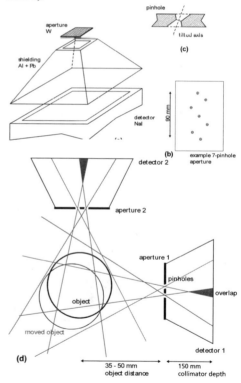

Figure 1 *A collimator used consists of a detector with shielding (a) and aperture (b) with one or more pinholes (c). An example setup of T-SPECT (d) using two detectors mounted perpendicular to one another. The object is translated through the FOV while the detectors remain stationary.*

Figure 2 *The object is moved in the field of view of one or more detectors (a), no rotation is required. A variety of detector arrangements (b) are possible using T-SPECT.*

2 T-SPECT System

We move the object through the field of view using three orthogonal linear axes (see figure 2(a)). The acceleration of the motion is minimal, thus tissue displacement is negligible. As a result of the strong position dependency of the point spread function (PSF) for a pinhole imaging system combined with the tilted axes of the pinholes, translation of the object through the field of view provides projections suitable for 3D reconstructions. Projections are acquired at various locations throughout the field of view using a step and shoot approach. Each projection is saved along with the corresponding location and relative angle. The number of projections used in a T-SPECT study is comparable to R-SPECT. Thus the measurement time is approximately the same.

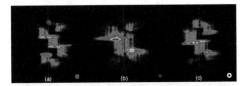

Figure 3 *Projections of a 40 mm Jaszczak phantom, through a 7-pinhole aperture, at three different positions (see colorcoded locations in figure 6). Projection (b) is closest to the first aperture and shows the greatest magnification. The overlap in this example varies from 5% to 20%.*

We reconstruct the object using multi-pinhole projections via Ordered Subset Expectation Maximization (OSEM). A dedicated physical model based on a ray-tracing technique has been developed to calculate the PSF for each voxel and pinhole. The reconstruction program operates on arbitrary multi-pinhole projections. Therefore, conventional R-SPECT can be viewed as a special case of the more general T-SPECT. Namely, R-SPECT has a fixed object location and varying angles while T-SPECT can vary both object and angle position.

Figure 3 shows the projections of a Jaszczak phantom at three different positions. The PSF changes significantly as a function of the distance from the pinhole and the angle of incidence. In particular, the magnification of the imaging system and the sensitivity levels change substantially for different locations in object space. The information gained by moving the object through the field of view of two orthogonal detectors provide sufficient information for reconstruction even though we do not rotate the detector. The performance of the system can be enhanced by using additional detectors (figure 2) or different setups. But we will show that two detectors are sufficient to perform high-resolution imaging.

3 Simulation Studies and Setup

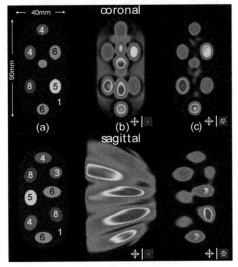

Figure 4 *A comparison of reconstruction results for a hot spot phantom (a) using a single detector and a 1-pinhole (b) or 7-pinhole (c) aperture.*

In this section we present the results of a small simulation study using a hot-spot phantom, containing 12 hot spots with various activities. The purpose of the study is to display the capabilities of T-SPECT when using 1 or 7 pinhole(s) and only one detector. When imaging with only a single detector, the 1-pinhole aperture lacks depth information. We show the results of such a setup in figure 4(b). We are able to distinguish the hot spots in the coronal slice, though "extra" hot spots are present due to the strong distortions visible in the sagittal slice. In the 7-pinhole case (figure 4(c)) a dramatic increase in depth information is introduced as a result of the tilted pinholes. We see that T-SPECT is able to distinguish the hot spots in both the coronal and sagittal slices in this case. The reason this distinction is possible is the capability of multi-pinhole apertures to view the object from different angles with only one detector. This simulation displays the strength of T-SPECT as compared to simple planar imaging performed with a similar setup.

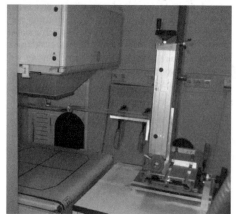

Figure 5   A picture of the T-SPECT setup using a commercial gamma camera equipped with a multi-pinhole detector and a 3D translation stage. The two detector setup was simulated by moving the detector.

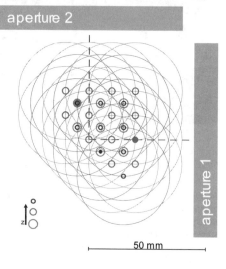

Figure 6   A diagram of the path used for reconstruction. The smaller circles represent the location of the center of the phantom while the larger circles correspond to the outer edge.

Imaging was performed using a translation table with three orthogonal axes, a seven pinhole detector placed in two perpendicular positions. The diameter of the pinholes was 2 mm and the collimator depth was 150 mm. The imaging path consisted of 30 stops located at three different heights and is displayed in figure 6. Example projections acquired using this system are displayed in figure 3.

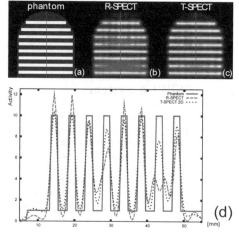

Figure 7   Additional simulation results showing a coronal slice of a Defrise phantom (a) with eight 2 mm slices of higher activity and 3 mm gaps in-between. Due to the axial movement in T-SPECT (c) axial structes are better reconstructed compared to R-SPECT (b). Profiles along the center of rotation are shown in (d).

4   Results

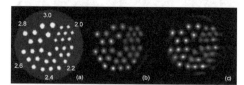

Figure 8   A transversal slice of a 40 mm Jaszczak-Phantom (a) imaged and reconstructed using R-SPECT with a radius of 50 mm (b) and a T-SPECT reconstruction using the path shown in figure 6. The rod diameters ranged from 2.0 to 3.0 mm. Both images were reconstructed using 60 projections. The two stationary detectors were located to the right and above the phantom as shown in figure 6.

In order to demontrate the viability of the T-SPECT imaging approach we performed studies of hot rod phantoms. In figure 8 we present reconstruction results of a 40 mm Jaszczak phantom imaged with a commercial gamma camera. Along with the 2x30 projections (1 minute in length) acquired in T-SPECT mode we also collected 60 conventional R-SPECT projections (30 seconds) for the sake of comparison. As shown in figure 8 we are able to reconstruct even the 2 mm rods of a Jaszczak phantom.

Furthermore, we have shown that the performance of T-SPECT is compareable to that of R-SPECT for this study. Note that the reconstructed resolution varies as a function of the distance to the detectors.

5 Conclusions and Future Work

T-SPECT is a novel multi-pinhole acquisition and reconstruction technique capable of simplifying the mechanical design of an imaging system, while still providing the performance needed for high-resolution small animal studies. Thus, it is a cost-efficient alternative to conventional R-SPECT.

Additionally it can be used for environments where the object is not easily accessable from all sides, such as human thyroid imaging or mammography. Notice that any combination of object and/or detector movement is allowed, all that is required is the information describing their relative position.

Given that T-SPECT is a generalization of R-SPECT, combinations of translation and rotation are possible. For instance, additional projection information can be obtained using an object holder that tilts the object. Such tilts allow the object to be viewed from additional directions without the disadvantages of rotation. We are also interested in combining R-SPECT and T-SPECT to allow for an arbitrary path partially or completely around the object with varying rotation radii optimized with regard for a specific object.

Pinhole SPECT Calibration

Jack Hoppin, Christian Lackas, Nils Schramm and Horst Halling
Zentralinstitut fuer Elektronik
Forschungszentrum Juelich GmbH
Leo-Brandt-Strasse
D-52425 Juelich

Abstract

In this work we discuss current approaches to SPECT calibration along with a new technique we have developed for calibrating our multi-pinhole SPECT systems. This new technique was developed in response to a need for more accurate estimates of system parameters as we image with increasing resolution, while maintaining a simple calibration procedure. The technique is described in full and results are shown indicating its success.

1 Introduction

The popularity of small-animal imaging as a research tool amongst biological scientists has increased dramatically in recent years with the development of higher resolution imaging systems. Specifically, nuclear medicine techniques such as SPECT and PET have provided researchers the ability to image biological function in small animals with sub-millimeter resolution. As we attempt to provide better and better resolution we have found that a major limiting factor in reconstructed resolution is the calibration of the imaging system. In this work, we present a new technique for calibrating our multi-pinhole SPECT systems.

Our multi-pinhole approach to SPECT imaging consists of upgrading commercial gamma cameras with specially designed collimators and multi-pinhole apertures [1,2]. We currently work with five such systems. We employ a reconstruction approach that models the point spread function (PSF) using a dedicated ray-tracing approach. Our approach requires five calibration parameters that define the system. In Figure 1 we present a schematic diagram defining these five parameters. The performance of our reconstruction algorithm is heavily influenced by even minor changes in these parameters, as will be shown in the next section.

We have developed a technique using a collimated beam source and a point-source phantom that accurately estimates these parameters. Our technique provides the user with a straightforward, robust and accurate approach to estimating the parameters required for an image reconstruction.

2 The Calibration Problem

As discussed in the introduction, even minor changes in the values of the estimated parameters of the imaging system have dramatic impacts on quality of the reconstructed images. In Figure 2 we present the results of a simulation study performed using a Jaszczak phantom in which we performed numerous reconstructions of the phantom with varying values of py. The figure displays the incredible sensitivity of the reconstruction to the value of pinhole shifts less than half a millimeter. As a result of such sensitivity, it is absolutely crucial to estimate these parameters accurately.

Many researchers have put forth techniques for estimating these parameters. The simplest approach is to reconstruct a phantom with known geometry using a wide range of parameters values. This approach, while very time consuming, is often very successful, especially when only one camera is used. Another approach to calibration is to measure the PSF directly by stepping a point source through object space. This

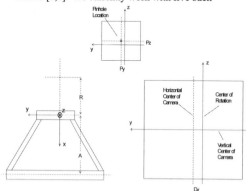

Figure 1 A schematic diagram of our imaging setup showing the five parameters of interest for our image reconstruction: collimator depth (A), radius of rotation (R), pinhole offset (py,pz) and center of rotation offset (dy).

approach is very time consuming, though ideal for cameras without a moving gantry [3,4].
Both of the previous techniques can take many days to perform. Given that the camera must be re-calibrated somewhat regularly it is desirable to have a simpler approach to the problem. Numerous groups

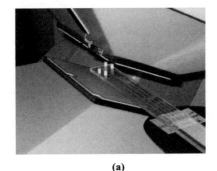

(a)

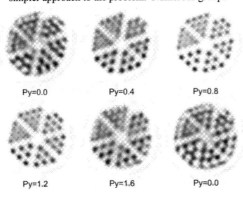

Figure 2 Reconstruction results of a Jaszczak phantom with bore diameters ranging from 1.2 to 1.7 mm. The values of A, R, pz and dy have been held constant while the value of py was varied from 0.0 to 2.0 mm. Notice the dramatic impact on image quality as the value of py is varied by only 2.0 mm.

have put forth attempts to solve this problem using only a static point source phantom or such a phantom in combination with another phantom translated through object space [5,6,7].

The approach of using a point-source phantom consists of first defining a model for a point-source projector for a single-pinhole aperture. Mathematically, such a projector can be represented by,

$$U = \frac{A(py - y)}{R - x} + py - dy$$
$$V = \frac{A(pz - z)}{R - z} + pz \quad (1)$$

where A, R, py, pz and dy are the parameters defined in Figure 1, x, y, and z are the coordinates of the point source, and U and V are the coordinates in the image plane. An example phantom and its projection are shown in Figure 3. Given this model, one can define a simple cost function between the real and model data as a function of the imaging parameters and the parameters defining the location of the phantom. For further detail, the reader is referred to [5].

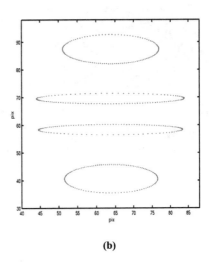

(b)

Figure 3 Our four-point calibration phantom (a) and its projection onto a detector over 120 angles (b).

3 Our Solution

The point-source solution is a proper solution to a well-posed problem and does work in simulation. We have found the approach to be successful when working with one of our cameras. The failings of this approach stems from the inherent difficulty one has separating the highly correlated pinhole offset py and the center of rotation shift dy. This correlation is no surprise given the relationship between py and dy shown in Equation (1). Yet another source of error in estimating these parameters lies in a detector tilt with respect to the axis of rotation. This tilt is also strongly correlated with py and dy, though rarely addressed in the literature.

Given that the point-source phantom approach was not working with some of our cameras we were forced to develop a new method, maintaining our goal of a straightforward calibration that does not require translation of any kind. The key to any calibration technique is the ability of said technique to separate the parameters py and dy. In order to separate these parameters we have introduced a collimated beam measurement in addition to the point-source phantom. In Figure 3 we show a photograph of our collimated beam phantom. The phantom is designed to image the perpendicular projection of the pinhole on the detector. Such a measurement yields the relationship given in the top equation of Equation (2). Using a technique outlined by Noo et. al [7], we are able to estimate the projection of the axis of rotation onto the detector. This added constraint is described and shown graphically in Figure 4.

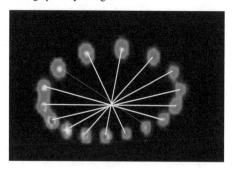

Figure 4 Diagram showing connected pairs of projected point sources separated by 180 degrees. The least-squares estimate of the intersection of these lines represents a point on the projection of the axis of rotation. This diagram yields the bottom equation given in Equation (2).

As discussed, the collimated beam and the estimate of the projection of the axis of rotation provide us with the set of linear equations,

$$py - dy = CB$$
$$py - dy = cont. - py\left(A/R\right)$$
(2)

where CB is the result of the measurement using the collimated beam and cont. is the estimate described in Figure 2. In order to solve this equation we must first estimate A and R using the standard point-source phantom approach. This approach has been shown to provide very reliable estimates of A and R independent of its estimation of py and dy. Thus combining the two approaches gives us accurate estimates of the system parameters.

In Figure 5 we show reconstructions of a Jaszczak phantom imaged with one of our cameras that clearly indicate the improvements gained using our technique of solving for py and dy.

4 Conclusions

We have presented a new technique for calibrating a pinhole SPECT system that combines a measurement with a point-source phantom and a collimated beam to allow for the separation of pinhole and axis of rotation offset. Our new technique has been shown to successfully estimate the parameters we require to perform our reconstructions.

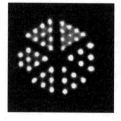 

Figure 5 Reconstructed Jaszczak phantoms using our method (left) and the point-source-phantom technique (right). The approach using only the point-source phantom was unable to properly separate py and dy.

3 References

[1] Schramm, N. U.; Ebel, G.; Engeland, U.; Schurrat, T.; Behe, M.; Behr, T. M.: High-resolution SPECT Using Multipinhole Collimation. IEEE Transactions on Nuclear Science, 50 (2003) 315-320

[2] Lackas, C.; Schramm, N. U.; Hoppin, J. W.; Engeland, U.; Wirrwar, A.; Halling, H.: T-SPECT: A novel imaging technique for small animal imaging. (to appear in IEEE Transactions on Nuclear Science)

[3] Furenlid, L. R.; Wilson, D. W.; Chen, Y.; Kim, H.; Pietraski, P. J.; Crawford, M. J.; Barrett, H. H.: FastSPECT II: A Second-Generation High-Resolution Dynamic SPECT Imager.

[4] Wilson, D. W.; Barrett, H. H.; Furenlid, L. R.: A new design for a SPECT small-animal imager. IEEE Nuclear Science Symposium Conference Record (2001)

[5] Beque, D.; Nuyts, J.; Bormans, G.; Suetens, P. ; Dupont, P. : Characterization of Acquistion Geometry of Pinhole SPECT. IEEE Transactions on Medical Imaging 22 (2003) 599-612

[6] Rizo, Ph.; Grangeat, P.; Guillemand, R. : Geometric Calibration Method for Multiple-Head Cone-Beam SPECT System.

[7] Noo, F.; Clackdoyle, R.; Mennessier, C. ; White, T. A.; Roney, T. J.: Analytic method based on identification of ellipse parameters for scanner calibration in cone beam tomography. Physics in Medicine and Biology 45 (2000) 3489-3508

Tomographische Bildrekonstruktion für die hochauflösende Positronen-Emissions-Tomographie (PET)

Patrick Musmann[1], Nils Schramm[1], Simone Weber[1]
[1]Forschungszentrum Jülich GmbH, Jülich, Deutschland

Kurzfassung

Mit der Positronen-Emissions-Tomographie (PET) lassen sich Stoffwechselfunktionen von Organismen "in vivo" untersuchen. Bei modernen, hoch-sensitiven PET-Geräten, wie dem ClearPET® Neuro, ist dabei der Trend zu beobachten, dass sie sich nicht mehr auf die Erfassung dreidimensionaler Projektionen beschränken, sondern zusätzlich eine Vielzahl von Eigenschaften für jedes Ereignis registrieren (List-Mode Format). Wird dieser Zugewinn an Information im Rekonstruktionprozess berücksichtigt, kann das zu einer deutlichen Verbesserung des Rekonstruktionsergebnisses führen.
Der hier vorgestellte Ansatz für eine List-Mode Rekonstruktion berechnet die Gewichte der Systemmatrix mittels Strahlverfolgung (Raytracing) durch den Scanner. Dadurch wird die Abhängigkeit der koinzidenten Antwortfunktion von der Position des Voxels innerhalb der Koinzidenzröhre und von der Orientierung der Detektormodule modelliert

1 Einleitung

Mit der Positronen-Emissions-Tomographie lassen sich Stoffwechselfunktionen von lebenden Organismen untersuchen. Die gefilterte Rückprojektion (FBP) ggf. mit "Rebinnning"-Methoden [1] kombiniert war dabei lange der Standard in der Bildrekonstruktion, insbesondere bei zweidimensionalen Messungen und im klinischen Anwendungsbereich. Ihr Vorteil liegt dabei primär in der schnellen Rekonstruktion. Allerdings neigt sie auch zur Artefaktbildung, da sie auf der analytischen Inversion rauschfreier, kontinuierlicher Projektionsdaten basiert.
Abhilfe versprechen hier die iterativen Rekonstruktionsverfahren, wie z.B. der "Maximum Likelihood - Expectation Maximization" Algorithmus. Diese algebraischen Verfahren bedürfen zwar eines höheren Rechenaufwandes, bieten aber die Möglichkeit der exakten Modellierung der dem Messvorgang zugrundeliegenden physikalischen Prozesse in der Systemmatrix. Moderne, hoch-sensitive PET-Geräte beschränken sich heute nicht mehr auf die alleinige dreidimensionale Erfassung der Messdaten, sondern registrieren eine Vielzahl von Parametern, wie z.B. Energie und Gantry-Position. Gespeichert werden die Messdaten dabei ereignisorientiert im sogenannten List-Mode Format. Das hat den Vorteil, dass die Messdaten in ihrer höchst-möglichen räumlichen und zeitlichen Auflösung vorliegen. Wird dieser Zugewinn an Information im Rekonstruktionprozess berücksichtigt, führt das zu einer deutlichen Verbesserung der Qualität des rekonstruierten Bildes, gegenüber der Rekonstruktion aus "gebinnten" Sinogrammen.
Allerdings kann eine List-Mode Rekonstruktion, abhängig von der Anzahl der Ereignisse, sehr zeitaufwendig sein. Deshalb ist eine gute Modellierung der Systemmatrix hier um so wichtiger, um die Zahl der nötigen Iterationen so gering wie möglich zu halten.
Der hier für das ClearPET® Neuro vorgestellte Ansatz für einen Rekonstruktionsalgorithmus ist ähnlich den Arbeiten von [2] und [3].

2 Datenbeispiel

Fast alle heutigen iterativen Rekonstruktionsverfahren basieren auf dem "Maximum Likelihood - Expectation Maximization" (MLEM) Algorithmus von Shepp und Vardi [4], der mittels Poisson Sta-tistik den physikalischen Detektionprozess beschreibt.

2.1 MLEM

Der EM-Algorithmus für den ML-Erwartungswert ist gegeben durch:
wobei

$$n_j^{k+1} = \frac{n_j^k}{\sum_{i=1}^{I} a_{ij}} \sum_{i=1}^{I} a_{ij} \frac{m_i}{q_i^k}$$

$$q_i^k = \sum_{j=1}^{J} a_{ij} n_j^k$$

die erwartete Aktivität der "Line of response" (LOR) i ist, wenn die Intensitätsverteilung des Objektes n_j^{k+1} war (bei der k-ten Iteration). Die Systemmatrix a_{ij} gibt dabei die Wahrscheinlichkeit an, dass eine Emission von Voxel j entlang der LOR i detektiert wird. I ist die Anzahl aller möglichen LORs im Scannersystem und J die Anzahl der Voxel im Rekonstruktionsvolumen.

2.2 List-Mode MLEM

Bei der List-Mode Akquisition werden keine Projektionen registriert, sondern nur die Detektorpaare, zwischen denen die koinzidenten Ereignisse aufgetreten sind. Das heißt, der Messwert m_i ist bei den List-Mode Daten für jede LOR nun implizit 1.
Die List-Mode MLEM ist somit gegeben durch [5]:

$$n_j^{k+1} = \frac{n_j^k}{\sum_{i=1}^{I} a_{ij}} \sum_{i=1}^{M} a_{ij} \frac{1}{q_i^k}$$

wobei M die Anzahl der koinzidenten Ereignisse ist.
Ein Nachteil ist, dass der Algorithmus für große M, wie es bei hochsensitiven Geräten, hohen Aktivitäten oder langer Messdauer der Fall ist, relativ langsam arbeitet.

2.2 OSEM

Der "Ordered subset" EM Algorithmus (OSEM) [6] ist eine Erweiterung des MLEM, der die Konvergenz beschleunigt. Hierbei werden die Projektionsdaten in S geordnete Teilmengen (Subsets) sortiert, die dann sukzessiv rekonstruiert werden. Jedes einzelne Subset verbessert somit die Schätzung des Objektvolumens. Eine komplette OSEM Iteration entspricht dabei der einmaligen Verwendung aller Subsets. Eine OSEM Rekonstruktion mit einem Subset (S=1) entspricht der normalen MLEM Rekonstruktion.
Solange die Anzahl der Subsets nicht zu groß ist, d.h. die Anzahl der Projektionen in einem Subset nicht zu klein gewählt wird, erreicht der OSEM Algorithmus eine Reduktion der notwendigen Iterationen um einen Faktor, der der Anzahl der Subsets entspricht.
Eine weitere Verbesserung der Rekonstruktion kann erzielt werden, wenn auch die Reihenfolge der Subsets so variiert wird, dass aufeinander folgende Subsets möglichst unterschiedliche (Winkel-) Informationen enthalten [7].

3 Berechnung der Systemmatrix

Der wichtigste Teil bei der Implementation eines iterativen Rekonstruktionsalgorithmus ist die Modellierung der Systemmatrix, gegeben durch die Koeffizienten a_{ij}.
Für die hochauflösende PET ist es von besonderer Bedeutung, dass die Wahrscheinlichkeit, dass ein im Voxel j emittiertes Photonen-Paar in der LOR i detektiert wird, stark vom Ort des Voxels und von der relativen Lage und Orientierung der Detektormodule zueinander abhängt.

3.1 Voxel-Strahl-Zuordnung

In der Literatur werden verschiedene Verfahren zur Berechnung der Koeffizienten a_{ij} beschrieben (1-0-, Schnittlängen-, Volumengewichtung), die sich in ihrer Genauigkeit und dem erforderlichen Rechenaufwand unterscheiden. Aufgrund der großen Anzahl (mehrere 10 Millionen) der LORs in einem Scannersystem wird bei der List-Mode Rekonstruktion meist ein einfaches Modell verwendet, wie der Siddon Algorithmus [8], der die geometrische Schnittlänge der LOR i mit dem Voxel j berechnet.

3.2 Berechnung der Detektionswahrscheinlichkeit eines Strahls

Zusätzlich zu der geometrischen Gewichtung müssen die intrinsische Antwortfunktionen (IRF) der Detektoren mit berücksichtigt werden. Die Detektionswahrscheinlichkeit P eines einzelnen Strahls hängt dabei von dessen Schnittlänge l mit dem absorbierenden Kristall ab:
wobei μ der lineare Abschwächungskoeffizient des

$$P_{absorption} = 1 - \exp(-\mu l)$$

Kristalls ist. Zusätzlich muss auch die Durchgangswahrscheinlichkeit aller K Kristalle, die der Strahl durchlaufen hat, ohne absorbiert zu werden, berücksichtig werden:
Die koinzidente Antwortfunktion (CRF) eines Detek-

$$P_{transmission} = \exp(-\sum_{k=1}^{K} \mu_k l_k)$$

torpaares ergibt sich dann aus dem Produkt aller beteiligten Transmissions- und Absorptionswahrscheinlichkeiten (vergleiche Bild 1). Im Ortsbereich entspricht das einer Faltung der beiden beteiligten IRFs.

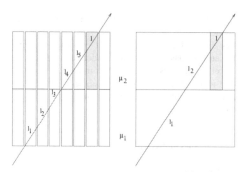

Bild 1: Berechnung der IRF am Beispiel eines Detektorblocks des ClearPET® Neuro. Rechts sind zur einfacheren Berechnung die einzelnen Kristalle zu einem Layer zusammengefasst.

3.3 Raytracing der Systemmatrix

Für die genaue Berechnung der a_{ij} ist es von besonderer Bedeutung, neben der geometrischen Gewichtung und der Detektionswahrscheinlichkeit auch die räumliche Ausdehnung der LOR zu berücksichtigen, d.h die "Line of response" wird zur "Tube of re-sponse" (TOR).

Zur Bestimmung der Voxel-Gewichte a_j für eine feste TOR i, werden nun n zufällige Strahlen durch die beiden an der Koinzidenz beteiligten Kristalle gelegt. Für jeden dieser Strahlen wird zunächst die CRF bestimmt. Dann werden mittels des Siddon Algorithmus die beteiligten Voxel-Indices bestimmt und die geometrischen Schnittlängen mit der CRF und n skaliert. Aus der additiven Überlagerung aller Strahlen ergibt sich die räumliche Verteilung der gesuchten Gewichte für die Vor- und Rückprojektion einer LOR.

Mit einer hinreichenden Anzahl von Strahlen erhält man somit eine Dichteverteilung, die dem statistischen Gewichtungsverfahren [9] unter Berücksichtigung der Detektionswahrscheinlichkeit entspricht.

Ein Vorteil dieses Ansatzes ist es, dass die kostenintensive Berechnung der Gewichte implizit nur für die Voxel geschieht, die auch einen Beitrag zur jeweiligen TOR liefern.

4 Ergebnisse

4.1 Beispiele CRF

Bild 2 (links) zeigt das Ergebnis der Gewichtsberechnung für ein Detektorpaar, deren Achsen parallel zu LOR orientiert sind. Die CRF hängt dabei nur von den Abständen zu den beiden Detektoren und vom Abstand zur X-Achse ab.

Ein wichtiger Nachteil dieser Methode wird ebenfalls deutlich: das Verfahren kann erst dann seine volle Wirkung entfalten, wenn die Voxelbreite kleiner als die Kristallbreite ist, da sonst alle Strahlen die selben Voxel durchstoßen.

Rechts im Bild 2 sieht man den Einfluss gegeneinander geneigter Detektorblöcke auf die Form der Antwortfunktion. Deutlich sichtbar ist die asymmetrische Form der CRF hervorgerufen durch den Parallaxeneffekt und die Transmission vor dem untersuchten Kristall.

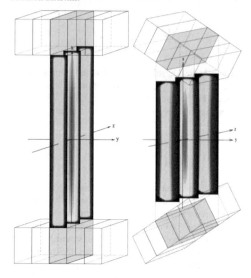

Bild 2: Die CRF eines Detektorpaares mit Detektorachse parallel zur LOR (links) und mit gegenüber der LOR geneigter Detektorachse.

4.2 Simuliertes Phantom

Die Qualität der Rekonstruktion wurde an einem simulierten Phantom getestet. Dazu wurde mit GATE [10] ein homogener Zylinder (Länge 4cm, Radius 1cm) im Zentrum des "Field of view" (FOV) simuliert. In Bild 3 ist oben das Rekonstruktionsergebnis für einen einzelnen Durchlauf des List-Mode Datensatzes mit $n=100$ Strahlen, 60 OSEM Iterationen und einer Voxelgröße von 1mm dargestellt. Das Ergebnis zeigt eine zufriedenstellende Homogenität. Allerdings sind auch noch kleinere Artefakte der Detektorgeometrie zu erkennen.

Zum Vergleich ist unten in Bild 3 eine Rekonstruktion mit nur einem Strahl ($n=1$) dargestellt (der Strahl verläuft dann stets durch die Zentren der Detektoren).

Hier sind deutliche Artefakte zu erkennen; durch die nicht vorhandene räumliche Ausdehnung der Koinzidenzlinie gibt es Bereiche im Objektvolumen, die von den Projektionen nur unzureichend abgedeckt werden.

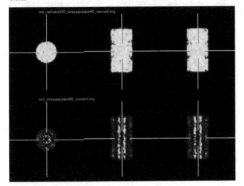

Bild 3: List-Mode Rekonstruktion eines homogenen Zylinders mit 60 Subsets und $n=100$ (oben) bzw. $n=1$ (unten).

5 Diskussion

Die Beispiele der Koinzidenten Antwortfunktionen in 4.1 zeigen, dass das Raytracing Verfahren geeignet ist, die Form und Dichteverteilung der Koinzidenzröhren zu berechnen. Die gewünschte Genauigkeit kann durch die Anzahl der zu berechnenden Strahlen festgelegt werden. Dabei ist jedoch zu beachten, dass die Anzahl der Strahlen hinreichend groß sein muss, um die notwendige Statistik zu gewährleisten.

Der daraus entstehende Nachteil liegt auf der Hand: will man das volle Potential des Verfahrens ausschöpfen, sollte zusätzlich die Voxelgröße klein sein. Somit ergeben sich in der Regel sehr lange Rekonstruktionszeiten. In der Praxis muss daher ein Kompromiss zwischen Genauigkeit und Geschwindigkeit gefunden werden.

Die Rekonstruktion des simulierten Phantoms zeigt, wie erwartet, eine deutliche Verbesserung der Bildqualität unter Berücksichtigung der CRF; bei komplexen Scannergeometrien, wie beim ClearPET® Neuro, lässt sich so die Auflösung im gesamten rekonstruierten Bild verbessern.

Der Wegfall der äußeren MLEM Iteration und die Beschränkung auf reine "innere" OSEM Iterationen, scheint in der Praxis akzeptabel zu sein. Allerdings steht die vollständige Evaluation des Rekonstruktionsverfahrens für den produktiven Einsatz am ClearPET® Neuro noch aus.

6 Literatur

[1] Oehler, M. et al.: Implementation of a "Fourier-Rebinning"-Algorithmus for a high resolution PET Scanner. Konferenzbeitrag f. Remagener Physiktage, 2004

[2] Reader, A. J.; et al.: Regularized One-Pass List-Mode EM Algorithm for High Resolution 3D PET Image Reconstruction into Large Arrays. IEEE Trans. Nucl. Sci. Vol. 49, pp. 693-699, 2002

[3] Levkovitz, R. et al.: The Design and Implementation of COSEM, an Iterative Algorithm for Fully 3-D Listmode Data. IEEE Trans. Med. Imag. Vol. 20, pp. 633-642, 2001

[4] Shepp, L. A.; Vardi, Y.: Maximum likelihood reconstruction for emission tomography. IEEE Trans. Med. Imag. Vol. 1, pp. 113-122, 1982

[5] Barret, H. H. et al.: List-mode likelihood. Journal Opt. Soc. Am. Vol. 14, pp. 2914-2923, 1997

[6] Hudson, H. M.; Larkin, R. S.: Accelerated image reconstruction using ordered subsets of projection data. IEEE Trans. Med. Imag. Vol. 13, pp. 601-609, 1994

[7] Herman, G. T.; Meyer, L. B.: Algebraic Reconstruction Techniques Can Be Made Computationally Efficient. IEEE Trans. Med. Imag. Vol. 12, No. 3, pp. 600-608, 1993

[8] Siddon, R. L.: Fast calculation of the exact radiological path for a three-dimensional CT array. Med. Phys. Vol. 43, pp. 252-255, 1985

[9] Terstegge, A. et al.: High Resolution and Better Quantifikation by Tube of Response Modelling in 3D PET Reconstruction. IEEE Nucl. Sci. Symposium, pp. 1603-1607, 1996

[10] Santin, G. et al.: GATE: a Geant4-based simulation platform for PET and SPECT. IEEE Trans. Nucl. Sci. Vol. 50, pp. 1516ff., 2003

Implementation of a Fourier-Rebinning Algorithm for a High-Resolution PET-Scanner

M. Oehler[1], S. Weber[2], P. Musmann[2] and T. M. Buzug[1]
1) RheinAhrCampus Remagen, Department of Mathematics and Technology, Suedallee 2, D-53424 Remagen
2) Forschungszentrum Jülich, Zentralinstitut für Elektronik, 52425 Jülich

Abstract

At present a high-resolution PET scanner (ClearPET Neuro) is constructed at the Research Centre Jülich. While human PET systems have a spatial resolution of 5-7 mm, high resolution PET needs a resolution of < 2 mm and a high sensitivity which is achieved e.g. by taking data in fully 3D mode. The **Fo**urier-**Re**binning algorithm (FORE) is a procedure that sorts the 3D data into a 2D data set. The 3D image is then recovered by applying a 2D reconstruction method to each slice, such as filtered back projection. FORE is approximate, but allows an efficient implementation, based on the 2D Fourier transformation of the data [1] and speeds up the reconstruction time significantly.

1 Introduction

1.1 Positron-Emission-Tomography (PET)

Positron emission tomography is a non invasive nuclear medical imaging modality which offers the possibility to study the metabolism of a living organism. The concentration and distribution of a radioactive tracer substance on the body is measured and the metabolism is made visible.

At present a new high resolution PET-scanner – the ClearPET Neuro [2] – is constructed at the Research Centre Jülich. The aim of this paper is to present the principle and the results of the Fourier Rebinning (FORE) algorithm implementation for this new PET scanner.

1.2 Sinogram Data of a Cylindrical PET Scanner

Let's have a closer look at a cylindrical scanner with radius R, length L, and the z axis defined as axial direction. Any plane which is normal to the z-axis is called transaxial plane (see Fig. 1).

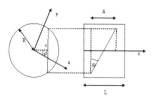

Figure 1: Geometry of a cylindrical PET scanner; left: transaxial view; right: axial view.

The distribution of the radioactivity is given by the function f(x,y,z) defined in a cylindrical *field-of-view* (FOV) Ω with radius $R_\Omega < R$ and with the same axis and length of the scanner. The integral of the function f(x,y,z) is measured via any direct line between two detectors, called *line-of-response* (LOR). Data measured in this way depend on the distance s between the z-axis and the projection of the LOR in the transaxial plane and the angle ϕ between the projection and the y-axis. Such a data representation is called sinogram [1].

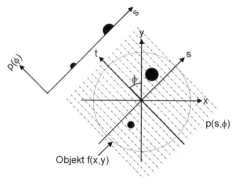

Figure 2: Projection of the distributed radioactivity.

The reconstruction problem is to recover the function f(x,y,z) from these data.

Fig. 2 shows the projection geometry of the distributed activity under one angle for the 2D situation. The ClearPET Neuro has an open septa-less cylindrical geometry, which allows to measure coincidences between all pairs of detectors, not only between detectors which lie within one transaxial plane (2D), such that an entire 3D data set is acquired. In 3D the line integral is characterized by the following parameters:

$$p(s,\phi,z,\delta) =$$
$$\int_{-\infty}^{+\infty} f(\cos\phi - t\sin\phi, s\sin\phi + t\cos\phi, z + t\delta)dt \quad (1)$$

with

- $\delta = \tan\Theta$ (δ is called ring difference, because it is proportional to the difference of the indices between the two rings in coincidence),
- Θ is the angle between the LOR and the transaxial plane,
- the integration variable t is defined along the projection of the LOR in the transaxial plane and no longer along the LOR, and
- z is the axial coordinate which determines the point midway between the two detectors.

The data set which belongs to one fixed data pair (z,δ) is called *oblique* sinogram. If $\delta = 0$ it is called *direct* sinogram.

A PET scanner with N rings in which every pair of ring is allocated by one data pair (z,δ) consists of N^2 sinograms including N direct and N(N-1) oblique sinograms, if a 3D data set is acquired. If the scanner operates in 2D mode (with interslice septa), so that only a coincidence in one transaxial plane is allowed, the data set will consist of only N direct sinograms $(z,0)$. The 2D data set is then described by Eq. (1) but with $\delta = 0$:

$$p_{2D}(s,\phi,z) = p(s,\phi,z,0) \quad (2)$$

2 Rebinning Methods

Usually the 3D data, measured with a multi-ring scanner, are reconstructed by using the 3D reprojection algorithm (3D-RP algorithm) [3]. Due to the high number of measured LOR's in this 3D data set, it is clear that the 3D-RP algorithm takes much more time than a 2D slice by slice filtered backprojektion (FBP), which is used for reconstruction of a 2D data set. To reduce the reconstruction time several options have been tested.

One possibility is to use faster hardware to speed up the implementation of the 3D-RP algorithm, which is in most cases very expensive. Another option is to decrease the sampling rate of the 3D data to reduce the number of LORs which must be reprojected by grouping them.

The third alternative is to 'rebin' the 3D data into a 2D data set. The main concept of rebinning consists in converting the 3D data set into its equivalent stack of 2D data sets. Now this can be reconstructed with less expenditure of time by using every established 2D reconstruction method such as the FBP. Another advantage of rebinnig is the better measuring statistics of the 3D data set in comparison to a 2D data set,

which is quite interesting because of a better signal to noise ratio [1]. In the next sections the different rebinning methods are described.

2.1 Single-Slice Rebinning (SSR)

Single-slice rebinning is the oldest (1987) and easiest rebinning method [4]. In this case the coincidence between two different slices *Z1* and *Z2* is attached to the slice *Z_new* which lies midway between the two slices.

This method makes sense when the measured distribution of the activity is near to the centre of the FOV or when the axial dimension of the tomograph is very small.

For large axial acceptance angles or the measurement of larger activity distributions like the complete brain, the use of other methods should be considered [5] since the SSR leads to some loss in resolution and image quality.

2.2 Multi-Slice Rebinning (MSR)

The multi-slice rebinning [6] is an extension of the single-slice rebinning. Coincidences between two slices are distributed equally among every slice that lies between these two.

This also leads to some axial loss of resolution which increases with a greater axial acceptance angle. In comparison to the SSR it is more accurate but less stable in the presence of noise [7].

2.3 Fourier-Rebinning (FORE)

Both the SSR as well as the MSR yield very inexact redistributions of the 3D data to a 2D data set. For this reason Michel Defrise et al. [1] 1997 developed an exact rebinning algorithm that is shortly described in the next section. For detailed information see [1].

First the 2D-Fourier transformation of the 3D sinogram data (1) is calculated with respect to s and ϕ:

$$P(\omega,k,z,\delta) = \int_{0}^{2\pi}\int_{-R_\Omega}^{R_\Omega} p(s,\phi,z,\delta) \cdot e^{(-ik\phi - i\omega s)} ds d\phi \quad (3)$$

with $\omega \in R$ (the radial frequency) and $k \in Z$ (the Fourier index). The conversion into cylindrical coordinates and a 1D-Fourier transformation leads to

$$\wp(\omega,k,\omega_z,\delta) = \int_{-\infty}^{\infty} P(\omega,k,z,\delta) \cdot e^{-i\omega_z z} dz \quad (4)$$

and further to

$$\wp(\omega, k, \omega_z, \delta) =$$

$$\int_0^{R_\Omega} \int_0^{2\pi} e^{-ik\beta} \int_0^{2\pi} F(\rho, \beta, \omega_z) \cdot e^{(-ik\phi - i\omega_z \rho \cos\phi - i\omega_z \rho\beta\sin\phi)} d\phi d\beta \rho d\rho$$

Here ω_z is the axial frequency and $F(\rho, \beta, \omega_z)$ is the 1D axial Fourier transformation of $f(x, y, z)$ in cylindrical coordinates.

After the Fourier transformation along the z-axis, the variable ϕ is no longer present in the argument $F(\rho, \beta, \omega_z)$. Therefore, the ϕ integral can be factored out. Using the standard integral representation of Bessel functions [8] one has:

$$\int_0^{2\pi} e^{-ik\phi - i\omega\rho\cos\phi - i\omega_z \rho\delta\sin\phi} d\phi$$

$$= e^{-ik \arctan\left(\frac{\delta\omega_z}{\omega}\right)} \cdot 2\pi(-i)^k J_k\left(\rho\omega\sqrt{1+\frac{\delta^2\omega_z^2}{\omega^2}}\right) \quad (5)$$

Finally we get a relation between the 3D-FT of direct and oblique sinograms:

$$\wp(\omega, k, \omega_z, \delta)$$
$$= e^{-ik\arctan\left(\frac{\delta\omega_z}{\omega}\right)} \cdot \wp\left(\omega\sqrt{1+\frac{\delta^2\omega_z^2}{\omega^2}}, k, \omega_z, 0\right) \quad (6)$$

This relation can be called „exact rebinning formula". Based on this formula Defrise et al. developed the approximate Fourier rebinning algorithm, which leads to a much faster implementation as the exact rebinning algorithm does.

The Fourier rebinning equation is an approximation of the exact rebinning formula. It can be obtained by considering the Taylor expansion on the exact rebinning formula.

The zero order approximation of the Taylor expansion is the SSR approximation. The first order is the Fourier rebinning approximation which is given by equation (7):

$$P(\omega, k, z, \delta) \approx P(\omega, k, z - \frac{\delta k}{\omega}, 0) \quad (7)$$

This is the main result from the approximation. Equation (7) shows that the relation between a 2D Fourier transform of an oblique sinogram and a 2D Fourier transform of a direct sinogram is given by a frequency dependent offset:

$$\Delta z = -\frac{k\delta}{\omega} \quad (8)$$

The original derivation of the Fourier rebinning equation (7) was based on the frequency-distance relation discovered by Edholm et al. [9]. It states that the value of P at the frequency (ω, k) of a 2D Fourier transformed sinogram receives contributions mainly from sources located at a set distance $t = -k/\omega$ along the LOR.

The second approximation of the Taylor equation yields information about the accuracy of the approximate Fourier rebinning.

The equation (7) is only valid for high frequencies [1], hence the frequency space (ω, k) is divided into 3 areas and a different method is applied in every region to estimate the 2D sinogram data (2).

Region 1: High-frequency region: Calculation via Fourier rebinning.
Region 2: The consistency condition $R\Omega > |k/\omega|$ is not satisfied and hence the sinogram data are zero in this area.
Region 3: Low-frequency region: calculation via SSR.

3 Implementation of the Fourier-Rebinning Algorithm

Defrise et al. divide the implementation of the FORE-algorithm into 5 steps [1]:

1) Initialise a stack of 2D Fourier transformed sinograms: $P_{2D}(\omega, k, z) = 0$

2) Consider sequentially each pair of oblique sinograms (z, δ) and $(z, -\delta)$. For this step both sinograms were added by using the symmetry: $p(s, \phi + \pi, z, \delta) = p(-s, \phi, z, -\delta)$

   a. If necessary pad in s-direction and ϕ-direction to get the right dimensions for the calculation of the 2D-FFT.
   b. Calculate the 2D-FFT with respect to s and ϕ by using equation (7) to get $P(\omega, k, z, \delta)$. The FFT-algorithm which was used in this work is FFTW [10].
   c. For each sample (ω, k) in region 1
   $$(|k/\omega| < R_\Omega; |k| > k_{\lim} or |\omega| > \omega_{\lim})$$
   calculate: $z_{new} = z - \delta k / \omega$.
   If $|z_{new}| \leq L/2$ calculate the nearest two sampled slices by using linear interpolation
   $z_1 < z_{new} \leq z_2$
   Add $(z_2 - z_{new}) \cdot P(\omega, k, z, \delta)$ to $P_{2D}(\omega, k, z_1)$.

Add $(z_{new} - z_1) \cdot P(\omega, k, z, \delta)$ to $P_{2D}(\omega, k, z_2)$.

d. If $\delta \leq \delta_{lim}$ calculate the samples (ω, k) in region 3 ($|k| \leq k_{lim}$ and $|\omega| \leq \omega_{lim}$)

Add $P(\omega, k, z, \delta)$ to $P_{2D}(\omega, k, z)$.

3) Normalise the rebinned data P_{2D} to take the variable number of contributions into account for each sample ω, k, z. The normalisation factors are calculated by applying the procedure 2c)-2d) to unit data $P(\omega, k, z, \delta) = 1$.

4) Take the inverse 2D-FFT of $P_{2D}(\omega, k, z)$.

5) Reconstruct each slice by 2D FBP.

4 Results of the Image Reconstruction

Generally two reconstruction methods are differentiated: analytical and algebraic (iterative) reconstruction methods. The aim of both is to reconstruct images of the distribution of the activity from the measured projection data (sinograms). A difference between these two methods consists in the fact that the algebraic reconstruction method handles the data as discretised data right from the beginning, while the analytical methods discretise the data after establishing the reconstruction regulation. In this work only the analytical reconstruction methods and from these in detail the FBP [11,12] is used for the reconstruction.

The decision for the FBP is based on the fact that it is the mostly used reconstruction method which is easy to implement. Also it represents, in combination with the FORE-algorithm, an alternative to the 3D-RP algorithm.

For the visualisation of the images a tool named "MPI"-Tool [13] is used. This program offers the possibility to display the three cut planes next to each other. The first image shows a cut through the x-y-plane of the scanner, i.e. a cut orthogonally to the scanner axis (transversal). The second image shows the cut parallel and horizontal to the scanner axis (through the x,-z-plane (coronal)). Image number three shows the cut through the y-z-plane (parallel to the axis of the scanner vertical (sagittal)). Furthermore this tool offers the option to display several reconstructed images among each other which makes them well comparable.

Because the ClearPET Neuro scanner wasn't finished yet, all the reconstructed images are based on GEANT3 [14] simulated 3D data. Figure 3 shows a sequence of images reconstructed from simulated 3D sinogram data of an array consisting of many different line sources with different intensities. At the top of fig. 3 the simulated data were reconstructed by using the 3D-RP algorithm. The second part of fig. 3 displays the reconstruction via the 2D FBP after rebinning the 3D-sinogram data into its equivalent 2D data set by using FORE. The difference between these two reconstructed sequences is demonstrated later in this text.

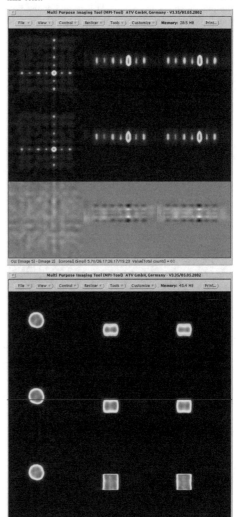

Fig. 3: a) top: 3D-RP, b) middle: FORE+ 2D-FBP and bottom: difference between a) and b).

Altogether, fig. 3 shows very well that the results achieved with FORE in combination with the FBP and those of the 3D-RP algorithm are very well comparable. The advantage of the FORE-algorithm in combination with the 2D FBP consists in a definite saving of time, which constitutes 90% of what is spent for the 3D-RP algorithm.

Considering the difference between SSR, MSR and FORE, always in combination with the 2D FBP, FORE yields the most accurate results of the image reconstruction as reflected in fig. 4.

In this figure simulated 3D sinogram data of a cylindrical tube were used for the different reconstruction methods. The top of fig. 4 displays the SSR+2D-FBP, the middle shows the MSR+2D FBP and at the bottom FORE in combination with the 2D FBP was used for reconstruction.

Only in the last case of the reconstruction with FORE + 2D FBP the cylinder fills the whole length of the scanner, which reflects the reality. The other two cases display the cylinder too short.

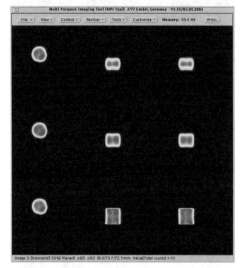

Fig. 4: Top: SSR+2D-FBP, middle: MSR+2D-FBP and bottom: FORE+2D-FBP

5 Discussion

As a first step, the FORE algorithm was only implemented for a cylindrical ring scanner geometry (**fig. 5**). But the real scanner geometry of the ClearPET Neuro shows some differences to a ring scanner geometry (**fig. 6**).

Fig. 5: Ring scanner

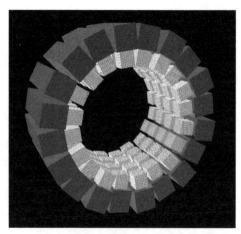

Fig. 6: ClearPET-Neuro (3D)

Fig. 6 illustrates the spaces between the detectors of the scanner along the axial size of the scanner as well as transaxially. To compensate them partially every second detector element inside one ring is pushed back ¼ of a detector width.

These openings within the ClearPet Neuro are causing problems in the form of spaces inside the 3D sinogram data in both directions, axially and transaxially. The 3D sinogram data of three line sources with different activities, which are lying parallel to the scanner axis, shown in fig. 7, are one of the first, and still uncorrected, data acquired with the ClearPET Neuro. The spaces can be clearly seen within the 3D data along the scanner axis, here marked by the orange arrows.

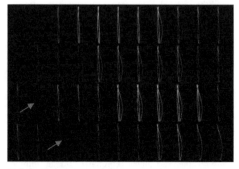

Fig. 7: Segment zero of the 3D-sinogram data set of the three line sources

Using the FORE-algorithm in combination with the 2D FBP on these 3D sinogram data and comparing it to the reconstruction via the 3D-RP algorithm, leads to the following result (see **fig. 8**)

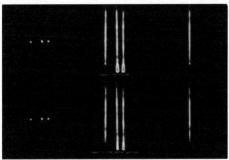

Fig. 8: top: 3D-RP and bottom: FORE+2D-FBP

At the top of fig. 8 the result of the reconstruction of the sinogram data with the 3D RP-algorithm and at the bottom the reconstruction with the FORE-algorithm in combination with the 2D FBP (in the three different views) is displayed. In both cases the artefacts in z-direction are caused by the missing sinogram data (middle of each). Both methods have the problem that the missing sinogram data caused by the spaces within the scanner are still not compensated and so the data have defects. But in a direct comparison the results of FORE + 2D FBP are very similar to the 3D-RP so that this method can be used as a reasonable and fast alternative.

A future task will be the estimation of the missing sinogram data e.g. by making a similar estimation as it is made within the 3D-RP-algorithm, before using the FORE-algorithm.

For the ClearPET Neuro, a listmode based iterative 3D reconstruction is also under development [15] which will model the system matrix precisely. This method will yield high quality images but will be extremely time consuming, so that both the FORE based and the 3D methods will complement each other.

5 References

[1] M. Defrise, P.E. Kinahan, D.W. Townsend, C. Michel, M. Sibomana and D.F Newport, "Exact and Approximate Rebinning Algorithms for 3-D PET Data", IEEE Trans. Med. Imag., vol. 16, 1997, pp.145-158.

[2] S. Weber, "Development of a high resolution positron emission tomograph", Remagener Physiktage, 2004.

[3] P. E. Kinahan, J. G. Rogers, "Analytic 3D image reconstruction using all detected events", IEEE Trans. Nucl. Sci., vol. 36, p. 964, 1989.

[4] M. E. Daube-Witherspoon and G. Muehllehner, "Treatment of axial data in three-dimensional PET", Journal of Nuclear Medicine, vol. 28, pp. 1717-1724, 1987.

[5] P. E. Kinahan and J. S. Karp, „Figures of merit for comparing reconstruction algorithms with a volume-imaging PET scanner, Physics in Medicine and Biology, vol. 39, pp.631-642, 1994.

[6] R. M. Lewitt, G. Muehllehner and J. S. Karp, „Three-dimensional image reconstruction for PET by multi-slice rebinning and axial image filtering", Physics in Medicine and Biology, vol. 39, pp. 321-340, 1994.

[7] M. Defrise, A. Geissbuhler and D. W. Townsend, „A performance study of 3_D reconstruction algorithms for PET", Physics in Medicine and Biology, vol. 39, pp. 305-320, 1994.

[8] I. N. Bronstein, „Taschenbuch der Mathematik", Verlag Harri Deutsch.

[9] P. R. Edholm, R.M. Lewitt and B. Lindholm, "Novel properties of the Fourier Decomposition of the sinogram" in Int.Workshop Physics and Engineering of Computerized Multidimensional Imaging and Processing, Proc. SPIE, 1986, vol.671, pp. 8-18.

[10] http://www.fftw.org/

[11] Th. M. Buzug, „Einführung in die Computertomographie", Springer Verlag, Heidelberg, 2004.

[12] N. Schramm, „Entwicklung eines hochauflösenden Einzelphotonen-Tomographen für kleine Objekte", Bericht des Forschungszentrums Jülich, ISSN 0944-2952, Zentralinstitut für Elektronik Jül-3841.

[13] MPI-Tool, Multi Purpose Imaging; Advanced Tomo Vision Gesellschaft für medizinische Bildverarbeitung mbH, Am Anger 18, Erftstadt.

[14] Application Software Group, Computing and Network Division. GEANT Detector Description and Simulation Tool. CERN, Genf, Schweiz.

[15] P. Musmann, N. Schramm, S. Weber, „Tomographische Bildrekonstruktion für die hochauflösende PET", Remagener Physiktage, 2004.

Non-Standard Medical Imaging Modalities

Thorsten M. Buzug, Dietrich Holz, Matthias Kohl-Bareis and Georg Schmitz
Department of Mathematics and Technology, RheinAhrCampus Remagen, Südallee 2, 53424 Remagen

Abstract

Enormous efforts are made to achieve advances in image quality and acquisition time of today's standard medical imaging modalities as X-ray fluoroscopy, computed tomography, magnetic resonance and ultrasound. However, new principal diagnostic insights cannot be expected from the linear extension of the physical principles of these methods. In this paper a review is given on non-standard imaging modalities that include new physical principles as terahertz and diffuse optical imaging or well known principles – that had been rejected some years ago and are recently raised to interest by detector advances – as thermography. Another class of non-standard modalities is the combination of standard scanners with novel diagnostic agents or different physical concepts as Overhauser imaging, US-based molecular imaging and MR elastography, respectively.

1 Introduction

X-ray fluoroscopy, computed tomography (CT), ultrasound (US) and magnetic resonance imaging (MRI) are today's standard diagnostic imaging modalities. Together with the nuclear imaging techniques Single Photon Emission Computed Tomography (SPECT) and Positron Emission Tomography (PET) insights to morphology and functional behaviour of the human body have been dramatically improved. In the last decades enormous progress in acquisition time and image quality of these modalities has been achieved. However, all the achievements are based on gradual developments. With new X-ray detectors a dose reduction for CT can be realized and there is a strong technology push towards cone-beam geometries. For MRI a paradigm change allows for 3 T scanners and new coding principles in US give a better overall image quality. Although – to some extent – these standard diagnostic tools can also be used to acquire functional information, the modalities are limited by their physical principles and no revolutionary progress can be expected.

In this paper a spot light on the non-standard modalities is given. There are two lines of interest:

- due to recent technology advances in imaging modalities like thermography, terahertz imaging, diffuse optical imaging and impedance tomography a new diagnostic window has been opened,
- the established modalities summarized above can be enhanced with additional physical principles e.g. Overhauser imaging based on hyperpolarized materials in low-field MR, MR elastography or US-based molecular imaging.

These novel principles are functional or molecular imaging opportunities to gain new insight to the human body.

2 Non-Standard Modalities

2.1 Thermography

Undoubtedly, thermography is an effective medical screening modality. Recently, this has been manifested since fevering individuals potentially infected with SARS had been identified in a huge number of persons in airport security areas by their facial temperature profile. However, since its invention it has been conjectured that medical thermography is a functional imaging modality that can be used to visualize pathologically increased metabolism by its temperature signature. Especially for some type of cancer the detection of tumours follow the thermographic paradigm, that the strong growth of malign tumours is necessarily accompanied by an increase of metabolism leading to a conspicuous temperature signature.

Very often thermography is the interpretation of subcutaneous processes from the cutaneous temperature distribution. Thanks to a human skin-emission coefficient of $\varepsilon = 0.98 \pm 0.01$ for $\lambda > 2$ μm this is indeed a promising approach. On the other hand, the human core temperature is held constant for depths larger than 20 mm, and, therefore, the thermographic paradigm holds for near-to-skin processes only. Consequently, medical diagnosis based on thermography can be expected to yield sensible results in cases of inflammation processes, arthrosis, rheumatism, circulatory disturbances, any cases of allergy with skin symptomatic as well as burning and scalding, frost bites and monitoring the success of skin transplantation etc. As an example of a prominent circulatory disturbance Fig. 1 shows the effective cooling of a hand of a person smoking a cigarette. This qualitative example shows the vascular narrowing and the subsequent reduction in blood perfusion associated with smoking. For this image sequence a simple line-array detector scanner (NEC TH5104 Thermo Tracer) is used.

 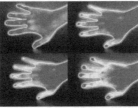

Figure 1: Cigarette-smoke induced vascular narrowing resulting in a cooling down of fingers. The NEC TH5104 Thermo Tracer camera gives an image of 255×223 pixels and is based on a thermo-electrically cooled HgCdTe line-array detector. The camera has a nominal thermal resolution of δT = 100 mK.

As mentioned above, a strong emphasis is given to the detection of malign tumours. Many research groups find out indications for breast cancer. But it seems that the spatial cutaneous temperature signature of a breast tumour can be detected in a late cancer state only. Therefore, the cancer-diagnosis application carried out in our group exclusively focuses on malign skin processes. In the following the instrumentation of a thermographic imaging study on malign melanoma is presented. First clinical trials are carried out with a FLIR SC3000 thermographic camera. The FLIR SC3000 scanner is based on a Sterling-cooled quantum well infrared photodetector (QWIP) with focal-plane-array (FPA) technology. The IR sensitive chip consists of 320 x 240 pixels having a high sensitivity and a nominal thermal resolution of δT = 30 mK.

The QWIP-detection principle can be explained by quantum mechanics. In general, IR photons promote electrons across the interband gap ΔE from the valence to the conduction band producing a photocurrent (see Fig. 2a). By tuning the gap ΔE, the spectral response of the detector can be customized. The material $Hg_{1-x}Cd_xTe$ of the NEC line-array detector introduced above is an example where the gap width can be determined by varying x.

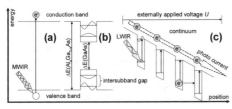

Figure 2: Energy-level system of IR photon detection with QWIP.

However, detection of very long wavelength infrared radiation requires extremely low band-gap materials that are difficult to process [1]. Therefore, intersubband transitions in multi-quantum-well structures (equivalent to the particle in a box problem) made of large bandgap semiconductors are suggested [2]. Fig. 2b shows a schematic energy diagram of a quantum-well situation. The wavelength of the highest response and the cutoff frequency can be customised by the variation of the layer thickness giving the quantum well width and the barrier composition giving the barrier height. The $GaAs/Al_xGa_{1-x}As$ material system allows for a continuous tailoring of these characteristics.

Fig. 2c shows the principle of a so called bound-to-quasibound QWIP. The depth and width of the quantum box can be adjusted that the first excited energy level fits to the well top. Long wavelength infrared radiation (LWIR) can promote electrons from the ground state to the first excited state. An externally applied voltage tilts the overall energy level such that the excited electrons flow down the resulting potential pitch producing a photocurrent [3,4]

$$I \propto \int_{\lambda_1}^{\lambda_2} \varepsilon(\lambda,T)\rho(\lambda,T)d\lambda, \qquad (1)$$

where $\varepsilon(\lambda,T)$ is the emission coefficient and $\rho(\lambda,T)$ is Planck's radiation law. Eq. (1) is the band limited version of the Stefan-Boltzmann law. For a review of QWIP and thermography physics see [4] and [5], respectively.

For the screening of suspicious moles in our study a clinical protocol is proposed that consists of a thermographic sequel after a provocation of the skin area by cooling. Following the thermographic paradigm malign melanoma should compensate the provocation faster than healthy tissue due to enhanced metabolism and an increased vascularization of the tumour.

2.2 T-Ray Imaging

Since a couple of years there is a scientific race to set up practical systems for the generation of terahertz radiation. The frequency band between 0.1-10 THz is a previously unused imaging range that came into focus with advances in T-ray generation and detection technology. However, due to the earlier unavailability of the corresponding wavelengths between 30 μm and 3 mm this band is still called the terahertz gap [6]. All research prototype systems are working very inefficiently yielding an output power in the range of nano to micro watt. For a review of current systems see [7]. Interestingly, the first terahertz quantum cascade laser (demonstrated in 1994) is based on the same technology of periodic layers of two semiconductor materials as explained for the QWIP thermography-detector principle in Fig. 2c. Köhler's laser [8] consists of repetitions of GaAs quantum wells separated by $Al_{0.15}Ga_{0.85}As$ barriers. For broadband T-rays detector systems are based on helium cooled silicon, germanium and InSb bolometers.

However, up to now there are only a few commercially available T-ray systems. Very recently, advances in ultra-fast pulsed laser technology have led to the generation and detection of broad bandwidth terahertz radiation for the first time. This dramatic advance was made possible by applying new concepts in semiconductor physics to these commercially available laser systems [9].

Figure 3: TPI™ probe system (courtesy TeraView Ltd. [9]) equipped with a remote head attached to an articulated arm.

The TeraView system shown in Fig. 3 scans a maximum 25x25 mm area at its highest resolution in approximately 200 seconds.

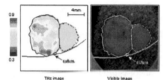

Figure 4: Appearance of basal-cell carcinoma in the T-ray regime [9].

First trials with T-ray imaging focus on oncology and oral health care. Fig. 4 shows that in laboratory experiments healthy and malign skin tissue can be distinguished.

Figure 5: Visible (left) and terahertz absorption image (right) of a tooth decay (marked with an array) taken with a TPI System [9].

For the images shown here the terahertz pulses were generated by femto-second pulses from a RegA 900 laser (Coherent Inc). By scanning the terahertz beam in the x and y planes, an image was obtained from the reflected pulse. In Fig. 5 an example of dental T-ray imaging is given. Dental material types can be differentiated by refractive-index differences revealing early stages of caries in the enamel layers of the human teeth [9].

2.2 Diffuse Optical Imaging

Over the last years a number of methods have been developed based on diffuse optical imaging. The foundation of all these methods is that the absorption coefficient of tissue is low (< 0.02 mm^{-1}) for light in the near-infrared wavelength range between 650 and 1000 nm resulting in a sufficiently high penetration depth and long optical pathlengths. As for this part of the spectrum the scattering coefficient is larger by roughly 2 orders of magnitude compared to the absorption coefficient, the transport of the photons can be described in terms of diffusion processes. The main chromophores giving contrast are the oxygenated and deoxygenated haemoglobin components (oxy-Hb, deoxy-Hb), lipids and water with additional and much smaller contributions from e.g. cytochromes. By and large, the applications proposed are dominated by studies of brain and muscle metabolism as well as breast tumour diagnosis.

Soon after the introduction of blood oxygen level dependent contrast (BOLD) in functional MRI of the brain it was recognized that the underlying neurovascular coupling, i.e. the localised response of flow, volume and oxygenation of blood following a cortical stimulus, can be measured optically based on the different absorption spectra of the haemoglobin components [10].

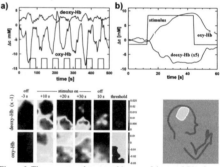

Figure 6: Time course and topographic map of haemoglobin changes during a finger tapping exercise

The example of a functional study [11] shown above is based on a system integrating 16 low power laser diodes at 760 and 850 nm and 8 detectors. Light delivering optical fibers were arranged in a grid over the somatosensory cortex of a volunteer with a distance of 2.5 cm between source and detector fibers. The time courses calculated from changes in attenuation following a finger tapping paradigm emphasize the stimulus correlated increase in oxy-Hb and decrease in deoxy-Hb (a: for seven stimuli; b: average). The different measurement positions allowed a topographic map of the haemoglobin changes to be calculated with a clear localised vascular response in the cortex.

Similar studies in adult volunteers have shown activation maps of i.e. the visual, auditory or frontal cortex [e.g. 12, 13]. Coregistration of the optical haemoglobin signal and the BOLD-contrast in fMRI proved a good correlation [14].

Though the temporal resolution of these measurements is good, the diffuse nature of the photon transport limits the resolution to about 1 cm^2 which is rather poor when compared to fMRI. However, the clear advantage of the optical method is the undemanding instrumentation with the potential of a true bedside monitoring e.g. in stroke monitoring [15].

Most applications of optical detection of brain functions rely on the measurement of changes in light intensity at different wavelengths for the calculation of changes in haemoglobin. Technically more challenging is the measurement of the temporal spread of a short light pulse by the multiple scattering in the tissue which permits an absolute quantification of chromophore concentrations.

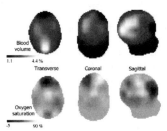

Figure 7: 3D - optical images of a newborn infant.

The ultimate aim of an optical three-dimensional imaging of blood volume and oxygenation is pursue by a few research groups with an example given in Fig. 7 [16]. The images of the head of a newborn infant were generated with a 32-channel time-resolved system based on the analysis of photon time-of-flight with ps-resolution and a finite-element model for image reconstruction. The cerebral haemorrhage located near the left ventricle could be resolved when blood volume and blood oxygen saturation were calculated from data at 780 and 815 nm. This study clearly highlights the 3D - potential of diffuse optical imaging.

2.3 Overhauser Imaging

In the design of standard MRI scanners generally a compromise has to be made between patient comfort and patient accessibility on one side and field strength and homogeneity on the other side. The latter strongly influence image quality and spatial as well as temporal resolution in the images. Therefore the freedom in magnet geometry is limited by field strength.

A possible solution is using the Overhauser effect in MR imaging that allows for high signal to noise ratio at very low field strength. This is achieved by transferring polarization of electrons, which is much stronger than that of protons at a given field strength, to the protons and thereby enhance the proton signal by a factor of up to 660.

The precondition is that electrons are in close vicinity to protons, that can be realized in biological systems by the application of dedicated contrast agents that provide: (1) free electrons as i.e. nitroxide radicals, (2) biochemical stability and (3) no toxicity. In case of weak scalar coupling between protons and electrons the coupled system of electrons and protons has four energy levels as shown in Fig. 8 [17, 18].

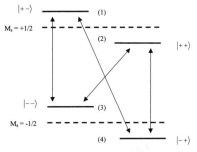

Figure 8: Energy-level diagram of the coupled electron-proton system.

In the ket representation $|+-\rangle$ the first sign refers to the orientation of the magnetic moment of the electron with respect to the external field, the second to the proton. Transition $1 \rightarrow 3$ and $2 \rightarrow 4$ are flips of the electron spin only (EPR transitions, single quantum transition) and $1 \rightarrow 2$ and $3 \rightarrow 4$ are transitions of the proton spin only (NMR transitions, single quantum transition), the transition $1 \rightarrow 4$ and $2 \rightarrow 3$ involve both (zero or double quantum transition). Weak scalar coupling favours transitions $1 \leftrightarrow 4$ and all others are much less probable. Let $W(1 \rightarrow 4)$ and $W(4 \rightarrow 1)$ denote the probabilities for transitions in both directions between levels (1) and (4). Then the number of transitions is given by

$$N_+^p \cdot n_-^e \cdot W(1 \rightarrow 4) = N_-^p \cdot n_+^e \cdot W(4 \rightarrow 1) \quad (2)$$

In thermal equilibrium the probabilities follow Boltzman's law:

$$\frac{W(1 \rightarrow 4)}{W(4 \rightarrow 1)} = e^{\frac{E(1)-E(4)}{kT}} = e^{\hbar \frac{\omega(1 \rightarrow 3)+\omega(1 \rightarrow 2)}{kT}} \quad (3)$$

If the transition $1 \rightarrow 3$ and $2 \rightarrow 4$ are saturated, then $n_e^+ = n_e^-$ and

$$\frac{N_+^p}{N_-^p} = \frac{n_+^e \cdot W(4 \rightarrow 1)}{n_-^e \cdot W(1 \rightarrow 4)} = \frac{W(4 \rightarrow 1)}{W(1 \rightarrow 4)}$$

$$= e^{\hbar \frac{\omega(1 \rightarrow 3)+\omega(1 \rightarrow 2)}{kT}} \approx 1 + \hbar \frac{\omega(1 \rightarrow 3)+\omega(1 \rightarrow 2)}{kT} \quad (4)$$

This means that the difference in the population of the proton levels is considerably increased compared to the situation in standard MRI since the electron resonance frequency exceeds the proton resonance frequency by about 660 times. In case of dipolar interaction similar considerations are valid but the enhancement factor is about 330.

Overhauser enhanced imaging can be considered to consist of a standard MRI sequence preceded by a dedicated spin polarization phase (see Fig. 9).

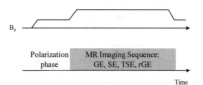

Figure 9: Field cycling and timing in Overhauser Imaging.

Electromagnetic waves with very high frequencies have only a limited penetration depth in biological tissue. This is also true for Overhauser Imaging and requires a reduction of the magnetic field strength during the polarization phase. Our Overhauser system cycles the field strength between 8 mT corresponding to 227 MHz electron resonance frequency during polarization and 15mT corresponding to 625 kHz proton resonance frequency during imaging.

Although this seems to be the solution to many MR problems, one severe difficulty is left: to synthesize a suitable contrast agent that is tolerated by the biological system.

2.4 MR Elastography

Magnetic Resonance Elastography (MRE) is a new imaging technique based on Magnetic Resonance Imaging (MRI) that provides information about elastic properties of body tissue. This information is presented in tomographic images as in standard MRI.

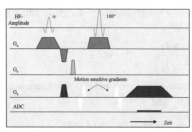

Figure 10: Motion sensitive MR sequence used for MRE.

The principle behind MRE is the sensitivity of MRI to motion such as heart beat and breathing which in many clinical applications is adverse and reduces image quality considerably. However, already in MR angiography, one of today's routine MR applications, the movement of blood is used to depict the vessel system.

In MRE an external mechanical oscillator is used to induce an acoustic wave into the body tissue. The propagation of this wave through the body is monitored by use of a motion sensitized MR sequence depicted in Fig. 10. This type of sequence may be repeated with motion sensitizing gradients applied in all three spatial directions providing the complete vector displacement field [19, 20] of a tissue voxel.

The motion sensitizing gradients encode the displacement of spins in the tissue in the phase of the MR signal. The amplitude of displacement on the other hand depends on the local elastic properties of the tissue. Thus the final MRE image provides information about the spatial distribution of tissue elasticity and its anisotropy.

The most challenging part of the MRE technique is the physical model of wave propagation in biological tissue and the mathematical algorithms to calculate Youngs's modulus and the Poisson's ration from the MR data that depict the displacement field. The relation between the external force applied and the spatial distribution of the mechanical wave is given by

$$\rho \underbrace{\frac{\partial^2 u_i}{\partial t^2}}_{\text{external force}} + \gamma \underbrace{\frac{\partial u_i}{\partial t}}_{\text{attenuation}} = \lambda_{iklm} \frac{\partial^2 u_m}{\partial x_k \partial x_l} \qquad (5)$$

with λ being the elasticity tensor, γ parameter describing the attenuation, u_i the i-th component of the displacement vector and ρ the tissue density.

A clinical application currently under investigation by several groups is MRE applied in mammography [21].

2.5 US-based Molecular Imaging

Specific ultrasound contrast agents based on molecular biology techniques may become an important alternative to MRI-based molecular imaging [22]. To create image contrast, different mechanisms are under discussion: most commercial contrast agents consist of microbubbles and use the strong contrast of the difference in acoustic impedance of a gas-blood interface. These microbubbles have approximately the diameter of red blood cells and therefore some pass the lungs. They are used to enhance the contrast in perfused areas and to increase the signal used for Doppler flow measurements. Additionally, the microbubbles can be destroyed by a strong enough diagnostic ultrasound pulse which allows using bubbles filled with medication for targeted drug delivery under ultrasound imaging surveillance. Alternatively, liposomes are investigated [23]. Here, most likely chemical processes create gaseous inclusions that cause the contrast. Liposomes are smaller than microbubbles and simplify the lung passage as well as the coupling to molecular ligands.

A new sort of contrast medium is created by perfluorocarbon nanoparticles of up to 250 nm diameter [24]. They show a much smaller impedance mismatch to blood and thus a low contrast as long as they are not bound to a target structure. As soon as the particles attach to a target structure by means of certain ligands, they create a continuous layer which shows a clear contrast against blood [24].

Fig. 11 shows an example of a targeted contrast medium for the detection of arteriosclerotic plaque [24, 25]. The binding mechanism is the strong avidin / biotin coupling often used in molecular biology. Biotin is bound to a fibrin antibody which couples to the fibrin found in plaque. When this medium is in circulation it binds to arteriosclerotic plaque.

The avidin is injected and couples to the biotin. In a last step perfluorocarbon nanoparticles coupled to biotin are applied and build a continuous contrast layer on the plaque that can be imaged with ultrasound. So far the detection of contrast media is based only on the image brightness. A more specific detection of contrast media using their frequency dependent scattering would be of great interest and will be the topic of future research.

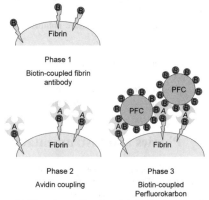

Figure 11: Three phases of imaging arteriosclerotic plaque by a target contrast medium with perfluorocarbon nanoparticles. (from [25]).

3 Conclusions and Perspectives

Currently there is no break through in the diagnostic capabilities of non-standard medical imaging modalities. However, it is obvious that a gain following the established imaging lines will result in relatively small image quality enhancements – simply because it has been achieved so much. Therefore, it seems important to find new imaging principles that will substantially improve the diagnostic insights. Thermography, T-ray imaging and diffuse optical imaging as well as combinatory modalities as Overhauser imaging, magnetic resonance elastography and US-based molecular imaging are candidates, but today any user is faced with serious disadvantages of these imaging modalities concerning e.g. resolution or imaging depth. On the other hand, none of the modalities described in this paper concentrate on better resolution for deeper insights to morphology. All the modalities focus on functional imaging supplementing the standard modalities with additional information.

Acknowledgement

We appreciate the clinical T-ray images supplied by TeraView Ltd, Cambridge UK.

References

[1] S. D. Gunapala and S. V. Bandara, *Quantum Well Infrared Photodetector (QWIP) Focal Plane Array (FPA)*, Semiconductors and Semimetals **62** (1999) 197.

[2] L. Esaki and H. Sakaki, IBM Tech. Disc. Bull. **20** (1977) 2456.

[3] B. F. Levine, J. Appl. Phys. **74** (1993) R1.

[4] S. D. Gunapala et al., *640x512 Pixel Four-Band, Broad-Band, and Narrow-Band Quantum Well Infrared Photodetector Focal Plane Arrays*, preprint.

[5] N. Schuster and V. G. Kolobrodov, *Infrarotthermographie*, Wiley-VCH, 2000.

[6] C. Sirtori, *Bridge for the Terahertz gap*, Nature **417** (2002) 132.

[7] M. Pospiech, *Terahertz Imaging*, Script of the University of Sheffield (2003).

[8] R. Köhler et al., *Terahertz Semiconductor Heterostructure Laser*, Nature **417** (2002) 156.

[9] TeraView Ltd, Cambridge UK, http://www.teraview.co.uk

[10] A. Villringer and B. Chance *Non-invasive optical spectroscopy and imaging of human brain function*, Trends Neurosci. **20** (1997) 435.

[11] M. Kohl et al., *Topographic Imaging of Cortical Activation*. Proc Polish Acad. Sci. 'Laser–Doppler flowmetry and near infrared spectroscopy in medical diagnosis' (2002) 57.

[12] A. Maki et al. *Spatial and temporal analysis of human motor activity using noninvasive NIR topography*. Med. Phys. 22 (1997) 2005.

[13] E. Watanabe et al. *Non-invasive assessment of language dominance with near-infrared spectroscopic mapping*. Neurosci. Lett. **256** (1998) 49.

[14] A. Kleinschmidt et al. *Simultaneous recording of cerebral blood oxygenation changes during human brain activation by magnetic resonance imaging and near-infrared spectroscopy*. J. Cereb. Blood Flow Metab. **16** (1996) 817.

[15] H. Kato et al. *Near-infrared spectroscopic topography as a tool to monitor motor reorganization after hemiparetic stroke: a comparison with functional MRI* Stroke **33** (2002) 2032.

[16] J. Hebden et al., *Three-dimensional optical tomography of the premature infant brain* Phys. Med. Biol. **47** (2002) 4155

[17] C.P. Slichter, *Principles of Magnetic Resonance*, Springer, 1990.

[18] A. Abragam, *Principles of Nuclear Magnetism*. Oxford University Press, 1961.

[19] R. Muthupillai et al, *Magnetic Resonance Imaging of Transverse Acoustic Strain Waves*, MRM, **36** (1996) 266-274.

[20] R. Muthupillai et al, *Magnetic Resonance Elastography by Direct Visualization of Propagating Acoustic Strain Waves*, Science **269** (1995) 1854-1857.

[21] E. Sondermann et al., *MR-Elastographie (MRE) der Mamma: Entwicklung und Validierung des Verfahrens im klinischen Setting*, Deutscher Röntgenkongress, Book of abstracts 2002.

[22] P. A. Dayton., K. W. Ferrara, *Targeted imaging using ultrasound*, Journal of Magnetic Resonance Imaging, vol. **16** (2002) 362-377.

[23] H. Alkan-Onyuksel, S. M. Demos, G. M. Lanza et al., *Development of inherently echogenic liposomes as an ultrasonic contrast agent*, Journal of Pharmacological Science, vol. **85**, (1996) 486-490.

[24] J. N. Marsh, C. S. Hall, M. J. Scott et al., *Improvements in the ultrasonic contrast of targeted perfluorocarbon nanoparticles using an acoustic transmission line model*, IEEE Trans. Ultrason., Ferroelect., Freq. Contr., vol. **49** (2002) 29-38.

[25] C. S. Hall, G. M. Lanza, J. H. Rose et al., *Experimental determination of phase velocity of perfluorocarbons: applications to targeted contrast agents*, IEEE Trans. Ultrason., Ferroelect., Freq. Contr., vol. **47** (2000) 75-84.

Thermography-Based Detection of Skin Cancer

S. Schumann[1], L. Pfaffmann[1], U. Reinhold[2], M. Marklewitz[3], J. Ruhlmann[1)3)] and T. M. Buzug[1)#]

1) Department of Mathematics and Technology, RheinAhrCampus Remagen, Suedallee 2, D-53424 Remagen
2) Dermato-onkologische Praxisklinik Bonn, Friedrich-Breuer-Str. 72-78, D-53225 Bonn-Beuel
3) Medical Center Bonn, Spessartstraße 9, D-53119 Bonn

Abstract

Annually 133.000 people world-wide get sick on malign melanoma, tendency increasing. The purpose of this study is the early diagnosis of malignant skin cancer. At the moment the dermatologists are screening for anomalies at the relevant lesion by examining the skin area with a microscope. To determine changes, another scan has to be taken in a follow-up session after a time period of about 15-20 weeks. Today's visual diagnostic decision is based on the pragmatic ABCD-approach (**A**symmetric, **B**order, **C**olour, and **D**iameter). However, there is no adequate and sound non-invasive way to find out, if a skin spot is either malign or benign. If the visual approach corroborates a suspicion of skin cancer, histology is needed to make explicit diagnosis. To avoid unnecessary surgeries (on false positive alarm) and to initiate necessary surgeries in early stages a new diagnostic screening approach is presented here. Based on the fact that malign melanoma have higher metabolism as well as increased blood flow, it has been conjectured that malign melanoma have slightly higher temperature compared to the healthy skin that can be measured by high resolution thermographic imaging.

1 Introduction

In the last 20 to 30 years the number of patients who were diagnosed with skin cancer has increased dramatically. However, there is no appropriate and sound way to decide non-invasively, if a skin tumour is benign or malignant. Generally, the diagnosis is made with Stolz's traditional ABCD rule of dermatoscopy based on the four main criteria or lesion parameters: **A**symmetric, **B**order, **C**olour and **D**iameter, with a semi-quantitative score system [1,2]. Frequently, this method is improved by computerised scanning methods based on polarised light surface microscopes [3]. With both methods a suspicious spot has to be observed over a period of time to obtain a reliable result, i.e. the evolution of the spot is important which refines the ABCD method into the ABCDE method [2]. However, for the time being the only way to get an accurate diagnostic finding is an invasive histological examination. To avoid redundant excise of tissue and to detect malignant melanoma in an early stage, a new method is proposed here. Thanks to the facts that malignant melanoma have higher consumption of glucose caused by a higher level of metabolism and an augmented branching of blood vessels, the so called angiogenesis, the temperature should be different to its environment, e.g. the surrounding healthy tissue.
Associated with the increased demand of energy, it is conjectured that malignant melanoma shows a higher temperature (2-4 K) than its surrounding skin [4]. Infrared imaging has a high potential to detect the beginnings of angiogenesis, when cancer cells first try to develop their own blood supply, which is a necessary step before they can grow rapidly and metastasize. To substantiate the conjecture an infrared imaging devices of the latest generation is used.

2 Physics and Instrumentation

Thermographic imaging makes use of the infrared spectral band. The physical formulas describing this topic are Planck's law, Wien's displacement law and the Stefan-Boltzmann law. With the Planck's law

$$\rho(v)dv = \frac{8\pi \cdot v^2}{c^3} \frac{hv}{e^{\frac{hv}{kT}}-1} dv \quad (1)$$

the relation between the temperature and the wavelength or frequency, respectively, is described (see figure 1).

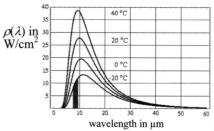

Fig. 1: Spectral distribution of electromagnetic radiation for a so called black body at 4 different temperatures.

Contact: buzug@rheinahrcampus.de, phone: +49 (0) 26 42 – 932 – 318, fax: +49 (0) 26 42 – 932 – 301.

The higher the temperature is, the shorter is the wavelength. Wien's displacement law

$$T\lambda_{max} = 2.898 \text{ mm K} \quad (2)$$

states that there is an inverse relationship between the wavelength of the peak of the emission of a black body and its temperature. The Stefan-Boltzman law helps to calculate the total radiant emittance.
The thermography camera measures and images the emitted infrared radiation from an object. The fact that radiation is a function of object surface temperature makes it possible for the camera to calculate and display temperature variations. This is done via the sensitive Stefan-Boltzman law

$$I \propto \int_{\lambda_1}^{\lambda_2} \varepsilon(\lambda,T)\rho(\lambda,T)d\lambda \quad (3)$$

formulated for a narrow wavelength interval (see figure 1 for illustration of the interval). However, the radiation measured by the camera does not solely depend on the temperature of the object but is also a function of the emissivity $\varepsilon(\lambda,T)$. Radiation captured with the camera may also originate from the surroundings and is reflected from the object of interest. In the application discussed here the human skin-emission coefficient is $\varepsilon = 0.98 \pm 0.01$ for $\lambda > 2$ μm [5], therefore, spurious radiation not generated by the skin can be neglected.

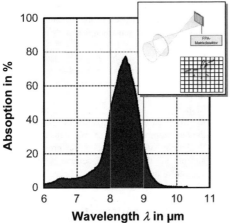

Fig. 2: Sensitivity of the latest FLIR thermographic camera based on QWIP detector technology [6].

The small integration bandwidth is achieved by a so called quantum well infrared photodetector (QWIP) of the latest focal plane array (FPA, see figure 2) thermography camera SC 3000 of FLIR.
While the physical principles of infrared imaging are clear there are a variety of problems one has to cope with in medical applications of thermographic imaging. Typical sources of errors in diagnostic imaging are given in the following list.

o Complexity of an exact model of the medical thermo-regulation process. Even the simple bio-heat equation of Pennes [7]

$$\rho C \frac{\partial T}{\partial t} = \nabla \cdot (K\nabla T) + w_b C_b \rho_b (T_a - T_v) + Q_m \quad (4)$$

(where T is the local temperature, ρ, C and K are the tissue density, specific heat and heat conductivity respectively, Q_b is the Pennes perfusion term, Q_m is the metabolic heat production) is an ill-posed problem,

o patient-depended variability of the thermo-regulation process due to different bio-feedback time constants,

o spectral specification as well as accuracy and resolution of the infrared camera,

o there is no definition of a standard in active thermography concerning the methodology of thermal excitation or provocation, respectively,

o vagueness of determination of thermal characteristic of skin cancer, i.e. a variation of the emissivity coefficient of the suspicious moles,

o only the surface temperature is imaged,

o unknown relation between thermo-regulative time constants and amount of applied energy in active thermography,

o inhomogeneity and speed variation of energy transfer in active thermography,

o spurious reflection of background radiation,

o insufficient patient acclimatisation before imaging (patient induced uncertainty: E.g. time of patient's most recent warm meal, as well as ambience induced: E.g. instable thermoregulation of the examination room, etc.).

For a discussion of some of these points see [8]. However, in the past decades strong emphasis is given to the detection of malign tumours especially mamma carcinoma with thermographic imaging.
Many research groups find out indications for breast cancer [9-11 and the papers cited therein]. But it seems that the spatial cutaneous temperature signature of a breast tumour can be detected in a late cancer state only. Therefore, the cancer-diagnosis application carried out in this paper exclusively focuses on malign skin processes.
And – it is worthwhile to focus on this application of thermography, because medical infrared imaging is the only diagnostic method that is purely passive and, therefore, inherently is without any dose limitation. Additionally, it is a low-cost system – compared to MR or PET – that yields functional information.

3 Infrared Imaging Procedure

To be accepted in clinical routine a simple, reproducible measurement protocol of the thermographic image acquisition and patient preparation has to be set up. In the first trials shown in this paper an active thermography strategy has been established. The main point of protocol is that the skin spot under suspicion is investigated dynamically.

To get the characteristics or temperature signature, respectively, the relevant skin area has to be provoked. Generally, two directions are possible. On one hand the skin can be warmed up and, on the other hand, the skin can be cooled down. One major risk of the warming-up method is that the denaturising process of the proteins starts when the skin temperature exceeds 42°C. For that reason the cooling-down method is used in our experiments that produce a substantial temperature difference within a few minutes.

The cooling is carried out using direct contact with cooled gel packs. An area of about 10 cm by 10 cm is cooled down to 20° C. After this patient-preparation step the signature of the thermo-regulation process is recorded by the FLIR SC 3000 camera with a temperature resolution of 0.03 K.

A sequel of 300 images is taken in a total time interval of five minutes. However, in practice that interval depends on the level of cooling and the type of skin lesion. As shown in figure 3 the camera is placed directly in front of the lesion using a macro lens.

Fig. 3: Patient situation during recording in clinical practice.

A difficulty in the evaluation of the thermo-regulation process is that a spot cannot be automatically detected on the basis of the temperature image alone, because at the starting point of thermo regulation the entire skin region of interest has a homogeneous temperature distribution.

To overcome this problem a marker is attached (see figure 4) to the skin and a normal digital photo is taken prior to the thermographic session.

By comparing the marker of each image with the digital image motion of the patient can be compensated. This is important because a small motion in an image acquisition with a macroscopic lens can cause errors in the correlation between the frames of the thermographic sequel. Without stable motion compensation an automated comparison of the temperatures of skin and spot is not possible.

Fig. 4: Marker attached to the skin.

4 Image Processing

A few elementary image processing steps are required for the automated evaluation of the temperature signature of the thermo-regulation process:

o   Detection of bore-holes of the fiducial marker (that can be seen even in the thermographic sequel.) using the generalized Hough transformation.
o   Estimation of motion parameters based on the homologous landmarks obtained in the first step.
o   Identification of suspicious skin spot in the digital photo using an active contour.
o   Mapping of spot boundary, i.e. the active contour, to each of the thermographic frames using the motion model of the second step.
o   Evaluation of thermo-regulation process inside and outside the skin area defined by the active contour of the third step.

4.1 Hough Transformation

The Hough transformation [12,13] is basically used to detect mathematically defined objects in an image. Historically, the Hough transformation is developed to detect lines given in parametric form by

$$x \cos(\varphi) + y \sin(\varphi) = l, \qquad (5)$$

where l is the length of a normal of the line measured from the origin (0,0) and φ is the angle between the normal line and the x-coordinate.

Plotting all possible values l and φ for a point (x,y) in an image will result in a sinusoidal curve in the polar Hough space (l,φ). It is well known, that the Hough transformation is a relative to the Radon transformation.

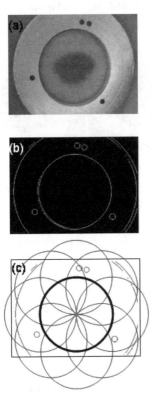

Fig. 5: Illustration of the circular Hough transform (CHT).

In this Hough parameter space points that are collinear in the Cartesian image become visible as maxima, because they yield sinusoidal curves which intersect at a common (r,φ) point. The transformation is implemented by quantisation of the Hough space into finite accumulator cells. However, to detect the drills of the fiducial marker the 2D-Hough transformation must be extended to the 3D circle Hough transform (CHT). The CHT can be used to find circular patterns of a given radius r (see figure 5a). A first step in the Hough transformation evaluation edges of the image must be detected. This is done by application of an appropriate edge-filter (e.g. Sobel filter) which can be seen in figure 5b.
The parametric equation of a circle is

$$r = \sqrt{(x-a)^2 + (y-b)^2}, \quad (6)$$

where (a,b) is the centre point and r is the radius of the circle, respectively.
Each edge point contributes a circle of radius r to the accumulator space (see figure 5c). Intensity maxima in the accumulator space are detected, where these circles overlap at the centre of the original circle. In figure 6 the resulting 3D accumulator space is illustrated.

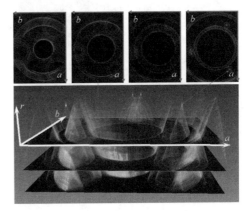

Fig. 6: 4 Slices of different radii of the 3D accumulator space of circle Hough transformation. Below: The four fiducial landmarks (compare figure 5a) can easily be detected in the upper part of the rendered 3D accumulator space.

After detecting the drills in both images, the digital photo can be transformed by affine transformation.

4.2 Motion Compensation

As mentioned above an appropriate transformation has to be applied for motion compensation. Due to the fact, that the digital photo potentially yields a rotated, scaled, shifted and slightly sheared image of the area of interest the parameters of an affine transformation

$$\begin{bmatrix} x' \\ y' \end{bmatrix} = \begin{bmatrix} a_{11} & a_{12} \\ a_{12} & a_{22} \end{bmatrix} \begin{bmatrix} x \\ y \end{bmatrix} + \begin{bmatrix} t_x \\ t_y \end{bmatrix} \quad (7)$$

have to be estimated.
To calculate the parameters at least three non-collinear landmarks are needed (see figure 7; step 1). Applying this very transformation on an active contour that has been used to identify the border of the lesion in the digital photo (see figure 7; step 2), the area of interest can be analyzed without motion distortion.

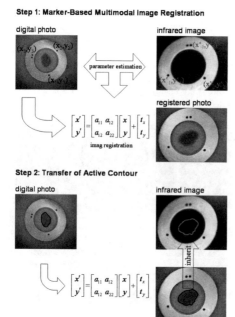

Fig. 7: Two major steps of motion compensation: Estimation of parameters of affine transformation and mapping of active contour via the affine transformation.

5 Results

The clinical study is still ongoing. However, two examples of the first trials are given in this paper.

5.1 Basaliom

In the first case, a basal-cell carcinoma, the skin has been cooled down to 27 °C. The subsequent thermoregulation process is observed to 5 minutes taking infrared frames in intervals of 1 s. The healthy skin regulates its temperature to 36 °C within 5 minutes. Figure 8 shows the digital photo and the corresponding infrared image after completes thermo regulation.
The signature of the curves is exponential. However, focusing on the thermo regulation of the spot it can be seen that there is a synchronous warm up to 32 °C in the first 90 s.
As can be seen in figure 9 the temperature of the basaliom is increasing slower than the surrounding healthy skin when the temperature exceeds 32 °C.
The basaliom is not visible at the temperature image in the first 90 s of the recording it is clearly recognizable in the end.

One possible explanation for the lower temperature is based on the physiological characteristics of a basaliom.
A basaliom is created from the cells which can produce an isolation layer, like it is already mentioned in a study from Maleszka et al [5] who saw this effect for psoriatic arthritis.

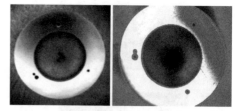

Fig. 8: Original and temperature image of a basaliom.

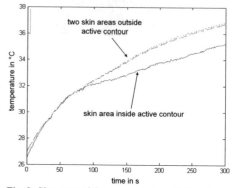

Fig. 9: Signature of thermo regulation of a basaliom.

5.2 Dysplastic Nevus

In the second case, a dysplastic nevus, the skin lesion is invisible in the temperature image during the entire recording sequence. Figure 10 shows the digital photo and the corresponding infrared image after a time interval of 5 min.

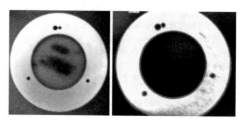

Fig. 10: Original and temperature image of a dysplastic nevus.

In this case the dysplatsic nevus is semi-malignant as histology proves.

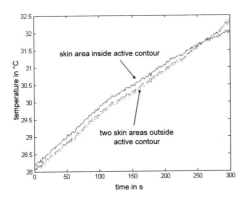

Fig. 11: Signature of thermo regulation of a dysplastic nevus.

6 Conclusion

During our first clinical trials the method of thermography-based evaluation of skin lesions turns out to be promising. Especially in the case of a basal-cell carcinoma the method yields a clear diagnostic result. It seems that the well known principles of thermography, that had been rejected some years ago and is recently raised to interest by detector advances, gives a powerful tool for dermatologist's diagnosis.

However, the recording protocol needs an improved standardisation due to patient-individual variation of the acquisition conditions and, in addition to the novel procedure, the traditional ABCD lesion features should support the thermographic evaluation.

7 References

[1] W. Stolz, D. Hölzel, A. Riemann, et al., *Multivariate analysis of criteria given by dermatoscopy for recognition of melanocytic lesions*, Book of Abstracts, Fiftieth Meeting of the American Academy of Dermatology, Dallas, Texas, December 1991.

[2] R. A. Fiorini, G. Dacquino and G. Laguteta, *A New Melanoma Diagnosis Active Support System*, Proceedings of the 26th Annual International Conference of the IEEE EMBS San Francisco, CA, September 1-5, 2004, 3206-3209.

[3] G. Zouridakis, M. Doshi, and N. Mullani, *Early Diagnosis of Skin Cancer Based on Segmentation and Measurement of Vascularization and Pigmentation in Nevoscope Images*, Proceedings of the 26th Annual International Conference of the IEEE EMBS San Francisco, CA, September 1-5, 2004, 1593-1596.

[4] O. Dössel, *Bildgebende Verfahren in der Medizin*, Springer, Berlin, 2000.

[5] N. Schuster, V. G. Kolobrodov, *Infrarotthermographie*, Viley-VCH, Berlin, 2000.

[6] FLIR Inc., SC 3000 Handbook.

[7] H. H. Pennes, *Analysis of tissue and arterial blood temperatures in the resting human forearm*, Journal of Applied Physiology, vol. **1**, no. 2, 1948, 93–122.

[8] A. Z. Nowakowski, *Limitations of Active Dynamic Thermography in Medical Diagnostics*, Proceedings of the 26th Annual International Conference of the IEEE EMBS San Francisco, CA, September 1-5, 2004, 1179-1182.

[9] R. N. Lawson, *Implications of Surface Temperatures in the Diagnosis of Breast Cancer*, Can. Med. Assoc. J. **75**, 1956, 309.

[10] M. Gautherie and C. Gros, *Breast thermography and cancer risk prediction*, Cancer, **45**, 1980, 51.

[11] N. Scales, C. Herry, M. Frize, *Automated Image Segmentation for Breast Analysis Using Infrared Images*, Proceedings of the 26th Annual International Conference of the IEEE EMBS San Francisco, CA, September 1-5, 2004, 1737-1740.

[12] P. V. C. Hough, *A method and Means for recognizing complex pattern*, US Patent, No. 3069654, 1962.

[13] B. Jähne, *Bildverarbeitung*. 5.Aufl, Heidelberg: Springer-Verlag, 2002.

[14] A. Zalewska, B. Wiecek, A. Sysa-Jedrzejowska, G. Gralewicz, G. Owczarek, *Qualitative thermograhic analysis of psoriatic skin lesions*, Proceedings of the 26th Annual International Conference of the IEEE EMBS San Francisco, CA, September 1-5, 2004, 1192-1195.

[15] R. Maleszka, M. Rosewicka, M. Parafiniuk, A. Kempinska, D. Mikulska, *Trial of thermographic investigations application in patient with psotriatic arthritis*. (Polish) Dermatol Klin **5**, 2003, 11-15.

Bildgebung / Medical Imaging

Simulation von Detektorsystemen für die medizinische Physik

D. Krücker, M. Khodaverdi, J. Perez[(*)], H. Herzog, U. Pietrzyk[(*)]
Institut für Medizin, Forschungszentrum Jülich GmbH
[(*)] und Bergische Universität Wuppertal - Fachbereich C - Physik

Kurzfassung

Simulationen spielen seit Jahren eine wesentliche Rolle in der Entwicklung von Geräten und Methoden in vielen Bereichen der Naturwissenschaften. Anhand zweier Fallbeispiele (µCT und PET) wird der Einsatz von GEANT4 und GATE für nuklearmedizinische Fragestellungen vorgeführt.

1 Einleitung

Die Einsatzmöglichkeiten von Computersimulationen sind vielfältig. In der Entwicklungsphase eines Strahlungsdetektors helfen sie bei der Optimierung des Designs und erlauben Voraussagen über Leistung und Qualität des geplanten Systems. In der Anwendung des Detektors helfen Simulationen bei der Entwicklung von Methoden zur Kalibrierung, oder dem besseren Verständnis von Untergrundquellen und anderen Fragen, die sich bei der Auswertung experimenteller Daten ergeben.

Es existiert eine große Zahl von Programmen zur Simulation von Detektoren, wie sie in der medizinischen Physik eingesetzt werden. Neben Programmen, die auf die Simulation von PET- und SPECT-Geräten spezialisiert sind [1], finden sich Programmbibliotheken, die es erlauben allgemeine Fragestellungen aus der Nuklearmedizin zu untersuchen [2]. Für die Simulation vollständiger Detektorsysteme ist sicherlich GEANT4 [3] das umfangreichste Programmpaket. GEANT (GEometry ANd Tracking) wurde für die Simulation von Experimenten in der Hochenergiephysik verwendet und wurde vor allem am Forschungszentrum CERN in Genf über Jahrzehnte hinweg entwickelt. Die derzeit aktuelle Version GEANT4 ist das Ergebnis eines internationalen Projektes, das das gesammelte physikalische Know-how in einem objektorientierten Ansatz verfügbar macht. Frühzeitig wurden dabei auch nuklearmedizinische Fragestellungen berücksichtigt. Dazu gehören PET- und SPECT-, ebenso wie Dosimetrieanwendungen.

Insbesondere für den Einsatz in der Emissionstomographie wurde kürzlich die Simulationsplattform GATE [4] als benutzerfreundliche Schnittstelle zu GEANT4 vorgestellt. An der Entwicklung von GATE war auch das Institut für Medizin am Forschungszentrum Jülich beteiligt, wo GEANT4 und GATE in verschiedenen Bereichen eingesetzt werden; u.a. für Untersuchungen zum Einsatz atypischer Positronenstrahler in der PET, für die Kalibration und Normierung des ClearPET, sowie für das Design eines µCT. Hier sollen zwei typische Anwendungsbeispiel vorgestellt werden: eine Designstudie für einen Mikro-Computertomographen (µCT), die mit GEANT4 durchgeführt wurde, sowie die GATE-Simulation eines Positronen-Emissions-Tomographen (PET).

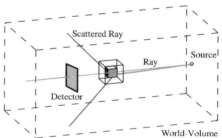

Detektor	Röntgenröhre (ggf. mit Al-Filter)
Szintillator: GdOS, CsI	Wolfram
Halbleiter: aSe, CdZnTe	Molybdän

Bild 1 Aufbau der µCT-Simulation

2 Designstudie für ein µCT

Das erste Beispiel stammt aus einer Doktorarbeit zum Design eines µCT, die im Rahmen des ClearPET-Projektes durchgeführt wurde. Es handelt sich also um einen Röntgentomographen, der zusammen mit einem Kleintier-PET verwendet werden soll. Es sei erwähnt, dass die zugrunde liegende Arbeit [5] das Problem detaillierter behandelt, als es hier dargestellt wird. An diesem Beispiel lassen sich die typischen Schritte einer Simulationsstudie vorführen.

Aufgrund der verschiedenen praktischen Randbedingungen ergibt sich eine **Vorauswahl** an Komponenten und Materialien, die als Ausgangspunkt der Studie dienen. Das geplante System sollte aus einer Mikrofokusröntgenröhre und einem Festkörperdetektor bestehen. Für die Röntgenröhre standen Wolfram und Molybdän als Targetmaterialien zur Auswahl und es bestand die Möglichkeit, Al-Filter verschiedener Stärke einzusetzen. Als Detektor kamen verschiedene Szintillator- oder Halbleitermaterialien in Betracht (siehe Tabelle zu Bild 1). Die **Fragestellung** war nun, für welche Kombination von Mikrofokusröntgenröhre und Festkörperdetektor die besten Ergebnisse zu erwarten sind. Zuerst wurde ein einfaches **Modell des Systems** erstellt (siehe Bild 1), wobei die Simulation hier auf die wesentlichen Elemente beschränkt wurde. Der nächste Schritt bestand in der **Validierung der Simulation**. Das Spektrum der Röntgenröhren mit verschiedenen Al-Filtern konnte in einem Messaufbau überprüft werden. Die Massenschwächungskoeffizienten der Detektormaterialien wurden der NIST-Datenbank [6] entnommen. In beiden Fällen ergab sich ein sehr gute Übereinstimmung mit den simulierten Daten. Um nun die Simulation für die Designstudie einzusetzen, musste ein **Qualitätskriterium** definiert werden. In der vorliegenden Arbeit war dies u.a. der Kontrast, der zwischen verschiedenen Gewebetypen im rekonstruierten CT-Bild zu beobachten ist. Zu diesem Zweck wurde ein Kontrastphantom zwischen Röntgenröhre und Detektor eingefügt, in dem gleichzeitig 5 verschiedene Materialien simuliert wurden: Wasser, Luft, sowie Fett-, Knochen- und Hirngewebe. Die Simulation von je 20700 Photonen wurde für 180 Projektionsrichtungen durchgeführt und die so gewonnenen Daten mittels gefilterter Rückprojektion rekonstruiert. Die Rechenzeit für die Simulation von je 180 Projektionen betrug auf einem 800 MHz PIII Linux-Rechner etwa 1 Woche. Der letzte Schritt besteht in der **Variation der Parameter**. An dieser Stelle verhinderte der Zeitbedarf der Simulation ein Durchprobieren aller Materialkombinationen bzw. Filterstärken. In der vorliegenden Studie wurde deshalb eine Vorauswahl anhand der Simulationsergebnisse für das einfache Röntgenbild getroffen und die vollständige CT-Simulation nur für 4 Kombinationen von Detektormaterial und Röntgenquelle durchgeführt.

Die Simulation erlaubt es hier also, das Design mit Blick auf eine komplexe Größe, die nicht einfach analytisch zugänglich ist, nämlich den Kontrast im rekonstruierten CT-Bild, zu optimieren. Man sieht hier allerdings auch, dass die Simulation bildgebender Systeme anspruchsvoll in Bezug auf die Rechenzeit ist.

3 GATE

Für das vorangegangene Beispiel wurden etwa 2500 Programmzeilen geschrieben. Dafür waren, einschließlich Test- und Debuggingphasen, ca. 6 Monate Arbeit nötig. Im Vergleich dazu ist die detaillierte Simulation eines vollständigen PET- oder SPECT-Systems um Größenordnungen aufwendiger. Allerdings ähnelt sich die Programmierarbeit für jedes simulierte Detektorsystem zu großen Teilen. Da außerdem nicht alle potenziellen Nutzer aus dem Bereich der Nuklearmedizin über die notwendigen Kenntnisse (und die Zeit) verfügen, eine derartige Simulationssoftware zu entwickeln, wäre ein Programmpaket wünschenswert, das dem Benutzer möglichst viel Funktionalität gleichsam in „Fertigbauweise" bereitstellt. Mit GATE steht ein solches System zur Verfügung.

Die OpenGATE-Kollaboration hat sich im Jahre 2001 mit der Zielsetzung gegründet, eine Simulationssoftware für SPECT- und PET-Anwendungen zu entwickeln. Das Programm sollte modular sein und sich leicht an spezielle Bedürfnisse adaptieren lassen. Es sollte zudem ohne C++-Kenntnisse und ohne Erfahrung in der GEANT4-Programmierung benutzbar sein. Seit Sommer 2004 ist GATE öffentlich verfügbar [4]. Zwei Konzepte sind wesentlich für das Programm: die System-Templates und die Makro-Programmierung. Innerhalb von GATE stehen verschiedene Systeme zur Verfügung, wie *Scanner*, *CPET*, *ECAT*, *SPECTHead* u.a. Mit diesen Systemen verbunden ist eine bestimmte Struktur der Geometrie und der Detektorauslese, die einer Familie von Scannern gemeinsam ist. Dabei werden auch gerätetypische Dateiformate bereitgestellt (z.B. ECAT7). Innerhalb des gewählten Systems kann der Anwender die Details der Simulation mit Hilfe von Makrodateien frei definieren. GATE integriert sich dazu in die GEANT4-Benutzerschnittstelle. Weiterhin stellt GATE eine große Anzahl von Modulen für die Simulation der Auswerteelektronik zur Verfügung wie *quantumEfficiency, blurring, crosstalk, threshold, deadtime* u.v.m. Diese Module können ebenfalls mittels Makrobefehlen in die Simulation eingefügt werden.

Darüber hinaus zeichnet sich GATE durch die Fähigkeit aus, zeitabhängige Systeme zu simulieren. GEANT4 kann während der Simulation die Geometrie nicht verändern. GATE umgeht diese Einschränkung, in dem es Bewegungen schrittweise simuliert. Der simulierte Zeitraum kann dafür in eine beliebige Anzahl diskreter Intervalle zerlegt werden. So wird es möglich, z. B. die Bewegung eines Detektorkopfes um einen Patienten zu beschreiben. Die Zeitabhängigkeit von radioaktiven Quellen wird ebenfalls berücksichtigt, wobei die Quellenstärke kontinuierlich aktualisiert wird.

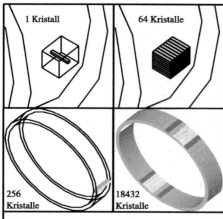

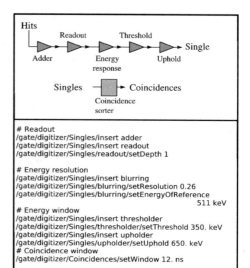

Bild 2 Vier Schritte zur Definition der Kristallgeometrie

Bild 3 Aufbau einer Auslesekette

Mit GATE wurden verschiedene kommerzielle und nichtkommerzielle PET und SPECT-Geräte simuliert. Eine Liste von Referenzen zur Validierung findet sich in [4]. Messung und Simulation für Größen wie Sensitivität, Auflösung, Streuanteil u.a. stimmen typischerweise innerhalb von 5% überein.

3.1 Simulation eines PET-Scanners

Anhand eines ECAT EXACT HR+ von CPS/Siemens soll die Leichtigkeit demonstriert werden, mit der auch komplexe Detektorsysteme innerhalb von GATE definiert werden können. Bild 2 zeigt, wie die 18432 BGO-Kristalle dieses Scanners mit wenigen Zeilen innerhalb des Makros definiert werden. In ähnlicher Weise kann die Digitalisierungskette festgelegt werden (Bild 3). Detaillierte Ergebnisse zur Simulation des HR+ finden sich in [7].

4 Ausblick

Die beiden Beispiele machen deutlich, dass Simulationen ein wichtiges Werkzeug zum Verständnis von Strahlungsdetektoren sind. Die Ergebnisse stimmen auch für komplexere Systeme gut mit den experimentellen Werten überein. Ein wesentliches Problem ist allerdings der hohe Rechenzeitbedarf bei der Simulation bildgebender Systeme. Das vorige PET-Beispiel benötigt im Mittel 17 ms für die Simulation des Systems bei einem ^{18}F-Zerfall auf einem handelsüblichen Rechner (2.4 GHz P4). Für die Simulation einer realen Messung, die typischerweise 10^9 ... 10^{10} Zerfälle enthält, ergeben sich dann schnell Rechenzeiten im Bereich von 10^2 ... 10^3 Tagen. Methoden zur Beschleunigung zu finden ist also eine wesentliche Aufgabenstellung bei der Simulation solcher Systeme. Dabei sind zwei Strategien möglich: Parallelisierung und Effizienzsteigerung. Die hier beschriebenen Simulationen sind trivialerweise parallelisierbar. Das einzelne Zerfallsereignis oder das einzelne Röntgenphoton ist unabhängig von allen anderen. Die Simulation kann also praktisch auf eine beliebige Anzahl von Prozessen aufgeteilt werden. Allerdings ist dabei ein gewisser Verwaltungsaufwand nötig. Innerhalb der OpenGATE-Kollaboration beschäftigt sich eine eigene Arbeitsgruppe damit, dem Benutzer einheitliche Strukturen für die Parallelisierung bereitzustellen. Bei der Verbesserung der Effizienz der Simulation sind verschiedene Methoden zur Varianzreduktion in der Diskussion. Solche Ansätze sind insbesondere in Situationen wichtig, wo in Phantomen oder Kollimatoren ein Großteil der simulierten Photonen absorbiert werden, ohne ein Detektorsignal zu erzeugen. Hier sind verschiedene Strategien denkbar, die Wahrscheinlichkeit zu manipulieren, dass ein Photon den Detektor erreicht. Für solche Algorithmen stellt die aktuelle GEANT4 Version die notwendigen Programmstrukturen bereits zur Verfügung.

5 Danksagung

Diese Arbeit wurde zum Teil von der Deutschen Forschungsgemeinschaft (DFG) innerhalb des Projektes HE 3090/2-1 gefördert.

6 Literatur

[1] I. Buvat und I. Castiglioni, Q. J. Nucl. Med. 46 (2002) 48
[2] F. Verhaegen und J. Seuntjens, Phys. Med. Biol. 48 (2003) R107
[3] S. Agostinelli et al., Nucl. Instr. and Meth. A506 (2003) 250
[4] GATE: a simulation toolkit for PET and SPECT, S. Jan et al., Phys. Med. Biol. 49 (2004) 4543, http://www-lphe.epfl.ch/GATE
[5] M. Khodaverdi, Designstudie eines μCT-Zusatzes für einen hochauflösenden Positronen-Emissions-Tomographen, Forschungszentrum Jülich GmbH, 2004
[6] National Institute of Standards and Technology, http://www.nist.gov
[7] S. Jan et al., Monte Carlo Simulation for the ECAT EXACT HR+ system using GATE, IEEE Trans. Nucl. Sci., to be published

Multimodale Abbildung der Haut mit hochfrequentem Ultraschall und Optischer Kohärenztomographie

M. Vogt[1,5], A. Knütel[2,5], T. Grünendick[3,5], K. Hoffmann[4,5], P. Altmeyer[4,5], H. Ermert[1,5]

[1]Lehrstuhl für Hochfrequenztechnik, Ruhr-Universität Bochum, IC 6/133, 44780 Bochum
[2]Isis optronics GmbH, Innstr. 34, 68199 Mannheim
[3]Visomed AG, Universitätsstr. 160, 44801 Bochum
[4]Klinik für Dermatologie und Allergologie, Ruhr-Universität Bochum, Gudrunstr. 56, 44791 Bochum
[5]Kompetenzzentrum Medizintechnik Ruhr (KMR), Bochum

Kurzfassung

Hochfrequenter Ultraschall (HFUS) und Optische Kohärenztomographie (OCT) sind Techniken für die hochaufgelöste und nichtinvasive Abbildung von Geweben, die u.a. in der Dermatologie für die Abbildung der Haut zum Einsatz kommen. Beide Techniken arbeiten auf Grundlage vergleichbarer Konzepte, wobei jedoch unterschiedliche Wellenphänomene genutzt werden. Die diagnostische Aussagekraft kann daher durch die Kombination von HFUS und OCT in einem multimodalen Ansatz gesteigert werden. In diesem Beitrag werden neben einem systemtheoretischen Vergleich beider Techniken Ergebnisse vergleichender Messungen an technischen Objekten zur Quantifizierung des Auflösungsvermögens und Ergebnisse von Messungen in-vivo präsentiert. Das Potential einer multimodalen Bildgebung der Haut mit HFUS und OCT wird diskutiert.

1 Einführung

In der dermatologischen Diagnostik hat die visuelle Befundung von Hautveränderungen durch den Mediziner eine große Bedeutung, da die zu untersuchenden Läsionen unmittelbar zugänglich sind [1]. Auf den Erfahrungen der Dermatologen basierend hat sich die „ABCD-Regel" der Dermatologie etabliert, in deren Zusammenhang Merkmale (,features') für die Erkennung tumoröser Hautveränderungen angegeben werden. Die Asymmetrie (,**a**symmetry'), die Begrenzung (,**b**order'), die Farbgebung (,**c**olor') und der Durchmesser (,**d**iameter') sind dabei zu quantifizierende Parameter. Ein wesentliches Problem besteht darin, dass die Ergebnisse einer Klassifikation auf Grundlage der ABCD-Regel stark von den individuellen Erfahrungen des die Untersuchung durchführenden Mediziners abhängen. Dieser Umstand hat zu der Entwicklung computergestützter Bildverarbeitungssysteme geführt, bei denen die Hautläsionen mit Hilfe eines digitalen Kamerasystems abgebildet, die Bilddaten automatisch analysiert und eine Klassifikation durchgeführt werden. Derartige Implementierungen von Mustererkennungssystemen werden auf Grundlage großer Datenbanken mit Fallbeispielen trainiert, bei denen histologische Befunde von Pathologen die Referenz, d.h. den „Goldstandard", darstellen.

Die beschriebene Art der Diagnostik beruht lediglich auf der Einschätzung *oberflächlicher* Hautstrukturen, wobei Informationen über die Ausdehnung und Gestalt von Hautläsionen über der Tiefe nicht zur Verfügung stehen. Tomographische Abbildungsverfahren, d.h. Schnittbildverfahren, wie die Sonographie erlauben darüber hinaus die nichtinvasive Abbildung von Geweben über der Tiefe [1]. Für die Ultraschallabbildung in der allgemeinen medizinischen Diagnostik wird Ultraschall im Frequenzbereich von etwa 1 MHz bis 10 MHz eingesetzt. Für die Abbildung der Haut ist das dabei gegebene Ortsauflösungsvermögen jedoch ungenügend. In der Dermatologie kommt daher hochfrequenter Ultraschall (HFUS) im Bereich oberhalb von 20 MHz zum Einsatz [1-3]. Die Optische Kohärenztomographie (OCT, ,Optical Coherence Tomography') stellt ein zur hochfrequenten Sonographie analoges Verfahren dar, bei dem jedoch Licht im nahen Infrarotbereich anstelle von Ultraschallwellen in das Gewebe eingestrahlt wird [4]. Bei beiden Verfahren werden die rückgestreuten und reflektierten Wellenfelder über der Laufzeit analysiert, und es werden die Rückstreueigenschaften des Gewebes abgebildet. Mit der unterschiedlichen Physik der eingesetzten Ultraschallwellen bzw. des Lichtes liefern die beiden Abbildungsverfahren unterschiedliche Bildinformationen [4-6].

Im Folgenden werden HFUS und OCT aus systemtheoretischer Sicht und anhand von Messungen an technischen Objekten und in vivo hinsichtlich ihrer diagnostischen Aussagekraft für Anwendungen in der Dermatologie miteinander verglichen. Die Zielsetzung einer *multimodalen* Bildgebung liegt darin, nicht-redundante Informationen aus der Fusion der erhaltenen Bildinformationen zu gewinnen.

2 Abbildung der Haut mit HFUS und OCT

2.1 Eigenschaften diagnostischer Abbildungssysteme

Diagnostischer Abbildungsverfahren ermöglichen die nichtinvasive Abbildung von Gewebestrukturen. *Tomographische* Abbildungsverfahren liefern dabei *Schnittbilder*, in denen physikalische Gewebeeigenschaften einer Schicht punktweise in einer Bildebene repräsentiert sind. Eine wesentliche Zielsetzung ist dabei die Realisierung eines möglichst guten Ortsauflösungsvermögens, d.h. benachbarte Gewebestrukturen sollen in der Abbildung möglichst gut voneinander unterschieden werden können. Die *Punktbildfunktion* (PBF) eines diagnostischen Abbildungssystems ist das von dem System gelieferte Bild für den Fall der Abbildung eines einzelnen, isolierten und *punktförmigen* Bildobjektes. Im folgenden werden zweidimensionale (2D) Abbildungsverfahren betrachtet, bei denen Gewebeeigenschaften über der Tiefe (axiale Ortskoordinate) und transversal dazu (laterale Ortskoordinate) abgebildet werden. Als axiale und laterale Ortsauflösungen werden dabei die räumlichen Ausdehnungen der PBF entlang der beiden Ortskoordinaten bezogen auf einen Abfall auf die Hälfte des Maximums (FWHM: ‚full width half maximum') betrachtet. Die Messung der PBF eines vorliegenden Abbildungssystems kann in der Praxis mit Hilfe punktförmiger Testobjekte („Phantome") erfolgen, deren Abmessungen kleiner als die zu erwartende Auflösung sind. Für die Realisierung derartiger Messungen können u.a. Fadenphantome, reflektierende Oberflächen und isolierte Punktstreuer eingesetzt werden.

2.2 Hochfrequenter Ultraschall (HFUS)

Bei der Ultraschallabbildung werden typischerweise Puls-Echo-Messungen realisiert. Dazu wird ein elektroakustischer Ultraschallwandler mit einem pulsförmigen Sendesignal angesteuert, und die an *akustischen* Gewebeinhomogenitäten diffus rückgestreuten bzw. spiegelnd reflektierten Ultraschallwellen werden von demselben Schallwandler in ein elektrisches Echosignal umgesetzt. Unter Annahme einer konstanten Schallgeschwindigkeit innerhalb des Gewebes können den Echosignalen über der Laufzeit axiale Entfernungen in Schallausbreitungsrichtung zugeordnet werden. Für die Bildgebung wird dabei ein schmaler Schallstrahl geformt, entlang dessen sich die Schallwellen ausbreiten, und dieser Schallstrahl wird für die Realisierung eines linearen Scans transversal zur Schallausbreitungsrichtung verfahren. Im B-Bild (‚brightness') wird die Hüllkurve des bandbegrenzten Echosignals nach Demodulation über der axialen und lateralen Ortskoorinate dargestellt. Damit wird die *akustische* Rückstreuung bzw. die Reflektivität des Gewebes qualitativ abgebildet.

Das Ortsauflösungsvermögen hängt von der Mittenfrequenz f_0 und der Bandbreite B des Ultraschallabbildungssystems an. Mit größer werdender Mittenfrequenz wird bei gegebener Apertur des Ultraschallwandlers eine stärkere Fokussierung (schmalerer Schallstrahl) und damit ein besseres laterales Ortsauflösungsvermögen realisiert. Mit größer werdender Bandbreite wird die Pulsdauer rückgestreuter Echosignale kleiner und damit das axiale Ortsauflösungsvermögen verbessert [2-4].

Während in der konventionellen, niederfrequenten Sonographie für die Bildgebung im wesentlichen Ultraschallwandlerarrays (Zeilen von 64 bis 250 Elementen), deren Einzelelemente einzeln elektronisch angesteuert werden, zum Einsatz kommen, stehen derartige Arrays für hochfrequente Ultraschallanwendungen nicht zur Verfügung. Für die Bildgebung werden daher sphärisch fokussierte Einzelelementschallwandler eingesetzt. Diese weisen einen Schallstrahl auf, der sich vom Schallwandler ausgehend verjüngt und in einen Fokus mündet. Im dahinterliegenden Fernfeld des Schallwandlers weitet sich das Schallfeld auf. Für die Bildgebung wird der Fokus des Schallwandlers in die abzubildende ROI (‚region of interest') gelegt. Eine Abtastung entlang der lateralen Ortskoordinate erfolgt durch mechanisches, motorgesteuertes Verfahren des Einzelelementschallwandlers.

Am Lehrstuhl für Hochfrequenztechnik haben wir ein hochfrequentes Ultraschallabbildungssystem mit einem im 100 MHz-Bereich arbeitenden Einzelelementschallwandler realisiert [2-4]:

Bild 1   HFUS-System: Applikator

2.3 Optische Kohärenztomographie (OCT)

Die OCT arbeitet analog zum HFUS, wobei jedoch Licht anstelle von Ultraschall zum Einsatz kommt. Das eingestrahlte Licht wird an *optischen* Gewebeinhomogenitäten diffus rückgestreut bzw. spiegelnd reflektiert. Aus der Analyse rückgestreuten bzw. reflektierten Lichtes kann somit die *optische* Rückstreuung bzw. Reflektivität des Gewebes qualitativ abgebildet werden. Auf Grund der gegenüber der

Schallgeschwindigkeit (Mittelwert 1540 m/s) viel größeren Lichtgeschwindigkeit (Mittelwert $2{,}14 \cdot 10^8$ m/s) im Gewebe sind die Laufzeiten des Lichts sehr wesentlich kleiner als bei der Sonographie. Unmittelbare Laufzeitmessungen aus der Analyse rückgestreuten bzw. reflektierten Lichtes über der Zeitachse können damit technisch nur sehr schwierig realisiert werden.

Für die OCT wird daher ein interferometrischer Ansatz gewählt. Das Licht aus einer *niedrigkohärenten* Lichtquelle (LED, ‚light emitting diode') wird über einen optischen Koppler in einen Referenzpfad und einen Messpfad eingekoppelt. Im Referenzpfad wird das Licht an einem Spiegel reflektiert, während das Licht im abzubildende Gewebe, das innerhalb des Messpfades liegt, rückgestreut bzw. reflektiert wird. Rückgestreute und reflektierte Lichtanteile durchlaufen den Koppler und überlagern sich an einer Fotodiode im Empfänger [4]:

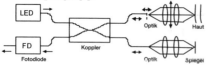

Bild 2 OCT-Abbildungskonzept; Blockschaltbild

Die Fotodiode liefert ein elektrisches Signal, das proportional zur Intensität des Lichtes ist. Bei dem vorliegenden Michelson-Interferometer erfolgt damit eine Korrelation beider Lichtsignalanteile. Hinter der Fotodiode liegt dann ein Messsignal an, wenn die optischen Weglängen in Referenz- und Messpfad gleich groß sind. Im Umkehrschluss bedeutet dies, dass durch Änderung der optischen Weglängendifferenz eine axiale Abtastung realisiert werden kann. Die laterale Abtastung erfolgt durch mechanische Verfahren der abbildenden Optik, siehe Bild 2.

Für die im folgenden vorgestellten Messungen ist ein kommerzielles OCT-Abbildungssystem (‚SkinDex300', Isis optronics GmbH, Mannheim) eingesetzt worden, das bei einer Wellenlänge von 1300 nm arbeitet. Bei diesem System liefert eine integrierte CCD-Kamera optische Übersichtsbilder der Hautoberfläche, mit deren Hilfe die abzubildende ROI auf einfache Art und Weise in den Scanbereich des Systems gebracht werden kann:

Bild 3 OCT-System (Isis optronics GmbH)

3 Multimodale Abbildung der Haut

3.1 Multimodale Bildgebung

Bei der *multimodalen* Bildgebung besteht das Ziel darin, nicht-redundante Informationen Bildinformationen zu gewinnen und diese miteinander zu fusionieren [5,6]. Zu diesem Zweck kommen verschiedene, auf unterschiedlichen physikalischen Konzepten basierende Abbildungsverfahren zum Einsatz, mit deren Hilfe unterschiedliche physikalische Gewebeeigenschaften abgebildet werden. Im Rahmen einer nachgeschalteten Bildanalyse erfolgen eine Registrierung, d.h. räumliche Zuordnung, der Bilddatensätze zueinander und die Repräsentation in einem gemeinsamen Koordinatensystem.

3.2 HFUS und OCT als komplementäre Abbildungsverfahren

HFUS und OCT sind beides tomographische Abbildungsverfahren, die ortsaufgelöste Schnittbilder liefern. Die Eindringtiefe der eingesetzten Ultraschallwellen bzw. des eingesetzten Lichtes ist jedoch stark unterschiedlich. Mit der in der Dermatologie bereits gut etablierten 20 MHz-Sonographie besteht die Möglichkeit, die gesamte Haut und das darunter liegende subkutane Fett abzubilden. Die hier eingesetzte 100 MHz-Sonographie hingegen ist auf die hochaufgelöste Abbildung der obersten Hautschichten, der Epidermis, und der Dermis ausgerichtet. Die Eindringtiefe in das Gewebe ist mit etwa 2-3 mm kleiner als bei der 20 MHz-Sonographie [2,3].

Das bei der OCT eingesetzte Licht erfährt eine sehr starke Streuung im Gewebe, und mit dem interferometrischen Ansatz kann gleichzeitig ein großer Dynamikumfang realisiert werden. In der Konsequenz können mit Hilfe des OCT insbesondere die obersten Hautschichten hochaufgelöst abgebildet werden, die Eindringtiefe ist auf einige hundert Mikrometer begrenzt [4].

4 Ergebnisse

4.1 Durchgeführte Untersuchungen

Zur Quantifizierung der Abbildungseigenschaften des eingesetzten HFUS- und OCT-Systems sind zunächst Messungen an technischen Objekten durchgeführt worden. Die Oberfläche einer Glasplatte ist als gut definiertes Einzelobjekt zur Bestimmung der axialen Ortsauflösung abgebildet worden. Mit Hilfe von Messungen an einem Fadenphantom (Wolframdraht mit 7 µm Durchmesser) ist die laterale Ortsauflösung gemessen worden.

In einem weiteren Schritt sind vergleichende Messungen in-vivo durchgeführt worden, um das Potential beider Systeme für klinische Anwendungen auszuloten [4].

4.2 Messungen an technischen Objekten

In Bild 4 sind HFUS- und OCT-Abbildungen einer Glasplatte zu sehen:

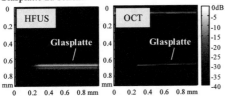

Bild 4 HFUS-Abbildung (links) und OCT-Abbildung einer Glasplatte

Als Ergebnis dieser Untersuchungen ist für das HFUS-System eine axiale Auflösung von 9,3 µm gemessen Mit einer Messung an einem Drahtphantom ist weiterhin eine laterale Auflösung von 60 µm gemessen worden. Axiale und laterale Auflösung des OCT-Systems wurden mit 5,8 µm und 4,1 µm gemessen.

4.2 Messungen in-vivo

Hinsichtlich dermatologischer Anwendungen ist es von großem Interesse zu analysieren, wie sich die geschichteten Strukturen der Haut abbilden. In Bild 5 ist ein HFUS-Bild gesunder Haut am Handgelenk zu sehen:

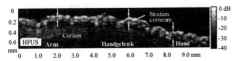

Bild 5 HFUS-Abbildung: Gesunde Haut, Übergang am Handgelenk

In der Ultraschallabbildung ist die Hautoberfläche als ein erstes, echoreiches Band erkennbar. Darunter ist das stratum corneum in der Epidermis als ein echoarmes Band zu sehen. Diese Schichtstruktur weitet sich beim Übergang von der am Unterarm vorliegenden Felderhaut hin zur Leistenhaut an der Hand, wie in Bild 5 gut erkennbar ist. Weiter darunter ist die Dermis (Corium) als echoreiche Struktur zu sehen.

In Bild 6 sind OCT-Abbildungen zu sehen, die an derselben Stelle aufgenommen worden sind. Da der maximale Bildbereich des OCT-Systems (0,9 mm / 1,0 mm in axialer / lateraler Richtung) wesentlich kleiner ist als der maximale Bildbereich des HFUS-Systems (3,2 mm / 10 mm in axialer / lateraler Richtung), sind drei Einzelaufnahmen entlang der lateralen Ortskoordinate realisiert worden:

Bild 6 OCT-Abbildung: Gesunde Haut, Übergang am Handgelenk

In der OCT-Abbildung ist die Schichtstruktur der Haut, die sich in der HFUS-Abbildung gezeigt hat, ebenfalls erkennbar. Beim Übergang vom Arm auf die Hand ist deutlich die Aufweitung des stratum corneums zu sehen. Gleichzeitig ist erkennbar, dass die Ortsauflösung besser ist als bei der HFUS-Abbildung, während die Eindringtiefe in die Haut kleiner ist.

Ergebnisse eines weiteren Vergleichs sind in Bild 7 und Bild 8 zu sehen. Abgebildet ist dort die Nagelplatte und die Haut an der Nagelfalz an der Hand eines Probanden:

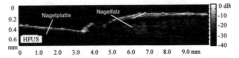

Bild 7 HFUS-Abbildung: Fingernagel

In der Ultraschallabbildung ist auf der rechten Seite die Haut an der Nagelfalz sichtbar, während die eigentliche Nagelplatte echoarm erscheint. Die vergleichende OCT-Abbildung weist hingegen neben rückgestreuten Lichtes in der Nagelfalz auch rückgestreute Anteile aus der Nagelplatte selbst auf:

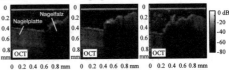

Bild 8 OCT-Abbildung: Fingernagel

Im Gegensatz zu HFUS kann mit der OCT die Nagelplatte aufgrund ihrer optischen Transparenz gut abgebildet werden.

4.3 Fusion von OCT und Histologie

Für die Klassifikation tumoröser Hautveränderungen stellen histologische Schnittbildfolgen, die postoperativ von einem Histologen befundet werden, den Goldstandard dar. In Hinblick auf den Entwurf bildbasierter Diagnosesysteme ist es daher wichtig, Histologien auf die Bilddatensätze der tomographischen Abbildungsverfahren zu registrieren, d.h. räumlich zuzuordnen. In Bild 9 ist exemplarisch das Ergebnis einer solchen Registrierung zu sehen:

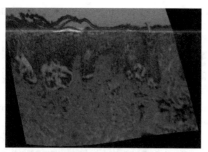

Bild 9 OCT-Abbildung und überlagerte Histologie: Basalzellkarzinom (Basaliom)

In diesem Fall ist preoperativ ein OCT-Bild aufgenommen worden, das in Bild 9 überlagert zu einer postoperativ gewonnenen Histologie dargestellt ist. Bei der Entnahme und Weiterverarbeitung der Gewebeprobe sind geometrische Verzerrungen unvermeidlich. Um diese zu korrigieren, ist eine nichtrigide Koordinatentransformation zwischen den Bilddatensätzen auf Grundlage einer landmarkenbasierten Registrierung mit manuell platzierten Landmarken erfolgt. Im Ergebnis können die beiden Bilddatensätze nach vollzogener Koordinatentransformation räumlich einander präzise zugeordnet werden.

5 Diskussion und Ausblick

Im Rahmen dieser Arbeit sind die hochfrequente Ultraschallbildgebung (HFUS) und die Optische Kohärenztomographie (OCT) als tomographische Abbildungsverfahren für die hochaufgelöste Gewebeabbildung mit Anwendungen in der Dermatologie miteinander verglichen worden. HFUS und OCT stellen zueinander analoge Verfahren dar, bei denen aus dem Gewebe rückgestreute und reflektierte Ultraschallwellen bzw. Licht über der Laufzeit aufgelöst abgebildet werden. Bei beiden Konzepten erfolgt gleichzeitig eine Fokussierung entlang der lateralen Ortskoordinate, womit eine Bildgebung ermöglicht wird. Bedingt durch die unterschiedliche Physik der eingesetzten Wellenfelder werden unterschiedliche physikalische Gewebeeigenschaften abgebildet.

Mit Hilfe von Messungen an technischen Objekten ist das Ortsauflösungsvermögen der eingesetzten Systeme gemessen worden. Das eingesetzte OCT-System weist dabei eine um einen Faktor 1,6 bessere axiale und um einen Faktor 15 bessere laterale Ortsauflösung auf. Messungen in-vivo zeigen andererseits, dass die Eindringtiefe in die Haut und der maximale Messbereich bei HFUS größer sind als bei der OCT. Beide Verfahren ermöglichen die Abbildung von Hautstrukturen in-vivo. Es sind ferner erste Untersuchungen zur Fusion von OCT-Bilddatensätzen und histologischen Schnittbildfolgen durchgeführt worden.

Unsere weiteren Arbeiten zielen auf die Sammlung klinisch relevanter Bilddatensätze mit beiden Abbildungsmodalitäten, deren Fusion und die Implementierung eines Systems zur Klassifikation tumoröser Hautveränderungen hin.

6 Danksagung

Ein Projekt des Kompetenzzentrums Medizintechnik Ruhr (KMR), Bochum. Gefördert durch das Bundesministerium für Bildung und Forschung (BMBF), Nr. 13N8079.

7 Literatur

[1] Altmeyer, P., Hoffmann, K., Stücker, M.: Bildbasierte Diagnostik der Haut, Biomedizinische Technik., vol. 46 (Ergänzungsband), pp. 32-35, 2001

[2] Ermert, H., Vogt, M., Paßmann, C., el Gammal, S., Kaspar, K., Hoffmann, K., Altmeyer, P.: High frequency ultrasound (50-150 MHz) in dermatology. In: Skin Cancer and UV Radiation, (Eds. P. Altmeyer, K. Hoffmann, M. Stücker), pp. 1023-1051, Springer, Berlin, Heidelberg, New York, 1997

[3] Vogt, M., Kaspar, K., Altmeyer, P., Hoffmann, K., El Gammal, S.: High frequency ultrasound for high resolution skin imaging, Frequenz, vol. 55, no. 1-2, pp. 12-20, 2001

[4] Vogt, M., Knüttel, A., Hoffmann, K., Altmeyer, P., Ermert, H.: Comparison of high frequency ultrasound and optical coherence tomography as modalities for high resolution and non invasive skin imaging, Biomedizinische Technik, vol. 48, no. 5, pp. 116-121, 2003

[5] Vogt, M., Ermert, H.: Biomicroscopy of the skin utilizing high frequency ultrasound in a multi modal approach. Proceedings: Workshop on Ultrasound in Biomeasurements, Diagnostics and Therapy, pp. 69-73, September 7-9, 2004, Gdansk-Sobieszewo, Poland

[6] Vogt, M., Scharenberg, S., Scharenberg, R., Grünendick, T., Knüttel, A., Hoffmann, K., Altmeyer, P., Ermert, H.: Multimodal imaging of the skin with high frequency ultrasound, optical coherence tomography and epiluminescence microscopy. Proceedings: 38. Jahrestagung der DGBMT, 22.-24.09.2004, Ilmenau, Germany, Biomedizinische Technik, Band 49, Ergänzungsband 2, Teil 2, S. 854-855, Sept. 2004

3D-Reconstruction of Microscopic Translucent Silicate-Based Marine and Freshwater-Organisms

D. Schmitz, R. Herpers, D. Seibt[1], W. Heiden

Bonn-Rhein-Sieg University of Applied Sciences, Department of Computer Science, St. Augustin, Germany
[1] DLR-Institute of Aerospace Medicine, Cologne, Germany

Abstract

The objective of the presented approach is to develop a 3D-reconstruction method for micro organisms from sequences of microscopic images by varying the level-of-focus. The approach is limited to translucent silicate-based marine and freshwater organisms (e.g. radiolarians). The proposed 3D-reconstruction method exploits the connectivity of similarly oriented and spatially adjacent edge elements in consecutive image layers. This yields a 3D-mesh representing the global shape of the objects together with details of the inner structure. Possible applications can be found in comparative morphology or hydrobiology, where e.g. deficiencies in growth and structure during incubation in toxic water or gravity effects on metabolism have to be determined.

1 Introduction

At the DLR-Institute of Aerospace Medicine of the German Aerospace Center, the influence of gravity on the development of living organisms is studied. One question to be answered is for instance, whether zero or low gravity has effects on the building of shells by aquatic micro organisms. Therefore, the fine structure of these organisms has to be measured and quantified to enable an detailed evaluation of changes of the morphology. However, morphological changes or changes of the shape of micro organisms have to be evaluated for the entire 3D shape of the organisms. The standard approach for 3D-reconstruction of micro organisms is based on a confocal laser scanning microscopy [8] which however, proved to be not applicable for the use in outer space. Problems do occur during out-of-space missions, where the lenses of the finely tuned confocal laser scanning microscopes will be misplaced by shocks and hyper gravity forces during lift-off. Therefore, a software based 3D-reconstructiuon process for image sequences acquired with more robust light microscopes was needed.

In light microscopy, rays emitted by the microscopic specimens are distorted by the lenses. According to the Gaussian law for lenses [5], rays of a point representing a structure of a specimen converge to a point rather than an area on the focal plane only if the distance between the lense and the strutucture matches exactly the focal depth. Recording microscopic images, the part of the structures of the specimen which is in focus is dependent on the magnification of the lenses used. Here, it is assumed that structures in a certain volume corresponding to a thickness of a hypothetic slice defined by [4] as the depth-of-focus, δ (Mag), are sufficiently in focus so that images of structures in this limited volume occur sharp. This means, that at high magnification levels, structures of the specimen exceeding in-depth beyond δ (Mag) the level-of-focus plus half the slice thickness are out of focus. Thus, parts of the image are blurred due to non-convergence of the light rays on the focal plane ([2], [3], [4]). Blurred image areas appear as overlapping circular patches with low sharpness. This particular feature is exploited for our 3D-reconstruction method motivated by the "Shape from Focus" approach of Nayar et al. [6]. Figure 1 shows the depth of focus, δ (Mag), expressed as a volume in which image rays assumed to converge sufficiently. Thus, the specimen is divided into several levels-of-focus or volumetric slices [1], [4].

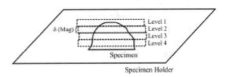

Figure 1: δ (Mag), the depth of focus at a given magnification, expressed as a volumetric slices [4].

Image sequences recorded at different levels of focus were used to reconstruct the 3D-shape of the specimen. All out-of-focus regions in each image layer were removed while the remaining edge structures were connected and a 3D-mesh was built. By doing so, a reconstruction of the 3D-shape of the specimens from planar slices was achieved.

In the next chapter the material used in this study is introduced and the image acquisition in presented. In chapter 3 the several steps of the developed 3D-reconstruction approach is discussed. After the processing of the reconstruction procedures, the developed method yields a 3D-mesh representing the global shape of the specimens together with details of the inner structure.

2 Material and methods

2.1 Specimens

The reconstruction method was developed for and tested on various types of freshwater and marine forms of micro organisms from the kingdoms of plants and animals in microscopic preparations [7] provided by Johannes Lieder Inc. Among those were:

- Radiolarians
- Foraminiferians
- Sponge needles
- Diatomeans
- Phytomonadina (*Volvox*)
- Larvae of sea-urchins (*Psammechinus miliaris*)

The specimens were chosen due to details and contrast of inner as well as outer structure. Furthermore in this first step of this study relatively large specimens were investigated to evaluate the performance and principle correctness of the developed approach.

2.2 Image acquisition

The microscope used in this study was a Leitz „Laborlux S" with a video camera Agfa-tv model HR 480 (1/2''-chip) equipped with objectives of the type PL-Fluota with 10-, 25-, and 40-times magnification (numerical appertures: 0.30, 0.60, and 0.70 respectively). Since the magnification of the ocular was by factor 15, the maximal possible overall magnification was 15x40 with an estimated depth-of-focus, δ (Mag) of 5 µm at this level [4]. This magnification was used to record the sequences of all investigated specimens which are in average about 20µm in size for the radiolarian and about 100µm for the *Volvox* measured using a Thoma-chamber (fig. 2) with intersections at 5 µm [7].

Figure 2: Thoma-chamber at 15x40 magnification [7] used to determine the size of the specimen.

Sample raw images of a sequence of a radiolarian (focus levels 1, 4, 5, and 6) are shown in fig. 3. In level 1 (top image of fig 3) central surface areas of the radiolarian are in focus and the included structures are sharp and clearly visible. In deeper layers surface areas which are in focus are visible as a ring of darkly outlined and sharp structures surrounding the central area of the specimen. These regions grow broader while increasing the level-of-focus, thus descending to the midsection.

In average, 20 slices were recorded for all specimens, while the maximum and minimum amount of slices varied between 8 and 42 slices dependent of the size of the specimen. At the maximum number of 42, however, it turned out that the overlap of structures in consecutive image pairs was too large to allow for a reliable 3D-reconstruction.

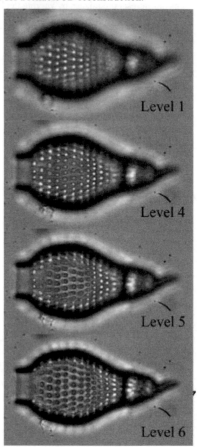

Figure 3: Sample raw images of a sequence of a radiolarian at different levels-of-focus (image size 350x200 pixel). In the top image (level 1) central surface areas of the specimen are in focus and its structures are sharp and clearly visible. In deeper layers surface areas which are in focus are visible as a ring of darkly outlined and sharp structures surrounding the central area of the specimen.

3 3D-Reconstruction

The 3D-reconstruction process is subdivided into the following subprocesses:

1. Edge-detection
2. Thresholding
3. Skeletonisation and artefact removal
4. Vectorisation
5. Edge-refinement and meshing
6. Rendering

The different subprocesses are described in more detail in the following subsections.

3.1 Edge detection

Edge detection is performed by applying Kirsch-filtering with standard kernels for various edge-orientations (horizontal, vertical, diagonal) ([3], [9] [10]). The filtering results were inverted and rescaled to 8 bit gray values. Thus, the gray values correspond to the "sharpness" of the edge structures detected in the image (after inversion) while dark areas represent sharp edges (see fig 4).

3.2 Thresholding

The approach of Nayar et al. for 3D-reconstruction [6] was based on iterating vertically over consecutive images from topmost to lowest, detecting the "sharpest" (in this case, darkest) pixels in each column and removing the remaining ones, thus yields only parts of the images which are in focus. This approach, however, is limited to opaque objects only. In our approach, fine inner structures that are blurred due to convolution with structures from overlaying image-levels would be lost when applying this method. Therefore, an interactive thresholding process has been developed with the scope to preserve also inner structures of the translucent specimens. The cut-off-level of the gray values was chosen interactively and adapted to the depth of the slice. By applying this method all out-of-focus regions could be reliably removed while preserving inner structures. The results of this process are shown in figure 5. It should be noted that the tubuli exceeding the surface towards the midsection of the specimen/inner structures are preserved, as deeper layers come into focus. Due to impurities of the lenses, some artefacts are always in focus on the focal plane and cannot be removed by pure thresholding.

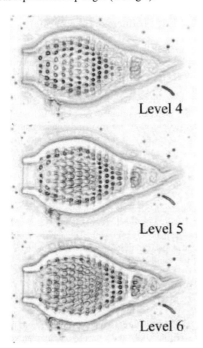

Figure 4: Consecutive images from fig. 1 after application of Kirsch edge detection, inversion and rescaling. Deep dark image structures represent surface areas which are in focus to be used for the following 3D-reconstruction process.

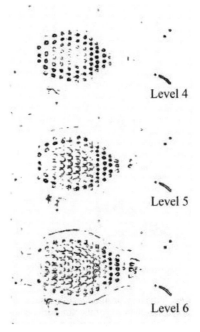

Figure 5: Consecutive images from fig. 1 after thresholding. Remaining surface areas per image layer are connected in depth as well as within the image plane to establish a 3D-mesh. Artefacts visible as dark blobs outside the main body of the specimen were removed interactively for subsequent processing steps.

3.3 Skeletonisation and artefact removal

For skeletonisation a recursive thinning algorithm has been developed and applied by using template based morphological operations for opening and closing of structures as described by Jähne and Haußecker [3]. This yields sequences with edge structures limited to a maximum width of one pixel. Artefacts caused by contamination of the microscopic preparations and impurities of the lenses were removed interactively prior to the next processing step.

3.4 Vectorisation

The spatial orientations of connected edge structures were evaluated and classified in four classes (fig. 6). Horizontal, line shaped edge structures were marked blue, vertical green and diagonal yellow while intersections, junctions, and corners were coloured teal. The pixels of edge segments of the same orientation or class type on each image level were connected and vectorised during a first vectorisation iteration.

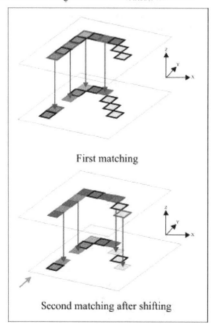

Figure 6: Vertical reconstruction of the structure by connecting edge elements of the same orientation or class type (the spatial orientation is colour-coded).

The interlayer distance was calculated of the length-to-depth ratio of the specimens and the level and depth of focus. Consecutive image pairs were explored for in-depth edge structures. Edge structures of the same orientation or class type which were exactly neighboured were connected by an iterative approach and vectorised. This is computed by iterating over the reference image of each pair and comparing the orientation of each image position left, representing the surface structure of the specimen with the orientation at the same position in the consecutive image layer (test image). If the orientations/classes match, a vector is computed and the image position is assigned as visited (red). Matching was allowed between identical orientations of neighboured image positions. Nodes and diagonal edge structures were considered for possible matches as well if and only if the edge structure of one layer is horizontal or vertical oriented.

In order to detect more connected edge structures, similarily oriented edges in close spatial vicinity were considered for matching and evaluated as well (fig. 6). The test image is shifted relative to the reference image one pixel in each of the four directions (horizontal, vertical, and left/right diagonal). The matching process was repeated as described above for all computed shifts. Edges marked as visited and matched before were removed from further considerations (fig. 6). The iteration process was terminated, when no more matches could be found or computed. Image positions which could not be assigned to any other neighbouring structure at all remained unmatched and were not considered further.

3.5 Edge refinement and meshing

Since the matching process described above results in a solid surface description of the specimen when all image positions are connected to their neighbours, a thinning algorithm has to be applied to obtain a thin mesh, which is needed for visualisation. For that, all depth oriented vectors which were too close to each other are removed. This results in a sparse 3D-mesh with a minimum distance of one pixel between adjacent depth vectors (fig.7).

3.6 Rendering

The computed edge vectors representing parts of the mesh of the 3D-object reconstructed were rendered in an OpenGL environment. The environment was embedded in a graphic user interface faciliating the procedures and algorithms described above.

4 Results

The 3D-shape of three sequences of different radiolarians (figs: 7,8,9) and one phytomonadian of the *Volvox* type (fig. 10) have been processed. The results of the developed 3D-reconstruction method of a sample radiolarian is shown in figure 7 at different viewing angles, in which the shape of the organism can be seen as well as details of the inner structure like the tubuli exceeding from the surface to the midsection of the radiolarian.

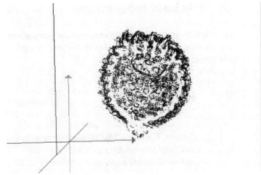

Figure 9: 3D-reconstruction of a third radiolarian

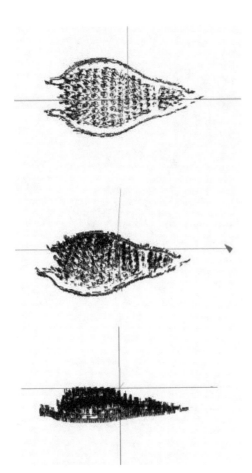

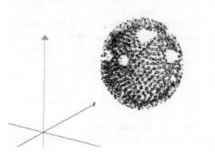

Figure 10: 3D-reconstruction of a *Volvox*. The white blobs represent subcolonies of cells which could not be resolved due to the higher density.

5 Discussion

The proposed 3D-reconstruction method exploits the connectivity of similarly oriented and spatially adjacent edge elements in consecutive image layers. Problems may occur during the reconstruction process in image layers beyond the midsection due to accumulated blurring effects in depth. In order to overcome these difficulties, rotatory symmetry of the specimens could be faciliated. Thus, the entire 3D-shape of the specimen might be built from two halves.

Due to the chosen methodology the presented approach is limited to translucent silicate-based marine and freshwater-organisms (e.g. Radiolarians or Volvox), since reflections caused by calcitic shells cannot be resolved.

Possible applications can be found in comparative morphology or hydrobiology, where e.g. deficiencies in growth and structure during incubation in toxic water or gravity effects on the building of shells have to be determined.

Figure 7: Results of the 3D-reconstruction process for a radiolarian at different viewing angles rotating around the X-axis. The shape of the radiolarian can be seen as well as details of inner structure like the tubuli exceeding from the surface to the midsection.

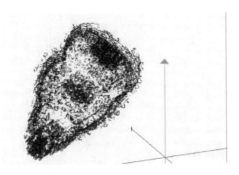

Figure 8: 3D-reconstruction of another type of radiolarian of a bigger size.

6 Acknowledgements

D. Schmitz gratefully acknowledges the support by the members of the DLR- Institute of Aerospace Medicine.

7 References

[1] Bonton, P. et al. (2002) *Colour Image In 2D and 3D Microscopy For The Automation Of Pollen Rate Measurement*. Image Anal Stereol, 21:25-30.

[2] Castleman, K. R. (1996) *Digital Image Processing*. New York: Prentice Hall Inc.

[3] Jähne, B., Haußecker, H. (2000). *Computer Vision and Applications*. London, New York: Academic Press.

[4] Knappertsbusch, M. W. (2002). *Stereographic Virtual Reality Representations Of Microfossils In Light Microscopy*. Paleontologica Electronica 5(3):11pp.

[5] Grehn, J., vonHessberg, A., Holz, H.-G., Krause, J., Krüger, H., Schmidt, H.K., (1992). Metzler Physik. Hannover: Schroedel.

[6] Nayar, S. K., Nakagawa, Y. (1994). *Shape from Focus*. IEEE Transactions On Pattern Analysis And Machine Intelligence, Vol.16, No. 8.

[7] Nultsch, W. (1968). *Mikroskopisch-Botanisches Praktikum*. Stuttgart, New York: Thieme.

[8] Packroff, G., Lawrence, J. R., Neu, T. R. (2002). *In Situ Confocal Laser Scanning Microscopy of Protozoans in Cultures and Complex Biofilm Communities*. Acta Protozool. 41:245-253.

[9] Parker, J. R. (1996). *Algorithms for Image Processing and Computer Vision*. New York: John Wiley&Sons.

[10] Zimmer, W. D., Bonz, E. (1996). *Einführung in die digitale Bildverarbeitung*. München Wien: Carl Hanser.

Development of a high resolution positron emission tomograph

Simone Weber, Forschungszentrum Jülich, Zentralinstitut für Elektronik, D-52425 Jülich

Abstract

High resolution positron emission tomography (PET) is a non-invasive tool which is used e.g. in preclinical drug development and biological research. To study mice or rats, which are the most important species for the respective applications, a spatial resolution of 2 mm or less is required, while at the same time the scanner sensitivity has to be high enough to allow data acquisition in short periods of time. Considering the ClearPET® Neuro which has been developed at the Forschungszentrum Jülich as example, it is shown how both, high spatial resolution as well as high sensitivity, can be reached simultaneously.

1 Introduction

High resolution PET scanners dedicated for studies on small animals gain increasing interest with regard to their application in preclinical drug development and experimental studies of animal models of various diseases [1]. Such systems permit researchers to perform animal studies by repeated measurements in a single animal. The design of dedicated experimental tomographs has in the past often been optimized to achieve a spatial resolution of 2 mm or less. Since there is a trade-off between spatial resolution and sensitivity, a high spatial resolution was, to a certain extent, pursued at the expense of scanner sensitivity.

While the image resolution is important to distinguish between structures, a better spatial resolution will not necessarily lead to a better image quality. An increase in spatial resolution requires smaller, and consequently more voxels covering the field-of-view (FOV) to be reconstructed. Increasing the spatial resolution of a scanner by a factor of 2 requires an increase in sensitivity by a factor of 2^3 to get the same statistical error per volume element. So the required signal-to-noise ratio per image pixel has to be considered as well.

2 The ClearPET® Neuro small animal PET scanner

2.1 Scintillator crystals

Detectors for small animal PET scanners usually consist of scintillating crystals coupled to position sensitive photomultiplier tubes. Crystal cuboids with a cross section of 2 mm or less are required to achieve a high spatial resolution. The crystals have to be bright and fast to ensure a correct readout of the scintillation light. For high sensitivity, crystals with a high density are required which are about 20 mm long to allow the 511 keV gamma quanta to interact in the crystal with a high probability.

While in the past BGO ($Bi_4(GeO_4)_3$) (**table 1**) was the scintillator of choice for human PET systems, it is now more and more replaced by LSO ($Lu_2(SiO_4)O:Ce$) since its discovery in the early 1990th [2]. Another interesting candidate is the newly developed Lutetium Aluminium Perovskite (LuAP) crystal, possibly with some Yttrium additive (LuYAP) ($Lu_xY_{1-x}AlO_3:Ce$) which is brighter and has a faster decay time than BGO [3].

	(Lu-Y)AP	LSO	BGO
density [g/cm³]	> 7.7	7.4	7.13
light yield* (photons/511 keV)	1418	5007	1299
scintillation decay time [ns]	20.5 (65%) 160 (35%)	40	300
energy resolution	19.6 %	12.6 %	22.8 %

* for a 2 x 2 x 10 mm³ crystal

Table 1 Properties of scintillator crystals used in PET

2.2 Depth-of-interaction estimation

When using long scintillator cuboids with a small cross-section, the spatial resolution may suffer from parallax errors due to penetration effects (**fig. 1 a**). When emitted some distance outside of the center of the FOV, the gamma quanta may impinge obliquely onto the crystals and hence penetrate some crystals before interacting in the scintillator. The line-of-response (LOR) which is estimated may differ significantly from the true gamma emission line, which may lead to a significant deterioration of spatial resolution depending on the position within the FOV.

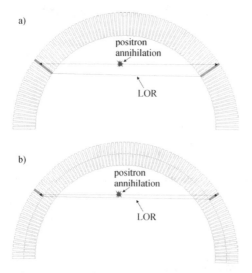

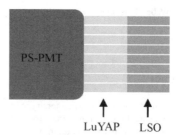

Figure 2 Two crystal layers coupled to a position sensitive PMT – side view.

Figure 1 Depth-of-interaction estimation; a) parallax error due to crystal penetration; b) reduced parallax error with detectors consisting of two crystal layers.

The parallax error can be reduced by estimating the depth-of-interaction (DOI) within a crystal. One approach to estimate DOI is to use two crystal layers instead of one single crystal (**fig. 1 b**), the so-called phoswich configuration. The two layers have to be read out separately or alternatively have to be differentiated e.g. by means of different scintillator properties. The ClearPET® Neuro scanner uses 2 x 2 x 10 mm LSO and LuYAP crystals, with a crystal pitch of 0.3 mm, in phoswich configuration. While in the first scanner version the DOI estimation is not yet implemented, the layer of interaction will be identified in the future from the scintillation decay time of the corresponding scintillator material [4]. From the two scintillation decay times of LuYAP, only the slow component will be used to distinguish between the two layers.

Figure 3 Detector head, consisting of two layers with 64 crystals (covered with white reflector) each, coupled to a position sensitive photomultiplier tube.

2.3 Data processing

A high sensitivity scanner requires a fast readout of the data as well as a low deadtime of the detectors. The ClearPET® Neuro uses position sensitive photomultiplier tubes (Hamamatsu 7600M64) for the readout of the scintillator cuboids. One detector head consists of 2 layers of crystals with 64 crystals each coupled to the 64 photocathode pixels of the photomultiplier tube (**fig. 2**, **fig. 3**).

The signal from each photomultiplier is sampled continously by a free running analog-to-digital-converter (ADC) [5,6]. The pulse provides the information about the energy of the gamma, the pulse shape and the time information which is required for coincidence detection. All necessary information can be extracted by software data processing. This method also allows some suppression of cross-talk between crystals or the suppression of detector scatter. The single events are stored as list-mode data. Coincidences finally are sorted from the singles data set by software data processing.

2.4 System design

The ClearPET® Neuro is designed for applications in neurosciences. Therefore, the main design goal, besides a spatial resolution of ~2 mm, is a sensitivity of the scanner which allows the user to take data of extensive activity distributions in ~30 sec time frames.

Four detector heads, arranged in line, build a module. 20 modules are arranged in a ring (**fig. 4**) with a ring diameter of 13.8 cm. The axial size of the ring is 11.2 cm so that a FOV diameter of ~9 cm can be attained.

An insensitive region at the border of the detector heads (see **fig. 3**) results in gaps between the detectors axially as well as transaxially. Therefore, the detectors

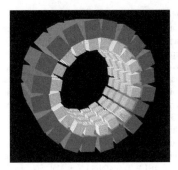

Figure 4 ClearPET® Neuro general design

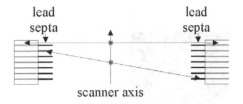

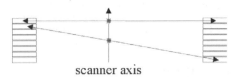

Figure 6 Top: data acquisition in 2D mode. Only events in a slice are acquired. Bottom: data acquisition in 3D mode. All events are acquired.

The ClearPET® Neuro takes 3D listmode data and, after sorting the data into 3D sinograms, uses an iterative reconstruction technique, the fully 3D „ordered subset maximum a posteriori one-step late" (OSMAPOSL) reconstruction algorithm [9].

Figure 5 ClearPET® Neuro detector ring

are rotating by 360 degree in step and shoot mode during data acquisition. Every second module is shifted ¼ of the detector width axially to compensate for the axial gaps.
The modules can be moved radially, so that the ring diameter can be expanded. **Fig. 5** shows the 20 modules of the scanner arranged in a ring with minimum diameter.

2.5 Data reconstruction

Besides using dedicated detectors and electronics to achieve a high sensitivity, another important factor is to use all available data for reconstruction. Human PET systems often use lead septa to allow coincidences only within a slice or between neighbouring slices of the scanner. In this case, 2D reconstruction techniques like 2D filtered backprojection (FBP) or 2D iterative techniques can be applied.
Without septa, all coincidences are acquired (**fig. 6**). The data then have to be reconstructed using dedicated 3D reconstruction techniques or have to be rebinned to the respective 2D dataset before reconstruction [7,8].

3 Results

A glass capillary with an inner diameter of 0.5 mm was filled with ^{18}F and was scanned at two positions within the FOV (center and 30 mm off-center, with ~2 hours time between both scans). During data acquisition, the scanner rotated in 1 degree steps over 360 degree and back. More than 28 million coincidence events were acquired in 12 min scan time (without time for detector rotation).
A coincidence window of 25 ns was used. Neither energy threshold nor further corrections of the data, like detector normalization or decay correction, were applied. The axial FOV is covered by 48 slices of 2.3 mm slice width. The coincidence data were binned into 3D sinograms with 80 projections and 81 tangential bins each for the direct slices and oblique slices with a slice difference up to +/- 15. The data are reconstructed using the OSMAPOSL algorithm. **Fig. 7** shows the reconstructed data of the two line sources together with a profile through the line source at the center of the FOV which is averaged over the length of the line source. The full-width at half-maximum (FWHM) of the profile is 1.6 mm. The inhomogeneity along the lines is mainly due to the missing detector efficiency normalization.

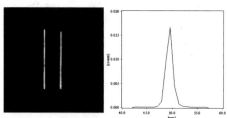

Figure 7 Left: Line sources at the center of the FOV and 30 mm off-center. Right: Profile through the line source at the center of FOV, averaged over the length of the line source.

4 Discussion

The ClearPET® Neuro is a high-resolution, high-sensitivity animal PET scanner which was developed recently at the Forschungszentrum Jülich. The data presented here are still preliminary. Though a thorough performance evaluation of the scanner has not yet been accomplished, and important corrections and calibrations are still missing, the scanner shows nevertheless a very promising performance in spatial resolution as well as sensitivity.

Up to now, general purpose reconstruction software is used for data reconstruction, but this software does not take into account the scanner characteristics. Due to the general design of the scanner and especially the gaps between the detectors, a dedicated reconstruction method is a very important topic [10] which will lead to a higher image quality.

5 Acknowledgements

The development of the ClearPET® Neuro was performed by a large team at the Forschungszentrum Jülich and within an international collaboration. I would like to thank especially A. Bauer, G. Brandenburg, D. Christ, M. Dehnhard, L. Fuß, B. Gundlich, U. Heinrichs, A. Hollendung, M. Khodaverdi, H. Larue, P. Musmann, M. Oehler, C. Parl, U. Pietrzyk, M. Streun and K. Ziemons, as well as the members of the Crystal Clear Collaboration, CERN.

6 References

[1] Weber, S.: Hochauflösende Positronen-Emissions-Tomographie – Entwicklung und Anwendung. Physikalische Methoden der Laser- und Medizintechnik, Fortschritt-Berichte VDI Reihe 17 Nr. 231. Düsseldorf: VDI Verlag, 2003

[2] Melcher, C.L., Schweitzer, J.S.: Cerium-doped lutetium oxyorthosilicate: a fast, efficient new scintillator. IEEE Trans Nucl Sci. Vol 39, 1992, pp. 502-505

[3] Weber, S., Christ, D., Kurzeja, M., et al.: Comparison of LuYAP, LSO, and BGO as scintillators for high resolution PET detectors. IEEE Trans Nucl Sci. Vol. 50, No. 5, 2003, pp. 1370-1372

[4] Streun, M., Brandenburg, G., Larue, H., et al.: Pulse shape discrimination of LSO and LuYAP scintillators for depth of interaction detection in PET. IEEE Trans Nucl Sci. Vol. 50, No. 3, 2003, pp. 344-347

[5] Streun, M., Brandenburg, G., Larue, H., et al.: A PET system with free running ADC's. Nucl Instr Meth A. Vol. 486, 2002, pp. 18-21

[6] Streun, M., Brandenburg, G., Larue, H., et al.: Coincidence detection by digital processing of free-running sampled pulses. Nucl Inst Meth A. Vol. 487, 2002, pp. 530-534

[7] Defrise, M., Kinahan, P.E., Townsend, D.W., Michel, C., Sibomana, M., Newport, D.F.: Exact and Approximate Rebinning Algorithms for 3-D PET Data. IEEE Trans Med Imag, Vol. 16, 1997, pp.145-158.

[8] Oehler, M., Weber, S., Musmann, P., Buzug, Th. M.: Implementation of a "Fourier Rebinning"-Algorithm for a high resolution PET-Scanner. Conference Record, Remagener Physiktage, RheinAhrCampus Remagen, Fachhochschule Koblenz, University of Applied Sciences, 29.09.-01.10.2004.

[9] Bettinardi, V., Pagani, E., Gilardi, M.C., et al.: Implementation and evaluation of a 3D one-step late reconstruction algorithm for 3D positron emission tomography brain studies using median root prior. Eur J Nucl Med, Vol. 29, No. 1, 2002, pp. 7-18

[10] Musmann, P., Schramm, N., Weber, S.: Tomographische Bildrekonstruktion für die hochauflösende Positronen-Emissions-Tomographie (PET). Conference Record, Remagener Physiktage, RheinAhrCampus Remagen, Fachhochschule Koblenz, University of Applied Sciences, 29.09.-01.10.2004.

Bildverarbeitung

Image Processing

Immersion Square -
a mobile platform for immersive visualizations

Rainer Herpers[1,2,3], Florian Hetmann[2], Axel Hau[1], Wolfgang Heiden[1]

[1] Bonn-Rhein-Sieg University of Applied Sciences, Department of Computer Science, Sankt Augustin, Germany
[2] Square Vision AG, Köln, Germany
[3] York University, Department of Computer Science, Toronto, Canada

Abstract

A mobile immersive visualization environment called Immersion Square is introduced which enables immersive visualizations for many application areas and offers a wide range of configuration flexibility. The standard setup of the projection environment consists of a three-wall-backprojection screen system. Immersion Square's hardware configuration is based on pure PC technology. The PC hardware components used are scalable in terms of their performance, starting from a single PC system equipped with Matrox Parhelia technology up to a multi processor or multi computer system wired by a LAN network.

1. Introduction

The first functional immersive Virtual Environments (VE) devices, based only on supercomputer technology, have been developed in the 1970's with BOOM [1], HMDs [2] and DataGloves [3] as examples. In the following decade a qualitative enhancement of body-based systems has been seen, as well as the development of first projection-based VE systems like the CAVE[TM] [4] or the Responsive Workbench [5]. Another decade later VE systems have migrated from experimental research to practical application. This migration, however, was accompanied by a tendency to use projective technology rather than ground or body based systems. The concept of "unplugged VE" proved much better suited for professional applications. Industrial requirements also pushed the development of distributed collaborative VE systems [6].
While VE applications used to be more or less restricted to large companies, there was an early tendency to develop PC- or even Internet-enabled VE systems. A first step was the development of VRML[7] as an Internet-ready standard file format and description language for 3D models and scenarios followed by some further approaches like X3D [8] or MPEG-4 BIFS [9] which however, weren't widely accepted. For the first years after the invention of VE technolgies applications of "virtual reality" (VR) on desktop computers (Desktop VR) was restricted to modeling and some simple interactive visualizations on standard 2D screens. With the immense growth of graphic performance of current personal computers, however, pushed forward by the 3D PC-games industry, the development of PC-based VR devices became possible. Now there are several PC-based systems replacing workstation-based VE technology, most of them offering some CAVE-type environment generated by a PC cluster. The Immersion Square visualization system belongs to this young family of pure PC-driven VE systems, which have been developed over the last few years [10,11].

In the following chapter some introductory information about perception in virtual environments are given. Thereafter, a detailed description of the hardware and software components of the Immersion Square visualization technology is presented. Some successfully realized applications are discussed in chapter 5 followed by some concluding remarks.

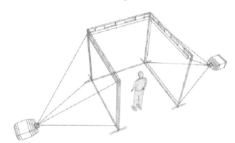

Figure 1: Sample setup of an immersive virtual environment. The field of view of the spectator is surrounded completely by projectable screens.

2. Perceptual concepts

Translating the CAVE concept to PC technology carries some general challenges, first of all the question how to reach an acceptable degree of immersion with considerably less computing power. In particular, the distribution of display modules performed by multi-pipe workstations in traditional VE systems has to be transferred to a cluster of individual computers. In the majority of current PC-based, CAVE-type VE systems this is realized with a cluster of Linux PCs. This concept is focused on stability and maximum performance rather than compatibility to standards, because in most cases individual rendering software has been developed.

The "acceptable degree of immersion" as mentioned above can be achieved by either producing computing (and graphical display) power comparable to graphics supercomputers by a highly connected PC cluster, which acts from outside as a single host, or by reducing the computing requirements to an amount that can be handled by a single host per wall. The former approach requires specialized hardware and software solutions, while the latter depends mainly on application dependencies and perception-related constraints rather applied to the application level.

Immersion in virtual environments depends mainly on the fulfillment of the following basic perceptual concepts:

- exclusion of real environment information from the user's perception channels (see fig. 1)
- presentation of visual (and auditory) depth
- consistent information through different perception channels
- fulfillment of (cognitive and, even more important, sub-conscious) perceptual expectations.

Despite a considerable amount of realism that can be achieved by today's multi-sensory VE systems, they still are all far from the ability to really fool human perception. We are referring here to all kinds of human perception, not only to the visual ones. However, if an application supports imagination (and helps in solving an actual problem), human perception is only too willing to forgive even dramatical visual flaws, as long as some well-known psychological dogmas of perception remain untouched.

3. Hardware components

3.1 Multi screen projection system

The Immersion Square visualization system is conceived as a CAVE type 3D visualization environment [11]. Several three wall immersive projection systems powered either by a multi computer system networked by a 100MB LAN, or by a single computer system, have been realized successfully. The configuration of the Immersion Square visualization system allows in principle for the operation of any number of screens, the only limiting factor is the speed and bandwidth of the connecting network for the network system configuration and the bus system for the single PC setup. The developed prototypes have been demonstrated several times with great success in the last few years (see figs. 2, 3, and 7).

Figure 2: Immersion Square attached with mirror modules presenting an entertainment scene. Presentation in the legislative building of the NRW state government in November 2002.

3.2 Three wall back projection system

The standard setup of the three wall projection system consists of a frame system carrying the seamlessly joined projection screen and three independent mirror system modules, which are attached from the rear-side to each wall (see figs. 2 and 3). The frame system is made of modular aluminium profiles, which can be easily assembled and disassembled. Each wall or screen has a visible size of 2.60 x 2.00 m (5.2 m²) surrounded by a 10 cm black frame.

Figure 3: Flexible projection screen configuration of the Immersion Square (here opened to 135°) demonstrated at the InfoComm Europe 2001. A civil construction scene is presented to be explored.

The back projection screen, manufactured as a single sheet of a size of (3 walls x 2.6 m width x 2.0 m height = 15.6 m²), is attached to the frame by a number of standard press buttons. The smooth and flexible material of the high quality back projection screen allows one to stretch the screen free from creases on the frame system. Furthermore, an optimal solution has been found to obtain sharp vertical edges where the walls join each other [10]. For that the screen material is folded and welded, enabling edge-free back-projection for a seamlessly continuing image around the corners, regardless of the setup angle between the walls. Flexible hinges allow the three wall projection system to be flexibly setup in a continuous range of 90

Figure 4: Demonstration of one mirror module of the Immersion Square. In the left image the independent frame setup carrying the back projection screen and the frame system carrying the mirror is presented. The mirror can be seen in the back of the projection space reflecting a computed view of the Louvre in Paris. The right most image shows the complete covering with a fire resistant cloth.

up to 135 degree configuration angle (inner joints)(see figs. 2 and 3).

The total disassembly time for the frame system including the back-projection screen is 20 min assuming two persons working. The flexible setup, as well as the mobility design in conjunction with the solution of the seamlessly joined projection screens that ensures defined and sharp edges at the joints of the walls, has been patented [10].

3.3 Mirror system

Although the Immersion Square system can be operated without any mirror system in general, the projection distances for the realisation of back projection are quite large even using high performance wide angle optics for the projection machines. So, mirror modules have been developed to reduce the projection distance for each wall by folding the projection (see Figure 4). One mirror module consists of a modular frame system carrying the projection machine on the one side and a foil mirror on the other. Each mirror module is independent and is attached to each projection wall. The frame system is covered by a black fire resistant cloth to darken the projection space in the rear area and to prevent the system from unwanted light sources. By realising these mirror systems combined with 1.1:1 optics for the projection machines, the space requirements for operating the Immersion Square can be reduced from 70 m^2 to 30 m^2 which is a reduction of more than 55%. The average time to disassemble one mirror module is 45 min for two persons working.

The entire equipment of the three wall back projection system and all three modules of mirror systems can be stored in few flight cases which easily fit into a single van. Thus, this technology offers for the first time a really portable solution of a virtual environment which enables a fair amount of immersion.

4. Software design

The general underlying principle of the Immersion Square software development is to support standard rendering tools and to extend these available 3D-visualization solutions to applications in VEs. Therefore, patches for several rendering engines have been developed, which enable a distributed and more importantly a synchronised visualization on multiple screens. Sample solutions for the Virtools [12] engine joined with an application of the Matrox Parhelia graphics card family as well as for the engines of blaxxun Contact 5.1 [13], Half-Life latest version [14], and Cortona 4.1 [15] have been successfully realised and tested. The general concept of the network version of the Immersion Square software system is that each screen's information is rendered independently using one or several disjunct hosts. So a classical client-server or client-master architecture has been realised, in which the master is controlling the spatial and temporal synchronisation of the clients. Since the computation load of the master process is rather small in comparison to the clients, the master thread can be physically computed on the same host as one client. So, for a three wall system just three physical hosts are necessary, while one host is additionally assigned as master. Every client receives a complete set of scene data from the master's hard-disk prior to each session start and then renders each frame for its particular perspective independently. The purpose of the master thread is solely to control the exact visualization time of the different views and to assign the part of the scene to be rendered by each client, according to user navigation. The Immersion Square software has been developed in C++ on a Windows platform. It is therefore completely Windows compatible. The network (LAN) uses TCP/IP to transfer synchronisation messages between the master and the clients.

4.1 OpenGL wrapper

While using a single PC system equipped with Matrox Parhelia graphic processor technology three different VGA outputs can be addressed, so that the visual output of a 12:3 aspect ration can be distributed to three different screens. However, in the standard configuration it is assumed that all three screens share the same perspective. For angled setups of three wall screen systems used for immersive visualisations, however, the perspective of the side views may vary between ±90° and ±135° dependent on the setup of the screen system (see fig. 5). To enable also corrected perspectives for the side views in angled setups an OpenGL wrapper has been developed. The OpenGL wrapper accesses graphic function calls being exchanged between the operating system and the application. The application's graphic function calls are evaluated and a perspective correction is computed for both side views. All three visualisations are finally processed and rendered by the subsystem before execution.

Figure 5: Three screen projection setup with (above) and without (below) perspective correction. The perspective corrections are easily visible at the straight lines of the edges of the side walk.

Research in this area is still in progress. It is intended that the approach should be adapted to PC systems equipped with the new PCI-Express bus system as soon as the technology is available, to be able to run two graphic cards simultaneously. This configuration would scale up rendering performance while maintaining adequate bandwidth on the interconnection between graphic processors and main memory.

4.2 Stereo projections

The impression of visual depth can be achieved through different visual effects. Although stereo projection systems are possible also for PC systems, the *Immersion Square* does not usually provide these since motion and 3D occlusion have been proven to provide sufficient 3D-cues [19]. However a shutter-based stereo projection system can be offered as an optional add-on to the Immersion Square system. It has been shown that for many typical applications, the advantage of more intensive visual depth is overcompensated by the disadvantage of incommodity that goes along with wearing shutter glasses for a longer period in VEs or frequently adding and removing the stereo glasses when entering or leaving VEs.

4.3 Visualisation of panoramic images

In general, the existing software modules can be adapted to other rendering solutions following individual needs and requirements. Furthermore, special solutions for the immersive visualisation of 360° or panoramic images have been developed to enable an integration of QuickTime VR [16] and even more advantageous of Spheron VR [17] high resolution data (see fig. 6). A sample solution for the visualization of synchronised video data has been realised as well based on a Windows Media Player platform.

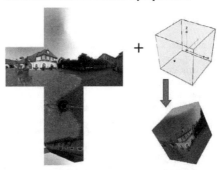

Figure 6: Concept of the computation of a 360° surround image from Spheron VR data. The high resolution data enable even visualizations on large screens.

Based on the consequent application of standard PC hardware components, all known PC compatible interaction devices can be used to interact with the system. Best experiences have been made with the cordless 3D mouse called Gyro Mouse [18].

Figure 7: Demonstration of the Immersion Square at the fair MEDICA 2002 presenting a 3D model of a beating heart.

Based on the above-mentioned special solution of the computation of 360° or panoramic views and the solution for the visualization of synchronised video data it was obvious to look for an easy-to-compute approach to record multi channel video sequences from different viewing angles on the one hand side and to provide a solution for wide field visualizations of these data comparable to those of power wall installations on the other hand side. One goal of this approach is to realize an approach which computes a common image plane also for the other video data recorded under different viewing perspectives [20].

In summary, the Immersion Square technology currently supports the following general application areas:

- immersive VR in interactive scenarios with real 3D geometry, including 3D games,
- quick interactive visualization of panoramic images projected on simple 3D geometries,
- immersive video display.

5. Application areas

High-performance graphics applications are migrating from specialized workstations to PC hardware. Although large-scale virtual scenarios (> 1 million polygons) equipped with many 3D details and expensive textures will currently not be able to be rendered with an acceptable performance on standard PC hardware, there are an increasing number of 3D data which can be already computed and visualised without the need of high performance graphics workstations. In many industrial and commercial applications 3D data are routinely produced or designed using standard CAD tools.

Most of these tools are equipped with import and export functions, which allow an easy transfer of the 3D data to standard formats. Furthermore, the manual construction of 3D-models and the acquisition of high quality textures is often time consuming and expensive. Therefore, the visualisation of video data and 360° or panoramic images is advantageous.

For more widely spread industrial and commercial applications of immersive visualization techniques, low cost solutions have to be developed that include a variety of mandatory features such as compatibility to standard data formats used routinely in work flows and accessibility within a standard local network. Furthermore the entire system must be easy to handle. In particular, the computer system should be easy to administrate. All those constraints and considerations will lead to visualization systems which are PC based and/or apply widely used standards.

In this context the Immersion Square technology has already been proven to work well in several application areas, such as:

- Architecture and civil construction (see figs. 3, 4),
- Biotechnology and Medicine (see figs. 7, 8),
- Mechanical and automotive engineering, and
- Entertainment and art (see fig. 2).

Figure 8: Visualisation and exploration of a lymphocyte model in the Immersion Square.

6. Conclusions

An immersive visualization system has been introduced which is fully modular and therefore mobile and easy to transport. The Immersion Square technology primarily aims at scalability and portability, making use of hardware and software standards in order to take advantage of progress being made in any of these technologies. This concept is thought to open wide application areas for marketing, construction and exploration tasks of a considerable number of 3D and VR data.

Relying only on standard hardware components, the Immersion Square hardware system can easily be configured according to user requirements. Thus, 3D scenes of an average size of up to 500.000 polygons (referring to current technology — with any new hardware development pushing the barrier) are possible to render depending on the performance of the rendering engine (see above) and of the processor/graphic card and bus system.

A reduction of the projection distance by using a mirror system is crucial to enable applications outside standard scientific environments. Furthermore, by covering the mirror modules the system gains independence since it can be placed almost in every environment (without the necessity of special external lighting conditions). The general design of the system is open, which means that more projection screens can easily be added. In summary, the Immersion Square technology is scalable in terms of display quality, spatial requirements, hardware configuration, rendering performance, and price.

Initial success has been made at developing and integrating multi channel video data, 360° and standard panoramic images into immersive visualization environments. The development of an alignment approach for multi channel video sequences, as well as the synchronization of the Windows Media Player, are some examples. The common underlying goal of this research is to provide low cost alternatives in the developing market of PC based VR and VE systems.

Acknowledgements

The authors gratefully acknowledge the financial support of the TRAFO program project "Synko-Square".

References

[1] Bolas, M.: Human factors in the design of an immersive display, IEEE Computer Graphics and Applications, January 1994

[2] Sutherland, I. E.: A head-mounted three dimensional display, Proc. AFIPS Fall Joint Computer Conference, 33, pp 757-764, 1968

[3] Sturman, D. J., Zeltzer, D.: A Survey of Glove-based Input. IEEE Computer Graphics and Applications, pp 30-39, January 1994

[4] Cruz-Neira, C., Leight, J., Papka, M., Barnes, C., Cohen, S.M., Das, S., Engelmann, R., Hudson, R., Roy, T., Siegel, L., Vasilakis, C., DeFanti, T.A., Sandin, D.J.: Scientists in Wonderland: A Report on Visualization Applications in the CAVE Virtual Reality Environment, Proc. IEEE Symposium on Research Frontiers in VR, pp 59-66, 1993

[5] Krüger, W. et al.: The Responsive Workbench, a Virtual Work Environment, IEEE Computer, July 1995

[6] Lehner, V., DeFanti, T.: Distributed Virtual Reality: Supporting Remote Collaboration in Vehicle Design, IEEE Computer Graphics and Applications, 17, 2, pp 13-17, 1997

[7] VRML'97 specification, On-line document, accessed on 09-Feb.-2004, available at: http://www.web3d.org/Specifications/VRML97/.

[8] X3D specification, On-line document, accessed on 09-02-2004, http://www.web3d.org/x3d.html.

[9] Battista, S., Casalino, F., Lande, C.: MPEG-4: A Multimedia Standard for the Third Millennium, Part 1, IEEE Multimedia, On-line document, accessed on 09-Feb.-2004, http://www.computer.org/ multimedia/articles/mpeg4_1.htm.

[10] R. Herpers, F. Hetmann, Dreiwandprojektionssystem, Deutsches Patentamt, Az. 101 28 347.4, 2001.

[11] F. Hetmann, R. Herpers, W. Heiden: "The Immersion Square - Immersive VR with Standard Components", VEonPC02 Proceedings, Protvino, St. Petersburg 2002.

[12] Virtools, http://www.virtools.com, accessed on 09-Feb.-2004.

[13] blaxxun interactive homepage, http://www.blaxxun.de accessed on 09-Feb.-2004.

[14] Sierra Studios, "The Official Half-Life website", http://www.sierrastudios.com/games/half-life accessed on 09-Feb.-2004.

[15] ParallelGraphics, http://www.parallelgraphics.com accessed on 09-Feb.-2004.

[16] Apple Computer Inc., http://www.apple.com quicktime accessed on 09-Feb.-2004.

[17] SpheronVR AG, http://www.spheron.com accessed on 09-Feb.-2004.

[18] GyroMouse pro, Gyration, Inc. Saratoga CA 95070 USA, http://www.gyration.com accessed on 09-Feb.-2004.

[19] Harris L.R., M. Jenkin, D. Zikovitz, F. Redlick, P. Jaekl, U. Jasiobdzka, H. Jenkin, R.S. Allison, Simulatine self motion I: Cues for perception of motion, Virtual Reality, 6, pp 75-85, 2002.

[20] Balsys M., Alignment of image data from a rectangular dual vision system for the visualisation in the immersive environment 'Immersion Square', Diploma Thesis, Department of Computer Science, Bonn-Rhein-Sieg University of Applied Sciences, Sept. 2003.

Bildverarbeitung │ Image Processing

An Open-Source System for Radiation Therapy

O. Grünwald[1], M. Oehler[1], S. Ostrowitzki[2], U. Altenburger[2], C. Haller[2], J. Ruhlmann[1) 2)] and T. M. Buzug[1]
1) Department of Mathematics and Technology, RheinAhrCampus Remagen, 53424 Remagen
2) Medical Center Bonn, Spessartstrasse 9, 53119 Bonn

Abstract

Although several industrial radio-therapy planning solutions presently exist, most of them do not cover the customer needs sufficiently. The aim of this project is to develop an open source radio-therapy planning system that would offer the ground work for individual customer solutions. During the first stage of the project we concentrate on the development of a visualisation module, since it is required for other parts such as dose calculation and beam modelling. The visualisation module will feature 2D and 3D visualisation of CT-, MRI- and PET DICOM images, registration of the images using Normalised Mutual Information, 2D and 3D visualisation of the matched images as well as 2D and 3D user interaction. In this paper an outline for our system is presented as well as a registration module, which is the first part of the visualisation module.

1 Oncology Management System

1.1 Oncology Management System

Oncology is a multi-faceted medical subject that involves cooperation of physicians from very different areas of expertise. The concept of cancer includes more than a hundred malignant diseases, which in turn can show varying characteristics. Thus, oncology is also a very complex medical field [2]. An Oncology Management System integrates all domains of oncology in one system and assumes administration and maintenance duties. This can be implemented in the form of an electronic patient record, which is available to all users over a central data base, inside the diagnostics department as well as inside the radiotherapy and chemotherapy departments. All relevant data, such as findings, images, prescriptions and others could then be recorded inside the electronic record, immediately becoming visible to all departments and leading to a decrease in administration efforts.

1.2 A System for Radiotherapy Planning

In this chapter an outline of a planning system for radiotherapy is presented, being an essential part of an Oncology Management System. In doing so the layout is limited to tele-roentgen therapy. The irradiation is carried out with the source of radiation positioned outside the patient. In most cases the radiation source is a Linear Accelerator. First, images from the diagnostics department are imported into the system. This is achieved best using a central data base, which stores all relevant patient data. Presently, the most common medical image format is DICOM (Digital Imaging and Communications in Medicine). It was developed by the American College of Radiology (ACR) and the National Electrical Manufacturers Association (NEMA). The Hounsfield values in the CT images are used to derive the density matrix and are therefore essential to dose calculation. Other modalities such as PET, SPECT and MRI introduce anatomical information that is mostly complementary to CT and are increasingly used in planning. During the next stage of the planning procedure the physician has to define the target volume as well as the organs at risk. This process is called segmentation and is described in detail in section 1.2.1.

1.2.1 Segmentation

The entire target volume is composed of three volumes. The Gross Tumour Volume (GTV) is the tumour tissue, which is visible in the diagnostic images. This volume is enclosed by the Clinical Target Volume (CTV) that accounts for unobservable extensions of the tumour as well as surrounding metastases or lymphoma. The Planning Target Volume (PTV) adds a safety margin to the CTV, allowing for physical errors caused by inaccurate position of the patient or the patient's movement during the irradiation. During treatment the PTV is supposed to receive the therapeutically effective dose whereas the dose inside the organs at risk has to be as small as possible. [1]
The segmentation of the PTV and organs at risk can be carried out manually or automatically. Manual segmentation is very time consuming since all slices of an image have to be handled separately. The most common methods are freehand drawing and polygon drawing. In most cases automatic segmentation is carried out using edge detection and region growing algorithms. [1] Figure 1 shows a CT slice with the PTV segmented using freehand drawing (red colour).

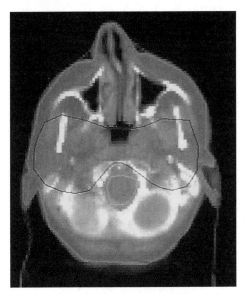

Figure 1. Planning Target Volume

The demonstration of the images is a very important aspect to segmentation. One of the possibilities is the display of one slice at a time involving an interactive scroll function. In addition to that axial and sagittal views can be interpolated from the data set resulting in a so called 2½D display.

As mentioned before, other modalities are often used in planning along with CT. To make the simultaneous display and processing of different modalities possible, the coordinate systems of the images have to be matched accordingly. This problem is called image registration and is further addressed in section 2 of this paper.

1.2.2 Adjustment of Irradiation Parameters and Beam Modelling

During the design of the plan the physicist has to utilise the technical facilities at hand to achieve an optimal dose distribution during treatment. The radiation type is one of the parameters that need to be set. In most cases roentgen rays are used, because of their penetration characteristics, being suitable for superficial treatments as well as treatments at higher depths, depending on the energy of the radiation source. Electron beams are used for superficial treatments because of their limited penetration depth. Recently there has also been research on treatment with protons and neutrons [1,2].

The beam direction, formed by the position of the treatment table and the gantry, is changed several times during treatment. The individual configurations for each irradiation interval have to be chosen carefully, taking the spatial position of the PTV and the organs at risk into account. Several convenient display methods are essential to this task. The Beam's Eye View (BEV) displays the beam settings on a surface model derived from the images or, in the simplest case, a cross section model. The point of view equals the current position of the radiation source as shown in Figure 2 [1].

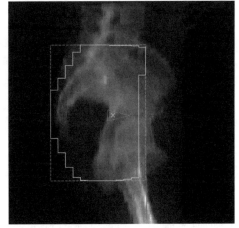

Figure 2. Beam's Eye View

The contour of the beam is plotted into the image (yellow in Figure 2) and is then adjusted interactively to equal the projection of the PTV. The Observer's View offers a three-dimensional view of the anatomy including all beams from any point of view, thus helping to limit the intersection volume of the beams to the PTV and minimise the exposure of the organs at risk. The Spherical View is another display method, which helps to find possible beam directions in the first place. The globular orbit of the radiation source, resulting from table and gantry movements is displayed as a spherical surface where gantry and table rotations represent the meridians and the parallels, respectively. The organs at risk are projected onto the surface so that the possible positions of the beam are to be found in between. [1]

Another parameter that needs to be evaluated is the beam intensity. There are several different approaches to this task. Earlier wedge filters were used to create a linear intensity distribution, decreasing constantly in one direction. Presently, virtual wedge filters are used where a blind is driven constantly over the field aperture thus decreasing the exposure time linearly [1,2].

Presently, the Multi Leaf Collimator (MLC) technology is becoming a standard in planning. The MLC consists of a number of opposed pairs of thin wolfram blades that can be computer operated to form arbitrary field contours. The configuration of the MLC can be changed for every irradiation interval during treatment, thus leading to a better alignment of the beam profile to the target volume. The Multi Leaf Collimator can also be adjusted successively or dynamically, leading to varying intensities inside the lateral beam plane (IMRT: Intensity Modulated Radiation Therapy) [1].

Since the plans for many tumour locations do not vary much for individual cases, a planning system should offer a range of standard plans stored in a data base. It should allow individual configuration of the provided standard plans as well as the generation of user defined standard plans.

1.2.3 Dose Calculation

The characterisation of an optimal plan is a consequence of certain clinical criteria. These can be the tolerated dose deposed in the organs at risk or the homogeneity of the dose distribution inside the PTV. In practise a compromise has to be made between the accuracy of the calculated dose distribution and the computing time. Therefore a planning system should offer several algorithms for dose calculation allowing the physicist to choose the most appropriate method depending on the individual oncological case. In this section the most common dose calculation algorithms are introduced.

Pencil Beam is counted among the convolution algorithms, the central principle of which is the calculation of energy deposition of a primary beam in water resulting in a corresponding energy dose kernel. In the Pencil Beam method the primary beam is an infinitely thin beam that impinges the surface of an atom. During the dose calculation the Linac spectrum is treated as a sum of separate primary beams (primary fluence) meaning that the dose in any point is the sum of the contributions of all energy kernels. Assuming that there are no inhomogeneities in the plane that is lateral relative to the beam direction, the applied dose can be calculated as the convolution of the primary fluence and the dose kernel. Convolution algorithms are very fast and can be used for dose calculation of irregularly formed fields. The correction of inhomogeneities inside the lateral plane is achieved by scaling the dose according to depth, relative to water, and is therefore very inaccurate in regions with high tissue inhomogeneities [1].

In superposition algorithms the dose kernel of the Pencil Beam is reduced to a small point so that it equals the dose distribution caused by absorption of a primary photon in water. Tissue inhomogeneities are taken into account using a three-dimensional scaling of the point kernel with the relative electron densities.

The total dose distribution is then the superposition of all point kernels. Superposition algorithms are more accurate than the Pencil Beam, but they are also very time-consuming [1].

The Monte Carlo method has proved to be the most accurate one. Here, the propagation of the particles and their interaction with tissue are considered as a series of random events. Since the probabilities of the aforementioned events are known, the trajectory and the energy loss of the radiation can be calculated using random numbers. The total dose distribution is the average value energy transfers of many particles (up to 100 Millions). The computing times for this algorithm are still very high, however, hindering its clinical use [1,3].

1.2.4 Demonstration of Dose Distribution

After the dose distribution is calculated it has to be displayed in a way that allows the physician to review it and choose the best plan in case there are alternatives. Two- and three-dimensional views provide a quick overview. Equal doses are displayed as lines, bands or semi-transparent clouds [1]. Figure 3 shows a CT slice with the areas of equal dose enclosed in colour coded lines.

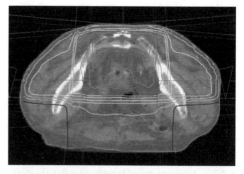

Figure 3. Equal dose display

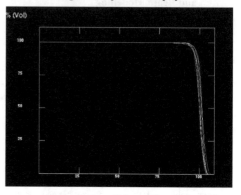

Figure 4. Dose Volume Histogram

Dose Volume Histograms are also very important tools in plan review. They are less complex than three-dimensional dose demonstrations and help to discover overdoses and inhomogeneities in dose distribution comparably easily. Figure 4 shows a Dose Volume Histogram with the volume portion in percent plotted against the relative dose applied to it [1].
Summation and Subtraction tools are also very important for Dose Volume Histograms in order to compare alternative plans.

1.2.5 Inverse Planning

In inverse planning the program uses input specifications in order to find the optimal plan parameters. These input specifications can be tolerance dose restrictions for organs at risk, the desired dose distribution inside the target volume as well as the corresponding weighting factors, which denote the importance of the specifications relative to each other [1].
The optimisation is based on objective functions, which can be physical or biological. It is usually carried out iteratively using a gradient algorithm or a stochastic method called Simulated Annealing. The gradient algorithm scans the objective function successively and finds a minimum or a maximum by comparison of the values on the right side to the values on the left. Although the gradient algorithm tends to be fast it has the disadvantage of being instable regarding local optima. The Simulated Annealing algorithm calculates the search directions randomly. If the new parameters result in a better dose distribution, the new settings are accepted and the search is continued. In case the new dose distribution is inferior to the previous one, it is also accepted being weighted with a probability, which becomes smaller as the optimisation goes on. Thus, the algorithm does not stop after reaching a local optimum, which is a great advantage as opposed to the gradient algorithm. Since many iteration steps are required to find the optimum, the algorithm is highly time consuming [1].

1.3 Industrial Solutions

In this section we introduce several commercial Oncology-Management-Systems and radiotherapy planning systems.

1.3.1 LANTIS, Siemens

LANTIS is a Siemens Oncology Management System that provides a range of information management tools implemented in a modular design.
The *Commander* module provides the groundwork for building a paperless electronic patient record. Administrative tasks such as the management of patient, stuff and location schedules are carried out by the *Supervisor* module. *Digital Photos and Facts* helps to verify patient identification and field data. The *Image RT* module allows the integration of all patient information, including diagnostic and treatment images, in one management system. The *Clinical Assessment* module helps to track and evaluate the treatment progress providing the possibility to make individual adjustments to the treatment process. The *Reporter* module provides more than 90 standard reports for documentation. Other modules are *RTP Link* for the transfer of plans from the planning system to the electronic patient record, *Gateway* for the interface between the Oncology Management System and the Hospital Information System, *Transcriber* for document management and others.

1.3.2 VARiS Vision, Varian

VARiS Vision is a Varian Oncology Management System. One of the aims of this system is to rationalise and accelerate the clinical treatment process. VARiS Vision also uses an electronic patient record stored in a central data base.
The module *Patient Manager* is an entry point for all information inside the electronic record. It includes a summary page with patient identification information, overviews for appointments and tasks, prescription and treatment data, physician information, insurance information, diagnosis and staging information and summaries for radiation therapy and medical oncology. From within the Patient Manager the user has access to the Radiation *Therapy Chart*, which provides the management for the complete treatment process. The *Time Planner* module manages resources and activities. *Activity Capture* allows the user to capture financial data for cervices rendered and activities executed within the system. The *Document Manager* module can be used for dictating, typing, importing and storing of clinical documentation. The module *Vision* allows management of images and treatment plans.

1.3.3 Eclipse, Varian

Eclipse is a Varian planning system for radiotherapy. This system offers a wide range of components implemented in separate modules.
The *Virtual Simulation* module converts the medical image data into a patient model that can be used for target delineation and field design. The segmentation module *Contouring* offers advanced drawing and editing tools as well as automatic segmentation tools including region growing. The *Field Setup* module offers a number of standard plans, which can be adjusted for individual cases using a *Wizard*. There are several algorithms provided for dose calculation,

which can be carried out in the background. *Interactive IMRT Planning* is a highly interactive module, which is used for inverse planning and allows interactive change of plan specifications. *Plan Evaluation* is plan review module that allows locking the displays of alternative plans together, so that the same view is used for each plan in evaluation. This module also offers several 2D and 3D views as well as Dose Volume Histograms with additional summation and subtraction tools.

1.3.4 XiO, CMS

XiO is a CMS radiotherapy planning system. The segmentation module offers drawing and editing tools as well as patented tools for automatic segmentation. Display possibilities during planning and plan review include the Beam's Eye View, Observers View, transversal, coronal and sagittal plane views as well as displays of equal dose lines/areas and Dose Volume Histograms. XiO offers several algorithms for dose calculation. The Fourier Transform Convolution algorithm and the Multi-Grid Superposition algorithm are used for photon dose calculation. Electron dose calculation is carried out using the Pencil beam algorithm. A module for inverse planning for IMRT is also provided.

2 Registration Module

2.1 Registration Using Geometry of PET-CT Appliance

The images were imported using the open source DICOM Toolkit by Kuratorium OFFIS e. V. and registered using the geometry information of the PET-CT appliance. During the registration process the PET image is interpolated to match the CT image using trilinear interpolation.

2.2 Normalised Mutual Information

In information theory, entropy can be considered as a quantitative measure of uncertainty and, therefore, it provides a convenient measure of information content of a signal. This concept may be applied to imaging as well, including medical imaging. In this case, the intensity of each voxel represents a stochastic variable X, which can have a value x_i, and the whole image, being a set of voxels, provides information on human anatomy. The amount of information provided by an image is proportional to the number of different voxel intensities in it and the diversity of their probabilities of occurrence [5,6].
The Shannon-Wiener entropy measure $H(X)$, which is defined as the expectation E_X of the logarithm of the probability $p(X)$ of a stochastic variable is the most common definition of entropy:

$$H(X) = -E_x[\log(p(X))]$$
$$= -\Sigma\, p(X = x_i) \cdot \log(p(X = x_i))$$

If we consider A and B as the sets of values of the images a(x) and b(x), where x denotes the image position, then the average entropy of A and B can be calculated from their histograms as follows:

$$H(A) = -\sum_{a \in A} p(a) \cdot \log(p(a))$$
$$H(B) = -\sum_{b \in B} p(b) \cdot \log(p(b))$$

where $p(a)$ and $p(b)$ are the marginal probabilities provided by histograms, and $H(A)$ and $H(B)$ are the marginal entropies of A and B, respectively [5,6].
The joint entropy $H(A,B)$ can be used as a measure to show how well the two images are aligned. It is calculated from the joint histogram, where the members of the set A are along the x-axis and the members of the set B are along the y-axis, as shown in figure 5. Then,

$$H(A,B) = -\sum_{a \in A}\sum_{b \in B} p(a,b) \cdot \log(p(a,b))$$

where $p(a,b)$ is the joint probability. In the case of best alignment, the joint entropy is minimal [5,6].

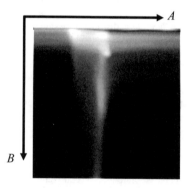

Figure 5. Joint histogram

To maximise the information content in the overlap, the marginal entropies must be maximised and the joint entropy must be minimised. In order to make the similarity measure independent to the amount of overlap, we use the normalised mutual information (NMI) as our similarity measure, defined as:

$$NMI(A,B) = \frac{H(A) + H(B)}{H(A,B)}$$

2.3 Examination of NMI

In order to understand and analyse the performance of the optimisation algorithm the behaviour of the Normalised Mutual Information was investigated using a PET-CT data set. Figure 6 shows the results for horizontal translations from -10mm to 10mm, assessed in 0.5mm steps. The results for the in-plane rotations from -5° to 5°, assessed in 0.5° steps, are plotted in Figure 7.

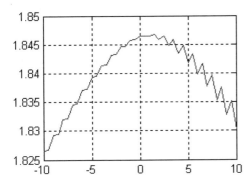

Figure 6. NMI, plotted versus horizontal translations.

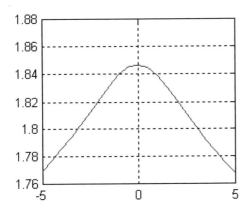

Figure 7. NMI plotted versus in-plane rotation.

2.4 Optimisation

The registration of the images achieved using the geometry information of the combined PET-CT appliance is optimised using an iterative gradient descent algorithm. At this stage we concentrate on the optimisation of three of the six degrees of freedom of a rigid body transformation. These are the in plane rotation and the horizontal and vertical in plane translations. The algorithm finds the optimal transformation by comparing the values of Normalised Mutual Information for transformations achieved with each parameter separately and choosing the parameter which increases the NMI value the most. A definite step size of several millimetres is set for the initial transformation and reduced several times during the optimisation process, acting as a multiple resolution approach.

The data set used in this study was optimised with an initial step size of 5mm for the translations and 5° for the rotation. The step size was reduced four times during optimisation down to 0.3125mm and 0.3125°, respectively. The resulting parameter changes were a horizontal translation of 0.9375mm, a vertical translation of 1.25mm and a rotation of 0°. The Normalised Mutual Information was increased from 1.84643 to 1.84709 during optimisation. Figure 8 shows a section of a slice of the matched image prior to optimisation on the left side and after optimisation on the right side.

Figure 8. PET-CT image (CT: blue, PET: red), left: prior to optimisation, right: after optimisation.

3 Future Work

Subsequent work will include a robustness test of the optimisation algorithm and an upgrade of the module for registration of images acquired with separate diagnostic appliances.
In further development the registration module will become a part of a visualisation module which will feature 2D and 3D visualisation of CT, MRI and PET DICOM images, 2D and 3D visualisation of the matched image as well as 2D and 3D user interaction.

4 References

[1] Medizinische Strahlenphysik, W. Schlegel, J. Bille, Springer 2002
[2] Grundlagen der Strahlentherapie, E. Richter, T. Feyerabend, Springer 2002
[3] Entwicklung schneller Algorithmen zur Dosisberechnung für die Bestrahlung inhomogener Medien mit hochenergetischen Photonen, C. Schulze, 1995, Ruprecht-Karls-Universität Heidelberg
[4] BEAM: a Monte Carlo Code to Simulate Radiotherapy Treatment Units, Medical Physics, Ausg. 22, Nr. 5, 1995
[5] An Overlap Invariant Entropy Measure of 3D medical Image Alignment, C. Studholme, D.L.G. Hill, D.J. Hawkes, Pattern Recognition, Ausg. 32, S. 71-86, 1999
[6] Medical Image Registration, Joseph V. Hajnal, Derek L.G. Hill, David J. Hawkes, CRC Press, 2001

Hardware Accelerated 2D/3D Image Segmentation for Real-Time Medical and Industrial Applications

P.H. Dillinger[1], J.-F. Vogelbruch[1], J. Leinen[1], S. Suslov[1], M. Ramm[1], R. Patzak[1], K. Zwoll[1], H. Winkler[2], K. Schwan[2], H. Halling[1]

[1]Central Institute for Electronics, Research Center Jülich, D-52425 Jülich
[2]Array Electronics GmbH, Ehamostr. 27, D-85658 Egmating

Abstract

Real-time image segmentation is one of the most demanding tasks in image processing in the future. The applications comprise scientific tasks, e.g. segmenting volume data sets for medical inspection, as well as for industrial use, e.g. 2D surface inspection. The new Grey-Value-Structure-Code (GSC) algorithm meets the quality requests and is perfectly parallelisable on a suitable hardware platform presented here. This new FPGA-board features an up-to-date Virtex-II-Pro architecture, two large and independent DDR-SDRAM channels, two fast and independent ZBT-SRAM channels, a PCI bus, CameraLink interface, and two 32 Bit extension connectors. The hardware implementation can speed up the complex segmentation processing of 4096^2 images up to 7 fps.

1 Introduction

One of the elementary steps in scientific and industrial image processing is image segmentation. Neighbouring pixel or voxel values will be considered identical if a common feature is similar. For most applications their grey value will be used as feature.

Many works describe numerous different techniques and strategies of image segmentations (see [1-5]):

- point oriented techniques (e.g. threshold methods) [6]
- edge detection [7] and watershed techniques [8]
- region growing techniques (e.g. region growing/merging [9], split & merge [10], pyramid linking [11]
- deformable models [12]
- direct in-image-data classification methods [13]

After the segmentation the new combined areas are classified into application specific objects like bones or tissue in medical applications, or homogenous / distorted regions in industrial quality assurance tasks. The quality of various segmentation algorithms differs in many ways, so that for every application the best suitable one should be selected to match the requirements. Not every algorithm can be accelerated by an FPGA or DSP processor. Rehrmann [14] proposes a 2D segmentation algorithm (CSC) based on a hierarchical approach of Hartmann [15] to include local precision by keeping a global overview.

Vogelbruch [16-18] extended this algorithm to 3D (3D-GSC) by identification of an appropriate 3D island structure and proving this to be unique in 3D. Both approaches are perfectly parallelisable because the island itself and islands of each hierarchy level can be processed independently. The only disadvantage is the high memory amount and bandwidth needed to keep all the parallel processing units busy and make them most efficient.

In the past 10 years many special hardware solutions (e.g. CNAPS [19], IMAP [20], SYMPHONIE [21], CC-IPP [22], and SIMD architectures [23]) have been proposed but none has been successfully brought to market. Today image processing tasks of low complexity are implemented on graphic controllers with DSP units (e.g. Matrox Genesis). For the solution of more complex tasks a significant trend to use standard FPGA architectures [24-26] is visible. But the available hardware architectures still do not meet the complex requirements of innovative algorithms used here. The proposed hardware platform described in this paper satisfies the needed memory requirements of the 3D-GSC algorithm. It contains two large banks of SDRAM and two banks of fast SRAM for caching purposes. The memory word width is set to 128 Bit per bank to match the island structure of the GSC algorithm. Due to the use of an FPGA as processing core the board is suitable for many other image or high memory demanding algorithms.

2 2D/3D Segmentation Algorithm

The 3D-GSC [16] merges local precision and global view by re-evaluating homogeneity decisions taken on a lower hierarchy level on the basis of the global view through a subsequent splitting of contiguous but non-similar regions. The basis of the 3D-GSC is a newly developed 3D island structure (shown in Figure 1) with the following capabilities explained in a simplified way for the 2D case (in Figure 2):

- homogeneous periodical lattice ⇒ efficient algorithmic realization
- covering of all lattice points ⇒ complete region linkage
- central symmetrical islands ⇒ isotropic region linkage
- complete simple overlapping ⇒ unique connectivity + splitting
- simple hierarchy ⇒ multi-scale approach, recursive implementation

Fig. 1 3D hierarchical island structure

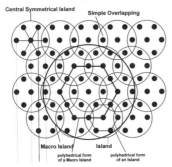

Fig. 2 2D hierarchical island structure

It has been shown that only the 14-neighbourhood of a rhombic dodecahedron can satisfy the above mentioned requirements. But due to the inhomogeneous neighbourhood structure of the rhombic dodecahedron not all overlapping points in a macro island can be considered for region linkage. This demands for explicit splitting of those overlappings.

The generation of the 3D-GSC takes place in the following phases:

- In the **coding phase** neighbouring and similar voxels are combined to local regions of the lowest hierarchy level.
- During the following **linking phase** (Fig. 3) these regions are linked hierarchically to global segments up to the highest hierarchy level. A region of one hierarchy level consists of contiguous and similar regions of the hierarchy level underneath.

The grey value of this region is calculated by the grey value mean of the participating regions.

- During the linking phase two regions can be non-similar, but overlapping. Therefore, in order to obtain a disjoint segmentation result, the overlapping area must be separated afterwards. This procedure is carried out recursively down to the lowest hierarchy level during the **splitting phase** (Fig. 4), which is initiated immediately after the linking of an island.

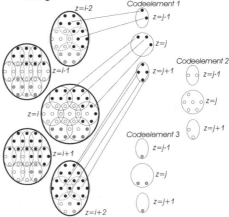

Fig. 3 Three dimensional linking

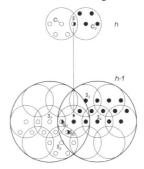

Fig. 4 Two dimensional splitting

3 Hardware Acceleration

For segmenting 2D images up to 4096^2 pixel and 3D images up to 512^2 voxel (16 bpv) in real-time a hardware acceleration is required. This will be realized with a PCI compliant extension board to support standard PC systems. The core of this coprocessor board is a Xilinx Virtex II Pro FPGA processor ($> 10^6$ gates) equipped with sufficient local memory organized in at least two separate channels. The latter is needed for concurrent access to overcome the memory bandwidth bottleneck of the FPGA and to enhance parallelism.

Industrial applications (e.g. for quality assurance) can be accomplished by connecting an external camera directly to the FPGA board or via the CameraLink. Image data can also be transferred from any other source via the PCI bus (64 Bit at 66 MHz). For future purposes two 32 Bit extension connectors are available. In later versions it is planned to replace the PLX chip with PCI-Express capabilities.

A user front-end software running on the host PC controls the FPGA board and takes over any required preprocessing of the image data. The FPGA segmented image data is transferred back to the application software where further processing, monitoring, and storing takes place.

3.1 128 Bit Coprocessing Board

The FPGA coprocessing board contains at least two independent memory banks (Fig. 5). These two banks will be socketed to support DDR SDRAM modules (up to 1 GB) with 266 MSamples/sec (PC2100). Because of the slower FPGA design clock, these modules will be operating at 220 MHz to obtain doubleword data (128 Bit) at 110 MHz on the FPGA ports (about 1.6 GByte/sec per channel). Two additional and smaller SRAM memory banks (up to 8 MB) will be supplied for caching purposes. These ZBT (zero bus turnaround) banks for fast random access are also designed to be independent and operate at 110 MHz and 128 Bit data width. All four banks achieve an overall data bandwidth of about 6.4 GByte/sec. The use of this 128 Bit processor board is not limited to this 2D/3D segmentation algorithm, but can also be used in many other applications.

3.2 Acceleration Strategies

The acceleration model used for the 2D/3D segmentation algorithm can be entitled as *Dual Channel with Half/Full Caching*. The SDRAM memory is used separately as input/output data and the SRAM is partly used for caching purposes to obtain linear access to the SDRAM which improves its timing characteristic. There are at least two other possibilities to use this memory arrangement on the FPGA board.

Combined Channel with Half/Full Caching
In this configuration the two SDRAM banks are combined to access either 256 Bit data width with normal speed or 128 Bit with double speed (for Dual Port capability). A similar setup is used in current Intel-based mainboards, where a high memory bandwidth is gained by dual channel access to the memory. An arbiter is necessary to distribute memory access for best performance. Some part of the data can be cached using the SRAM for linear and faster SDRAM access.

Single Channel with or without Caching
For Virtex II Pro Chips where two PowerPC processor cores are integrated, these cores can have its own independent SDRAM memory. The SRAM can be used to cache the access or to extend the total of independent memory banks to four. This is possible not only for the CPU cores but for all independent working designs inside the FPGA. Some kind of interconnect protocol is needed to exchange data between autonomous FPGA designs.

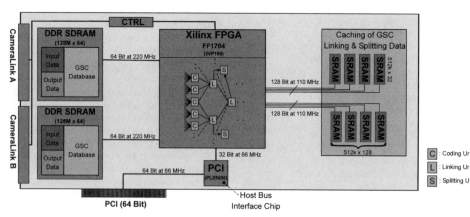

Fig. 5 Block diagram of the FPGA board

4 Hardware Implementation

4.1 HW-Adapted Data Structure

The data structure has been modified for memory access reduction in hardware:
- 128 bit word length for single access of all regions of an island of hierarchy level
  $\Rightarrow$ less memory access operations
- relative indices instead of absolute addresses for database entries
  $\Rightarrow$ lower memory consumption
- Single-linked tree structure (Subcode pointers in the software version will be replaced by position computing and comparison of father indices)
  $\Rightarrow$ regular size of region entries (no size fields)
  $\Rightarrow$ omission of indirect addressing via GSC Key Table

4.2 Coding Modules

For the coding phase 20 different algorithms (10 SW, 10 optimised for HW) have been examined to determine the best implementation (Fig. 6). The verification criteria have been:
- Sensitivity: ability to generate different numbers of regions over a broad threshold interval
- Singularity: ability to avoid singularities (pixels which can not be linked to regions)
- Coverage of reference patterns determined by the average of a set of manual segmentations of 20 characteristic patterns.

For each pattern the best 10 algorithms receive a ranking score for each criteria (bar graph in Fig. 6) weighted with its importance to obtain a comprising measure (upper curve in Fig. 6).

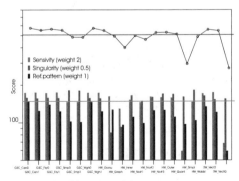

Fig. 6 Algorithm comparision

Under these conditions the best HW algorithms are "HW Wobble" and "HW No.2". Because of the lower FPGA resource consumption (600 LUT compared to 1200 LUT) "HW No.2" has been chosen for the implementation of a coding module (CM).

The "HW No.2" algorithm is operating in 4 steps (Fig. 7). In each step the edges are examined for similarity to form intermediate regions (3 comparisons in one step). For the next step the corresponding voxels are updated with the intermediate region mean value.

Fig. 7 Region detection in 4 edge phases for the "HW No.2" algorithm

For the parallel detection of regions in islands in the scope of a macro island at most 20 CMs are required (Fig. 8). These regions form the basis for the subsequent linking phases. The phases in Fig. 8 correspond to the macro island positions in an image shown in Fig. 9.

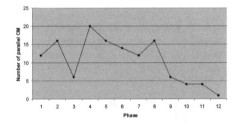

Fig. 8 Number of required parallel CM for different phases in an image

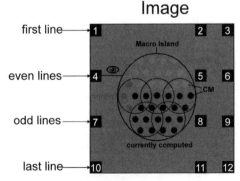

Fig. 9 Phases in an image

5 Performance Enhancement

5.1 Software Runtime Optimisation

Runtime optimisation has been performed both by manual code improvements (man. opt.) and using compiler and processor-specific optimisations (comp. opt. and add. µProc opt.). Fig. 10 compares the runtimes on an Intel-PC with Pentium4@2.66 GHz and 1 GB of DRAM.

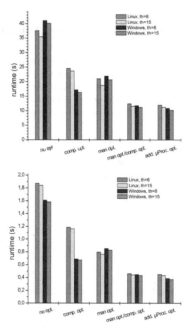

Fig. 10 (a) generation of a 256^3 label dataset with 3D-GSC, (b) generation of a 1024^2 label dataset with 2D-GSC

5.2 Hardware Performance Estimation

Fig. includes the performance estimation for the 2D case based on memory accesses for the specially designed HW architecture presented in Fig. 5. The worst case is given when both the image data and the database are accessed totally sequentially leading to a speedup of approximately 28. Exploiting the 4 memory banks for parallel access and avoiding overhead the best case can be obtained resulting in a speedup of approximately 50.

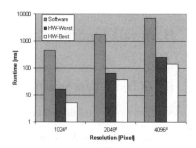

Fig. 11 Performance estimation in hardware

5.3 Segmentation Results

Fig. 12 shows an image segmentation example of a simulated 3D MRI brain dataset [28]. On the basis of the segmentation results 99,985% / 98,5 % of all voxels in a noised ellipsoid phantom / simulated MRI data set respectively are detected properly as class members using a simple n-to-1 classifier.

Fig. 12 (a) original image (b) initial region codings (c) intermediate codings, (d) completely segmented (e) classified with n-to-1 classifier

6 Acknowledgement

This work is supported by the BMBF grant No. 01 IR C01 A.

7 References

[1] L.P. Clarke, R.P. Velthuizen, M.A. Camacho, J.J. Heine, M. Vaidyanathan, L.O. Hall, R.W. Thatcher, M.L. Silbiger: MRI segmentation: methods and applications. Magnetic Resonance Imaging 13 No. 3 (1995) 343-368

[2] R.C. Gonzalez, R.E. Woods: Digital Image Processing. 2nd~Ed. Addison-Wesley (1993)

[3] B. Jähne: Digitale Bildverarbeitung. 4th~Ed. Springer. (1997)

[4] N.R. Pal, S.K. Pal: A review on image segmentation techniques. Pattern Recognition 26 (1993) 1277-1294

[5] J.C. Russ: The Image Processing Handbook. 3rd~Ed. CRC Press LLC. (1999)

[6] S.U. Lee, S.Y. Chung: A comparative performance study of several global thresholding techniques for segmentation. Computer Vision, Graphics and Image Processing 52 (1990) 171-190

[7] D. Ziou, S. Tabbone: Edge detection techniques - An overview. Technical Report 195 Departement de math et informatique. Universite de Sherbrooke (1997)

[8] L. Vincent, P. Soille: Watersheds in digital spaces; an efficient algorithm based on immersion simulations. IEEE Transactions on Pattern Analysis and Machine Intelligence 13 (1991) 583-597

[9] R.K. Justice, E.M. Stokely: 3D segmentation of MR brain images using seeded region growing IEEE Engineering in Medicine and Biology Society 18 (1996) 1083-1084

[10] S.L. Horowitz, T. Pavlidis: Picture segmentation by a tree traversal algorithm. Journal of the Association for Computing Machinery 23 (1976) 368-388

[11] P.J. Burt: The pyramid as a structure for efficient computation. Multiresolution Image Processing and Analysis. A. Rosenfeld (ed.). Springer-Verlag (1984) 6—35

[12] T. McInerney, D. Terzopoulos: Deformable models in medical images analysis: a survey. Medical Image Analysis 1 No. 2 (1996) 91-108

[13] Y. Shimshoni: Introduction to Classification Methods. School of Mathematical Sciences. Tel-Aviv University (1995) http://www.math.tau.ac.il/~shimsh/DM-bib.htm

[14] V. Rehrmann: Stabile, echtzeitfähige Farbbildauswertung. Koblenzer Schriften zur Informatik 1 (1994)

[15] G. Hartmann: Recognition of hierarchically encoded images by technical and biological systems. Biological Cybernetics 57 (1987) 73-84

[16] J.-F. Vogelbruch: Segmentierung von Volumendatensätzen mittels dreidimensionaler hierarchischer Inselstrukturen. Schriften des Forschungszentrum Jülich/Informationstechnik 3 PhD Thesis RWTH Aachen (2002)

[17] J.-F. Vogelbruch, P. Sturm, R. Patzak, L. Priese, H. Halling: 3D Segmentierung mittels hierarchischer Inselstrukturen. Proc. Bildverarbeitung für die Medizin, Leipzig (2002)

[18] J.-F. Vogelbruch, R. Patzak, H. Halling: 3D Segmentierung und Visualisierung von Volumendatensätzen. Proc. Remagener Physiktage (2002)

[19] D.W. Hammerstrom, D.P. Lulich: Image Processing using One-dimensional Processor Arrays. Proc. IEEE 84 No. 7 (1996) 1005-1018

[20] M.W. van der Molen, S. Kyo: Documentation for the IMAP-VISION image processing card and the 1DC language. NEC incubation Center (1997)

[21] D. Juvin, Ch. Gamrat, J.F. Larue, Th. Collette: SYMPHONIE: Calculateur Massivement Parallele, Modelisation et Realisation. Journees Adequation Algorithmes Architectures. Toulouse (1996)

[22] P.P. Jonker, J. Vogelbruch: The CC/IPP, an MIMD-SIMD Architecture for Image Processing and Pattern Recognition. Computer Architecture for Machine Perception (1997) 33-39

[23] T. Le, W. Snelgrove, S. Panchanathan: SIMD processor arrays for image and video processing: a review. Multimedia Hardware Architectures 3311 S. Panchanathan, F. Sijstermans, S.I. Sudharsanan (eds.) (1998) 30-41

[24] C. Dick: Computing multidimensional DFT using Xilinx FPGAs. Signal Processing Algorithm and Technology 8 (1998)

[25] K. Compton, S. Hauck: Reconfigurable Computing: A Survey of Systems and Software. ACM Computing Surveys 34 No. 2 (2002) 171-210

[26] B. Draper, R. Beveridge, W. Böhm, C. Ross, M. Chawathe: Implementing Image Applications on FPGAs. Pattern Recognition (2002)

[27] P. Dillinger, J.-F. Vogelbruch, J. Leinen: 3D-RETISEG Datenbank-Struktur. Technical Documentation. 3rd Draft (2003)

[28] D.L. Collins, A.P. Zijdenbos, V. Kollokian, J.G. Sled, N.J. Kabani, C.J. Holmes, A.C. Evans : Design and Construction of a Realistic Digital Brain Phantom, IEEE Transactions on Medical Imaging, vol.17, No.3, p.463-468, June 1998

Segmentierung individueller Hirnatlanten

G. Wagenknecht[1], H. Belitz[1], L. Wischnewski[1], J. Castellanos[1], H. Herzog[2], H.-J. Kaiser[3], U. Büll[3], O. Sabri[4]

1   Zentralinstitut für Elektronik, Forschungszentrum Jülich GmbH, D-52425 Jülich
2   Institut für Medizin, Forschungszentrum Jülich GmbH, D-52425 Jülich
3   Klinik für Nuklearmedizin, Universitätsklinikum der RWTH Aachen, D-52057 Aachen
4   Klinik und Poliklinik für Nuklearmedizin, Universität Leipzig, D-04103 Leipzig

Kurzfassung

Die Segmentierung von MRT-Bilddaten zur Generierung individueller Hirnatlanten basiert im wesentlichen auf der Kombination einer Gewebeklassifikation mittels neuronalem Netz und wissensbasierten Ansätzen. Dieses Grundkonzept wird zur Zeit um Schritte zur Bildvorverarbeitung sowie um eine modellbasierte Segmentierung subkortikaler Strukturen und eine semiautomatische Komponente zur Flexibilisierung erweitert. Schnittstellen zu kommerzieller Software (z.B. PMOD) erlauben die Anwendung der Atlanten zur regionenbasierten Funktionsanalyse.
Evaluationsergebnisse werden anhand von Simulationsdaten sowie realen Daten präsentiert. Auf Basis realitätsnaher Kopf-Software-Phantome werden Bildstörungen, wie Rauschen und Inhomogenitäten, simuliert und die Segmentierungsergebnisse in Abhängigkeit dieser Artefakte dargestellt. Applikationsbeispiele zeigen die Ergebnisse der atlasbasierten Quantifizierung von PET-Daten.

1   Einleitung

Das Gehirn ist das wichtigste und in Struktur und Funktion komplexeste Organ des menschlichen Körpers. Die in einer Hirnregion ausgeübte Funktion, wie z.B. Sprache, Motorik etc., spiegelt sich in der neuronalen Aktivität und den hiermit verknüpften biochemischen Prozessen wieder. Dreidimensionale Abbildungen des Stoffwechsels, der Durchblutung oder der Verteilung von Rezeptorliganden können mit der Positronenemissionstomographie (PET) gewonnen werden. Mit der Magnetresonanztomographie (MRT) können die Hirnstrukturen, insbesondere die verschiedenen Gewebearten, dreidimensional abgebildet werden.
Zur Funktionsanalyse in anatomischen 3D-Regionen des Gehirns ist die vorherige Segmentierung dieser Regionen erforderlich. Hierzu existieren im wesentlichen folgende Ansätze: 1. Interaktive Segmentierung der interessierenden Regionen (ROI) [1]; 2. Segmentierung durch Registrierung mit einem Referenzatlanten [2-4]. Während interaktive Methoden sehr zeitaufwendig sind, besteht bei den Registrierungsmethoden das Problem der Handhabung der interindividuellen Variabilität menschlicher Gehirne. Diese wird in [4] durch die Generierung sogenannter „Wahrscheinlichkeitskarten" berücksichtigt.
Durch die automatisierte Segmentierung individueller 3D-Hirnregionen (3D-IROI-Atlanten) auf Basis T1-gewichteter MRT-Bilddaten des jeweiligen Patienten werden interindividuelle Unterschiede implizit berücksichtigt. Da die zugrundeliegenden Methodiken in [5-6] bereits ausführlich beschrieben wurden, werden diese hier lediglich in ihren wesentlichen Elementen wiedergegeben (Kap. 2.1). In Kap. 2.2 werden die zur Zeit in der Entwicklung befindlichen Erweiterungen der Methodik dargestellt. Den Schwerpunkt des Beitrages bilden die Beschreibung von Segmentierungsergebnissen (Kap. 3) sowie der Applikation zur atlasbasierten Quantifizierung von PET-Bilddaten unter Berücksichtigung der verschiedenen Auflösung von MRT und PET (Kap. 4).

2   Methoden

Das Grundkonzept der Atlas-Extraktion besteht aus einer Gewebedifferenzierung durch Klassifikation und einer hierauf aufbauenden wissensbasierten Analyse zur weitergehenden Differenzierung anatomischer Regionen. Die extrahierten Atlanten werden zur atlasbasierten Quantifizierung funktioneller Daten in ihrer Auflösung und abhängig vom Partialvolumeneffekt modifiziert.
Dieses Grundkonzept wird zur Zeit erweitert um eine Vorverarbeitung zur Rauschunterdrückung, eine modellbasierte Basalgangliensegmentierung sowie eine semiautomatische Komponente zur Flexibilisierung der Regionenextraktion.
Zur Nutzung der Segmentierungsergebnisse in kommerziellen Programmen zur Funktionsanalyse (z.B. PMOD) wurde eine Schnittstelle auf Basis eines Konturverfolgungsalgorithmus geschaffen.

2.1 Grundkonzept

2.1.1 Gewebedifferenzierung

Die Differenzierung des Hirngewebes in graue (GM) und weiße Substanz (WM) sowie des Liquorraums (CSF), des Fettgewebes (SB) und des Hintergrundes (BG) erfolgt ausschließlich auf Basis des zu segmentierenden individuellen T1-gewichteten MRT-Datensatzes. Die hierzu entwickelte vollautomatische überwachte Klassifikation besteht aus einer automatischen Stichprobenextraktion, die auf der randomisierten Auswahl geeigneter Stichprobenelemente aus vorsegmentierten Bereichen beruht, und einem neuronalen Netz [7], das auf Basis dieser Stichprobe trainiert wird und den gesamten MRT-Datensatz klassifiziert.

Die überwachte Klassifikation liefert für jedes Voxel ein Klassenlabel, das die anatomische Zugehörigkeit widerspiegelt. Dieses Wissen ist Voraussetzung für den zweiten Schritt der Atlas-Extraktion, die wissensbasierte Analyse der klassifizierten Bilddaten.

2.1.2 Regionenextraktion

Die Grenzen dreidimensionaler anatomischer Regionen verlaufen teilweise entlang bereits extrahierter Gewebegrenzen, zum Teil aber auch innerhalb eines Gewebetyps. Die entwickelten Algorithmen zur weiteren Differenzierung anatomischer Regionen nutzen das anatomische Wissen über die hierarchischen Beziehungen und Lagerelationen der Hirnregionen, das in Form attributierter relationaler Graphen repräsentiert wird. Hierdurch ist die Reihenfolge der Extraktion festgelegt und die Suche nach neuen Regionen erfolgt in Relation zu bereits extrahierten Regionen und Regionengrenzen.

Die grobe Detektion einer neuen Region erfolgt durch den Vergleich mit einem einfachen z.B. rechteckförmigen Template. Die genauen Grenzen werden dann durch Regionenwachstum in Richtung der Gewebegrenzen und morphologische Operationen in Richtung der übrigen Grenzen bestimmt. Das Erreichen einer Gewebegrenze beendet den Regionenwachstumsprozess automatisch. Die Anzahl der Iterationen für die morphologischen Operationen kann weitestgehend standardisiert werden.

2.1.3 Auflösungsanpassung

Die Modifikation der Atlanten zur Anpassung an die vergleichsweise geringe Auflösung funktioneller Verfahren (z.B. PET: 4-6 mm FWHM[1], SPECT: 8-12 mm FWHM) dient der Berücksichtigung von Partialvolumeneffekten in der atlasbasierten Quantifizierung funktioneller Bilddaten [8]. Der Partialvolumeneffekt kann z.B. für den 2-5 mm breiten Kortex zu einer „Verfälschung" der dort regional gemessenen funktionellen Parameter führen. Die Faltung des 3D-IROI-Atlanten mit der idealisierten „point spread function" des PET-Systems führt zu einer Überlagerung der Regionen. Durch einen Schwellwert $\Delta V(IROI)$ wird ein minimal erforderliches Partialvolumen festgelegt, das mindestens erforderlich ist, um einem Voxel die Klasse IROI zuzuweisen. Ziel dieser partialvolumenabhängigen Regionenmodifikation ist die Steuerung des Einflusses benachbarter Atlasregionen auf das Quantifizierungsergebnis der betrachteten Region.

2.2 Erweiterungen

2.2.1 Bildverbesserung

Abhängig von der Qualität des MRT-Datensatzes ist eine Vorverarbeitung zur Bildverbesserung erforderlich. Hierzu wird eine automatisierte adaptive Methode zur Rauschunterdrückung auf Basis anisotroper Diffusionsfilter realisiert [9].

2.2.2 Modellbasierte Segmentierung

Da die Basalganglien aufgrund ihrer Grauwertcharakteristik mit rein bildbasierten Methoden nur schwer abgrenzbar sind, wird eine Verbesserung der Segmentierung durch das Einbringen von topologischem und geometrischem Modellwissen angestrebt. Dieses Vorwissen sowie die Bildinformation steuern eine aktive Oberfläche, die gegen die Segmentierung des gesuchten Objektes konvergiert [10].

2.2.3 Sulcus-Segmentierung

Die bisher realisierte Detektion der Sulci diente im wesentlichen der Abgrenzung der Hirnlappen. Durch die Bestimmung weiterer Sulci auf Basis von Skelettierungsmerkmalen soll die Segmentierung des Kortex verfeinert werden.

2.2.4 Semiautomatische Segmentierung

Die Entwicklung semiautomatischer Segmentierungsverfahren dient der Ergänzung der automatisch generierten Atlanten durch flexibel definierbare Regionen auf Basis der Kortexoberfläche. Bei dem zugrundeliegenden Live-Wire-Verfahren wird mittels der Methode der dynamischen Programmierung ein optimaler Konturverlauf zwischen zwei vom Benutzer definierten Stützpunkten berechnet [11,12].

---

[1] Full width at half maximum

2.2.5 Konturverfolgung

Kommerzielle Programme zur Quantifizierung funktioneller Bilddaten können meist nur 2D-Konturen verarbeiten, da eine interaktive Regionendefinition vorausgesetzt wird. Um eine Schnittstelle zu diesen Programmen (z.B. PMOD) zu schaffen, wurde ein Kon-turverfolgungsalgorithmus entwickelt, der aus dem 3D-Atlas 2D-Konturen extrahiert und in dem spezifisch benötigten ROI-File-Format ablegt.

3 Ergebnisse

Da die Segmentierung der anatomischen Regionen und die anschließende auflösungsangepasste Modifikation der Atlanten auf der Gewebedifferenzierung aufbauen, beeinflusst deren Genauigkeit die Atlasextraktion essentiell. Daher werden zunächst quantitative Evaluationsergebnisse zur Gewebedifferenzierung anhand simulierter Kopf-Software-Phantome dargestellt. Anschließend werden exemplarische Ergebnisse der weitergehenden Segmentierung anatomischer Regionen anhand realen Datenmaterials beschrieben.

Die Klassifikationsergebnisse wurden mit folgender Parametereinstellung des dreilagigen neuronalen Feed-Forward-Netzes gewonnen: nichtlineare Sigmoid-Funktion als Aktivierungsfunktion; $N_{in}=1$ Eingangs-knoten (da lediglich der Grauwert als Merkmal zur Klassifikation genutzt wird); $N_{out}=5$ Ausgangsknoten [da fünf Klassen (GM, WM, CSF, SB, BG) differenziert werden]. Die Anzahl versteckter Knoten entspricht der doppelten Anzahl der Ausgangsknoten, also $N_{hid}=10$. Die Lernrate wurde mit $\varepsilon = 0.1$, der Momentumterm mit $\alpha=0.9$ eingestellt. Der Trainingsalgo-rithmus ist ein Back-Propagation-Algorithmus und die Anzahl der Trainingszyklen beträgt $CYC=10^6$. Mit dieser Parametereinstellung wurden 50 realitätsnahe Kopf-Software-Phantome (KSP) untersucht. KSP1-KSP50 sind mit additivem gleichverteiltem weißen Rauschen behaftet. Von KSP1 bis KSP9 steigt die Amplitude des Rauschens. KSP10-KSP20 sind mit einem linearen Inhomogenitätsfeld überlagert, wobei die maximale Amplitude dieses Feldes von KSP10 bis KSP20 zunimmt. KSP21-KSP31 sind mit einem quadratisch ansteigenden Inhomogenitätsfeld behaftet, wobei auch hier der Maximalwert von KSP21 bis KSP31 zunimmt. Von KSP32 bis KSP50 nimmt die Standardabweichung der zugrundeliegenden Grauwertverteilungen zu, so daß sich die Verteilungen der zu differenzierenden Klassen zunehmend überlagern.

Die Fehlklassifikationsrate err_V wird durch den Vergleich des Klassifikationsergebnisses K mit dem korrekten Klassifikationsergebnis K_{orig} für jedes Voxel ermittelt, wobei N_V die Anzahl der Voxel des untersuchten Datensatzes beschreibt (Gl. 1).

$$err_V = \frac{1}{N_V} \sum_{n_V=0}^{N_V-1} \left(K_{orig}(n_V) \neq K(n_V) \right) \quad (1)$$

Berechnet wurde der Mittelwert m, die Standardabweichung std sowie Minimum min und Maximum max der Fehlklassifikationsrate für das gesamte Datenkollektiv KSP1-KSP50 sowie die Teilkollektive KSP1-KSP9, KSP10-KSP20, KSP21-KSP31 und KSP32-KSP50.

Die Ergebnisse für das Gesamtkollektiv mit eine mittleren Fehlklassifikationsrate von 3.587%, einer Standardabweichung von ±0.466% und einem Bereich von 2.726%-4.927% zeigen die Güte und Stabilität des neuronalen Klassifikationsverfahrens. In **Tabelle 1** sind die Ergebnisse für die vier verschiedenen Phantomgruppen separat dargestellt.

	$m \pm std$	$min - max$
KSP1-KSP9	3.665% ± 0.43%	3.204% - 4.528%
KSP10-KSP20	3.562% ± 0.563%	2.975% - 4.927%
KSP21-KSP31	3.668% ± 0.629%	2.726% - 4.914%
KSP32-KSP50	3.519% ± 0.318%	2.896% - 4.232%

Tabelle 1 Fehlklassifikationsraten err_V für die vier Phantomgruppen: $m \pm std$ und Bereich $min - max$.

Die höchsten Fehlerraten wurden für die Kopf-Software-Phantome ermittelt, die am stärksten durch Inhomogenitätsartefakte beeinträchtigt sind. Dies sind KSP20 mit $err_V(KSP20)= 4.927\%$ sowie KSP31 mit $err_V(KSP31)= 4.914\%$. Die Variabilität der Ergebnisse, ausgedrückt durch die Standardabweichung std, ist für die beiden mit Inhomogenitäten behafteten Teilkollektive mit ±0.563% (KSP10-KSP20) und ±0.629% (KSP21-KSP31) ebenfalls größer als für die beiden anderen Gruppen mit ±0.43% (KSP1-KSP9) und ±0.318% (KSP32-KSP50). Verglichen mit den anderen Parametermodifikationen wurden die Klassifikationsergebnisse durch Überlagerung der Inhomogenitätsfelder am stärksten beinträchtigt.

Die benötigte Rechenzeit zur Klassifikation eines Datensatzes der Matrixgröße 256×256×256 beträgt inklusive aller Schreib-Lese-Prozeduren 4 min 56.5 sec auf einer SUN Ultra60 und <1 min für die meisten zur Regionenextraktion (s. Abschnitt 2.1.2) entwickelten Algorithmen.

Im folgenden werden exemplarische Ergebnisse der Regionenextraktion bildlich dargestellt. In **Bild 1** und **Bild 2** ist die Segmentierung des MNI-Datensatzes zu sehen. Dieser Datensatz stammt vom Montreal Neurological Institute und stellt eine Mittelung von 27 T1-gewichteten Akquisitionen eines Probanden dar.

Der Datensatz besitzt somit ein sehr gutes Signal zu Rauschverhältnis, durch die Mittelung der Scans jedoch leicht unscharfe Grenzen zwischen den Gewebearten (siehe **Bild 1**, obere Zeile). Der Datensatz wurde aufgrund seiner Bedeutung für die Neurowissenschaften bearbeitet.

In der mittleren Zeile von **Bild 1** sind die Ergebnisse nach der Trennung cerebraler und extracerebraler Regionen, dem sogenannten „Peeling", dargestellt. Der „Peeling"-Algorithmus ist der erste wissensbasierte Algorithmus zur Regionenextraktion und arbeitet somit auf den durch das neuronale Netz klassifizierten Datensätzen. Der zweite Schritt ist die Hemisphären-Segmentierung, die in der unteren Zeile von **Bild 1** dargestellt ist. Da die Hemisphärengrenze als Referenzstruktur zur Orientierung in allen weiteren Extraktionsschritten genutzt wird, ist deren Segmentierung von besonderer Bedeutung.

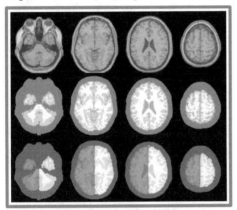

Bild 1 Oben: Schichten des MNI-Datensatzes; Mitte: Segmentierung des Hirngewebes (GM, WM) und CSF, „Peeling" des Gehirns; Unten: Segmentierung der Hirnhemisphären.

In **Bild 2** ist der MNI-Datensatz noch einmal angeschnitten in einer 3D-Darstellung zu sehen. Auf der linken Seite ist die cerebrale Region nach Anwendung des „Peeling"-Algorithmus dargestellt. Zur Orientierung wurde die basale Schicht des Datensatzes mit eingeblendet. Auf der rechten Seite ist der Originaldatensatz gezeigt, wobei hier die Grenze der weißen Substanz in der linken Hemisphärenhälfte eingeblendet wurde. Voraussetzung hierfür ist wiederum die Hemisphären-Segmentierung.

In **Bild 3** sind Schnitte durch verschiedene individuelle 3D-IROI-Atlanten dargestellt. Die zugrundeliegenden T1-gewichteten MRT-Bilddaten wurden einem Patientenkollektiv entnommen, das im Rahmen einer klinischen Studie akquiriert wurde [13]. Die Schnittbilder sind in verschiedenen anatomischen Orientierungen im Vergleich zu korrespondierenden Schnitten aus einem Anatomieatlanten [14] dargestellt. Die Güte der Regionenextraktion wird durch diese vergleichende Darstellung sehr deutlich. In der ersten Zeile von **Bild 3** ist die Segmentierung des Kleinhirns, in der mittleren Zeile die Segmentierung der Basalganglien, des Hirnstamms und des Kleinhirn dargestellt. Die untere Zeile zeigt schließlich die segmentierte Gyrus cinguli - Region, die Inselregionen sowie die Hirnlappen.

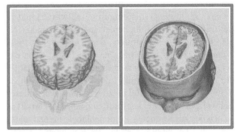

Bild 2 Links: „Peeling" des Gehirns; 3D-Ansicht des MNI-Datensatzes. Rechts: Segmentierte Grenzkonturen der WM-Region in der linken Hemisphäre dem originalen MNI-Datensatz überlagert.

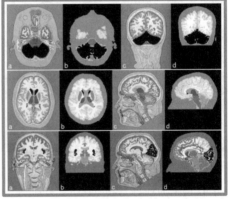

Bild 3 3D-IROI-Atlanten extrahiert aus unterschiedlichen Patientendatensätzen; b,d: Ausgewählte Schichten individueller Atlanten in verschiedenen anatomischen Orientierungen; a,c: Korrespondierende Schichten des Kretschmann-Atlanten [14].

4 Applikationen

Die atlasbasierte quantitative Analyse koregistrierter funktioneller Bilddaten wird anhand einer exemplarischen in vivo - Applikation beschrieben. Bei dieser PET-Studie handelt es sich um eine Aktivierungsstudie mit einer Wortrepititionsaufgabe (WR) als Ruhebedin-gung und einer Wortassoziationsaufgabe (WA) als Aktivierungsbedingung [13]. Der individuelle Atlas wird entsprechend Abschnitt 2.1.3 modifiziert. In Abhängigkeit vom Schwellwert $\Delta V(IROI)$ kann dann der Einfluss benachbarter Atlasregionen auf das Quantifizierungsergebnis der betrachteten Region gesteuert werden.

In **Bild 4** sind drei ausgewählte Schnitte aus PET-Datensätzen verschiedener Patienten dargestellt, wobei in den ersten beiden Zeilen die Ergebnisse aufgrund der Aktivierungsbedingung, in der letzen Zeile ein Beispiel bei Ruhebedingung dargestellt ist. In der ersten Zeile ist die Frontallappenregion (FL), in der zweiten Zeile die Gyrus cinguli - Region (CI) und in der dritten Zeile sind die beiden Insel - Regionen (IN) als Regionengrenzkonturen dem jeweiligen PET-Schnitt überlagert. Von links nach rechts ist - von der originalen Atlas-Region ausgehend - die Modifikation der Regionen aufgrund immer höher gewählter Schwellwerte ΔV(IROI) zu sehen. Da dieser Schwellwert die Mindestforderung an das Partialvolumen des zugeordneten Regionenlabels im betrachteten Voxel beschreibt, werden die Regionen bei steigendem Schwellwert immer kleiner, da nur noch wenige Voxel den Anforderungen an das Partialvolumen genügen.

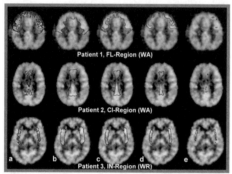

Bild 4 Aktivierungsstudie Wort-Repetition (WR) / Assoziation (WA): a. Original 3D-IROIs (FL, CI und IN) in ausgewählten PET-Schnitten; b.-e. Modifizierte 3D-IROIs mit ΔV(IROI): 10%, 30%, 50%, 70%.

Die Variation des Schwellwertes erfolgte im Bereich 10-80% [8] (siehe **Bild 5**). Die Quantifizierungsergebnisse zeigen eine klare Abhängigkeit von diesem Schwellwert. Im Bereich 10-60% ist der Anstieg der Ergebnisse in Abhängigkeit von ΔV(IROI) für die Aktivierungs- und Ruhebedingung sehr ähnlich. Dies gilt für jede der untersuchten Regionen und ist in **Bild 5** für die Frontallappenregion (FLL), die Gyrus cinguli – Region (CIL) und die Insel-Region (INL) der linken Hemisphäre dargestellt.

Dies ist ein sehr wichtiges Ergebnis und hat Bedeutung für die Wahl einer geeigneten Methode zu Auswertung solcher Aktivierungsstudien. Da differenzenbasierte Methoden weniger von der Schwellwertwahl und damit vom Partialvolumeneinfluss benachbarter Regionen beeinflusst sind, sind diese Methoden quotientenbasierten Verfahren vorzuziehen.

Für Schwellwerte >60% bleiben nur wenige „Restregionen" übrig. Die Ergebnisse sind daher stark von diesen noch verbleibenden Teilregionen abhängig und daher wenig aussagekräftig. Diese Reduktion ist insbesondere im Falle kortikaler Regionen erheblich und bedeutet, dass nahezu jedes Kortex-Voxel stark vom Partialvolumeneffekt beeinflusst ist. Die Wahl des Schwellwertes ΔV(IROI) kann daher nur einen Kompromiss zwischen dem noch vertretbaren Partialvolumeneinfluss aus benachbarten Regionen und der Reduktion der zu untersuchenden Region darstellen.

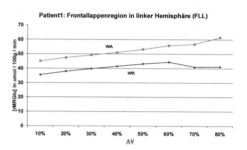

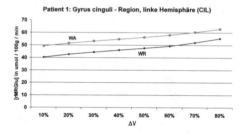

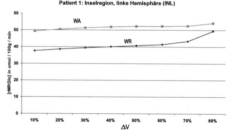

Bild 5 Aktivierungsstudie Wort-Repetition (WR) / Assoziation (WA): Quantifizierungsergebnisse für ΔV(IROI): 10-80% verdeutlichen den Einfluss partial-volumenbedingter Effekte für drei unterschiedliche Regionen (Patient 1). ΔV(IROI): 0-60%: Ähnlicher Anstieg der Ergebnisse für WA und WR. ΔV(IROI) >60%: Die Ergebnisse sind abhängig von wenigen „Restregionen" und daher wenig aussagekräftig.

MRT-basierte individuelle Regionenatlanten des Gehirns sind nicht auf die beschriebene Applikation beschränkt, sondern können auch in anderen Feldern der Diagnostik, Therapie und Hirnforschung Anwendung finden. Beispiele sind morphometrische Analysen des Gehirns oder die Unterstützung der bildgestützten Neurochirurgie.

5 Diskussion und Ausblick

Der Zeitaufwand rein interaktiver Segmentierungsverfahren kann abhängig von der medizinischen Fragestellung und damit der Regionenanzahl und geforderten Genauigkeit erheblich sein (> 1 Tag). Daher werden die Regionen oft nur grob und zudem auf Basis von 2D-Schnitten bestimmt. Eine Genauigkeit, wie sie mit einer automatisierten 3D-Segmentierung möglich ist, wird in der Praxis mit rein manueller Segmentierung daher kaum erreicht. Diese Genauigkeit ist aber Voraussetzung für viele Anwendungen, insbesondere auch für die Berücksichtigung von Partialvolumeneffekten im Rahmen funktioneller Analysen.

Dem Problem der interindividuellen Variabilität des Gehirns wird bei Verwendung von Referenzatlanten häufig durch "Vergröberung" der Referenz-Regionen, z.B. durch Glättung oder Mittelung mehrerer Datensätze, begegnet. Auch dies beeinflusst die quantitative Analyse der registrierten PET-Daten.

Die automatisierte Extraktion individueller 3D-Regionenatlanten erfordert keine interaktive Delineation von Regionengrenzen. Die nichtlineare Zuordnung zu einem anatomischen oder statistischen Hirnatlanten ist ebenfalls nicht erforderlich. Durch die beschriebene Methode zur Auflösungsanpassung können partialvolumenbedingte Effekte in funktionellen Bilddaten in erster Näherung berücksichtigt werden.Die dargestellten Ansätze und Methoden werden künftig weiterverfolgt und die beschriebenen Erweiterungen in das Gesamtsystem integriert.

Danksagung: Diese Arbeit wurde durch das IZKF"ZNS" des Universitätsklinikums der RWTH Aachen und das Tenure-Track-Programm des Forschungszentrums Jülich GmbH gefördert.

7 Literatur

[1] O. Sabri, D. Hellwig, et. al., Correlation of Neuropsychological, Morphological and Funcional (Regional Cerebral Blood Flow and Glucose Utilization) Findings in Cerebral Microangiopathy, J Nucl Med (1998), Vol. 39, No. 1, pp. 147-154.

[2] D.L. Collins, C.J. Holmes, T.M. Peters, A.C. Evans, Automatic 3D model-based neuroanatomical segmentation, Hum Brain Map (1995), Vol. 3, No. 3, pp. 190-208.

[3] K.J. Friston, J. Ashburner, et. al., Spatial registration and normalization of images, Hum Brain Map (1995), Vol. 2, pp. 165-189.

[4] K. Amunts, K. Zilles, Advances in cytoarchitectonic mapping of the human cerebral cortex,. Neuroimaging Clinics of North America (2001), Vol. 11, No. 2, pp. 151-169.

[5] G. Wagenknecht, Entwicklung eines Verfahrens zur Generierung individueller 3D-„Regions-of-Interest"-Atlanten des menschlichen Gehirns aus MRT-Bilddaten zur quantitativen Analyse koregistrierter funktioneller ECT-Bilddaten, Dissertationsschrift, Shaker, Aachen, 2002.

[6] G. Wagenknecht, MRT-basierte individuelle Regionenatlanten des menschlichen Gehirns: Ziele und Methoden, In: TM. Buzug et. al. (Hrsg.). Physikalische Methoden der Laser- und Medizintechnik. Fortschritt-Berichte VDI, Reihe 17 Nr. 231, VDI Verlag, Düsseldorf (2003), S. 58-64.

[7] D.E. Rumelhardt, and J.L. McClelland, Parallel Distributed Processing, Vol. 1: Foundations, MIT Press, Cambridge, 1989.

[8] G. Wagenknecht, HJ. Kaiser, O. Sabri, U. Buell, MRI-based individual 3D region-of-interest atlas of the human brain: Influence of the partial volume threshold on the quantification of functional data, Eur J Nucl Med (2002); Vol. 29, pp.157.

[9] Castellanos J, Wagenknecht G, Tolxdorff T. Advances of a 3D noise suppression method for MRI data, VDE Verlag (im Druck).

[10] H. Belitz, G. Wagenknecht, H. Müller. Modellbasierte Segmentierung von MRT-Daten des menschlichen Gehirn,. VDE Verlag (im Druck)

[11] L. Wischnewski, G. Wagenknecht. Semiautomatische Segmentierung dreidimensionaler Strukturen des Gehirns mit Methoden der dynamischen Programmierung. In: T. Tolxdorff, et. al. (Hrsg.). Bildverarbeitung für die Medizin (2004), Springer-Verlag, Berlin, S. 55-59.

[12] L. Wischnewski, G. Wagenknecht, R. Dillmann, Ein semiautomatisches Verfahren zur Segmentierung von Hirnregionen in MRT-Daten, VDE Verlag (im Druck).

[13] M. Schreckenberger, E. Gouzoulis-Mayfrank, O. Sabri, et. al., "Ecstasy"-induced changes of cerebral glucose metabolism and their correlation to acute psychopathology. An 18-FDG PET study, Eur J Nucl Med (1999), Vol. 26, pp. 1572-1579.

[14] HJ. Kretschmann, W. Weinrich, Klinische Neuroanatomie und kranielle Bilddiagnostik, Thieme Verlag, Stuttgart, 1991.

Brain Tumour Segmentation with Quantitative MRI

Sabrina Wildenburg, Heiko Neeb and Nadim J. Shah
Institute of Medicine, Forschungszentrum Jülich GmbH, Germany

Abstract

One of the challenges met in brain tumour diagnostics is associated with the precise separation between tumour and normal healthy tissue for biopsies and anti-cancer therapy. We present a new method for tumour tissue segmentation based on extensions to multivariate clustering analysis of quantitative T_1, T_2^* and H_2O content maps acquired with MRI. Additional parameters that are sensitive to the local magnetic field offset were also included in the segmentation algorithm. First results from a current clinical study for the differentiation of tumour, oedema, and surrounding healthy tissue are shown. Furthermore, results from the objective evaluation of anti-oedema cortisone therapy are presented demonstrating the clinical power of quantitative MRI in combination with multivariate image analysis.

1 Introduction

Imaging cerebral neoplasias is a challenging task because human brain tumours are in general not homogenous objects that are well separated from the surrounding healthy tissue. It is therefore of the utmost importance for correct diagnosis, treatment strategy evaluation, and the treatment itself to collect as much information as possible about the tumour and its structure *in vivo*. Histological results obtained from biopsy samples e.g. depend on the correct delineation of the viable and growing part of the neoplasia. Knowledge of the presence and the extent of regions with decreased oxygen supply, e.g., can alter therapy decisions because such regions are in general less responsive to standard radiation treatment strategies [1]. Current methods used in clinical routine for cerebral tumour imaging are almost exclusively based on invasive procedures requiring the injection of either paramagnetic or radioactive contrast agents [2][3]. Even though both techniques may result in good delineation of tumour borders, substructures are typically not identifiable.

We present a method for tissue segmentation based on multivariate clustering analysis and newly developed methods for quantitative MR imaging. This approach makes full use of the advantages offered by the quantitative nature of the acquired data including the possibility for objective monitoring of treatment effects. In addition, the multivariate algorithm effectively uses all available information such as the correlation structure within the data.

First results from a current clinical study for the differentiation of tumour, oedema, and surrounding healthy tissue are presented. Furthermore, a quantitative evaluation of anti-oedema cortisone treatment is presented. Based on the high precision of the new MR measurement sequences in combination with the precise segmentation process and a total measurement time of approximately 21 minutes, the presented method clearly demonstrates its relevance for use in clinical studies as well as routine diagnosis.

2 Materials and Methods

Data Acquisition

T_2^* was measured using the QUTE-EPI sequence [4] with the following parameters: TR=138ms; TE=4ms; α=90°; 17 slices; 64 time points; echo spacing=2ms; matrix size=256^2; FOV=220mm; slice thickness=5mm; rf-spoiling employed; 10 preparation scans. T_1 mapping was performed using TAPIR [5][6]. The following parameters were employed: TR=15ms; TE_1: TE_2: TE_3=2.8: 5.1: 7.5ms; α=25°; 17 slices; 20 time points; matrix size=256^2; FOV=220mm; slice thickness=5mm; sequential excitation; delay time τ=2s. The first point for the first slice on the recovery curve was sampled 10ms after inversion. Inefficiencies of the inversion pulse were corrected using the procedure described in [7]. Based on the results from T_2^* and T_1 mapping, water content was quantitatively determined using the procedure described in [8]. In addition, the mean absolute deviation of the measured signal decay curve R_{Mean} from an exponential fit to the measured points was included in the clustering process. R_{Mean} is large in regions with large field inhomogeneities or significant bi-exponential signal behaviour. In order to perform quantitative H_2O content mapping in regions of large field inhomogeneities, the signal decay curve was fitted with a 3rd order polynomial function and the fit was constrained to a number N_p of measured points. The linear slope $α_1$ extracted from the polynomial fit and N_p were used as additional variables for the multivariate image segmentation process.

Quantitative measurement of all the described parameters was performed in a 28 year old patient with oligodendroglioma in the frontal area and a 53 year old patient with a cerebral pulmonary metastasis before and 2 weeks after treatment with methylpresdnisolone.

Image Segmentation

Clustering using the "Zoom-In" approach

Multivariate image segmentation was based on an extension to the standard agglomerative hierarchical and k-Means clustering approach [9]. Instead of assigning all image points only once to a given number of clusters, a tree-like approach called Zoom-in-Clustering (ZIC) was developed in order to reduce the number of false cluster assignments.

Zoom-in-Clustering is based on the interactive and iterative combination of conventional clustering algorithms like k-Means. During the first step, clustering is performed for a given number $N_{c,1}$ of clusters. $N_{c,1}$ is either externally given or based on the Sum of Squares Within (SSW) criterion (see below). Depending on the results obtained it is possible to either stop the process or to define one or more clusters 1..k (k< $N_{c,1}$) that should be re-clustered again in a second step. The second step includes the same conventional clustering algorithm as before but works only on a subset of points (the "zoom-in" region) which where assigned previously to one of the clusters 1..k. The number of clusters in the second step $N_{c,2}$ is again either externally given or automatically extracted from the SSW criterion. In this case, the Sum of Squares Within will be determined based only on the points used for clustering in the second step. In addition, variables different from the ones used in the first step can be used for clustering. This offers the advantage of including variables known to be only sensitive or specific within the sub-region to be re-clustered again. The whole process can be repeated as many times as desired.

Cluster determination and assignment

Zoom-in-Clustering was performed in a pre-selected image of a transverse slice. Due to memory constraints resulting from the large pair-wise distance matrices required for clustering, the resolution of the slice was reduced by a factor of 2. The region-of-interest used for clustering was manually chosen and included the complete hemisphere where a tumourous area was visible in the corresponding T_1 map.
Mean values $z(C_i)$ of the quantitative MR parameters used in the clustering process were calculated for each cluster C_i separately. Image points from all slices including the slice used for clustering were then assigned to their most likely cluster based on the multi-dimensional Euclidian distance between each point and the respective cluster mean. Points were colour-coded and displayed as a cluster image using MATLAB V6.5, Release 13 (The Mathworks Inc., Natick/MA, USA).

The determination of the most likely number of clusters in each step relied on a frequently-used criterion, the so-called Intra-Class-Variance. The Intra-Class-Variance is based on the sum of the averaged squared distances of the elements x within a cluster from its centre, $z(C_i)$ summed over all N clusters

$$SSW(C) = \sum_{i=1}^{N} \sum_{x \in C_i} (x - z(C_i))^2 \quad [1].$$

The SSW was determined for different cluster numbers N and plotted against N. Based on this plot, the Elbow criterion was used to determine the approximate number of clusters present in the data [9].

Segmentation evaluation

The clusters were solely determined by the quantitative MR parameter patterns without taking the spatial distribution of the resulting cluster structure into account. If points surrounding a given point i were all classified as tissue-class C, then there is a large prior probability for this point to belong to the same class C. These local changes in prior probability were not taken into account by the algorithm presented and can therefore be used to evaluate the quality of the clustering results. In order to check the spatial homogeneity of the resulting cluster structure, the entropy

$$H(P) = -\sum_{i=1}^{N} p_i \log p_i \quad [2]$$

was calculated for each voxel as a measure of local coherence. $P=(p_1,...p_N)$ is the distribution of clusters in a 3x3 matrix of points surrounding the voxel for which the entropy is calculated and N represents the number of clusters.

The weighted mean entropy for each slice was calculated taking the different number of points assigned to different clusters into account:

$$<H> = \frac{\sum_{i=1}^{M}(M - k_{Ci}) H_i(P)}{\sum_{i=1}^{M}(M - k_{Ci})}. \quad [3]$$

Here, k_{Ci} is the number of points assigned to cluster C_i and M represents the total number of clustered points.

In addition, principal components analysis (PCA) was performed to reduce the dimension of the dataset. The first three principal components were colour coded according to the cluster assignment and plotted against each other in order to visualize the cluster structure in a lower dimension.

In vivo experiments

Two-step Zoom-in-Clustering based on both k-Means and Complete-Linkage was performed for the patient with oligodendroglioma. Five clusters were selected during the first step and zoom-in was performed with $N_{c,2}=5$ in a region comprising 2 clusters where tumour and oedema were identified. Water content, T_1 and T_2^* were used for the first clustering step and were extended by R_{Mean}, N_p and the linear term from the polynomial fit for the second step. To compare results with those obtained by a conventional approach, one-way k-Means and Complete-Linkage clustering with 8 clusters was performed using the same data as in the second step of Zoom-in-Clustering. The entropy for each image point and each of the four different clustering approaches was determined according to Eq. 2 and the weighted mean entropy was plotted as a function of slice position.

Data from the patient with cerebral pulmonary metastasis were processed with two-way Complete-Linkage Zoom-in-Clustering based on H_2O, T_1 and α_1 in both steps. Based on the SSW criterion, 4 clusters were chosen during the first and 3 clusters were chosen during the second step of ZIC. Data from the 2-week follow-up measurement of this patient were assigned to their most likely cluster as described above based on the cluster means determined from the first measurement. Changes in absolute water content as a result of anti-oedema therapy were determined based on the segmented oedematous region.

3 Results

Primary Tumour Substructure Definition

Results for the entropy calculation using k-Means and Complete-Linkage clustering both conventionally and with Zoom-In extension are shown in Fig. 1 for the patient with oligodendroglioma. Eight clusters were used for the calculation of all entropy maps in order to obtain comparable results. Boundaries between different cluster classes and therefore tissue types are clearly visible as they are expected to have large entropy. The tumourous region shows a heterogeneous entropy distribution regardless of the algorithm used which demonstrates the possible presence and identification of tumour sub-structures. For the different approaches presented, the slice dependence of the weighted mean entropy <H> is shown in Fig. 2. k-Means results in a larger <H> compared to the Complete-Linkage algorithm for both conventional and Zoom-In Clustering. The cluster structure determined

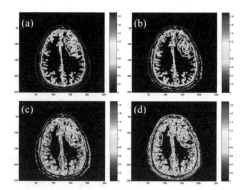

Figure 1: Entropy maps for Complete-Linkage ZIC (a), Complete-Linkage One-Way (b), k-Means ZIC (c) and k-Means One-Way.

by the Zoom-In approach is spatially more homogeneous than One-Way clustering resulting in a lower mean entropy. No clear dependence on slice position is observed including the slice used for clustering.

Fig. 3a shows the cluster maps resulting from the first step of Zoom-In Clustering. The tumourous region is well identified but partly overlapping with other brain structures. The overlap reduces when zooming into the yellow and light green region as shown in Fig 3b. The corresponding mean values for all 8 clusters are given in Tab. 1 showing the relevance of each individual parameter in defining the observed structure in Fig 3b. Results from principal component analysis are shown in Fig 3c-d where points assigned to a common cluster are displayed in the same colour. Plotting the first versus the second principal component clearly reveals the cluster structure also in the reduced space of the first two components for both steps of Zoom-In-Clustering.

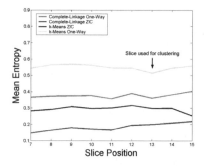

Figure 2: Weighted mean entropy as a function of slice position for the different clustering algorithms.

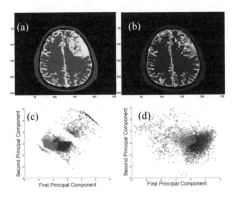

Figure 3: (a) Cluster assignment after the initial step of clustering for the patient with oligodendroglioma (b) as in (a) but after Zoom-In clustering. (c) First versus second principal component after the initial clustering step. (d) First versus second principal component after Zoom-In-Clustering.

This structure is also visible in the 3-dimensional representation of the dark blue region as shown in Fig. 4c. Results from the assignment of measured data points to clusters after a two-week cycle of methylpresdnisolone therapy clearly reveals the therapeutic effect of a reduced amount of peri-tumoural oedema (see Fig. 4d).

Based on the segmented images and the quantitative H_2O measurement used for clustering, it is furthermore possible to calculate the absolute water content within the selected cluster before and after therapy. Compared to normal white matter containing approximately 70.8% water in a 59 years old male [10], the increase of water in the ROI (dark blue region in Fig. 4) is 16.64 ml before therapy and 2.6 ml after therapy. Respective changes in volume and number of voxels comprising this cluster are given in Tab 2.

Anti-Oedema Therapy Evaluation

Results from the assignment of image points to clusters for the patient with secondary brain metastasis after the initial clustering step are shown in Fig. 4a. Based on the elbow-criterion, 3-4 clusters should be chosen as shown in Fig 5a. The data presented in Fig. 4a were clustered with $N_{c,1}=4$ which resulted in a clear segmentation of white and grey matter and tumour-associated oedema.

Results from the second step of the approach after zooming into the dark blue region of Fig. 4a are given in Fig. 4b. The results were obtained with $N_{c,2}=3$ consistent with the elbow criterion as demonstrated in Fig 5d. The second step of Zoom-In-Clustering resulted in better differentiation between CSF and oedemateous region that were identified as a common structure after the first step. In addition to the obvious large oedema in the left frontal hemisphere, a small region of oedema in the occipital region of the apparently normal hemisphere becomes visible after cluster assignment.

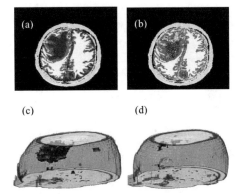

Figure 4: (a) Assignment of points to clusters after the initial clustering step for a transverse slice through the brain of a patient with pulmonary cerebral metastasis before the start of methylpresdnisolone treatment. (b) Results from ZIC for the same patient. (c) Oedema cluster (red structure) before (c) and after (d) two weeks cycle of methylpresdnisolone treatment.

Cluster Number	Cluster Colour	H_2O [%]	T_1 [ms]	T_2^* [ms]	α_1 [a.u.]	R_{Mean}[a.u.]	N_p	# Voxels
1	blue	76.18	950.51	64.984				16734
2	light blue	80.47	1764	152.09				636
3	green	69.21	1879.1	244.31				245
4	yellow	81.87	1653.9	71.896	-23.61	78.713	58.155	2003
5	orange	78.78	1802.2	45.738	-15.908	226.36	48.74	553
6	red	86.01	1838.8	71.246	-12.932	78.284	60.556	1113
7	dark red	66.73	1834	69.714	-24.217	356.21	37.547	294
8	brown	82.12	1504	106.67	-14.916	30.948	60.803	1701

Table 1: Mean values of variables used for clustering. Shown are the means for the eight different clusters identified by the Complete-Linkage Zoom-In algorithm as well as the number of points in each cluster.

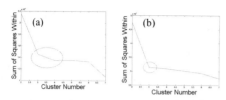

Figure 5: (a) Sum of Squares Within for the initial step of clustering and (b) after Zoom-In clustering. The red ellipse represents the region suggested by the Elbow-Criterion.

	Before Therapy	After Therapy
Number of voxels in oedema cluster	19687	5372
Mean H₂O content in oedema cluster	86 %	79 %
Absolute H₂O content in the oedema cluster	89.24 ml	22.40 ml
H₂O content increase	16.64 ml	2.6 ml

Table 2: Quantitative results from the anti-oedema therapy evaluation in a patient with pulmonary cerebral metastatic tumour.

4 Discussion

The results demonstrate that segmentation of images based on quantitative maps acquired with MRI is possible with the described Zoom-In-Clustering approach.

Clustering with k-Means and Complete-Linkage resulted in segmented images with good separation between grey matter, whiter matter and CSF. Even through the conventional approach enables good segmentation of non-pathological brains, they are in general less sensitive when pathological conditions such as cerebral tumours are present. This stems from the fact that tumours typically contain sub-regions not present in the normal healthy brain. Such regions include, e.g., necrosis, areas with a high and diffuse vascular activity or regions with altered oxygen supply [12]. When a slice through the human brain is clustered without taking the prior knowledge about the potential existence of such regions into account, all points have a finite probability to be classified as necrosis, e.g., even through necrosis may only exist in the tumour or peri-tumoural area. This results in a less homogeneous and less precise assignment of points that directly translates into a reduced quality of the segmented images.

This problem is strongly suppressed or even completely removed with the Zoom-In clustering algorithm presented. Images are segmented using an iterative and hierarchical approach. During a first step, a gross segmentation of the brain with only a few clusters is achieved using any conventional approach like k-Means or Complete-Linkage. Based on the first segmentation it is then possible to "zoom into" one or more clusters, and therefore areas, in order to define substructures only in this dedicated region. During the second step, only points in the selected region are re-clustered leaving all other pixels unchanged. The process of zooming into different regions can be easily repeated as many times as desired. Results from the Zoom-In step are directly displayed and can be evaluated online allowing interactive image processing.

In order to quantify the differences between the conventional and the Zoom-In approach, the spatial entropy distribution was determined. Entropy is a measure of uncertainty and therefore determines the spatial homogeneity of the cluster results obtained. The slice-dependent spatial entropy was calculated for k-Means and Complete-Linkage clustering both with the conventional and the Zoom-In approach. The results clearly revealed a reduced mean entropy and therefore increased spatial homogeneity associated with the Zoom-In algorithm. The entropy reduction was observed for both k-Means and Complete-Linkage based Zoom-In-Clustering compared to the conventional algorithm. The lowest entropy was observed for Complete-Linkage Zoom-In clustering which was even lower than the entropy for the respective k-Means ZIC. This effect may be caused by the different segmentation of grey- and white matter for both approaches. Complete-Linkage ZIC did not show a clear grey- white matter segmentation during the first pass of the algorithm in contrast to k-Means ZIC. This is also evident from the mean values for that cluster (see the first cluster in Tab. 1) which represents a mean T_1 and a mean H_2O content of grey and white matter. Due to the reduced number of tissue interfaces, the entropy is reduced without changing the effective homogeneity. In order to achieve good grey and white matter segmentation with Complete-Linkage, a Zoom-In step in this combined cluster has to be performed. Results from principal component analysis for both the first and the second step of Zoom-In clustering also revealed good separation between clusters obtained by this approach.

Zoom-In clustering allows semi-automated segmentation of images. Semi-automated clustering is possible and based on the extraction of the most likely cluster number relying on the determination of the Sum of Squares Within criterion. The approach is not limited to this specific measure so that different criteria can easily be incorporated into the current algorithm [11].

A further advantage offered by Zoom-In clustering is the possibility to include different variables at different steps of clustering. During the first step, only T_1, H_2O and T_2^* were used get a gross segmentation of the image which is sufficient for typical applications. In order to define tumour-associated substructures more efficiently, variables such as the mean deviation of the measured points from an exponential fit to these points were included. This variable is sensitive in regions with pronounced bi-exponential decay or with altered tissue oxygen extraction leading to a deviation from pure exponential behaviour [13]. These areas are typically present in tumours so that the inclusion of such variables provides valuable information and may result in more precise structure definition.

The Zoom-In clustering approach clearly demonstrated the presence of a tumour substructure in a patient suffering form an oligodendroglioma in the frontal area. Based on the available clinical information it is not possible to make any further statements about the histological origin of the identified structure as no matched histological information was acquired. Nevertheless, there is good evidence that the observed areas did not appear simply by chance because they were spatially homogeneous and were also identified at similar locations but in different slices. Both these facts demonstrate the statistical significance of the observed structure while the clinical significance has to be determined in future studies by comparing clustering results directly with histology.

The segmentation of the brain in a patient with a cerebral pulmonary metastasis based on Complete-Linkage Zoom-In clustering revealed a clear separation between grey and white matter, CSF and tumour associated oedema. The same structures were again identified in the follow-up measurement of the same patient after two weeks anti-oedema treatment with methylpresdnisolone. This clearly demonstrates one of the key advantages of the described clustering approach based on quantitative imaging. Once a cluster structure is defined, it can be used to classify the same patient again as many times as desired in longitudinal follow-up exams. This directly implies that objective therapy monitoring is feasible based on the quantitative nature of the data acquired in combination with an appropriate multivariate data analysis tool such as the described Zoom-In clustering approach. A clear reduction in water content and volume of the oedema was observed in the metastatic tumour patient indicating the success of the initiated treatment.

5 Conclusions

The results shown demonstrate the clinical relevance of quantitative imaging in combination with the described Zoom-In clustering approach. Zoom-In clustering offers clear advantages compared to conventional One-Way clustering as it results in spatially more homogeneous segmentations. In addition, it offers the maximum flexibility for inclusion of different information at different steps of the analysis as well as user defined Zoom-In strategies. Based on Zoom-In clustering of quantitative MR parameters, objective anti-oedema therapy monitoring was demonstrated in a patient with metastatic brain tumour. In addition, tumour substructure could be clearly and significantly identified in a patient suffering form an oligodendroglioma. Due to the missing information from histology, the clinical origin and significance of the identified areas remains to be determined in future studies.

6 References

[1] Lucignani G et al. Eur. J. Nucl. Med. Mol. Imaging 2004;3:1059-1063.
[2] Jacobs AH et al., Mol. Imaging 2002; 1(4):309-335.
[3] Barbier EL et al., J. Magn. Reson. Imaging 2001;
13:496-520.
[4] Dierkes T et al., ICS 2004;1265:181-185.
[5] Steinhoff S et al., Magn. Reson. Med. 2001; 46(1):131-140.
[6] Shah NJ et al., Neuroimage 2001;14:1175-1185.
[7] Shah NJ et al., German Patent Application 2000; No. 10028171.0.
[8] Zaitsev et al., Magn. Reson. Med. 2003;49(6): 1121-1132.
[9] Neeb H et al., ICS 2004;1265:113-123.
[10] Neeb H et al.,2004; in preparation.
[11] Brian S. Everitt, "Cluster Analysis"; Third Edition; Jon Wiley and Sons Inc. 1993.
[12] Nelson SJ et al., J Magn. Reson. Imaging 2002; 16:464-476.
[13] Yablonskiy D, Haake EM. Magn. Reson. Med. 1994;32:749-63.

Ein semiautomatisches Verfahren zur Segmentierung von Hirnregionen in MRT-Daten

L. Wischnewski[1], G. Wagenknecht[1], R. Dillmann[2]
1. Zentralinstitut für Elektronik, Forschungszentrum Jülich GmbH, D-52425 Jülich
2. Institut für Rechnerentwurf und Fehlertoleranz, Universität Karlsruhe, D-76131 Karlsruhe

Kurzfassung

Ziel des semiautomatischen Segmentierungsverfahrens ist die flexible Bestimmung von Regionen auf Basis der Hirnoberfläche. Zur Segmentierung einer auf der Hirnoberfläche liegenden Randkontur einer solchen „Region of Interest" wird der Live-Wire-Algorithmus auf den dreidimensional arbeitenden Fall erweitert. Komponenten des Live-Wire-Verfahrens sind die 3D-Merkmalsextraktion mittels eines Sobel- und Laplace-Filters, die hieraus bestimmte Kostenfunktion sowie die laufzeitoptimierte Graphsuche unter Einsatz von Prioritätswarteschlangen. Um Oberflächenpunkte innerhalb dieser Randkontur zu extrahieren, wird die Methode um eine automatische Komponente ergänzt, die den Live-Wire-Algorithmus mit einer hierarchischen Triangulation kombiniert. Die Güte des Verfahrens wird anhand von zwei Phantomkollektiven und zwei realen MRT-Datensätzen im Zwei- und Dreidimensionalen dargestellt.

1 Einleitung

Je nach Art der jeweiligen medizinischen Fragestellung werden in funktionellen Studien in den Neurowissenschaften unterschiedliche Gehirnregionen untersucht. Zur Darstellung des Gehirngewebes wird als bildgebendes Verfahren aufgrund des hervorragenden Kontrastes vornehmlich die Magnetresonanztomographie (MRT) eingesetzt. Die Analyse von funktionellen Parametern kann z.B. mit der Positronen-Emissionstomographie (PET) erfolgen.
Die Generierung dreidimensionaler Hirnatlanten stellt für die Analyse und Quantifizierung von Hirnregionen einen wichtigen Beitrag dar. Die Generierung von Hirnatlanten erfolgt in [1] mittels eines automatischen Segmentierungsverfahrens auf Basis des individuellen Bilddatenmaterials.
Durch den Einsatz eines semiautomatischen Verfahrens kann der verwendete Ansatz zur Generierung von individuellen Hirnatlanten ergänzt werden, indem neben automatisch segmentierten Regionen eine zusätzliche flexible Regionenbestimmung möglich wird. Im Folgenden wird das Konzept und die verwendeten Techniken zur flexiblen, semiautomatischen Regionenextraktion vorgestellt.

2 Methoden

Das entwickelte semiautomatische Segmentierungsverfahren basiert auf dem Live-Wire-Algorithmus. In [2, 3] wird der Live-Wire-Algorithmus zur Segmentierung von Objektkonturen im zweidimensionalen Bildbereich eingesetzt.

Bei der zweidimensionalen Variante des Live-Wire-Algorithmus spezifiziert der Benutzer Startpunkte auf der zu extrahierenden Objektkontur in einer Schicht des vorliegenden Datensatzes, die als Startpunkte in den automatischen Teil des Verfahrens eingehen. Im automatischen Teil wird eine graphbasierte Pfadsuche auf Basis der dynamischen Programmierung durchgeführt.
Zur Konstruktion des Graphen werden die Bildpunkte als Knoten des Graphen interpretiert. Auf dem zu segmentierenden Bild wird in einem Merkmalsextraktionsschritt eine lokale Kostenfunktion berechnet, die eine Linearkombination extrahierter Bildmerkmale darstellt. Die lokale Kostenfunktion wird zur Gewichtung der Kanten des Graphen eingesetzt. Die Pfadsuche berechnet vom Startpunkt zu allen anderen Punkten einen optimalen Pfad hinsichtlich der lokalen Kosten. Der optimale Pfad wird dann ausgehend vom Startpunkt zur aktuellen Mauscursorposition im Bild in Echtzeit angezeigt.
Zur Berücksichtigung der komplexen dreidimensionalen Topologie des Gehirns wurde das Live-Wire-Verfahren auf den dreidimensional arbeitenden Fall erweitert, indem neben der zweidimensionalen Pfadsuchtechnik eine Pfadsuche im dreidimensionalen Suchraum realisiert wurde. Eine schichtweise Segmentierung ist somit nicht notwendig.
Eine dreidimensional arbeitende Vorgehensweise ermöglicht die flexible Extraktion von 3D-Regionen des Gehirns auf Basis der Kortexoberfläche. Mit Hilfe des Live-Wire-Algorithmus lässt sich die Randkontur einer sogenannten „Region-of-Interest" im Volumendatensatz extrahieren, indem einzelne Pfadsegmente zu einer geschlossenen Randkontur verkettet werden. Das Live-Wire-Verfahren wurde um eine automatische

Oberflächenextraktion erweitert, um den Innenbereich der Randkontur auf der Kortexoberfläche zu bestimmen.

2.1 Lokale Kostenfunktion

In einem Merkmalsextraktionsschritt werden zur Berechnung der lokalen Kosten dreidimensionale Gradientenfilter eingesetzt. Die lokalen Kosten $l(p)$ für jedes Voxel p des Volumendatensatzes ergeben sich als Linearkombination des skalierten Gradientenbetrags l_G und des Laplaceschen Nulldurchganges l_Z [4]:

$$l(p) = \omega_G \cdot l_G(p) + \omega_Z \cdot l_Z(p) \quad (1)$$

Der Gradientenbetrag wird mittels eines 3D-Sobelfilters berechnet. Um Bereiche mit hoher Kantenstärke niedrige Kosten zuzuweisen, wird der Gradientenbetrag auf einen Wert zwischen 0 und 1 skaliert und mit einer linear abfallenden Funktion invertiert. Der Laplacesche Nulldurchgang l_Z wird durch einen 3D-Laplace-Operator und anschließender Binarisierung berechnet.

Neben einer linearen Abbildung des Gradientenbetrags auf einen Kostenfunktionswert kommt alternativ eine nichtlineare Funktion zum Einsatz. Der skalierte Gradientenbetrag wird mit einer modifizierten *Kotangens hyperbolicus*-Funktion auf einen neuen Kostenwert transformiert:

$$\coth_T(G) = \frac{e^{\lambda \tilde{G}} + e^{-\lambda \tilde{G}}}{e^{\lambda \tilde{G}} - e^{-\lambda \tilde{G}}} - 1,$$
$$mit\ \lambda \geq 1, \tilde{G} = \frac{G}{\max(G)} \quad (2)$$

Für stark gekrümmte Objektkanten kann der durch den Live-Wire-Algorithmus bestimmte Pfadverlauf über einen homogenen Bildbereich verlaufen, anstatt der Objektkontur zu folgen. Die nichtlineare Bewertung kompensiert diesen ungünstigen Pfadverlauf, da homogenen Bereichen besonders hohe lokale Kosten zugewiesen werden.

2.2 Graphsuche

Die Pfadsuche basiert auf dem Algorithmus von Dijkstra [5]. Für den aus den 2D-Schichten medizinischer Datensätze resultierenden zweidimensionalen Suchraum ist die vollständige Berechnung aller optimalen Pfade durchführbar. Eine vollständige 3D-Pfadsuche ist hingegen in medizinischen Volumendatensätzen nicht praktikabel.

Die Graphsuche erfolgt daher nach der in [6] verwendeten Pfadsuchtechnik *Live-Wire on the fly*. Hierbei werden nicht sämtliche optimale Pfade für den gesamten Graphen bestimmt, sondern nur der Teil des Graphen traversiert, der notwendig ist, um den optimalen Pfad vom Startpunkt zur aktuellen Mauscursorposition zu berechnen. Für die Graphtraversion stehen im Dreidimensionalen die 6er-Nachbarschaft, die 18er-Nachbarschaft und die 26er-Nachbarschaft zur Verfügung.

Eine effiziente Implementierung des Algorithmus von Dijkstra ist eine wesentliche Voraussetzung für die Anwendung des Verfahrens auf Volumendaten. Für eine effiziente Implementierung werden Prioritätswarteschlangen, z.B. ein binärer Heap oder ein Fibonacci-Heap, eingesetzt. Unter Verwendung des binären Heaps beträgt die Laufzeit des Graphsuchalgorithmus bei einem Graphen mit n Knoten und m Kanten $O((m+n) \log n)$ und unter Verwendung eines Fibonacci-Heaps amortisiert $O(m+n \log n)$.

2.3 Oberflächenextraktion

Eine „Region-of-Interest" wird auf der Kortexoberfläche bestimmt, indem der Benutzer zunächst die Randkontur der interessierenden Region mit Hilfe des Live-Wire-Algorithmus spezifiziert. Nach Extraktion dieser Randkontur erfolgt in einem automatischen Schritt die Bestimmung der Oberflächenpunkte, die innerhalb der Randkontur liegen.

Die automatische Bestimmung von Oberflächenpunkten erfolgt durch Kombination einer *hierarchischen Triangulation* [7] mit dem Live-Wire-Algorithmus. Eine hierarchische Triangulation verfeinert eine bestehende Triangulation sukzessive rekursiv durch weitere Triangulationen.

Die hierarchische Triangulation wird eingesetzt, um den von der benutzerdefinierten Randkontur begrenzten Innenbereich auf der Kortexoberfläche sukzessive in Dreiecke zu unterteilen. Das Verfahren bestimmt weitere Oberflächenpunkte, indem mittels des Live-Wire-Algorithmus automatisch die Randkontur des jeweiligen Dreiecks extrahiert wird.

Um die benutzerdefinierte Randkontur in Dreiecke unterteilen zu können, wird eine vereinfachte Modellannahme getroffen. Der durch den Live-Wire-Algorithmus resultierende variable Pfadverlauf im Dreidimensionalen wird auf ein Geradensegment in der Ebene abgebildet. Für die in **Bild 1** dargestellten Randkonturen ergibt sich in Abbildung 1-a ein konvexes Polygon und in Abbildung 1-b ein nicht konvexes Polygon.

Das entwickelte Verfahren löst das Problem der Oberflächenextraktion für den konvexen Fall einer benutzerdefinierten Randkontur. Als Eckpunkte des korrespondierenden Polygons dienen die vom Benutzer gesetzten Startpunkte, die bei der Randkonturextraktion durch den Live-Wire-Algorithmus zur Pfadberechnung verwendet werden.

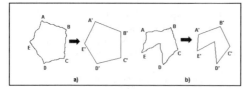

Bild 1 Abstraktion vom variablen Pfadverlauf auf Geradensegmente in der Ebene. Bild 1-a: Konvexer Fall einer benutzerdefinierten Randkontur; Bild 1-b: Nicht konvexer Fall einer benutzerdefinierten Randkontur.

Das Polygon wird durch eine Starttriangulation zunächst in eine Menge von Dreiecken unterteilt. Für die Starttriangulation werden sogenannte *Triangle-Strips* eingesetzt.

Die Dreiecke der Starttriangulation werden durch Anwendung einer quaternären Triangulation in weitere Dreiecke unterteilt, indem ein Dreieck in vier Dreiecke trianguliert wird (**Abbildung 2**). Wurden die Oberflächenpunkte der korrespondierenden Dreieckskante noch nicht extrahiert, wird der Live-Wire-Algorithmus unter Verwendung der Eckpunkte der Dreieckskante als Start- und Endpunkt für die Pfadsuche eingesetzt.

Zur Unterteilung eines Dreiecks in vier Dreiecke werden die Mittelpunkte der Dreieckskanten herangezogen. Für die Wahl des korrespondierenden Mittelpunktes der Dreieckskante wird der mittlere Pfadpunkt gewählt. Falls einer der Pfade der korrespondierenden Dreieckskanten eine Länge kleiner Drei aufweist oder die Mittelpunkte der Dreieckskanten aufgrund des variablen Pfadverlaufs zusammenfallen, wird die sukzessive Unterteilung des Dreiecks abgebrochen, da eine weitere Unterteilung des Dreiecks nicht möglich ist.

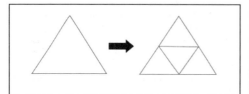

Bild 2 Unterteilung eines Dreiecks durch quaternäre Triangulation.

3 Ergebnisse

Im Rahmen einer Evaluationsstudie wurde der Algorithmus anhand von zwei künstlich generierten Phantomkollektiven untersucht. Es wurden zwei Phantome unterschiedlichen Typs mit einer Matrixgröße von (256x256x128) generiert. Der Vordergrund wurde mit einem Grauwert $g_1=170$ und der Hintergrund mit einem Grauwert $g_2=85$ im Intervall [0, 255] belegt. Der Vordergrund der Grundphantome wurde mit einer linear abfallenden Funktion, einer Stufenfunktion und einer geglätteten Stufenfunktion modifiziert, um den Kontrast an der zu extrahierenden Objektkontur variieren zu lassen. Die resultierenden Phantome wurden außerdem mit Gaußschem Rauschen unterschiedlicher Standardabweichungen verrauscht. **Abbildung 3** zeigt transversale Schnitte für jedes Exemplar der Kollektive.

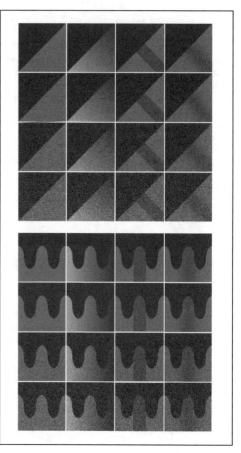

Bild 3 Konstruierte Phantomkollektive. Oben: Phantomkollektiv (I), Unten: Phantomkollektiv (II). 1.Spalte: Original Phantom (C), 2. Spalte: linear abfallende Funktion (L), 3. Spalte: Stufenfunktion (S), 4. Spalte: geglättete Stufenfunktion (G). Die Zeilen entsprechen den unterschiedlichen Rauschniveaus ($\sigma = 0$, $\sigma = 21.25$, $\sigma = 42.5$ und $\sigma = 63.75$).

Der Schwerpunkt der Untersuchung war auf den Einfluss der Merkmalsextraktion auf das Segmentierungsergebnis im Zweidimensionalen gerichtet. Außerdem wurde das Laufzeitverhalten der Graphsuche im Dreidimensionalen untersucht.

Zur quantitativen Bewertung der Segmentierungsergebnisse wurde die *Hausdorff-Distanz* [8] für die Referenzkontur und die vom Live-Wire-Algorithmus extrahierte Objektkontur berechnet. **Bild 4** stellt den ermittelten Interaktionsaufwand bei einer tolerierten Hausdorff-Distanz von 1.5 Pixel für das Phantomkollektiv (II) dar, der notwendig war, um den Vordergrund vom Hintergrund abzugrenzen. Bei Überschreitung der Schwelle wurde automatisch ein neuer Startpunkt auf der Objektkontur gesetzt. Aufgrund der Segmentierungsergebnisse sollte die Gewichtung des Gradientenbetrags größer als die Gewichtung des Laplaceschen Nulldurchganges sein. Der Interaktionsaufwand stieg deutlich an, wenn das Bildmaterial verrauscht war.

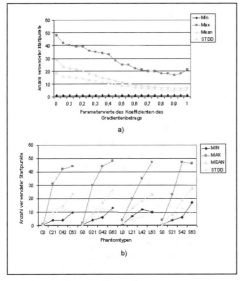

Bild 4 Evaluation der Merkmalsextraktion: a) Ermittelter Interaktionsaufwand für das Phantomkollektiv (II) bei einer tolerierten Hausdorffdistanz von 1.5 Pixeln bei unterschiedlicher Gewichtung ω_G des Gradientenbetrags; b) Ermittelter Interaktionsaufwand für das Phantomkollektiv (II) bei einer tolerierten Hausdorffdistanz von 1.5 Pixeln (MIN, MAX, MEAN, STDD über ω_G der Kostenfunktion).

Die Graphsuche wurde im Dreidimensionalen mit unterschiedlichen Prioritätswarteschlangen getestet. **Bild 5** zeigt das Laufzeitverhalten der Graphsuche für den binären Heap und den Fibonacci-Heap für die im Dreidimensionalen verwendeten Nachbarschaften. Die Laufzeit steigt stark an, wenn längere Pfadsegmente extrahiert werden (Bild 5-a). Bild 5-b und Bild 5-c stellen die Effizienz der verwendeten Prioritätswarteschlangen dar. Der binäre Heap ist dem Fibonacci-Heap bei den hier auftretenden Graphen vorzuziehen, da die Anzahl der Operationen pro Sekunde höher ist.

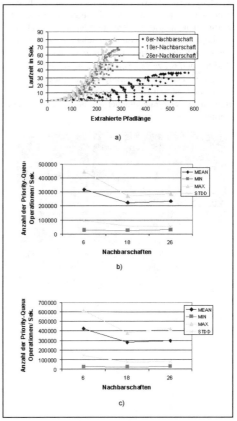

Bild 5 Laufzeitanalyse der Graphsuche im dreidimensionalen Suchraum: a) Laufzeit in Sekunden in Abhängigkeit der extrahierten Pfadlänge für die verschiedenen Nachbarschaften bei Verwendung eines binären Heaps; b) Anzahl der ausgeführten Operationen pro Sekunde des binären Heaps in Abhängigkeit der verwendeten Nachbarschaft; c) Anzahl der ausgeführten Operationen pro Sekunde des Fibonacci Heaps in Abhängigkeit der verwendeten Nachbarschaft.

Neben einer Evaluation an Phantomdaten wurden zwei reale Datensätze zunächst in 2D-Schichten segmentiert (**Bild 6**). Bild 6-a und Bild 6-d zeigen die Segmentierungsergebnisse einer Hirnaußenkonturextraktion bei einem über 27 Scans gemittelten, T1-gewichteten MR-Datensatz. Bild 6-b, 6-c, 6-e und 6-f zeigen Segmentierungsbeispiele eines zweiten T1-gewichteten MR-Datensatzes (1 Scan). Die Hirnaußenkontur konnte bei Einsatz einer linearen Bewertung des Gradientenbetrags mit 14 Startpunkten (Bild 6-a) und bei Einsatz einer nichtlinearen Bewertung mit 9 Startpunkten (Bild 6-d) segmentiert werden. Bild 6-b und 6-e stellen Problembereiche für die Anwendung des Live-Wire-Algorithmus dar. In Bereichen einer Objektkontur mit lokal dominantem Grauwertgradienten, wie der Kalotte oder im Bereich der Augen, kann der Pfad von einer schwächeren Objektkontur auf die stärkere Objektkontur überspringen. Bild 6-c und 6-f zeigen Segmentierungsergebnisse am Kleinhirn. Die Segmentierung konnte durch Einsatz der $coth_T$-Funktion (Bild 6-f) gegenüber der linearen Bewertung des Gradientenbetrags deutlich verbessert werden.

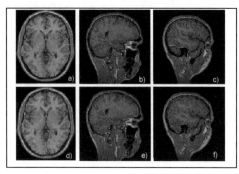

Bild 6 Segmentierungsergebnisse in transversalen und sagittalen Schichten T1-gewichteter MR-Datensätze.

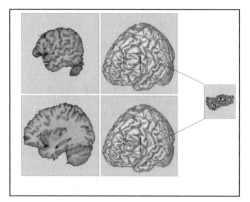

Bild 7 Beispiel einer Oberflächenextraktion: Linke Spalte: Setzen der Startpunkte in ausgewählten sagittalen Schichten eines automatisch segmentierten MR-Datensatzes; Rechte Spalte: Oberflächenextraktion. Die Randkontur (in rot gefärbt) wurde mit 5 Startpunkten erzeugt. Für die Pfadsuche wurde eine 26er-Nachbarschaft gewählt. Der extrahierte Oberflächenbereich besteht aus 203 Dreiecken.

Die Oberflächenextraktion wurde exemplarisch am gemittelten T1-gewichteten MR-Datensatz getestet (**Bild 7**). Der Datensatz wurde zunächst mit einem automatischen Segmentierungsverfahren [1] vorverarbeitet, mit dem das nichtcerebrale Gewebe aus dem Datensatz entfernt wurde. Zur Definition der geschlossenen Randkontur wurden in ausgewählten 2D-Schichten die Startpunkte für die Pfadextraktion durch den Benutzer ausgewählt (linker Teil der **Abbildung 7**). Auf den semiautomatischen Teil folgte die automatische Oberflächengenerierung, die den Innenbereich der Randkontur – dargestellt im rechten Teil der **Bild 7** - extrahierte. Aufgrund der Triangulation ergeben sich einige Lücken in der Oberfläche.

4 Diskussion und Ausblick

Das beschriebene Verfahren leistet einen Beitrag zur semiautomatischen Segmentierung, das zur flexiblen Extraktion von Bereichen auf der Kortexoberfläche eingesetzt werden kann. Der Live-Wire-Algorithmus wurde konsistent um eine Methodik erweitert, die die Pfadextraktion zur Oberflächengenerierung verwendet.

Die Evaluationsergebnisse zeigen, dass eine Weiterentwicklung der Methodik hinsichtlich des Live-Wire-Algorithmus zur Pfadextraktion und der automatischen Oberflächengenerierung sinnvoll ist. Der nächste im Projekt verfolgte Schritt zur Verbesserung der Pfadextraktion besteht in der Adaption der lokalen Kosten an die aktuell zu extrahierende Kontur (dem sogenannten On-the-fly-Training [2]). Durch die Adaption lassen sich auch schwächere Kanten segmentieren, weil ein Umspringen des Pfades auf eine Kontur mit einem höheren Gradientenbetrag verhindert werden kann. Außerdem soll der Einfluss anderer Bildmerkmale, wie die Grauwertinformation entlang der Objektkontur, und die Verwendung alternativer Kantendetektoren untersucht werden.

Zusätzlich zur partiellen Graphsuche sollte der Suchraum weiter eingeschränkt werden, um die Laufzeiten der Pfadsuche weiter zu reduzieren. Eine Beschränkung des Suchraums kann z.B. heuristisch mittels einer Strahlsuche erfolgen, bei der Knoten, deren lokale Kosten außerhalb eines Intervalls liegen, von der Graphsuche ausgeschlossen werden.

Zur flexiblen Oberflächenextraktion ist eine Erweiterung des Verfahrens auf den nicht konvexen Fall einer benutzerdefinierten Randkontur wünschenswert. Außerdem sind andere Triangulationsmethoden, wie z.B. die Delaunay-Triangulation, denkbar. Eine Alternative zum vorgestellten Ansatz stellt die in [9] beschriebene Vorgehensweise zur Oberflächengenerierung dar.

Das Verfahren wurde zur Segmentierung von MRT-Daten entwickelt. Weitere Anwendungsbereiche sind die Vorverarbeitung von PET- oder histologische Daten.

5 Literatur

[1] Wagenknecht G., Kaiser, H.J., Büll, U. et. al.: MRT-basierte individuelle Regionenatlanten des menschlichen Gehirns- Ziele, Methoden, Ausblick. In: Wittenberg T, Hastreiter P, Hoppe U, Handels H, Horsch A, Meinzer HP (Hrsg.). Bildverarbeitung für die Medizin 2003. Buchreihe Informatik Aktuell. Berlin: Springer-Verlag, 2003: 378-382

[2] Mortensen, E.N., Barrett, W.A.: Interactive Segmentation with Intelligent Scissors, Graphical Models and Image Processing 60: 349- 384, 1998

[3] Falcao, A.X., Udupa, J.K., Samarasekera, S., Sharma, S.: User-Steered Image Segmentation Paradigms: Live Wire and Live Lane, Graphical Models and Image Processing 60: 233 – 260, 1998

[4] Wischnewski, L., Wagenknecht, G.: Semiautomatische Segmentierung dreidimensionaler Strukturen des Gehirns mit Methoden der dynamischen Programmierung. In: Tolxdorff T., Braun J., Handels H., Horsch A., Meinzer H. P. (Hrsg.). Bildverarbeitung für die Medizin 2004. Buchreihe Informatik Aktuell. Berlin: Springer-Verlag: 55-59, 2004

[5] Dijkstra, E. W.: An note on two problems in connexion with graphs. In: Numerische Mathematik 1, 1959, S. 269- S. 271

[6] Falcao, A.X., Udupa, J.K., Miyazawa, F.K.: An Ultra-Fast User Steered Image Segmentation Paradigm: Live wire on the fly, IEEE Trans Med Imaging, Vol. 19, No 1, 2000

[7] L. D. Floriani and E. Puppo: Hierarchical triangulation for multiresolution surface description. ACM Transactions on Graphics, 14 (4) 363-411, 1995

[8] Chalana, V., Kim, Y.: A Methodology for Evaluation of Boundary Detection Algorithms on Medical Images, IEEE Trans Med Imaging. 16(5): 642-652, 1997

[9] Haenselmann T., Effelsberg, W.: Wavelet-based semi-automated live-wire segmentation: SPIE Human Vision and Electronic Imaging VII: 260-269, San Jose, 2003

Modellbasierte Segmentierung von MRT-Daten des menschlichen Gehirns

H. Belitz[1], G. Wagenknecht[1], H. Müller[2]
[1] Zentralinstitut für Elektronik, Forschungszentrum Jülich GmbH, D-52428 Jülich
[2] Lehrstuhl VII: Graphische Systeme, Fachbereich Informatik, Universität Dortmund, D-44221 Dortmund

Kurzfassung

In MRT-Bilddaten des menschlichen Gehirns sind die subkortikalen Bereiche aufgrund ihrer Grauwertcharakteristik kaum von ihrer Umgebung abgrenzbar. Rein bildbasierte, automatische Segmentierungsmethoden extrahieren diese Bereiche meist nur partiell. Durch Einbringen von topologischem und geometrischem Vorwissen ist jedoch eine automatische, modellbasierte 3D-Segmentierung denkbar. Ein System zur Segmentierung subkortikaler Regionen im Zweidimensionalen wird hier vorgestellt. Die bisherigen Ergebnisse werden anhand von realen Daten und Phantomdatensätzen präsentiert und eine Erweiterung des Modells auf Volumendaten diskutiert. Das sequentiell arbeitende Verfahren verwendet eine Gewebeklassifikation, mit einem GVF-Diffusionsalgorithmus nachbearbeitete Gradientenfelder, sowie aktive Konturmodelle, um den Bildraum bezüglich der gesuchten Strukturen zu partitionieren und diese dann zu segmentieren.

1 Einleitung

Für die Generierung individueller, dreidimensionaler Hirnatlanten des menschlichen Gehirns aus MRT-Daten [1] ist die Segmentierung subkortikaler Areale, insbesondere der Basalganglien, von Bedeutung.
In **Bild 1**, einer exemplarischen Schicht aus einem über 27 individuelle Scans gemittelten Datensatz [2], erkennt man deutlich, daß die Regionen der Basalganglien sich nur unzureichend von ihrer Umgebung abgrenzen. Dies ist insbesondere bei den Putamina und den Thalami der Fall. Automatische Segmentierungs-verfahren, die rein bildbasiert arbeiten, können aus diesem Grund die betreffenden Areale nur unzu-reichend oder unter Zuhilfenahme nicht vernach-lässigbarer manueller Interaktion extrahieren.
Bringt man jedoch Vorwissen über Form und Lage der gesuchten Strukturen mit ein, so ist auch eine automatische Segmentierung der Basalganglien denkbar. Dies führt zu den *modellbasierten Segmentierungsverfah-ren*. Dabei sind vor allem die sogenannten *aktiven Konturmodelle* hervorzuheben, die auf dem Snake-Modell von Kass et al [3] basieren.
Das primäre Ziel der Arbeit ist die Entwicklung eines dreidimensionalen Verfahrens, das den Volumendatensatz als Ganzes ohne nennenswerte Interaktion automatisch verarbeiten kann. Das hier vorgestellte modellbasierte Segmentierungssystem reduziert die Problematik zunächst auf den zweidimensionalen Fall einzelner Schichtbilder der medizinischen Datensätze.

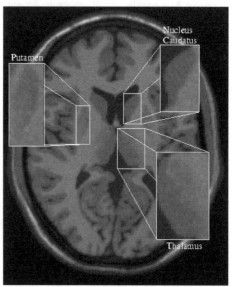

Bild 1 Schicht eines MRT-Datensatzes

2 Methodik

Das prototypische System setzt sich aus mehreren Modulen zusammen, die das Datenmaterial sequentiell verarbeiten. Diese Module werden im Folgenden kurz vorgestellt.

2.1 Vorverarbeitung

Im Vorverarbeitungsschritt kommt zuerst eine Artefaktreduktion zum Einsatz, sofern das Bildmaterial diese erfordern sollte. Diese Reduktion dient primär der Rauschunterdrückung und kann beispielsweise durch anisotrope Diffusionsfilter [4] erfolgen (vgl. **Bild 2a**).

Außerdem werden Informationen über die Gewebestruktur aus dem Bild extrahiert. Dabei kommen ein überwacht lernendes neuronales Netz und ein wissensbasierter „Peeling"-Algorithmus zur Anwendung, die alle Bildvoxel in Hintergrund, extrazerebrales Gewebe, Liquor, weiße und graue Substanz unterteilen [1] (siehe **Bild 2b**).

Durch diese Gewebeklassifikation können bereits grosse Teile des Bildes, zum Beispiel der extrazerebrale Liquorraum und die Schädelkalotte, verworfen werden, da sie für den Segmentierungsalgorithmus nicht von Belang sind. Desweiteren kann aus den Klassi-fikationsdaten mittels morphologischer Operatoren eine Maske gewonnen werden, die das graue Gewebe des Kortex heraus maskiert und auf diese Weise den Segmentierungsbereich weiter einschränkt (**Bild 2c**). Einzelne Gewebeklassen auszuschließen oder diese beispielsweise mittels eines Regionenwachstumsverfahrens zu segmentieren reicht für diesen Zweck nicht aus, da die subkortikalen Anteile grauen Gewebes meist mit dem Kortexgewebe verbunden sind. Die Beschränkung des Bildraums durch die Maske stabilisiert einerseits die spätere Segmentierung, andererseits verringert sie auch den zu durchsuchenden Bildraum.

In einem weiteren Schritt wird aus den Bilddaten mittels eines Canny- Operators ein Kantenbild erzeugt.

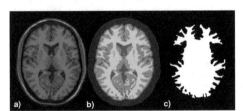

Bild 2 Ergebnisse der Rauschunterdrückung (a), Gewebeklassifikation (b) und Maskengenerierung (c)

2.2 Partitionierung

Die Partitionierung trennt einerseits die Regionen des Interesses (engl. *Regions of Interest – ROI*) voneinander, andererseits dient sie auch dem Zweck, für die Segmentierung uninteressante Regionen des Bildes von weiteren Arbeitsschritten auszuschließen. Wie schon bei der Maskengenerierung kann auch dieser Schritt die Laufzeit der Segmentierung positiv beeinflussen. Zudem erhöht die Trennung der interessanten Areale die Stabilität des Segmentierungsmodells. Die Partitionierung verwendet als Eingangsdatum den Gradienten des zuvor bestimmten Kantenbildes (siehe **Bild 3**, links). Dieses Vektorfeld wird mittels eines Gradienten-Vektorfluss-Verfahrens (GVF) [5] diffundiert (vgl. **Bild 3,** rechts). Die entstehenden Diffusionsfronten benachbarter Objekte treffen dabei aufeinander und bilden lokale Inhomogenitäten, d.h., stark voneinander abweichende Vektorenrichtungen, im resultierenden Feld.

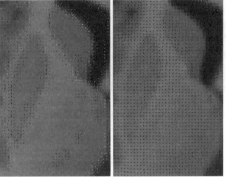

Bild 3 Gradient des Kantenbildes vor und nach der Anwendung der Gradientenvektorfelddiffusion

Mittels eines geeigneten Operators können diese Inhomogenitäten extrahiert werden. Der Operator wird definiert als

$$I(\mathbf{x}) = \max\left(\left|r(\mathbf{i}) - r(\mathbf{x})\right|\right), \mathbf{i} \in N(\mathbf{x}) \quad (1)$$

mit der Richtung des Vektors am Punkte $\mathbf{i}: r(\mathbf{i})$ und der 8er-Nachbarschaft des Punktes $\mathbf{x}: N(\mathbf{x})$.

Von der Antwort des Operators wird danach ein mit einem Gaußfilter kleiner Varianz geglättetes Kantenbild des ursprünglichen Bildmaterials abgezogen, da auch die Kanten des Bildes lokale Inhomogenitäten darstellen, diese für die nachfolgenden Schritte aber unerwünscht sind. Die so nachbearbeitete Operatorantwort ist in **Bild 4**, links, dargestellt.

Das entstehende Inhomogenitätsprofil reicht zur Partitionierung nicht aus, da es noch zahlreiche Bereiche enthält, die nicht von Interesse sind. Die sich gut abzeichnende zentrale Region kann mittels einer aktiven Kontur mit einer starken Gewichtung der äußeren Kräfte segmentiert werden. Die aktive Kontur schmiegt sich an den gesuchten Bildbereich an und trennt so die Putamina vom den übrigen gesuchten subkortikalen Regionen, die alle in der soeben segmentierten Region liegen. Um nun auch eine Region des Interesses für die Putamina definieren zu können, findet ein Abgleich mit der im Vorfeld erzeugten Maske statt. Durch eine Differenzenbildung von Maske und Partition erhält man eine neue Maske, die unter anderem auch die Putamina beinhaltet. Um deren Regionen des Interesses zu extrahieren, verwendet man ein morphologisches Closing, das alle irrelevanten Teile der neuen Maske verwirft. Das Ergebnis, also die zentrale Region und die Regionen der Putamina, ist in **Bild 4**, rechts, dargestellt.

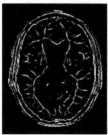 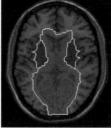

Bild 4 Antwort des Inhomogenitätsoperators und resultierende Partitionierung des Bildraums

2.3 Segmentierung

Die Segmentierung verwendet ein diskretes, deformierbares Modell, das auf der Arbeit von Viergever und Lobregt [6] basiert. Das Modell, ein *diskretes Partikelsystem*, wird als Polygon interpretiert, wobei die Stützpunkte des Polygons durch die Partikel gegeben sind. Das Modell ist *adaptiv diskretisierbar*, d.h., mit einer Veränderung der Polygongröße werden neue Partikel eingefügt oder Partikel gelöscht, so daß immer eine durchschnittliche Partikeldistanz und somit eine annähernd konstante Diskretisierungsrate gewährleistet wird. Die Stützpunkte des Polygons werden durch gewichtete Kräfteterme verschoben, was wiederum eine Deformation des gesamten Modells zur Folge hat. Bei der richtigen Parameterwahl konvergiert das Modell gegen die gesuchte Bildstruktur und liefert so die gewünschte Segmentierung.

Die Modellkräfte resultieren einerseits, in Form von *internen Kräften*, aus dem Vorwissen und sorgen für eine gewisse Glattheit der Kontur sowie für den Erhalt einer groben Form durch Abgleich mit einem zuvor definierten, einfachen Templatepolygon. Andererseits wird der Einfluß der Bildinformation durch *externe Kräfte* repräsentiert. Diese ergeben sich durch eine Gradientenbildung auf dem Kantenbild der jeweiligen Bildregion und die anschließende Diffusion dieser Felder mit dem GVF-Algorithmus. Das Ventrikelsystem wird zuvor mittels der Gewebeklassifikation heraus maskiert, da es sonst die kleineren Gradienten der subkortikalen Regionen überlagern könnte.

Nach Bestimmung der externen Kraftfelder können die subkortikalen Areale segmentiert werden. Die Partitionierung sowie die aus der Gewebeklassifikation hervorgehende Lage der Ventrikel ermöglichen die automatische Definition signifikanter Startpunkte für das aktive Konturmodell. Die Gewichtung der verschiedenen Kräfteterme ist abhängig von der gesuchten Region. So überwiegen beispielsweise bei der Segmentierung des Nucleus caudatus die externen Kräfte; die Vorgabe eines Templates ist hier in der Regel nicht von Nöten. Für den nur schwach abgrenzbaren Thalamus dagegen müssen die Templatekräfte sehr hoch gewichtet werden.

Wendet man das aktive Konturverfahren auf die gesuchten Bereiche an, so erhält man für die bisher verwendete Beispielschicht die in **Bild 5** dargestellten Ergebnisse.

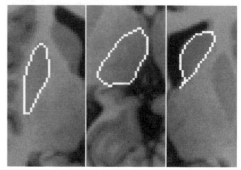

Bild 5 Segmentierung der gesuchten Regionen in der Beispielschicht des MRT-Datensatzes

3 Ergebnisse

Eine quantitative Evaluation des Verfahrens auf realem Datenmaterial erweist sich als schwierig. Da Goldstandards fehlen, ist der quantitative Fehler des Verfahrens nur grob, zum Beispiel durch Vergleiche mit manuellen Segmentierungen, abschätzbar. Um dennoch quantitative Aussagen über die Segmentierungsqualität machen zu können, wurden auf Basis von geometrischen Primitiven Phantome generiert, die exemplarische Schichten des realen Datenmaterials nachbilden. Durch die exakte, mathematische Definition der Phantome werden quantitative Aussagen möglich.

3.1 Konstruktion der Phantome

Die Schichtphantome wurden nach den Methoden der *Constructive Solid Geometry* [7] durch Mengenverknüpfungen von affin transformierten Ellipsen generiert. Das entstehende Modell wurde daraufhin im Bildraum diskretisiert. Um Bildartefakte zu simulieren, wurden diese Schichtphantome mit einem gaußschen Rauschen überlagert. Dabei wurden Standardabweichungen von 10, 20 und 40 in Bezug auf den Intensitätsbereich von [0,255] der Phantome gewählt. Desweiteren wurden die approximierten, subkortikalen Areale mit einem linearen Grauwertverlauf versehen, der die Phantomareale in die umgebende weiße Materie überführt. Die entstehenden Phantome sind für verschiedene Rauschniveaus in **Bild 6** dargestellt.

Bild 6 Erzeugtes Phantomkollektiv

3.2 Evaluation

Das entstandene Phantomkollektiv wurde verwendet, um das Verfahren quantitativ zu evaluieren. Dazu sind Diskrepanzmaße als objektive Kriterien subjektiven Goodness-Methoden vorzuziehen [8].

Für die Beurteilung der Segmentierungsqualität wurden Referenzkonturen A aus dem Phantom extrahiert und mittels verschiedener Diskrepanzmaße mit den Konturen der Segmentierungsergebnisse B verglichen. Dieses Verfahren wurde für verschiedene Diskretisierungsstufen, die sich im Modell durch die mittlere Distanz der Stützpunkte äußern, durchgeführt.

Für die Evaluation wurden drei verschiedene Diskrepanzmaße ausgewählt, die untereinander nur schwach korrelieren [9] und deshalb zusammen mit der Anzahl falsch klassifizierter Konturpixel gut zur Qualitätsbeurteilung heran gezogen werden können.

Das erste verwendete Maß ist die *Hausdorff-Distanz*

$$H(A,B) = \max\{h(A,B), h(B,A)\} \quad (2)$$
$$\text{mit } h(A,B) = \max_{a \in A} \min_{b \in B} d(a,b).$$

Sie beschreibt den maximalen Abstand der beiden verglichenen Konturen. Desweiteren wird der *Dice-Koeffizient*

$$C_D = \frac{2|A \cap B|}{|A| + |B|} \quad (3)$$

verwendet, welcher ein generelles Maß für die Übereinstimmung der Konturen ist. Die *mittlere Oberflächendistanz*

$$D_M = \frac{\sum_{a \in A} \min_{b \in B} d(a,b) + \sum_{b \in B} \min_{a \in A} d(b,a)}{|A| + |B|} \quad (4)$$

ist das dritte verwendete Maß. Sie mißt den mittleren Abstand der beiden Konturen. Diese drei Maße werden ergänzt durch ein viertes Maß, die Zahl der falsch segmentierten Konturpixel relativ zur Zahl der Pixel der Referenzkontur:

$$P_{FK} = (|A \cup B| - |A \cap B|)/|A| \quad (5)$$

Die Evaluationsergebnisse sind exemplarisch für die Putamen-Region in den **Bildern 7 - 10** dargestellt. Es zeigt sich, daß das Verfahren gegenüber schwachem Rauschen recht robust ist. Starkes Rauschen verschlechtert die Ergebnisse deutlich. Da der verwendete Grauwertverlauf in Verbindung mit hohem Rauschen eine eindeutige Konturerkennung aus dem Bildmaterial unmöglich macht, ist das Verfahren in diesem Fall sehr auf die gute Definition eines geometrischen Templates angewiesen. Erweist sich dieses als zu ungenau oder unpassend für die gesuchte Struktur, so kommt es zu eklatanten Segmentierungsfehlern.

Betrachtet man die Abhängigkeit von der Diskretisierung, so sind auch hier die Ergebnisse wieder zum Teil vom Rauschniveau des segmentierten Bildes abhängig. Bei keinem oder schwachem Rauschniveau ist eine hohe Diskretisierungsrate vorzuziehen, da diese eine genauere Annäherung an die gesuchte Kontur erlaubt. Bei hohen Rauschniveaus steigt jedoch auch die Anzahl der Partikel, die durch Rauschpixel im Bild falschen Kräften ausgesetzt sind. Diese Störeinflüsse können im schlimmsten Fall nicht mehr durch die internen Konturkräfte kompensiert werden. Dies führt zu einer Oszillation der aktiven Kontur um die Objektgrenzen. Eine Stabilisierung ist nur durch eine schlechtere Diskretisierung zu erreichen. Weiterhin ist zu beachten, daß auch für schwache Rauschniveaus bereits eine schlechtere Diskretisierung von Vorteil sein kann. So zeigen sich bei einem Rauschniveau von σ=10 und einer mittleren Stützpunktdistanz von 4 deutlich bessere Ergebnisse als bei kleineren Distanzen. Die Ursachen für dieses Verhalten sind noch genauer zu untersuchen.

Ebenfalls untersucht wurde der Einfluß der Intensitäts-Inhomogenitäten. Zu diesem Zweck wurden die

Segmentierungen der Phantome mit und ohne Grau
wertverlauf direkt verglichen. Dabei zeigte sich vo
allem, daß auftretende Inhomogenitäten die Rausch
empfindlichkeit signifikant erhöhen. Auf den Phan
tomen ohne Grauwertverlauf war auch für hoh
Rauschniveaus noch eine qualitativ akzeptable Seg
mentierung möglich. Zudem zeigte sich bei diesen
Phantomen konstanter Intensität eine geringere Ab
hängigkeit von der Diskretisierungsrate. Stark
Sprünge in den betrachteten Maßen, wie sie bei
spielsweise beim inhomogenen Phantom mit Rausch
niveau σ=10 zu beobachten sind, traten hier nicht auf

4 Ausblick

Das vorgestellte Verfahren zeigt grundsätzlich, daß
aktive Konturmodelle dazu geeignet sind, kontrastar
me, durch rein bildbasierte Ansätze nur unzureichen
abgrenzbare Regionen aus dem gegebenen Bildmate
rial zu extrahieren. Dazu ist jedoch eine Minderun
der Bildartefakte im Vorfeld zwingend notwendig.
Der nächste Schritt ist nun, das Verfahren auf de
dreidimensionalen Fall zu erweitern und so eine Seg
mentierung der Basalganglien auch in Volumendaten
sätzen zu ermöglichen. Dabei sind vor allem die kom
plexe Geometrie und die Konnektivität der Strukturen
untereinander zu beachten. Auch ist die automatische
Generierung von Gewichtsparametern für die verschiedenen Kräfteterme noch zu verbessern bzw. für
den dreidimensionalen Fall mit einer ganz neuen Methodik zu implementieren.
Weitere Ziele sind die Verbesserung der Segmentierungsqualität, z.B. durch Extraktion extremaler Bildpunkte oder durch eine intensitätsbasierte Kanten-
schätzung, sowie die Erhöhung der Segmentierungs-
geschwindigkeit. Zudem soll die Methodik der
Evaluation noch in der Hinsicht überarbeitet werden,
daß sie nicht nur Kontur-, sondern auch Regionenin-
formationen berücksichtigt, was unter anderem einen
direkten Vergleich mit anderen Segmentierungsmethoden erlauben würde.

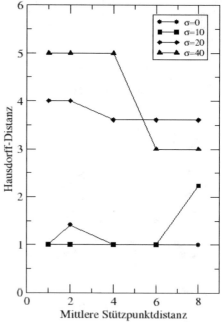

Bild 7 Putamen-Phantom: Hausdorff-Distanz

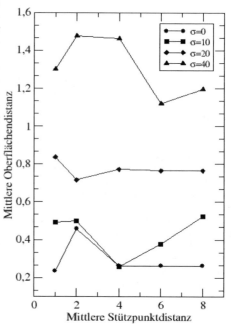

Bild 8 Putamen-Phantom: Mittlere Oberflächen-Distanz

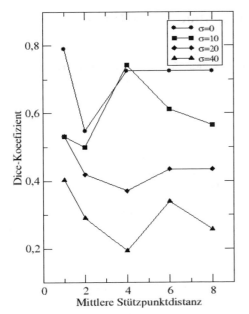

Bild 9 Putamen-Phantom: Dice-Koeffizient

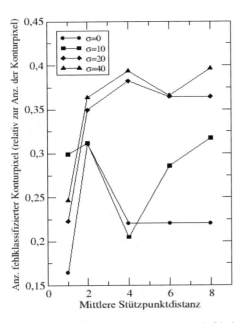

Bild 10 Putamen-Phantom: Relative Anzahl falsch segmentierter Pixel.

5 Literatur

[1] Wagenknecht, G.: Entwicklung eines Verfahrens zur Generierung individueller 3D-Regions of Interest-Atlanten des menschlichen Gehirns aus MRT-Bilddaten zur quantitativen Analyse koregistrierter funktioneller ECT-Bilddaten, Dissertation, RWTH Aachen, 2002

[2] Collins, D.L.; Neelin, P.; Peter, T.M.; Evans, A.C.; Automatic 3D registration of MR volumetric data in standardized talairach space. J. Comput. Assist. Tomogr., Vol. 18, No. 2, 1994, pp. 192-205

[3] Kass, M.; Witkin, A.; Terzopoulos, D.: Snakes: Active contour models. Int. J. Comput. Vision. Vol. 1, No. 4, 1988, pp. 321-331

[4] Castellanos, J.; Wagenknecht, G.; Tolxdorff, T.: Advances of a 3D noise suppression method for MRI data. VDE Verlag (im Druck)

[5] Prince, J.; Xu, C. : Generalized gradient vector flow external forces for active contours. Signal Processing - An International Journal. Vol. 71, No. 2, 1998, pp. 131-139

[6] Lobregt, S.; Viergever, M.A.: A Discrete Dynamic Contour Model. IEEE Transactions On Medical Imaging. Vol. 14, No. 1, 1991, pp. 12-24

[7] Abramowski, S.; Müller, H.: Geometrisches Modellieren. B.I. Wissenschaftsverlag, Mannheim, 1991

[8] Zhang, Y.J.: A survey on evaluation methods for image segmentation. Pattern recognition. Vol. 29, No. 8, 1996, pp. 1335-1346

[9] Heimann, T.; Thorn, M.; Kunert, T.; Meinzer, H.P.: Empirische Vergleichsmaße für die Evaluation von Segmentierungsergebnissen. Bildverarbeitung für die Medizin, Reihe Informatik Aktuell, Springer Verlag, Berlin, 2004, pp. 165-169

Advances of a 3D noise suppression method for MRI data

J. Castellanos[1], G. Wagenknecht[1], T. Tolxdorff[2]
[1] Central Institute for Electronics, Research Center Jülich, D-52425 Jülich
[2] Institute of Medical Informatics, Biostatistics and Epidemiology, Charité University Medicine Berlin, Benjamin Franklin Campus, D-12200 Berlin, Germany

Abstract

High resolution MR images are often affected by noise which results in undesired intensity overlapping. Dependent on the noise level, the segmentation and classification of different tissues can be difficult.
This work presents advances of a 3D automatic noise reduction method based on anisotropic diffusion filters. These filters are used because of their edge-preserving properties. The iterative procedure uses the distribution of the residual noise generated during the progression of the filtering process to adjust the filter parameters. Both simulated and real data are evaluated by means of similarity measures to illustrate the performance of the method.

1 Introduction

Traditional filters, such as Mean or Gaussian filters, are remarkable noise-reducing methods; they are easy to use and to implement. Unfortunately they do not respect the boundaries originated from regions with different intensities, eliminating sharp details and smoothing the edges. This result in blurred and diffused images (see **Figure 1**).

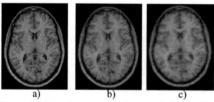

a)   b)   c)
Figure 1 Magnetic resonance image; a) Original image, b) processed image with a Gaussian filter after 3 and 6 iterations.

Anisotropic diffusion filters (ADF) overcome the major drawbacks of conventional linear algorithms because of the adaptive selection of diffusion strengths in accordance to the local characteristics of the image, encouraging the smoothing inside the regions and preventing the smoothing across the boundaries.
The aim of this project is to develop an adaptive method that automatically adjusts the two principal filter parameters, diffusion factor and number of iterations, according to the specific conditions of each data set. This method uses the differential information obtained from the original and processed data to define the feedback information required to control the process.
In particular this project concentrates in the processing of MR images of the brain, to support the automatic segmentation of its anatomical structures and the construction of individual atlases [2].
This paper is organized as follows. In Section 2 the principles of the anisotropic diffusion are presented. Section 3 presents some considerations about the noise and the general concept of our automatic method. In Section 4 some examples are analyzed and processed with the proposed method. And finally in Section 5 a brief summary of our work is presented.

2 Anisotropic Diffusion

The enhancement of an image can be seen as an evolution process where the image improves successively by means of convolving the original data with a defined kernel. In **Figure 1** the results obtained with the isotropic diffusion equation (eq. 1) are presented as an example. In this case the diffusion process takes place regardless of the image edges, produced between regions with different intensities.

$$\frac{\partial}{\partial t}I(\bar{u},t) = \mathrm{div}(\nabla I(\bar{u},t)) \qquad (1)$$

$I(\bar{u},t)$ represents the processed image at time t, $\bar{u}$ the image axes (e.g., x,y,z), ∇I the gradient of the image and t the iteration step (time).
Perona and Malik [3] modified this isotropic diffusion equation introducing the term $c(\bar{u},t)$:

$$\frac{\partial}{\partial t}I(\bar{u},t) = \mathrm{div}(c(\bar{u},t)\nabla I(\bar{u},t)) \qquad (2)$$

This new coefficient, c, was included to control the diffusion strength, according to an estimation of the boundaries position, stopping the diffusion when boundaries are found and favoring the diffusion in homogeneous regions.

2.1 Diffusion equations

The diffusion coefficient was defined as a monotonically decreasing function $c(\bar{u},t)=f(|\nabla I(\bar{u},t)|)$ varying as a function of the image gradient ∇I. The coefficient will get small values when the gradient's magnitude is large and values close to 1 when the gradient is near to zero. Perona and Malik proposed two of such functions:

$$c_1(\bar{u},t) = e^{-\left(\frac{|\nabla I(\bar{u},t)|}{k}\right)^2} \quad (3)$$

$$c_2(\bar{u},t) = \frac{1}{1+\left(\frac{|\nabla I(\bar{u},t)|}{k}\right)^2} \quad (4)$$

Additional functions were proposed by Black et. al.[4] (eq. 5) based on Tukey's error norm, and by Weickert [5] (eq. 6).

$$c_3(\bar{u},t) = \begin{cases} \frac{1}{2}\left(1-\left(\frac{\nabla I}{k}\right)^2\right)^2 & if \ |\nabla I| \leq k \\ 0 & if \ |\nabla I| > k \end{cases} \quad (5)$$

$$c_4(\bar{u},t) = \begin{cases} 1-e^{-\frac{Cm}{\left(|\nabla I|^2/k^2\right)^m}} & if \ |\nabla I| > 0 \\ 1 & if \ |\nabla I| = 0 \end{cases} \quad (6)$$

where m=2, 3 or 4 and Cm=2.3366, 2.9183 or 3.3148 respectively.

In all of these functions the parameter k, called the *diffusion constant or diffusion factor*, controls the final shape of the curves and determines the threshold where the functions begins to restrain the diffusion process. **Figure 2** shows the curves of the previous four diffusion functions plotted as a function of the gradient ∇I, normalized by this parameter k.

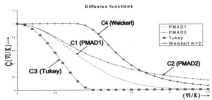

Figure 2 First (PMAD1) and second (PMAD2) Perona-Malik functions compared with Tukey's and Weickert (m=2) diffusion functions.

To better understand the effects of the diffusion function and of the diffusion constant, the flow function was introduced:

$$\phi(\bar{u},t) = c(\bar{u},t)\nabla I(\bar{u},t) \quad (7)$$

This flow function represents the diffusion strength produced by neighboring points at a defined location during the diffusion process (**Figure 3**).

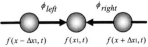

Figure 3 Representation of the diffusion in terms of the flow contribution in a 1D array. The intensity at $f(x_1,t)$ will be $I(x_1,t)=\phi_{right} - \phi_{left}$.

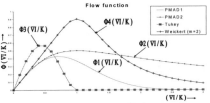

Figure 4 Flow curves of the diffusion functions.

Figure 4 presents the response from the flow function to the gradient of the intensity. The maximum flow produced by the second Perona-Malik function and Weickert function occur when the intensities of the gradients are close to the value of k. The first Perona-Malik flow function reaches its maximum near 70.7% of k, and Tukey's function when the gradient is almost half of k (approx. 0.45).

2.2 Noise Reduction

The right selection of the diffusion parameter k and of the number of iterations transforms these functions into effective de-noising filters. For each case, the diffusion constant k should be adjusted according to the following factors to produce the maximum flow at the intensities correspondent to the gradients of the noise. Used as de-noising filters, the corresponding diffusion factors are: $k_{PMAD1} \approx \sigma_e * \sqrt{2}$, $k_{PMAD2} \approx \sigma_e$, $k_{Tukey's} \approx \sigma_e * 2.22$, $k_{Weickert} \approx \sigma_e$, where σ_e is the estimation of the noise level.

The 2D transversal MR image of the brain (**Figure 1.a**) with an estimated noise magnitude $\sigma_e = 102$ was processed with each of the described functions. Details of its central part are presented in **Figure 5** as a 3D plot of their brightness.

Figure 5 shows that the different diffusion functions produce slightly different results. The PMAD1, Tukey and Weickert functions prefer high contrast edges to low contrast ones. This preference accurately preserves the position of the edges. In the case of the PMAD2 function, wider regions are favored over smaller ones, producing smoother results.

The number of employed iterations should be also considered, because some of the diffusion functions (PMAD1 and PMAD2) under certain conditions do not converge to a stable result. This results in a continuing with the diffusion process and modifying the image structures [6]. In contrast, if not enough number of iterations are considered the noise will not be completely suppressed.

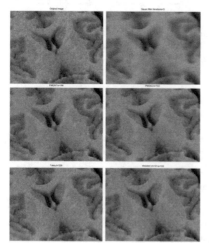

Figure 5 De-noising results: Original image (top left) and processed images (left to right, top to bottom) Gaussian filtered after 3 iterations, PMAD1 (*k=144*), PMAD2 (*k=102*), Tukey's (*k=228*) and Weickert (*k=102*). The images produced with the diffusion filters were obtained after 20 iterations.

The second important factor for a good performance of the de-noising process is the accurate determination of the noise gradient. This could be a complicated task, especially in cases such as MR images of the human brain, where different tissue textures and complex spatial interrelation of the different anatomical structures occur.

The next section presents the general approach taken to resolve these problems, searching for the optimal parameterization of the diffusion filters.

3 Automatic de-noising process

3.1 Operational conditions

The preprocessing character of this method imposes some restrictions to its design, limiting the sources of information to those exclusively contained in the initial image and those obtained during the process itself.

These conditions were established because of the necessity to preserve the flexibility and independence of the proposed method.

In this initial stage it was assumed that MR images were contaminated with additive Gaussian noise with mean zero. It was also assumed that the magnitude of the noise is the same across the 2D or 3D images. These conditions (the contamination with white Gaussian noise and the uniform distributed magnitude of the noise) are only approximations to the reality in order to simplify the problem.

In the following section, some characteristic of the noise that affect the studied images are presented.

3.2 Noise in MRI

Some brief comments about the nature of the noise present in MR images should be made in order to justify the chosen approach.

It is known that the measured complex data taken during the MR acquisition is corrupted with additive white noise having a Gaussian distribution. The sources of such perturbations are the ohmic losses in the receiving RF system and the noise properties of the electronic amplifiers [7]. Although the real and imaginary data obtained from the Fourier transformation are also Gaussian distributed, the non-linear computation of the magnitude image transforms this noise distribution into a Rice distribution [8].

Hence the Rican noise is signal dependent. In low intensity (dark) regions of the magnitude image the noise distribution tends to the Rayleigh distribution. In high intensity (bright) regions, the noise tends to a Gaussian distribution [9]. Frequently, the most important anatomical information is located between the magnitudes corresponding to the gray and white matter tissues, whose intensities are characterized to be medium or high, making our initial supposition of Gaussian noise contamination not far from reality.

3.3 Implementation

3.3.1 General concept

The proposed method incorporates the filtering process into a closed loop system, where the filter parameters are iteratively adjusted according to the intermediate results, searching for the maximum noise attenuation. The automatic method is divided in to three basic modules, *de-noising filters*, *control routines* and *evaluations methods* (**Figure 6**).

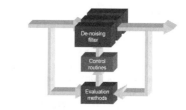

Figure 6 Methods schematic diagram.

The *De-noising filter* module contains the different diffusion filters implemented to process 2D or 3D images. They are defined currently as finite difference equations with the option to process four or eight neighbors for the 2D case, or 6, 18 and 26 neighbors for the 3D case.

The decision of which filter will be used and the values of the respective parameters correspond to the *Control routines*. These routines will receive the information from the *Evaluation methods* and the decisions are based on pre-programmed operations.

3.3.2 Evaluation

The *Evaluation methods* are the key component and currently one of the most complicated parts of the development. In fact, the entire control scheme is based on the way in which the results are evaluated.

In contrast to other techniques, such as the image compression methods, de-noising techniques do not have access to un-corrupted references, which could be used to control the filtering process by minimizing the error between the images. To overcome this difficulty a different approach was taken.

3.3.2.1 Differential information

MR images can be described as the addition between the intensity image of the examined tissues and the noise incorporated during the measurement. If an ideal filter were available, configured with optimal parameters, it would be expected that, after processing the initial image, the result would be the original image without any trace of noise.

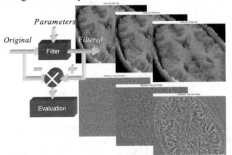

Figure 7 Differential information from the original and processed images used for monitoring the noise suppression.

Of course, such ideal filters do not exist, but instead, anisotropic diffusion filters are available which can be tuned to smooth structures with defined magnitudes (gradients). Then, the differential information obtained from the initial and processed images could be evaluated to determine the level of improvement achieved by analyzing the characteristics of this noise image. **Figure 7** illustrates this concept showing some processed images at different smoothing stages and the respective results of the subtraction.

3.3.2.2. Evaluation based on the local variance

The information contained in the differential image reflects the changes that occurred during the filtering process. If large texture variations are present in this image, this signifies that either the image is not filtered enough, because of the irregular noise distribution, or that the image was strongly smoothed and some anatomical structures have started to emerge in the differential picture. In the ideal case, a uniform pattern of white Gaussian noise across the image is expected.

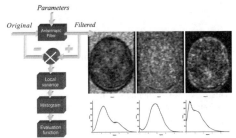

Figure 8 Schematic representation of the evaluation of the suppressed noise with some results from the local variance and histogram modules. The three examples correspond to a lightly smoothed, near optimum smoothed and heavily smoothed MR images.

The proposed evaluation method is divided in to three modules (**Figure 8**). The first is a local variance operator which produces a picture of the "noise", measuring the local variance with a 3x3 kernel at each point of the differential image. The local variance image is normalized to avoid biases in posterior operations.

The second module is a histogram responsible to extract the distribution information of the variance image. Its results are smoothed with a low pass filter to avoid discontinuities.

The final module is an evaluation function of the histogram results. This evaluation concentrates on some characteristics of the distribution of the variance picture providing the information needed for the feedback. The evaluation function is defined as:

$$Mp = (\max(h(v_i))) * \left(\frac{1}{FWHM}\right) * \left(\frac{1}{FW20\%M}\right) * \left(1 - \exp^{-\left(\frac{(t_{FWHM})^p}{0.2}\right)}\right) * \left(1 - \exp^{-\left(\frac{(t_{FW20\%M})^p}{0.2}\right)}\right) \quad (8)$$

where the first term (max(h)) represents the maximum value of the histogram, the inverse of the Full-Width at Half-Maximum (FWHM) and the inverse of the Full-Width at 20%-Maximum terms are indicators of the signal dispersion, and the last two exponential terms are functions that reflect the histogram symmetry at half and at 20% of the maximum.

This evaluation function produces the largest values when the local variance of the differential image has the maximum height and the minimum dispersion.

The two main parameters of the filter (i.e., factor k and number of iterations) can be now adjusted following the results of the evaluation function.

Figure 9 presents the results obtained after evaluating part of the parameter intervals with the proposed method.

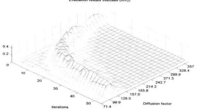

Figure 9 Results of the evaluation method using the Weickert (m=2) diffusion function.

This curve was generated, using *diffusion factor* values between 71.4 and 357 corresponding to 60% and 350% of the previously supposed "optimal" value for k (102). The iteration interval was defined between 2 and 52.

The pairs *Diffusion factor – number of iterations* corresponding to the maximum values in the **Figure 9** are considered to be close to the optimal parameters. These parameters will be automatically extracted and used to the final processing of the image.

In the next section some examples, using some of the "optimal" estimated parameters, are presented and their results are compared against the initial images.

4 Experimental results

In this section, we investigate the performance of the proposed method, using simulated and real data corrupted with additive Gaussian noise.

The first experiment used a Shepp-Logan phantom with an intensity interval of 0-256. The phantom was corrupted with noise with a standard deviation of 20. The second experiment uses a realistic magnetic resonance image taken from the Simulated Brain Database (*www.bic.mni.mcgill.ca/brainweb*) of the McGill University in Canada. This image was corrupted, using the online application, with 7% noise of the relative brightest tissue, keeping the intensity non-uniformity ("RF") equal to 0.

Both images were processed with the proposed evaluation method (section 3.3.2.2) using the Weickert (m=2) diffusion function as de-noising filter. Three estimated "optimal" parameter sets were selected from each of the produced evaluation curves. The de-noised images and the corrupted image were analyzed, using the original (un-corrupted) image as a reference. **Figure 10** presents some of the results obtained.

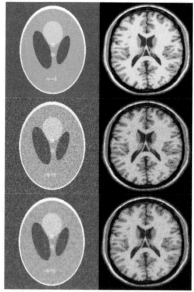

Figure 10 Results of the de-noising process, Shepp-Logan phantom (left) and simulated MR images (right). The original images are at the top, the corrupted images are at the middle and the processed images are presented at the bottom.

	Shepp-Logan				Brain Database			
Iterations	Noisy	10	15	20	Noisy	10	15	20
k factor	image	21.06	18.95	15.79	image	108	94.5	81
MAE	16.01	3.986	3.372	3.319	215.6	214.9	*215.8*	*215.8*
RMSE	20.07	5.684	5.051	5.558	255.8	242.6	243.7	244.7
SNR	-0.54	-0.02	-0.01	-0.02	1.763	1.868	1.868	1.857
PSNR	12.86	13.38	13.40	13.39	13.67	13.77	13.77	13.76
SSIM	0.936	0.994	0.995	0.994	0.961	0.964	0.964	0.964

Table 1 Results of the simulated data sets, using the Weickert (m=2) filter.

The results were evaluated by computing the mean-absolute error (MAE), the root-mean-square error (RMSE), the signal-to-noise ratio (SNR), the peak signal-to-noise ratio (PSNR) and the structural similarity index (SSIM) [10]. **Table 1** presents the results obtained with these indices.

These results show that a general improvement has been achieved with the suggested configurations. Only in two cases of the Brain Database experiment do the MAE values exceed the expected value of 215.6. In the Shepp-Logan case this improvement is clearer because of the sharp contrast between the different intensity regions and the monochromatic characteristic of each image region.

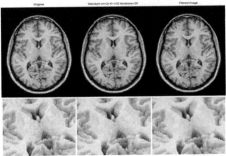

Figure 11 Results for the real MR image processed with the Weickert (m=2) filter, (left to right) original image, filtered image using the parameters according to the flow function (***k=102***) and 20 iterations, result using the parameters founded by the automatic function (***k=122.1075 and 20 iterations***).

	Weickert (m=2) filter						
	Flow curve		Automatic method				
Iterations	15	20	5	10	15	20	25
k factor	102	102	179.1	138.4	138.4	122.1	113.9
MAE	96.21	99.22	**95.69**	99.09	104.7	104.6	105.3
RMSE	110.7	116.1	**110.0**	**115.8**	125.4	125.4	126.5
SNR	0.012	0.014	0.016	0.020	0.020	0.019	0.019
PSNR	13.59	13.60	13.60	13.60	13.60	13.60	13.60
SSIM	0.996	0.996	0.997	0.996	0.996	0.996	0.996

Table 2 Results for the real MRI. The filtered images were compared against the original (corrupted) data.

Table 2 shows the results for the MR image. The first two columns contain the results using the parameters obtained from the flow estimation (sec. 2.3). The last five columns show the results generated with the automatic evaluation method. The results obtained with the automatic method are in general better than those obtained from the flow consideration. Only on three occasions (highlighted) do the results of automatic method extend below the values produced with the parameters from the flow estimation. Higher errors in MAE and RMSE in relation to the corrupted image denote a quality improvement. The SNR figures were also better for the automatic method, while the PSNR and the SSIM results were almost the same.

5   Conclusions

In this paper, the advances of an automatic noise-reduction method have been presented. This method uses the differential information, result from the subtraction between the initial and the intermediate images of filtering process, to overcome the problems related to the lack of adequate references. In this first version, an evaluation method has been proposed assuming some characteristics of the image noise. The results from simulated real MR images were included to demonstrate the viability of the method, obtaining similar or better results than simply using an estimation of the noise to determine the diffusion factor.

The project is in its initial stage and therefore exist significant chances of improvements, the evaluation of the local variance and the analysis of the differential information are only some examples. Extensive evaluations should be carried out to validate the method and to expose those weak points in order to correct them. This work represents an initial step but we believe that this step is in the right direction.

Acknowledgment: The authors want to thank the German academic exchange service (DAAD) for financial support to Mr. Castellanos under PN A/02/11312.

6   References

[1] Bankman, I., Handbook of Medical Imaging: Processing and Analysis, Academic Press; 1st edition, October, 2000.

[2] Wagenknecht, G., Entwicklung eines Verfahrens zur Generierung individueller 3D-„Regions-of-Interest"-Atlanten des menschlichen Gehirns aus MRT-Bilddaten zur quantitativen Analyse koregistrierter funktioneller ECT-Bilddaten, Shaker Verlag, Aachen, 2002.

[3] Perona, P. and Malik, J., Scale-space and Edge detection using anisotropic diffusion, IEEE Trans. In Pattern Analysis and Machine Intelligence, Vol. 12, No. 7, pp 629-639, Apr. 1990.

[4] Black, M., Sapiro, G., et. al., D., Robust Anisotropic Diffusion, IEEE Trans. Image Processing, Vol. 7, No. 3, pp 421-432, 1998.

[5] Weickert, J., Anisotropic Diffusion in Image Processing, ECMI Series, Teubner Verlag, Stuttgart, Germany, 1998.

[6] Nordström, N., Biased anisotropic diffusion - A unified regularization and diffusion approach to edge detection, In Proc. 1st European Conf. on Computer Vision, pp. 18-27, 1990.

[7] Vlaardingerbroek, M.T., Den Boer J.A., Magnetic Resonance Imaging: Theory and Practice, Springer Verlag; 2nd edition, Oct. 1999.

[8] Sijbers J., Signal and Noise Estimation from MR Images Ph.D. Thesis, Antwerp, 1998.

[9] Nowak R., "Wavelet-Based Rician Noise Removal for Magnetic Resonance Imaging." IEEE Trans. on Image Proc., 8, pp. 1408-1419, 1999.

[10] Wang, Z., and Bovik, A., Sheikh, H. R., Structural similarity based image quality assessment, Digital Video Image Quality and Perceptual Coding, Marcel Dekker Series in Signal Processing and Communications, 2004.

Bildverarbeitung   Image Processing

Hemodynamic Modeling of Abdominal Aorta Aneurysms – from Imaging to Visualization

U. Kose[1], M. Breeuwer[1], K. Visser[1], R. Hoogeveen[2], F. Laffargue[3], J-M Rouet[3], B.M. Wolters[4], S. de Putter[4], H. v.d. Bosch[5], J. Buth[6]

1 Philips Medical Systems, Medical IT - Advanced Development, Building QV 162, P.O. Box 10.000, 5680 DA Best, The Netherlands
2 Philips Medical Systems, MR Clinical Science, Building QR 0113, P.O. Box 10.000, 5680 DA Best, The Netherlands
3 Philips Medical Systems, MEDISYS Group, Building H4080, 51 Rue Carnot, 92150 Suresnes, France
4 Eindhoven University of Technology (TU/e), Faculty of Biomedical Technology, Building WH, P.O. Box 513, 5600 MB Eindhoven, The Netherlands
5 Catharina Hospital, Radiology, P.O. Box 1350, 5602ZA Eindhoven, The Netherlands
6 Catharina Hospital, Vascular Surgery, P.O. Box 1350, 5602ZA Eindhoven, The Netherlands

Abstract

Cardiovascular diseases are a major cause of death in the western society. The goal of the research presented is to improve the diagnosis and therapy selection and planning of cardiovascular diseases by patient-specific modeling of the blood circulation in the human cardiovascular system and its reaction to the circulation.

We investigate how to combine advanced imaging and analysis methods with sophisticated mathematical modeling approaches for the cardiovascular system. The research currently focuses on images of abdominal aorta aneurysms.

This research is a cooperation between Philips Medical Systems, the Eindhoven University of Technology and the Erasmus University Rotterdam, The Netherlands. The research is partially funded by the Dutch Ministry of Economic Affairs (Senter TS Program).

1 Introduction

In recent years, the assessment of patient-specific hemodynamic information of the cardiovascular system has become an important issue. It is believed that this information will improve the diagnosis and treatment of cardiovascular diseases.

We have started to investigate methods for patient-specific hemodynamic modeling of the blood flow in abdominal aortic aneurysms (AAA) and the resulting stress in the aorta wall.

Hemodynamic modeling of an AAA involves several steps:
1) 3D imaging of the abdominal aorta,
2) determination of the 3D geometry of the various AAA components from the acquired images (lumen, thrombus, wall, plaques, calcifications),
3) volume meshing, i.e. discretization of the 3D geometry,
4) formulation of the discrete equations that describe the blood flow dynamics and the wall-motion mechanics by means of a finite element method (FEM),
5) specification of the boundary conditions for these equations (e.g. input/output blood flow/pressure),
6) specification of the material properties of the components involved in the modeling,
7) solving of the equations by means of a computer,
8) comprehensive visualization and interpretation of the simulation results.

In the remainder of this paper the modeling steps are discussed into more detail and preliminary results of the different steps are shown. It will become clear that our hemodynamic simulations are still not completely patient specific especially due to the lack of patient-specific boundary conditions and material properties. We think that considerable effort is needed to realize true patient-specific modeling and to verify that the method works and is useful for improving diagnosis and treatment of AAA.

2 Methods

2.1 Imaging

Conventionally, CT Angiography (CTA) is used for visualizing the lumen, thrombus and calcifications of an AAA. However, CTA does not supply accurate information about the vessel wall, such as thickness and motion, which may be essential for patient-specific hemodynamic modeling and its evaluation. Therefore, investigations into the utility of Magnetic Resonance Imaging (MRI) for acquiring information on wall thickness, wall motion and also blood flow

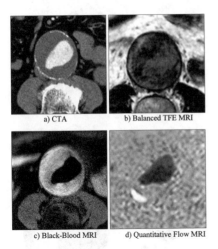

Figure 1 Examples of CTA and MRI images of an AAA patient (transversal slice)

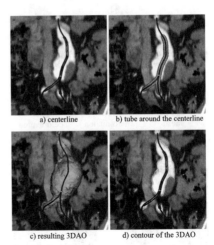

Figure 2 Example of 3DAO segmentation of the lumen of an AAA in CTA images

velocity have been performed. ECG-triggered MRI acquisition protocols have been optimized on healthy volunteers for the specific needs of this research and were recently evaluated on AAA patients at the Catharina Hospital Eindhoven. CTA images of each patient are also available to be able to compare results of the different imaging protocols and methods. **Figure 1** shows example MR and CTA images of which as much as possible the same slice position has been chosen to allow visual comparison.

2.2 Determination of the geometry

To determine the geometry of the various components of an AAA the technique of 3D Active Objects (3DAO) [1] is used which can be seen as the 3D extension of 2D Active Contours [2].
To segment the lumen with the 3DAO technique, first the centerline is tracked semi-automatically using user-defined start and end points (**Figure 2a**) [3]. Around the detected centerline an initial 3DAO – the surface of a tube – is placed (**Figure 2b**). This surface is then automatically iteratively deformed by shifting its nodes along their normals as long as the gray-values in the same direction in the image match the thresholds that limit the range of gray-values of the lumen (final result shown in **Figure 2 c-d**). For segmenting thrombus, wall and calcifications, similar approaches are under investigation.

2.3 Volume meshing

Special software has been developed for translating the surfaces of the AAA components acquired with the 3DAO technique into volume meshes that are needed to perform the hemodynamic simulations. Two different approaches for volume mesh generation are currently evaluated:
1) deformation of a standard hexahedral volume mesh into the patient-specific geometry of the lumen surface (**Figure 3a**)
2) division of the lumen volume covered by the lumen surface into tetrahedral volume elements using a local Delaunay tetrahedral meshing algorithm [4] that allows control over the size and the local quality of the resulting elements (**Figure 3b**).

2.4 Hemodynamic modeling

In the experiments that we have done so far, aortic blood flow is modeled with the Navier-Stokes equations. The blood velocity pattern used at the inflow plane is idealized. It has a uniform velocity across the inflow cross-section varying in time with a sine shaped pulse. The sine amplitude was chosen in accordance to physiologic values. The values for the pressure are also normal physiologic values.
For the wall motion simulations we assume that there is no thrombus present between lumen and wall and that the wall has a constant thickness around the lumen.

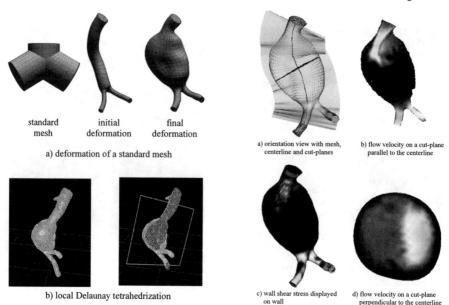

Figure 3 Examples of the evaluated volume meshing approaches

Simulations of local blood flow through the aorta as function of the phase in the cardiac cycle, the motion of the wall and its resulting displacement as well as the strain and stress in the wall were performed. All simulations were done with the Sepran Finite Element Modeling package from the Delft University of Technology, The Netherlands.

2.5 Boundary conditions and material properties

The boundary conditions for the blood flow simulations can be determined from Quantitative Flow MRI (see section 2.1). From this we get a more realistic velocity pattern and with that a more patient-specific simulation of the blood flow.

The determination of further boundary conditions, such as blood pressure or material properties (blood viscosity, wall elasticity), is far less straightforward and will therefore require substantial further research.

2.6 Visualization of results

Simulation of blood flow through an AAA and motion of the wall of an AAA results in a large amount of data. We have developed software that provides the user with visualization methods that allow for comprehensive display of the results (**Figure 4**). The visualization methods are implemented using the functionality of the Visualization Toolkit (VTK) from kitware [5].

Figure 4 Examples of visualizations of simulation results

3 Results

MRI has proven to be a powerful tool for visualizing most of the components of an AAA that are needed for a fully patient-specific hemodynamic simulation of blood flow and wall motion.

3D Balanced TFE (BTFE) MRI is found to be well suited for visualizing the lumen and thrombus as a function of phase in the cardiac cycle. In the most of the acquired BTFE images, a dark ring is visible that originates from partial volume effect of the transition between vessel wall and surrounding tissue. This ring is assumed to be a good feature for segmenting the outer vessel wall. The distinction between lumen and thrombus is, however, not always very clear in 3D BTFE images. Therefore we have also investigated the use of 2D BTFE imaging, since the contrast between lumen and thrombus is much better in these images (**Figure 5**).

Quantitative flow MRI is seen as a useful tool for visualizing and, more important for the simulations, quantifying the blood flow velocity through the vessel. Images acquired with the so-called Black Blood protocol appear to be a useful tool for visualizing the vessel wall and the thrombus. It even shows significant anatomical detail inside the thrombus (variation in pixel intensity) which may be of importance for the determination of tissue characteristics that are necessary for the FEM simulations. However, the distinction between vessel

wall and thrombus is not always possible.
With CTA some of the components of an AAA are also visible but mostly the distinction between different components or the anatomical detail of the component is not as good as with MRI. The distinction between lumen and thrombus, for example, is good in CTA but wall or further anatomical detail cannot be visualized with CTA. Another disadvantage of CTA is that flow imaging cannot be done. The most

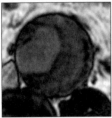 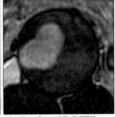

a) slice from 3D BTFE   b) slice from 2D BTFE

Figure 5 Comparison of 3D and 2D BTFE images (sections)

prominent advantage of CTA is that calcifications can be much better visualized with CTA than with MRI. In MRI signal voids occurring from calcification can only be observed when the calcification is large.

For being able to perform a completely patient-specific hemodynamic modeling, information about all components of an AAA is required. Seeing the advantages and disadvantages of CTA and MRI imaging we think that a combination of both modalities will be needed for acquiring this information.

For segmenting the different components of an AAA the technique of 3DAO seems to be very well suited. Segmentation of the lumen is no difficulty thanks to the very distinct feature of lumen as well in CTA as in 2D BTFE MRI. Segmentation of the thrombus from images of both modalities is still under investigation and first results look promising. Investigations into the segmentation and modeling of calcifications and wall are also ongoing.

First simulation results of blood flow show that the velocity distribution in an AAA varies significantly from that in healthy abdominal aortas. In the first wall-motion simulations we found that wall displacement, strain and stress vary substantially as function of the position on the wall.

4   Discussion and conclusions

In this paper we presented our research in the area of patient-specific hemodynamic modeling from image acquisition to visualization of simulation results and first results we made deduced from this research.

The first results of wall motion simulations show that the stress in the wall varies significantly over the position on the AAA wall what suggests that the geometry itself may play a role in rupture.

The first flow simulations show that the velocity distribution in an AAA in a plane perpendicular to the centerline varies much more than in a healthy aorta. This might play a role in the formation of thrombus and may therefore have an effect on the distribution of wall stress.

Up to now, our simulations are far from patient-specific, since we only use patient-specific geometry but no patient-specific inflow velocity distribution or other patient-specific material properties or boundary conditions. For the wall simulation further steps are to include thrombus and calcifications in the model geometry. In the first instance, "normal" values for the material properties will be used for these additional components but the final goal is to use patient-specific material properties for the simulations. For the blood flow simulations a next step is to use data acquired from Quantitative Flow MR images as inflow velocity distribution.

Another future task is to couple flow and wall-motion simulations which are currently done independently.

The goal is to use the simulations of the wall motion, strains and stresses for predicting rupture risk for the aortic wall. Therefore we have to validate whether the simulated hemodynamical parameters correlate with "clinical events" such as the position of rupture or the position of abnormal aneurysm growth.

5   Literature

[1] O. Gerard, A. Collet Billon, J-M. Rouet, M. Jacob, M. Fradkin, C. Allouche., "Efficient Model-Based Quantification of Left Ventricular Function in 3-D Echocardiography", IEEE Trans. on Medical Imaging, Vol. 21, No. 9, September 2001, pp. 1059-1068.

[2] S. Lobregt and M.A. Viergever, "A Discrete Dynamic Contour Model", IEEE Trans. on Medical Imaging, Vol. 14, No. 1, March 1995, pp. 12-24.

[3] O. Wink, W.J. Niessen, A.J. Frangi, B. Verdonck and M.A. Viergever, "3D MR Coronary Axis Determination using a Minimum Cost Path Approach", Magnetic Resonance in Medicine 2002, Vol. 47, p. 1169-1175.

[4] K. Ho-Le, "Finite Element Mesh Generation Methods: A Review and Classification", Computer Aided Design, Butterworth & Co. Ltd, Vol. 1, No. 20, Jan/Feb 1988, pp. 27-38.

[5] W. Schroeder, K. Martin, B. Lorensen, "The Visualization Toolkit, An Object-Oriented Approach To 3D Graphics", 3rd edition, ISBN 1-930934-07-6

Dreidimensionale Rekonstruktion zahnmedizinischer Präparate aus 2D-Schichtaufnahmen und µCT-Scans

Christoph Bourauel[1], Alireza Rahimi[1], Ludger Keilig[2], Susanne Reimann[1], Andreas Jäger[1], Thorsten M. Buzug[3]

[1] Poliklinik für Kieferorthopädie, Universitätsklinikum Bonn, Rhein. Friedrich-Wilhelms-Universität Bonn, Welschnonnenstraße 17, 53111 Bonn, Deutschland

[2] Abteilung für Zahnärztliche Propädeutik - Experimentelle Zahnheilkunde, Universitätsklinikum Bonn, Rhein. Friedrich-Wilhelms-Universität Bonn, Welschnonnenstraße 17, 53111 Bonn, Deutschland

[3] Fachbereich Mathematik und Technik, RheinAhrCampus Remagen, Südallee 2, 53424 Remagen, Deutschland

Zusammenfassung

Beim direkten Vergleich von Experiment und Finite-Elemente-Simulation ist eine detailgenaue Rekonstruktion der Präparatoberflächen anzustreben. Kommerziell erhältliche Software-Pakete zur vollautomatischen Modellgenerierung erfordern oftmals aufwendige manuelle Nachbearbeitungen der Modelle, da Strukturen nicht sauber erkannt werden. Für Untersuchungen in der dentalen Biomechanik sollten FE-Modelle basierend auf histologischen Serienschnitten, µCT-, CT- und MRT-Daten verschiedener Präparate erzeugt werden. Mit einem eigens entwickelten Algorithmus nach dem ‚Prinzip der geringsten Kosten' wurden die Schnitte segmentiert. Es folgte eine dreidimensionale Triangulation, Glättung der Oberflächen und eine Reduktion der Anzahl der Oberflächenelemente. Die so erstellten Modelle wurden in verschiedenen FE-Programmen (Marc, COSMOS/M) analysiert.

1 Einleitung

Automatische und halbautomatische Verfahren zur Bildanalyse und -segmentierung haben in den letzten Jahren durch eine Weiterentwicklung entsprechender Algorithmen sowie nicht zuletzt auch durch eine Steigerung der zur Verfügung stehenden Rechenleistung einen großen Aufschwung erfahren. Die hierbei verwendeten Methoden finden in immer mehr Bereichen Verbreitung und halten auch in der Medizin und Zahnmedizin Einzug.

Im Rahmen biomechanischer Untersuchungen in der Experimentellen Zahnheilkunde stellt sich oftmals die Aufgabe, das experimentell ermittelte Verhalten eines Präparates mit einer numerischen Simulation zu vergleichen. Die Rechnungen werden überwiegend mit Hilfe der Finite-Elemente-Methode (FEM) durchgeführt, wobei der Qualität und Realitätsnähe der entsprechenden FE-Modelle eine zentrale Bedeutung zukommt. **Bild 1** zeigt als Beispiel die Messung von Kraft/Auslenkungsdiagrammen an einem Rattenmolaren in einem frischen Unterkieferpräparat und das zu dem vermessenen Molaren entwickelte FE-Modell.

Die Untersuchungen wurden benutzt, um das mechanische Verhalten des Parodontalligamentes in experimentellen und numerischen Studien zu bestimmen. In den Experimenten zeigte sich ein nichtlinearer Zusammenhang, der in den FE-Simulationen gut nachvollzogen werden konnte (**Bild 2**, [2]). Andere Untersuchungen konzentrieren sich auf die Biomechanik der kieferorthopädischen Zahnbewegung, die Bestimmung von Knochenbelastungen bei prothetischer Versorgung mit Kronen, Brücken, Teil- oder Vollprothesen und die Auswirkungen von Dentalimplantaten auf den Alveolarknochen. Die Erkenntnisse können dazu verwendet werden, Behandlungsmethoden oder –elemente zu optimieren oder neue Geräte vor klinischem Einsatz zu simulieren. Um eine aussagekräftige Simulation durchführen zu können, werden präzise 3D-Oberflächenmodelle aller beteiligten Strukturen benötigt.

Bild 1 Biomechanische Messungen an einem Rattenpräparat in einem optomechanischen Messaufbau (links, Detailaufnahme Mitte [1]) und FE-Modell eines Rattenmolaren (rechts).

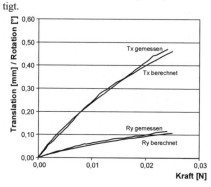

Bild 2 Messkurve und angefittetes nichtlineares Verhalten eines Rattenmolaren [2].

Ein wichtiger Schritt bei der Durchführung von Finite-Elemente-Analysen ist die Generierung des FE-Modells. Verschiedene 3D-Bildbearbeitungsprogramme bieten zwar die Möglichkeit, die rekonstruierte Geometrie auch in ein FE-Netz zu konvertieren, häufig besteht aber nur eine geringe Eingriffsmöglichkeit des Benutzers während der Modellerzeugung. Er ist somit nicht in der Lage, seine Kenntnisse über anatomische Strukturen in diesen Prozess einfließen zu lassen und es wird eventuell eine aufwendige Nachbearbeitung erforderlich.

Das hier vorgestellte Programm soll diese Probleme beseitigen helfen. Der Benutzer soll die Möglichkeit haben, sich interaktiv an der Erkennung relevanter Strukturen zu beteiligen. Um möglichst realistische Modelle der Präparate erstellen zu können, war es Ziel dieser Arbeit, ein Programm zur interaktiven dreidimensionalen und detailgenauen Rekonstruktion von Präparatoberflächen anhand verschiedenster Datenquellen zu entwickeln.

2 Material und Methoden

2.1 Verwendete Präparate

Bei den in experimentellen Studien eingesetzten Präparaten handelt es sich z.B. um ein- und mehrwurzelige humane Zähne, Schweine- und Rattenmolaren oder um Schweinekiefersegmente mit inserierten Implantaten. Diese Präparate sollen möglichst realistisch und mit hoher Genauigkeit rekonstruiert werden, um die experimentell gewonnenen Daten zur Validierung der numerischen Modelle nutzen zu können.

Als Grundlage für die Modellgenerierung sollen die Daten verschiedener bildgebender Verfahren eingesetzt werden, die in der (zahn-)medizinischen Anwendung Verbreitung gefunden haben. Dazu zählen mikroskopische Abbilder histologischer Serienschnitte, μCT-Scans sowie CT- und MRI-Scans verschiedener Präparate. Die Konturen der aus Sicht der Biomechanik interessanten Materialstrukturen (Parodontalligament, Dentin, Schmelz, Kortikalis, Spongiosa sowie gegebenenfalls Implantat) werden über interaktiv festgelegte Stützpunkte ermittelt. Beispiele verschiedener Präparate zeigt das **Bild 3**.

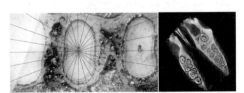

Bild 3 Histologisches Schnittbild eines humanen Unterkiefersegments im Prämolarenbereich (links) und μCT-Scan eine Oberkiefers einer Wistar-Ratte mit bis zu fünfwurzeligen Molaren (rechts).

2.2 Geometrierekonstruktion

2.2.1 Ablauf der Geometrierekonstruktion

Die Geometrierekonstruktion orientiert sich an den Umrissen der anatomischen Strukturen, die aus einzelnen Schichtaufnahmen der zu modellierenden Objekte extrahiert werden. In einem ersten Schritt werden diese Informationen aus jedem Schnitt extrahiert. Anschließend werden die Geometrieinformationen zweier benachbarter Schnitte zu einem Oberflächennetz erweitert und dann das vollständige Modell aus der Gesamtheit alle Schnitte dreidimensional rekonstruiert.

Bevor die Umrisse der beteiligten Strukturen aus den Schnittbildern extrahiert werden können, müssen in einer Bildanalyse eine Reihe von bildabhängigen Parametern, wie zum Beispiel die z-Koordinate des Schnittes sowie Informationen zu Farbverläufen innerhalb des Bildes, ermittelt werden. Optional können zusätzlich zwei Referenzpunkte zur Orientierung von digitalisierten histologischen Schnittaufnahmen angegeben werden.

2.2.2 Algorithmus zur Bildsegmentierung

Die eigentliche Bildsegmentierung erfolgt nach dem Dijkstra-Algorithmus [3] anhand eines sogenannten „Low-Cost"-Algorithmus, der unter Zuhilfenahme der bei der Bildanalyse berechneten Daten die kürzeste Verbindung zwischen zwei Punkten ermittelt [4-7]. Dazu wird eine Kostenfunktion verwendet, die jedem Pfad zwischen diesen beiden Punkten einen Wert

$$l(p,q) = w_Z \cdot f_Z(q) + w_G \cdot f_G(q) + w_D \cdot f_D(p,q)$$

zuweist, der umso kleiner ist, je besser der Pfad den Konturen des Bildes folgt. Die Kosten für einen Pfad werden ermittelt als Summe der Kosten zwischen allen benachbarten Pixeln auf dem Pfad, wobei die Kosten zwischen zwei benachbarten Pixeln als gewichtete Summe aus Laplacianwert, Gradientenwert und Gradientenrichtung ermittelt werden.

Die Gewichte $w_Z=0.45$, $w_G=0.45$ und $w_D=0.1$ wurden der Literatur entnommen [5]. Der Laplacewert

$$f_z(q) = \begin{cases} 0 & \text{für } I_L(q) = 0 \\ 1 & \text{für } I_L(q) \neq 0 \end{cases} \quad \text{mit } I_L = \begin{matrix} 1 & 1 & 1 \\ 1 & -8 & 1 \\ 1 & 1 & 1 \end{matrix}$$

eines Pixels q ist genau dann Null, wenn der Laplaceoperator angewandt auf die Grauwertefunktion in diesem Pixel eine Nullstelle hat [5]. Damit können strukturreiche Bildausschnitte hervorgehoben werden.

Der Gradient

$$f_G = \frac{\max(G') - G'}{\max(G')},$$

mit

$$G' = G - \min(G) \qquad G = \sqrt{I_x^2 + I_y^2}$$

$$I_x = \begin{matrix} -1 & 0 & 1 \\ -2 & 0 & 2 \\ -1 & 0 & 1 \end{matrix} \qquad I_y = \begin{matrix} -1 & -2 & -1 \\ 0 & 0 & 0 \\ 1 & 2 & 1 \end{matrix}$$

repräsentiert die Änderung der Grauwert-Intensität sowohl in x- als auch in y-Richtung. Er wird minimal, wenn ein Punkt auf einer Kante oder einer scharf abgegrenzten Struktur des Bildes liegt [5].

Die Gradientenrichtung

$$f_D(p,q) = \tfrac{2}{3\pi}\left\{a\cos(d_p(p,q)) + a\cos(d_q(p,q))\right\}$$

mit

$$d_p(p,q) = D'(p)L(p,p), \quad d_q(p,q) = L(p,q)D'(q)$$

und

$$D(p) = [I_x(p), I_y(p)], \qquad D'(p) = [I_y(p), -I_x(p)]$$

sowie

$$L(p,q) = \frac{1}{\|p-q\|}\begin{cases} q-p & \text{für } D'(p)(q-p) \geq 0 \\ p-q & \text{für } D'(p)(q-p) < 0 \end{cases}$$

repräsentiert die Richtung der maximalen Grauwertveränderung vom Punkt q zu all seinen Nachbarpunkten p [5].

Die Kostenfunktion extrahiert aus den einzelnen Bildern Polygone, die die verschiedenen Strukturen in der jeweiligen Ebene umranden. Die Polygone werden gegebenenfalls geglättet, und die Anzahl der Punkte wird reduziert.

2.2.3 Erzeugung des 3D-Netzes

Die Triangulation im dreidimensionalen Raum wird umgangen, um auf bewährte und schnelle 2D-Algorithmen zurückgreifen zu können: Zunächst werden jeweils die Polygone zweier benachbarter Ebenen in

Bild 4 Polygonzüge in zwei benachbarten Schnittbildern: Nach Projektion in eine gemeinsame Ebene werden alle Punkte miteinander als Dreiecke verknüpft. Anschließend müssen überzählige oder falsch verbundene Kanten gelöscht oder korrigiert werden (von links nach rechts).

eine gemeinsame Ebene projiziert und dort miteinander nach dem Prinzip der „Constrained Delaunay Triangulation" vernetzt [8]. **Bild 4** zeigt dies am Beispiel zweier Polygonzüge. Alle Punkte auf den beiden Polygonen werden so verbunden, dass die umschlossene Fläche vollständig mit Dreiecken ausgefüllt ist, wobei die existierenden Kanten der Polygone erhalten werden. Anschließend werden überzählige und falsch verbundene Dreiecke entfernt oder korrigiert, entstehende Lücken werden durch einen speziellen Algorithmus erkannt und nachträglich geschlossen.

Anhand dieser Triangulation werden dann jeweils benachbarte Schnittebenen im dreidimensionalen Raum miteinander vernetzt. Dies wird in aufeinanderfolgenden Schritten mit allen Schnittebenen durchgeführt, so dass ein vollständiges dreidimensionales Modell der auf den Schnittbildern dargestellten Strukturen entsteht. Dabei werden die verschiedenen Materialgruppen, wie Dentin, Zahnschmelz, Wurzelzement, Parodontalligament sowie kortikaler und spongiöser Knochen zu geschlossenen Körpern zusammengefasst, und in den jeweils letzten Schnitten werden die Körper z.B. an der Zahnkrone oder der Wurzelspitze automatisch geschlossen (vergl. **Bild 5**).

Ein besonderes Augenmerk bei der Vernetzung der segmentierten Schnittbilder liegt dabei auf der korrekten Zuordnung der Polygone auf den verschiedenen Ebenen [6], da überwiegend mehrwurzelige Zähne modelliert werden müssen. Auf jeder Ebene können die Bildsegmente, die ein zu modellierendes Material beschreiben, in verschiedene nicht zusammenhängende Bereiche zerfallen, so dass die Materialbegrenzung dieses Materials auf dieser Ebene durch mehrere Polygone beschrieben wird. Dabei kann sich die Anzahl der benötigten Polygone zwischen zwei benachbarten Ebenen unterscheiden, wenn sich zum Beispiel mehrere Materialbereiche vereinen oder ein neuer Bereich des Materials hinzukommt oder wegfällt.

Bild 6 illustriert diese Probleme anhand einiger Beispiele und zeigt die automatisch gefundenen Zuordnungen. Die Zuordnung der Polygone erfolgt über den Vergleich der Mittelpunkte der Polygone auf verschiedenen Ebenen sowie ihre Ausdehnung (Korrespondenzproblem): Liegen die Polygonmittelpunkte einer Ebene innerhalb der Polygone der zweiten Ebene, so werden diese beiden Strukturen als zueinander zugehörig definiert.

Für den Fall, dass zu einem Polygon auf einer Ebene zwei oder mehr zugehörige Polygone auf einer benachbarten Ebene gefunden werden (Verzweigungs-

Bild 5 Ausschnitt von etwa 10 Ebenen aus dem Wurzelbereich eines Rattenmolaren mit modelierter Pulpa, Wurzelzement und umgebendem PDL (links). Abschluss der Wurzelspitzen (rechts).

Bild 6 Korrespondenz- und Verzweigungsprobleme: Von der Zahnwurzel des Rattenmolaren teilen sich zunächst eine vordere Hauptwurzel und anschließend mehrere kleine posteriore Wurzeln ab (links und Mitte). Beim Übergang zur Zahnkrone sind die Aufteilungen in die Höcker zu modellieren (rechts).

problem), muss der Vernetzung dieser Polygone besondere Aufmerksamkeit gewidmet werden [6]. Um ein in der FE-Simulation möglichst gut handhabbares Modell zu erhalten, wird im Bereich der Verzweigung mittig zwischen den beiden Ebenen eine Hilfsebene eingezogen sowie auf dieser Ebene eine zusätzliche Polylinie erstellt, die in die Vernetzung einbezogen wird.

Die grafische Oberfläche des Programms wurde mit Qt 2.3 (Trolltech AS, Norwegen) unter Windows 2000 (Microsoft Corp., USA) erstellt. Zur Visualisierung der dreidimensionalen Oberflächen wurde die Grafikbibliothek Coin3D (Systems In Motion AS, Norwegen) verwandt. **Bild 7** zeigt als Beispiel die Erkennung der relevanten Strukturen eines mehrwurzeligen Rattenmolaren aus einem μCT-Scan sowie die daraus rekonstruierte Geometrie des Zahns mit Zahnhalteapparat im 3D-Viewer. Die einzelnen Strukturen des rekonstruierten Präparats müssen nunmehr an ein Finite-Elemente-Paket übergeben werden. Hierzu werden explizit die Knoten mit ihren Knotenkoordinaten sowie die Dreieckselemente der rekonstruierten Präparatoberfläche in eine ASCII-Datei exportiert. Diese Datei kann sowohl von dem FE-Programm MSC Marc/Mentat als auch von COSMOS/M importiert werden.

Im Finite-Elemente-Programm werden die Oberflächen durch entsprechende Vernetzungsalgorithmen in FE-Netze konvertiert, und die Strukturen werden mit den zugehörigen Materialparametern verknüpft. Für eine detailgenaue Überführung der Oberflächenmodelle mit ihrer Triangulation werden bevorzugt Tetraederelemente verwendet. Aber auch die Vernetzung mit Hexaeder-, sogenannten Solid- oder allgemeinen Volumenelementen ist möglich. Die in allen Rech-

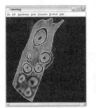 

Bild 7 Links: Schnitt durch das Segment eines Rattenoberkiefers im Wurzelbereich (μCT-Scan) und Erkennung der Strukturen Kortikalis, PDL, Dentin und Pulpa. Mitte und rechts: Daraus rekonstruiertes dreidimensionales Oberflächenmodell.

nungen verwendeten Materialparameter sind der **Tabelle 1** zu entnehmen. Sie wurden entweder selbst bestimmt [2, 9, 10] oder der Literatur entnommen [11].

3 Ergebnisse und Diskussion

Die Qualität und Realitätstreue der rekonstruierten Modelle hat unmittelbaren Einfluss auf die Genauigkeit, mit der die Simulationsergebnisse die Messungen darstellen. Selbstverständlich ist eine eingehende Sensitivity-Analyse durchzuführen, um die Einflüsse der verschiedenen Materialparameter, der Vernetzungsdichte und auch die Einflüsse der zugrunde liegenden Bildquellen zu ermitteln.

Dies soll hier nicht weiter dargestellt werden, da es den Rahmen dieses Beitrags deutlich sprengen würde. An dieser Stelle soll vielmehr die Eignung des entwickelten Programms zur Rekonstruktion verschiedenster Präparate in der dentalen Biomechanik aufgezeigt werden. Die folgenden drei Abschnitte zeigen die Rekonstruktionen und Ergebnisse von FE-Simulationen von Rattenmolaren, Schweineprämolaren und kieferorthopädischen Ankerimplantaten. Zu diesen Rechnungen existierten jeweils auch die entsprechenden Messungen, die mit den Simulationsergebnissen verglichen wurden.

3.1 Modellierung von Rattenmolaren

Nicht nur in der Medizin wird die Ratte häufig als Versuchstier verwendet. Auch in der Zahnmedizin gilt sie als adäquates Modell, um z.B. Grundlagen der kieferorthopädischen Zahnbewegung zu untersuchen. Im vorliegenden Fall wurden an Segmenten von Rattenoberkiefern die ersten Molaren mit entsprechend reduzierten Kraftsystemen belastet und die Auslenkungen gemessen. Um Aussagen über die Auswirkungen der kieferorthopädischen Belastung auf den Zahnhalteapparat treffen zu können, wurden die Präparate mit dem hier beschriebenen Programm in Finite-Elemente-Modelle umgesetzt [12]. **Bild 8** zeigt das Ergebnis der Rekonstruktion und die berechnete Auslenkung des Rattenmolaren bei einer Kraft von 0,1 N. Sie be-

Material	Elastizitäts-modul [MPa]	Querkontrak-tionszahl μ
Schmelz	80.000	0,3
Dentin	20.000	0,3
Kortikalis	15.000	0,3
Spongiosa	1.000	0,3
PDL (Schwein)	0,05 / 0,20	0,3 (bilinear)
PDL (Ratte)	0,12 / 0,60	0,3 (bilinear)
Implantat	100.000	0,3

Tabelle 1 Materialparameter von Zahn, Knochen und PDL. Das Verhalten des PDL von Ratten unterscheidet sich deutlich von dem von Nichtnagetieren.

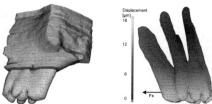
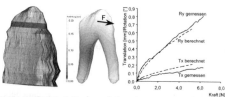

Bild 8 Finite-Elemente-Modell eines oberen Rattenmolaren mit umgebendem Parodontalligament und Alveolarknochen (links). Das Modell besteht aus etwa 60.000 10-Knoten Tetraederelementen. Bei einer einzelnen Kraft von 0,1 N beträgt die berechnete Auslenkung 18 μm (rechts).

Bild 9 FE-Modell eines Schweineprämolaren (links), berechnete Auslenkung bei einer Kraft von 6 N (Mitte) und Vergleich von gemessener und berechneter Zahnauslenkung zur Bestimmung von Materialparametern des Parodontalligaments (rechts).

trägt etwa 18 μm und stimmt gut mit den Messungen überein. Da die Rechnungen mit nichtlinearen Materialparametern durchgeführt wurden, und da große Deformationen zu erwarten sind, wurden nichtlineare Rechnungen durchgeführt. Außerdem wurde das 10-Knoten Tertraelement verwendet, da es bessere nichtlineare Eigenschaften aufweist als das 4-Knoten Tetraelement. Bild 8 zeigt die hervorragende Qualität der Geometrierekonstruktion, insbesondere im Bereich der Furkation und der Höcker.

Die berechneten Spannungen und Verzerrungen konnten mit histologischen Befunden um kieferorthopädisch bewegte Rattenmolaren korreliert werden [12]. Hierbei zeigte sich, dass in den Bereichen mit erhöhten Kompressionen und negativen Verzerrungen die Aktivitäten von Osteoklasten erhöht sind. Dies ist gleichbedeutend mit einer erhöhten Knochenresorption, einer Grundvoraussetzung für die kieferorthopädische Zahnbewegung. Insofern haben sich die FE-Modelle als hinreichend aussagekräftig für unsere experimentellen und theoretischen Untersuchungen in der dentalen Biomechanik erwiesen.

3.2 Bestimmung der Materialparameter von Schweinepräparaten

Das Verhalten des Zahnhalteapparates, insbesondere des PDL, ist von besonderem Interesse, da es nicht nur als Übermittler des mechanischen Reizes bei der kieferorthopädischen Zahnbewegung angesehen wird, sondern auch Dämpfungsaufgaben übernimmt, Auslenkungen der Zähne bis zu 0,2 mm im Rahmen der Kautätigkeit erlaubt und Zähne und Alveolarknochen vor Überlastungen schützt. Das biomechanische Verhalten des PDL ist äußerst komplex und weist sowohl hydrodynamische als auch nichtlinear-elastische Eigenschaften auf. Eine exakte Modellierung des PDL erfordert die Aufstellung eines mehrphasigen Materialgesetzes. Dies ist sehr aufwändig und Forschungsgegenstand mehrerer Arbeitsgruppen. Da hydrodynamische Phänomene wie Fließprozesse oder Dämpfung jedoch zeitabhängig sind und nach einer gewissen Zeitspanne abklingen, besteht auch die Möglichkeit, Kurzzeit- und Langzeitphänomene voneinander zu trennen. Dies war Gegenstand verschiedener Untersuchungen unserer Arbeitsgruppe, die sich für das Materialverhalten im Rahmen kieferorthopädischer Zahnbewegung interessiert [1, 2, 9, 10, 12, 13].

Die kieferorthopädische Zahnbewegung ist ein Langzeitphänomen, und so kommen praktisch ausschließlich die nichtlinear-elastischen Eigenschaften der Parodontalfasern zum Tragen. Bild 9 zeigt das FE-Modell eines Unterkiefer-Schweineprämolaren. An diesem Zahn wurden zunächst Last/Auslenkungsmessungen durchgeführt, anschließend wurde das Präparat mit den hier beschriebenen Methoden rekonstruiert. Durch Variation der Materialparameter des PDL konnte eine gute Rekonstruktion des Last/Auslenkungsverhaltens des Präparats erzielt werden, und es konnten die in Tabelle 1 gezeigten Parameter des PDL an über 10 Präparaten verifiziert werden.

3.3 Simulation der Belastung von kieferorthopädischen Implantaten

Kieferorthopädische Implantate werden im Rahmen orthodontischer Zahnbewegungen zur Verankerung von zahnbewegenden Kraftsystemen eingesetzt. Dabei ist von Interesse, inwieweit der Knochen um die Implantate belastet wird, und ob sich größere Auslenkungen an der Implantatspitze ergeben, da diese sofort belastet werden, also nicht osseointegriert sind. **Bild 10** zeigt das FE-Modell eines derartigen Ankerimplantats (TOMAS, Fa. Dentaurum, Pforzheim), das zunächst in ein Schweinekiefersegment eingeschraubt und dann in einem μCT-Scanner (SkyScan 1072, Aartselaar, Belgien) gescannt wurde. Die μCT-Bilder wurden wiederum genutzt, um mit den beschriebenen Methoden das FE-Modell zu generieren. Das Implan-

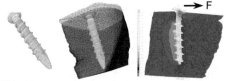

Bild 10 Finite-Elemente-Modell eines orthodontischen Verankerungs-Pins (links) mit umgebender Kortikalis und Spongiosa (Mitte). Die Simulation der Belastung mit 4 N resultiert in einem Ablösen des Implantats von der Knochenoberfläche (rechts) und einer überwiegenden Druckbelastung der Spongiosa.

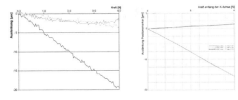

Bild 11 Gemessene (links) und berechnete (rechts) Auslenkung des Implantatkopfes bei einer Belastung mit 4 N. Die gemessene Hauptkomponente der Bewegung konnte in den Rechnungen gut reproduziert werden.

tat wurde anschließend im Rahmen einer Kontaktanalyse im FE-System Marc mit 4 N belastet.
In Bild 10 (rechts) ist zu erkennen, dass das Implantat bei Belastung mit 4 N bis in den Knochen hinein bogenförmig deformiert wird und sich dadurch von der Spongiosa auf der Zugseite ablöst. Dadurch erfährt der Knochen eine fast ausschließliche Druckbelastung auf der der Kraft entgegengesetzten Seite. Die dabei ermittelten Verzerrungen liegen mit maximal 2000 µstrain aber deutlich unter der physiologischen Grenze. **Bild 11** zeigt den Vergleich von berechneter und gemessener Auslenkung an der Spitze des Tomas-Pins. Bei einer Last von 4 N wurden für die Hauptkomponente der Bewegung in Richtung der X-Achse eine Auslenkung von 20 µm sowohl gemessen als auch berechnet. Nach den in Bild 10 dargestellten Ergebnissen kann diese Auslenkung überwiegend auf die Deformation des Implantats und seine Beweglichkeit in der Spongiosa zurückgeführt werden. Die gute Übereinstimmung von Messung und Rechnung zeigt auch hier, dass die generierten FE-Modelle von sehr guter Qualität sind und dass die Realitätsnähe der Aufgabenstellung angemessen ist.

4 Schlussfolgerungen

Die vorgestellten Beispiele zeigen, dass mit dem entwickelten ‚Low-Cost'-Algorithmus unterschiedlichste zahnmedizinische Präparate auf der Basis verschiedenster Bildquellen mit einer sehr guten Qualität rekonstruiert werden können. Die Realitätsnähe der entwickelten Modelle ist sowohl bei sehr kleinen (Rattenzähne) als auch bei größeren Objekten hinreichend für die gestellten Aufgaben in der dentalen Biomechanik. Abhängig von der Bildqualität und der Komplexität der zu modellierenden Strukturen (Vielzahl von Wurzeln, Kortikalis, Spongiosa, Pulpa, Dentin, Schmelz und Zement) kann die Rekonstruktion eines Schnittes bis zu 10 Minuten dauern, was im Wesentlichen auf die interaktive Erkennung und Markierung der einzelnen Strukturen zurückzuführen ist. Dadurch kann die Rekonstruktion eines Präparates bis zu 4-5 Stunden benötigen. Da anschließend keinerlei Korrekturen an dem FE-Modell mehr notwendig sind, kann die Erstellung eines vollständigen Rechenmodells in diesem Zeitrahmen als äußerst günstig angesehen werden.

5 Literatur

[1] Hinterkausen, M., Bourauel, C., Siebers, G., Haase, A., Drescher, D., Nellen, B.: In vitro analysis of the initial tooth mobility in a novel optomechanical set-up. Med. Eng. Phys. 20 (1998), 40-49

[2] Kawarizadeh, A., Bourauel, C., Jäger A.: Experimental and numerical determination of initial tooth mobilities and material parameters of the periodontal ligament in rat molar specimens. Eur. J. Orthod. 25 (2003), 569-578

[3] Dijkstra, E.W.: A note on two problems in connexion with graphs. Numerische Mathematik, 1 (1959), 269-270

[4] de Berg, M., van Kreveld, M., Overmars, M., Schwarzkopf, O.: Computational Geometry. Springer-Verlag (2000)

[5] Mortensen, E.N., Barrett, W.A.: Intelligent Scissors for Image Composition. Computer Graphics (SIGGRAPH `95), (1995), 191-198

[6] Klein, R., Schilling, A., Straßer, W.: Reconstruction and simplification of surfaces from contours. Pacific Graphics 1999, (1999)

[7] Lehmann, T., Oberschelp, W., Pelikan, E., Repges, R.: Bildverarbeitung für die Medizin. Springer-Verlag, (1997)

[8] Gopi, M., Krishnan, S., Silva, C.T.: Surface Reconstruction based on Lower Dimensional Localized Delaunay Triangulation. Eurographics 2000, 19 (2000)

[9] Vollmer, D., Haase, A., Bourauel, C.: Halbautomatische Generierung von Finite-Element-Netzen für zahnmedizinische Präparate. Biomed. Tech. 45 (2000), 62-69

[10] Poppe, M., Bourauel, C., Jäger, A.: Determination of the elasticity parameters of the human periodontal ligament and the location of the center of resistance of single-rooted teeth a study of autopsy specimens and their conversion into finite element models. J. Orofac. Orthop. 63 (2002), 358-370

[11] Abé, H., Hayashi, K., Sato, M. (Hrsg.): Data Book on Mechanical Properties of Living Cells, Tissues, and Organs. Springer Verlag, Tokyo, Berlin, Heidelberg, New York, 1996

[12] Kawarizadeh, A., Bourauel, C., Zhang, D., Goetz, W., Jäger A.: Correlation of stress and strain profiles and the distribution of osteoclastic cells induced by orthodontic loading in rat. Eur. J. Oral Sci. 112 (2004), 140-147

[13] Ziegler, A., Keilig, L., Kawarizadeh, A., Jäger, A., Bourauel, C.: Numerical simulation of the biomechanical behaviour of multirooted teeth. Eur. J. Orthod. (2004), in Druck

Approaches to Visualization in 3D Coronary MRA

Ulrike Blume[1], Matthias Stuber[2*], Meiyappan Solaiyappan[2], Dietrich Holz[1]
[1]Department of Mathematics and Technology, RheinAhrCampus Remagen, Suedallee 2, 32524 Remagen
[2]Johns Hopkins Medical Institutions, Department of Radiology, Baltimore, USA

Abstract

Today large volumetric 3D MRI datasets of the heart can be acquired and adequate visualization needs to be ensured. Therefore, we implemented an algorithm, which enables the 3D visualization of convex surfaces.

In a first step, a 3D triangulation of user-identified points on the surface of the heart is performed. In a second step, pixel values from within these triangles are projected on a 2D viewing plane (Figure). The orientation of this viewing plane can interactively be manipulated by the user.

Our software enables interactive visualization of coronary MRA datasets as obtained with contemporary 3D whole-heart MRI methodology.

1 Introduction

Coronary artery disease (CAD) is the most common form of heart disease in America and Europe. It is the leading cause of death among both men and women in the United States. Also two million Europeans die from coronary artery disease every year, death rates from CAD are higher in Northern, Central and Eastern Europe and lower in Southern and Western Europe.
Currently Cardiac catheterization is the golden standard for determining the presence of CAD. About 480,000 Germans undergo cardiac catheter examinations each year, and many of these patients are at the risk for coronary artery disease. Cardiac catheterization is an invasive procedure and it can be dangerous. It implicates a hospital stay and consequently an expensive procedure. X-ray coronary angiography involves ionizing radiation exposure for the patient and risks of different complications.
From this it follows that there is a big need for a non-invasive, safer, more convenient, and cost-effective technique. Coronary magnetic resonance angiography (MRA) can replace other cardiac tests such as the echocardiogram and the diagnostic cardiac catheterization. It is patient-friendly, non invasive, obtains a high resolution and uses good compatible contrast mediums.
For visualization of MRA datasets the "Soap-Bubble" Tool is used (**Fig. 1**). It was written by Alex Etienne and Matthias Stuber in 2002 [1]. This tool objectively enables to compare three dimensional coronary magnetic resonance angiographic (MRA) images between different scanning methods. It displays a planar reconstruction of multiple coronary branches in the resulting 2D image, which depends on the user-selected anatomy. Furthermore quantitative coronary analysis such as Vessel Length, Vessel Border, Sharpness, Diameter, signal-to-noise, and contrast-to-noise ratios are provided by the tool.

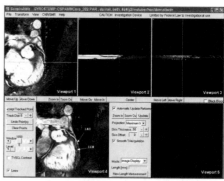

Fig. 1: Visualization of a Cardiac MR data set using the "Soap-Bubble" Tool.

The user interface of the tool consists of five view ports. View port 1-3 display three different orthogonal sections of the original 3D coronary MRA dataset. By using the mouse the user can selectively navigate through the dataset.
In order to display the desired coronary vessels the user has to mark them by attaching points with the right mouse button to the vessels slice by slice. After this the tool computes a surface by using a triangulation algorithm, which includes all points so that this surface also contains all coronary vessels, which were marked by the user. There are many possibilities to view the resulting reformatted image. The first one is displayed in view port 4; it is a planar reconstruction

---

* Contact: mstuber@mri.jhu.edu

in a 2D image. Furthermore the resulting image can be displayed as a movie or curved surface.
Today large volumetric 3D MRI datasets of the heart can be acquired and adequate visualization needs to be provided. The original "Soap-Bubble" Tool is not able to triangulate and to display a closed surface. After loading a whole-heart-dataset the result is displayed in view port 4 of **Fig. 2**.

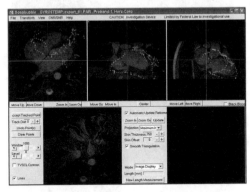

Fig. 2: Limitations of the "Soap-Bubble" Tool" in the visualization of 3D volume data sets with closed surfaces.

Therefore, we implemented an algorithm, which extends the "Soap-Bubble" Tool by the visualization of a 3D dataset as a convex surface.

2 Material and Methods

2.1 Analysis of default algorithm of the Soap Bubble Tool

For a 3D visualization of a convex hull, the Soap Bubble Tool needs to be extended.
The original tool used the TRI_SURF function, which is a provided function of IDL. It interpolates a regularly- or irregularly- gridded set of points with a smooth surface. It returns a two-dimensional floating-point array, which contains the triangulated and interpolated surface.
Unfortunately, this function cannot be used for a spherical triangulation or a triangulation of a closed surface.
Therefore, other provided functions and procedures in IDL have been analyzed which enable a spherical triangulation.

2.2 The TRIANGULATE and QHULL procedures

The TRIANGULATE procedure enables a spherical Delaunay triangulation of a set of points. A triangulation is a subdivision of a geometrical object into simplices (in our case triangles). The Delaunay triangulation algorithm of a set of points has the property that the circumsphere of every triangle does not contain any other points of the triangulation. If a set of points is given, it will be possible to create a vast number of triangulations. If T is the triangulation of these points, Theta (T) is assigned to the minimum angle measure and the smallest angle in all triangles. Via the Delaunay triangulation Theta (T) achieves its maximum value.

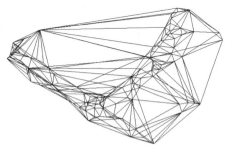

Fig. 3: Delaunay triangulated data set using TRIANGULATE procedure.

For triangulation a set of points is loaded. The TRIANGULATE procedure performs a Delaunay triangulation of the original points. This procedure is supported by IDL and introduce since version 4.0.
TRIANGULATE returns the connectivity list containing the adjacent nodes for each point in the Delaunay triangulation.
After the first triangulation the returned triangles are divided into very small triangles. The new created points are triangulated by a second procedure. This is performed by the default QHULL procedure, which is also based on the Delaunay triangulation and introduced since IDL 5.5. This procedure constructs a convex hull and gains a faster triangulation than the TRIANGULATE procedure. It returns also the indices of the connected triangles.
To allocate the right grey value to each triangle the original data set are loaded. The color of each triangle is allocated to the first named point in the connectivity list.
This user interface for visualization allows an interactive rotation of the image data by clicking and holding the left mouse button. Another possibility is to switch between the shading options 'flat' and 'grouraud'. The flat shading fills each triangle with a certain color

without any interpolation. The gouraud shading is a method for linear interpolation to simulate differing effects of color or light across the surface of a polygon. This very simple method is used to achieve smooth surface. The technique was invented by Henri Gouraud in 1971.

Other user-selected options are the style options wire and solid. The wire style displays only the mesh of triangles and the solid style fills the triangles with the right color. Furthermore the drag quality high and low is available.

The visualization of the object is based on the IDL graphic object called 'IDLgrPolygon' which needs the vertices and the connectivity list of the TRIANGULATE or QHULL procedure to draw a mesh of triangles. The color of each triangle is allocated to the first named point in the connectivity list.

The 'IDLgrPolygon' object includes one or more polygons through a given set of points and different illustration facilities. The vertices have to form a convex hull this means that a connectivity line between two vertices of the polygon cannot fall outside the polygon.

While the user is moving the heart by holding the left mouse button, the style option switch from solid to wire automatically to avoid time delays in displaying the new image.

2.3 Ray tracing/ 2D projection

To reduce the expenditure of time and improve a detailed visualization a new program was developed. It uses a Ray Tracing algorithm to visualize the coronary arteries of a whole heart MRA dataset. It combines the visualization with 'IDLgrPolygon' with a 2D projection from every point of view. This new program does not subdivide the triangles of the first triangulation and fills every triangle with a correct color instead of computing a 2D-projection which position was interactively defined by the user.

2.4 Texture mapping with OpenGL

To reduce the expenditure of time again, the application of the texture mapping method to the heart object was investigated.

Texture mapping is a method of adding an image (texture) to a computer generated graphic, resultant the amount of generating the scene is reduced.

In this application texture mapping was used to fill the triangles with the correct texture. It loads the connectivity list of triangles and the coordinates of the original points from the IDL program and adds the corresponding texture of the original dataset.

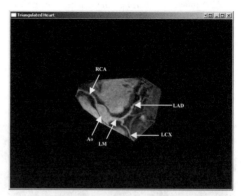

Fig. 4: Visualization of a 3D volume MRA data set using Ray Tracing/Texture Mapping method

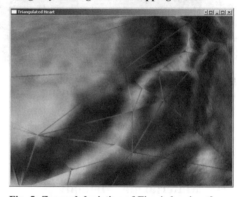

Fig. 5: Zoomed depiction of Fig. 4 showing the resultant triangles.

Fig. 4 shows the first result, the right and the left coronary arterial system are clearly visible. **Fig. 5** depicts the texture mapping method. The separate triangles can clearly be recognized. This method is very time effective, because the pixels in each triangle are not computed separately. The correct texture is copied from the original dataset and added to the correct triangle.

3 Results

3.1 The TRIANGULATE and QHULL procedures

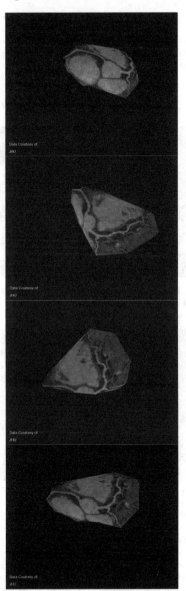

Fig. 6: Different views of a 3D MRA data set obtained using TRIANGULATE and QHULL procedures

Fig. 6 clearly illustrates the left and right coronary arterial system like ascending aorta, left main, left anterior descending, left coronary circumflex, and the right coronary artery from different view angles.

The advantage of this kind of visualization by using the default procedures TRIANGULATE and QHULL and the IDLgrPolygon object is the unique computation of the input data, the user selected points on the coronary vessels, and the loading of the original 3D data set which needs a lot of memory. The user can interactively define the point of view and the displayed image is represented without any time delay.

One disadvantage is the long computation time of the subdivision and the following triangulation.

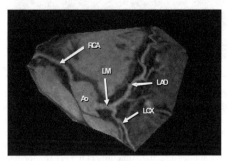

Fig. 7: Simultaneous visualization of a left and right coronary arterial system acquired using a contemporary whole-heart coronary Magnetic Resonance Angiography (MRA) method (Ao=ascending aorta, LM=left main, LAD=left anterior descending, LCX=left coronary circumflex, RCA=right coronary artery)

Another problem is the interpolation of the triangle edges. The resulting image does not show clear structures and at edges of the triangles of the coronary vessels are blurred. This is caused by the fact that one edge or connectivity line is part of two triangles. This caused of two different interpolations at one edge.

3.2. Ray tracing/ 2D projection

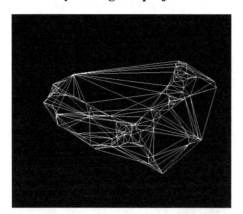

Fig. 8: Mesh of triangles showing during rotation

By clicking the left mouse button the user can rotate the mesh of the IDL graphic object (Fig. 8) on the display interactively. After he defined the new position of the surface, the program computes the difference in rotation and translation via matrix operations and applies the IDL function POLYFILLV for each triangle. This function returns a vector containing the subscripts of the pixel inside the triangle. It only designed for 2D so that very triangle has to calculate by itself.

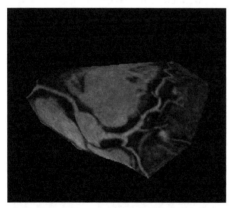

Fig. 9 Resultant visualization of the left and right coronary arterial system using Ray Tracing.

After the computation of the pixel inside the triangle a ray-tracing algorithm is used. Ray Tracing is a global illumination and point sampling algorithm to generate realistic images by computer. It traces the path of individual rays of light from the eye back to the object. This algorithm computes the smallest distance of the intersection point to the image plane and assigns the pixel of the image plane with the right color.

The visualization of the heart vessels in Fig. 9 is clearly improved. The calculation of each image does not need so much time as the TRIANGULATE procedure, because it does not use the subdivision of the triangles.
Otherwise it takes some time to calculate and display each image and the projection of each triangle on a 2D image plane with the function POLYFILLV. This function implicates a few inaccuracies, which are observable on the screen.

3.3 Texture mapping with OpenGL

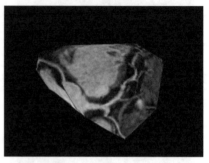

Fig. 10: Resultant visualization of the left and right coronary arterial system using Texture mapping with OpenGL

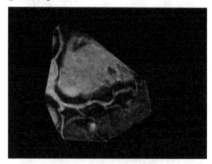

Fig. 11: Left coronary arterial system

The user can rotate and translate and scale the object interactively by using the keyboard and not the mouse.
The texture mapping method is a very fast way to visualize the coronary vessels on the triangulated closed surface. The program itself is still work in progress and not completed. The edges of the triangles are still visible and a few triangles have the wrong orientation of its texture, but this way of visualization looks very promising. Another question is still how to combine the tool, which is written in IDL with this visualization method, which is written in OpenGL and C++.

4 Conclusion and Future Work

The expenditure of time and computing, the resolution in the resulting 3D object, and the possibility of defining the point of view of the object by the user are always important factors for the different approaches in the visualization of volume datasets. A compromise between computation time and image quality must be found in order to provide diagnostic image quality within a reasonable short time. So far all results are showing the main arteries of the whole artery system like Ao (Aorta), RCA (right coronary artery), LM (left main), LCX (left coronary circumflex) and LAD (left anterior descending).

Furthermore the OpenGL program must be extended. The ambition is still to display the coronary vessels with OpenGL and to make the program still faster and to get a better resolution of the resulting object, so that the whole coronary vessel system is visible. To obtain this result it might be required to visualize the data by the hardware and not by software.

5 References

[1] Etienne A, Botnar RM, Van Muiswinkel AM, Boesiger P, Manning WJ, Stuber M "Soap-Bubble" visualization and quantitative analysis of 3D coronary magnetic resonance angiograms.
Magn Reson Med. 48: 658 – 666 (2002)

[2] IDL Online Help 6.0

Robotik

Robotics

Robotic-Guided Dental Laser Interventions

J. Bongartz[1], M. Ivanenko[2], P. Hering[2] and Th. M. Buzug[1],
[1] Department of Mathematics and Technology, RheinAhrCampus Remagen, Suedallee 2, D-53424 Remagen
[2] Holography and Laser Technology, Forschungszentrum caesar, Ludwig-Erhard-Allee 2, 53175 Bonn, Germany

Abstract

A novel sound-guided navigation concept for surgical laser interventions is proposed. The combination of image guidance and laser surgery is a promising approach e.g. in dental implantology. The main advantage compared to conventional drilling is the contactless ablation process that diminishes residual movements of the patient. However, the accuracy of the entire registration chain – from the CT imaging via optical navigation to the positioning precision of the robotic laser tool holder – has to be investigated in the course of the project. We will present the methodology for the error propagation estimation and a novel laser-based procedure to obtain a ground truth. The present paper describes parts of the cooperation between two principal projects of the German government at the science region Bonn, i.e. Caesar Bonn (center of advanced European studies and research) and the RheinAhrCampus Remagen. These institutions are compensation projects to change the face of Bonn after the movement of the Government to Berlin.

1. Introduction

The application of computer-aided planning, navigation and robot-guided laser surgery in dental implantology provides significant advantages [1,2] – compared to conventional practice – due to today's sophisticated techniques of patient-data visualization in combination with the flexibility and precision of novel robots and new laser-surgical instruments. However, a realisation of navigation and robot assistance in a field where only local anaesthesia is applied is a challenging task because of unavoidable patient movements during the intervention. In this paper we propose the combination of image-guided navigation, robotic assistance and laser surgery to improve the conventional surgical implantation procedure. A critical point for the success of the surgical intervention is the registration of the planned operation trajectory to the real trajectory in the OR. In the course of the project this point will be thoroughly investigated. As a preliminary study we have focussed on the registration error in a point-based registration.

The point-based registration is used in a large number of systems and is therefore chosen as a standard in our project. Different matching strategies as surface-surface or palpation based registration will be evaluated with respect to the point-based case. To evaluate the overall accuracy of the entire registration processing chain we have to estimate the localization error in the underlying image modality, the localization error of the fiducial or anatomical markers in the OR and the positioning error of the surgical tool. In our case this means a calculation of error propagation from the CT via the optical navigator to the robotic laser holder.

In chapter II a new methodology for the evaluation of fiducial and anatomical markers localization accuracy is presented. By using a holographic technique instead of the object itself the holographic image of the object is investigated. The advantage of this method is the contact-less localization of the markers since the holographic image is just a three-dimensional projection of the object. In chapter III a new algorithm for orientation detection of fiducial markers is presented. It is a detection method using a spherical coordinate transformation. From a CT-dataset 3D images of the fiducial markers are obtained. These 3D ROI-images and a corresponding 3D-image of a modelled marker in its precise spatial dimensions are transformed into spherical coordinate space. Using maximization of cross correlation as a similarity measure for the template matching the best matching position is obtained. In chapter IV the methodology of laser intervention is presented. Laser surgery is a smooth preserving intervention, which brings advantages whenever holes in small ridge-like bone structures have to be prepared. We developed a pulsed CO_2 laser system, which allows hard tissue removal with high precision and without thermal side effects. The strong absorption of the 9.6 or 10.6 µm laser light and the fast energy deposition by short pulses leads to a thermomechanical ablation process where no essential heat dissipation into neighboring tissue occurs. However, our main goal is to reduce the efforts of patient fixation. The laser "drilling" is a contactless process, and therefore, in comparison with the conventional drilling, the jaw is not subjected to forces. As a consequence, the online tracking and re-registration of the patient has to cope with small residual movements only, even in the case of non-invasive patient fixation. On the other hand, a major draw back in surgical laser drilling is the loss of tactile feedback, which in fact is the disadvantageous consequence when no forces are applied.

Thus, no direct information about the drilling depth and tissue type is available. This is a critical issue and it will be discussed in chapter V how this is treated in the project, because vital neighbouring structures as canalis mandibulae with the alveolar nerve must not be damaged during an intervention.

2. Ground Truth

We start the investigation with an anatomical phantom equipped with fiducial landmarks (see Fig. 1).

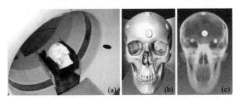

Fig. 1. (a) *GE Prospeed* CT scanner at RheinAhrCampus Remagen. A skull phantom equipped with fiducial markers is scanned with different protocols (different slice thickness in conventional modus and different pitches in spiral modus) as well as different phantom orientations in combination with different gantry tilts. One result for a pitch one spiral scan with 3 mm slice thickness is given as a 3D rendering in (b). 6 segmented fiducial markers can be seen in the semi-transparent visualization of the skull (c).

To obtain the ground truth a hologram of the phantom is taken with a holographic camera as shown in Fig. 2a. The GEOLA GP-2J camera system comprises a flash lamp pumped Nd:YLF laser oscillator with subsequent amplifycation and second harmonic generation.

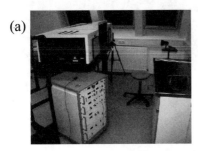

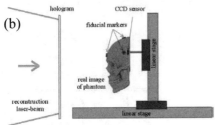

Fig. 2. (a) Holographic camera system for recording a hologram of the phantom (b). Reconstruction setup to localize the fiducial markers in the holographic real image.

The resulting laser-pulse has a wavelength of 526 nm, duration of 25 ns and maximum pulse energy of 2 J. Careful mode selection leads to a coherence length of above 6 m allowing recordings within a large volume of some cubic meters [7,8]. A typical object distance is approximately 60 cm.

Illuminating the resulting hologram with the complex conjugated reference beam of the recording setup by means of an equivalent continuous wave laser system (Coherent Verdi V-2) reconstructs the so called real image of the hologram, which corresponds to a three-dimensional image of the phantom with identical scale. The reconstruction set-up is sketched in Fig. 2b.

The real image is projected into the space in front of the hologram and its resolution depends on the aperture and therefore on the size of the hologram. Using holographic plates of 40 cm x 30 cm a resolution below 5 microns is achievable [8].

The real image is investigated by a CCD-array mounted on a three axis linear stage, which can be moved contact-less and with all degrees of freedom, since the real image is just a three-dimensional projection of the phantom. The CCD-array (Kodak KAI-4000) has 2048 x 2048 pixels of size 7.4 µm x 7.4 µm corresponding to an active area of 15.2 mm by 15.2 mm. The high resolution of the CCD-array allows the identification of microscopic features in the real image leading to a localization of suitable fiducial markers of the phantom with high precision [9]. We aim for an overall accuracy of marker localization below 15 microns. Fig. 3 shows a two-dimensional projection of the real holographic image ('slice') for a defined distance to the holographic plate.

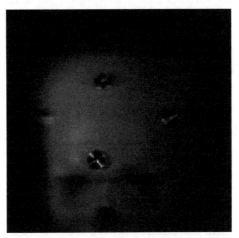

Fig. 3. A slice picture of the holographic phantom reconstruction.

A hologram is capable of recording all parts of the object, which are directed towards the holographic plate. To capture markers on the back of the phantom additional holograms of different perspectives are necessary to determine the relative positions between all markers.

3. Localization of fiducial markers and error propagation

A critical point for the success of the surgical intervention is the registration of the planned operation trajectory with the real trajectory in the OR. Here, we briefly describe the methodology of error-propagation investigation and a novel method to obtain an exact ground truth based on holography.

We used M fiducial markers as well as anatomical landmarks that are homologous point sets {p1, p2, p3, ... pM} in CT-space and {q1, q2, q3, ... qM} in OR-space to estimate the parameters of the rigid body transformation $q = A p - b$, where $A \in SO(3)$ is a rotation matrix and b a translation. Both sets of points are acquired with certain inaccuracies [1,2]. Fig. 1 shows a skull phantom with fiducial markers inside a GE Prospeed CT. To acquire data for an error propagation study several scans with different protocols, phantom orientations and gantry tilts have been performed. In Fig. 4 the result for a pitch one spiral scan with 3 mm slice thickness is shown. The variability of the positions determined in a segmentation step is mainly induced by partial volume averaging in the anisotropic CT data sets and can be expressed in terms of the covariance matrix C(p) for a certain marker at position p. However, the accurate segmentation process is an crucial step for the entire registration chain.

In many IGS applications cylindrically shaped fiducial markers are used for point-based registration strategies. These markers are distributed onto the skin of the patient in a way that the interventional target is lying in the centre of this distribution. For registration or transfer, respectively, of the planned intervention trajectory to the coordinates of the OR, a tracked pointer device is inserted to a centred milled cone of each marker. In that way an estimate of the marker centre is obtained. The required homologous set of points in the CT image must refer to these centre points.

The detection of the centre of each marker in the CT domain is confused by the typical quality problems of CT scans. Due to partial volume effects the marker boundaries cannot be accurately detected by a simple threshold operation [3]. Therefore, a template matching procedure is developed using object knowledge, i.e. precise geometrical shape knowledge of the fiducial. Generally, the problem in template matching strategies is the high dimensional search space for objects with rotational non-invariance.

Unfortunately, this is the case for the typically cylinder-shaped fiducial markers. Due to a variation of marker distribution from patient to patient, the orientation of a certain marker on the skull is a-priori not known. Therefore, we developed a template-matching procedure that is invariant to the orientation of the marker using a spherical coordinate transform. The entire procedure can be described as three steps.

A. Preparing the skull and CT imaging

The skull phantom is equipped with a number of fiducial markers of aluminum (to avoid additional problems with strong metal artifacts at this point of the project), in our experiments exemplarily chosen to be located three on the forehead, one on the top and two on the back of the head (see Fig. 1).

B. 3D reconstruction of the fiducial marker

Up to 11 slices from the CT image stack are used for the 3D reconstruction of the fiducial markers. This number depends on the CT scanning parameters and the orientation of the markers with respect to the axial CT slice. However, in principle the markers can be easily detected due to its high Hounsfield values. Figure 4 gives an overview of the marker positions on the phantom. This fact is used in a first detection step in which each slice is examined for high image intensity, to find those slices where the markers are included. From the knowledge of the exact geometrical shape a 3D ROI of (50x50x11) voxels is extracted around each centre of mass serving as preliminary marker centre. A 3D volume rendering of such a ROI is given in fig. 5. From that figure it can be seen, that the neighbouring bone is also included in the ROI. A threshold-based strategy to separate the marker from its bone would lead to an unacceptably high fiducial localization error (FLE) that is mainly caused by the partial volume effect as mentioned above.

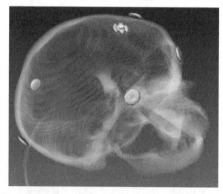

Fig. 4. The imaging quality of the fiducial depends on the position and orientation of the marker.

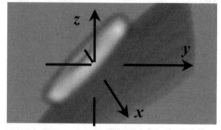

Fig. 5. Volume rendering of the 11-slice ROI including the marker and parts of the skull phantom.

C. Refinement of marker centre in the CT domain

A template-matching strategy is proposed for centre-point refinement in this algorithmic step. The main idea is an accurate segmentation of the marker from the bone using exact geometrical marker-shape knowledge. To handle the rotational non-invariance of the cylindrical marker the ROI is mapped to spherical coordinates using the preliminary centre point estimated in step B. For the definition of the polar and azimuth angle in the spherical coordinate transformation see fig. 6. The result of that transformation is shown in fig. 7.

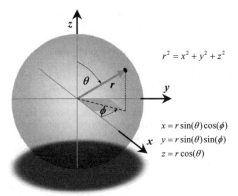

$$r^2 = x^2 + y^2 + z^2$$

$$x = r\sin(\theta)\cos(\phi)$$
$$y = r\sin(\theta)\sin(\phi)$$
$$z = r\cos(\theta)$$

Fig. 6. Definition of the spherical coordinates (r,ϕ,θ).

Fig. 7. Spherically transformed marker attached to skull phantom.

For the matching step an exact digital marker replica or template, respectively, is also mapped to spherical coordinates. The position of the marker centre in the CT image is refined stepwise by optimisation of the correlation similarity measure.

Unfortunately, the error of localization in the CT image is not the only source of inaccuracy in the registration process. The second error source is the localization error of the fiducial markers in the OR. We know from the technical booklet of the NDI Polaris navigation system that there is a system inherent inaccuracy of 0.35 mm.

However, this measure takes not into account the errors that are introduced by the surgical interaction. In Fig. 8 it is shown how the data of interaction inaccuracy are measured in the OR. The tip of the pointer device is positioned on a certain fiducial marker and the position data are acquired with 60 Hz while the pointer is moved around on a sphere.

As in the CT image the variability of the positions of the fiducial markers in the OR can be expressed in terms of the covariance matrix $C(\mathbf{q})$ for a certain marker at position $\mathbf{q}$. Fig. 8 shows that the covariance matrices can be geometrically interpreted as error ellipsoids.

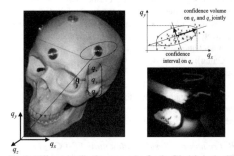

Fig. 8. Different localization accuracies for the fiducials in the CT domain (a). To measure covariance matrices for the fiducial markers in the OR domain (b) – that can generally be visualized as ellipsoids indicating confidence intervals – the tip of the pointer device points onto a certain marker. The position data are acquired with the navigation system when the pointer device is moved on a sphere.

Due to the fact that we are faced with errors in both coordinate systems, we have to estimate the transformation between the systems by minimizing

$$\chi^2 = \sum_{i=1}^{M}(\mathbf{q}_i - \mathbf{A}\mathbf{p}_i - \mathbf{b})^T (\mathbf{A}\mathbf{C}(\mathbf{p}_i)\mathbf{A}^T + \mathbf{C}(\mathbf{q}_i))^{-1}(\mathbf{q}_i - \mathbf{A}\mathbf{p}_i - \mathbf{b}) \cdot$$

4. Methodology of laser surgery

A second work package of the project deals with laser surgery. The laser procedure can bring advantages whenever holes in small ridge-like bone structures have to be prepared. Our main goal is, however, to reduce the efforts of patient fixation. The laser "drilling" is a contactless process, and therefore, in comparison with the conventional drilling, the jaw is not loaded with mechanical forces. As a consequence, the online tracking and re-registration of the patient has to cope with small residual movements only, even in the case of non-invasive patient fixation.

A basis for the development of a laser drilling procedure is provided by investigations on laser osteotomy with pulsed CO_2 lasers done at the last years (see [4, 5] and Refs. there). The goal of these works was a fast laser "processing" of hard tissue without thermal side effects. It can be reached if the deposited laser energy is confined in a very thin absorption layer and effectively used for a thermomechanical ablation of the hard tissue. The ablation process has to proceed at possibly low temperature and has to be faster than heat diffusion in the tissue.

A. Equipment and Experimental Setup

We use the CO_2 laser for hard tissue ablation because its wavelength (10.6 μm or 9.6 μm) is very strongly absorbed in the main bone tissue components, - hydroxylapatite, water and collagen. The light absorption depth in compact bone is only ~ 7 μm at 10.6 μm and ~ 5 μm at 9.6 μm.

Further important conditions of an effective ablation are high power density (J/cm^2) and intensity (W/cm^2) of the beam. That is why we use pulsed lasers and focus the beam strongly on the tissue. Furthermore, the modern pulsed CO_2 lasers possess a very good beam quality (focusability of the beam) and they are very reliable and relatively inexpensive. To prevent tissue parching and to cool it additionally we use a fine air-water spray (~ 0.5 ml H_2O/min). Other essential requirement is relatively low pulse repetition rate (< 100 Hz) or fast beam scanning over the tissue at the higher pulse repetition rates. The scanning technique allows to avoid unnecessary heating of the cut edges due to accumulation of the residual heat, which has not been exploited for the ablation.

B. Preliminary Results with a mini TEA CO_2 Laser

In experiments by the 'center of advanced european studies and research' in Bonn mainly two types of CO_2 lasers are used. The first one is a compact mini TEA CO_2 laser, which provides pulses of 1 µs duration and energies up to 40 mJ. The pulse composed of a spike of about 50 ns duration (FWHM) and a long tail, which contains about the half of the energy. The repetition rate is limited to 70 Hz. The beam can be focused down to the spot size of 260 µm without an optical breakdown in the water spray. As a result the energy density up to 80 J/cm^2/pulse and peak intensity up to 800 MW/cm^2 are possible. With this system we have demonstrated an extremely "clean" ablation of the compact bone tissue without any noticeable thermal side effects. Due to the very high peak intensity the ablation starts very quickly (time delay of ~10 ns) and the heat diffusion has no time to proceed (temperature relaxation time constant is about 100 µs). On the other hand, the material removal rate with these short pulses is below 10 µm/pulse. Only a part of the laser pulse energy is used for the ablation. Dense ablation debris absorbs "immediately" strongly laser light and thereby shields the tissue from the beam. Our experiments show, that this system can be used to produce deep (up to 10 mm) and narrow (200 µm) holes or incisions, as well as pits of different form in the bone (Fig. 9a). However, the drilling of relatively big holes for tooth implants will be rather time consuming. To remove 1 cm^3 of the compact tissue one need more than 10 kJ of the laser energy.

Fig. 9. 3-D removal of compact bone tissue (bull femur) with mini TEA CO_2 laser (a). Examples of deep holes drilled with the slab CO_2 laser under application of a special beam scanning procedure in compact bone (b). Irradiation parameters: λ = 10.6 µm, τ$_{1/2}$ = 80 µs, E = 80-90 mJ, f = 100-200 Hz, duration of the drilling up to 5 min, depth up to 16 mm.

C. Preliminary Results with a slab CO_2 Laser

Much faster processing of bone tissue is possible with a powerful "slab" CO_2 laser. It provides the energy up to 90 mJ in pulses of 80-µs-duration (FWHM). The repetition rate is up to 10000 Hz. For laser osteotomy the repetition rates of 50-1000 Hz are used.

The beam can be focused down to spot sizes of 120 µm and even smaller without an optical breakdown. The corresponding energy density amounts up to 600 J/cm^2/pulse by the peak intensity of 8 MW/cm^2. The system is coupled through an articulated mirror-arm and matching optic with a fast galvanic X-Y beam deflector, which is equipped with a plane-field focusing lens. The laser and the deflector are controlled with a PC. With this system we have reached for compact bone similar cutting rate like with usual mechanical surgical tools.

For example at average power of 70 W the cut rate of 70 mm/min by cut depth of 6 mm has been demonstrated [5]. The laser leaves no sign of carbonization in the tissue. Histological evidence of thermal effects can be found only in a 50-µm-broad zone adjacent to the cut boarder [6].

With the PC-controlled beam deflector an arbitrary geometry of tissue removal is possible. To produce round deep bores of variable diameter we have developed a special beam scanning procedure, which allows combining high laser power resulting in high drilling rates with very low thermal load on the tissue. A high aspect ratio (depth to diameter) is also possible with this scanning procedure (Fig. 9b).

5. Control of surgical laser drilling

A serious drawback of the contactless surgical laser drilling is the loss of tactile feedback. Thus, no direct information about the drilling depth and force moment, which correlates with type of tissue, is available. This is a critical issue, because vital neighboring structures as canalis mandibulae with the alveolar nerve must not be damaged during the intervention. Therefore some new method of laser drilling control must be developed.

We assume, that information about the drilling depth and type of the tissue can be extracted from the acoustic signal accompanying the laser ablation (Fig. 10). The main source of the acoustic signal is the ablation process at the bottom of the bore cave. The weaker signal come however, also from the side walls of the cave and from the very small tissue area around the bore entry, which experiences a petite but very fast thermal expansion after absorption of the light in sub-ablative wings of the focused Gaussian beam.

The acoustical diagnostic technique is under development by center of advanced european studies and research at present. In first experiments we used the pulsed TEA CO_2 laser (see above) and piezoelectric transducers for sound detection. Various bone specimens as well as reference materials were studied. For the determination of the ablation crater depth, we analyse the time delay between the laser-induced acoustic signal from the surface of the specimen and the bottom of the ablation crater. Results show a possibility of an accurate control of the cut depth in material with known acoustical properties. The measurements of the laser bore depth in homogeny tissue models have demonstrated precision up to ± 100 µm.

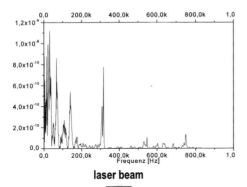

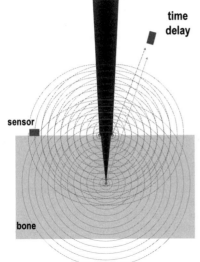

Fig. 10. Principle of acoustical depth control by laser tissue boring.

Other possibility to control the laser boring is acoustic tissue recognition (differentiation). For this we analyze acoustic spectra initiated by laser pulses in different samples: soft muscle tissue, compact and sponge bone, PMMA as a reference material. The recorded spectra show specific material-based features, especially in the region 50 kHz to 200 kHz. We work now on improvement and automation of the detection technique and on analyzing algorithm, which has to allow reliably recognize the tissue specific combination of the spectral signal futures independent on the laser pulse energy and bore depth. The technique will prompt surgeons with information about the transition from compact bone to sponge and/or soft tissue.

6. Summary

Two novel laser-based methods for computer-assisted surgery are presented. Based on a holographic experimental set-up a new methodology for ground truth in registration error propagation is obtained. In the framework of laser surgery a new navigation methodology is developed. This principle of sound guidance has an important advantage in comparison to the conventional drilling procedure, because these traditional methods measure the burr-hole depth only. The new method directly measures the distance-of-interest, i.e. the distance to neighboring organs-at-risk as the canalis mandibulae in dental implantology.

7. Acknowledgement

The project is embedded in the *Center of Expertise for Medical Imaging, Computing and Robotics – CeMicro –* which is a research center for medical image acquisition technology, medical signal processing and computation, and improvement of clinical procedures especially in image-guided interventions with robotic assistance. T.M.B. thanks the Ministry for Economy, Transport, Agriculture and Viniculture of Rhineland-Palatinate, Germany, for granting *CeMicro*.

8. References

[1] Th. M. Buzug, P. Hering, J. Bongartz and M. Ivanenko, A Novel Navigation Principle in Computer-Assisted Surgery, Proc. of EMBC 2004, San Francisco, September 2004.

[2] Th. M. Buzug, P. Hering, J. Bongartz, M. Ivanenko, U. Hartmann, D. Holz, J. Ruhlmann, G. Schmitz, G. Wahl and Y. Pohl, Acoustical Navigation of Laser Interventions in Implantology, Proc. of CAR 2004, Chicago, June 2004.

[3] Th. M. Buzug, Einführung in die Computertomographie, Springer Publishers, Berlin, (2004).

[4] M. M. Ivanenko, P. Hering. Wet bone ablation with mechanically Q-switched high-repetition-rate CO_2 laser.- Appl. Phys. **B 67**, 395 (1998).

[5] S. Afilal, M. Ivanenko, M. Werner, P. Hering. Osteotomie mit 80 µs CO2-Laserpulsen. In: Fortschritt-Berichte VDI, Biotechnik/Medizintechnik **17** (231), 164 (2003).

[6] M. Frentzen, W. Götz, M. Ivanenko, S. Afilal, M. Werner, P. Hering. Osteotomy with 80-µs CO_2 laser pulses - histological results. Lasers in Medical Science **18**, 119 (2003).

[7] J. Bongartz, D. Giel, P. Hering: "Living human face measurement using pulsed holography." Holography 2000. Bellingham, Wash.: SPIE proceedings 4149 (2000) 303-308.

[8] J. Bongartz, „Hochauflösende dreidimensionale Gesichts-profilvermessung mit kurzgepulster Holographie", PhD thesis, Heinrich-Heine Universitaet at Duesseldorf, 2002, http://deposit.ddb.de/cgi-bin/dokserv?idn=964966670.

[9] D.M. Giel, "Hologram tomography for surface topometry." PhD thesis, Heinrich-Heine-Universitaet Duesseldorf, 2003, http://deposit.ddb.de/cgi-bin/dokserv?idn=968530842.

Safety Analyses for Robotics:
An Example from Space Technology

Hendrik Schäbe, TÜV InterTraffic GmbH, Köln, Germany

Summary

Applications of robot systems need to be safe. Systematic methods for hazard identification, cause-consequence analysis, risk estimation and risk reduction are presented. An example from space technology is given.

The first step towards a safe system is hazard identification. The hazards and the event chains developing from them are systematically identified by a Hazard Analysis. A quantitative analysis can be carried out using the Consequence Tree Analysis. This kind of analysis studies consequences and their occurrence rates and allows to express the risk numerically.

The robotics system "Vital" serving as an example was planned to be used in a laboratory on board of the orbital station Mir. An experiment was planned to be performed with the experimenter being on the surface of the earth. For the experimenter, a virtual reality was generated using graphic computers and an eyephone that projects a stereographic picture of the laboratory environment for the operator. The operator is able to move inside the virtual reality and manipulate objects in the virtual reality using a dataglove. The manipulations carried out in the virtual reality can be transmitted in forms of command sequences to the laboratory onboard of the spacecraft. A robot transforms these commands into motions and actions and thus carries out the experiment.

1. Introduction

Robots are taking over more and more tasks formerly performed by human beings. Medical robotics are special applications. Without any doubt, the application of robots brings advantages, but also bears risks in it. The present paper is dedicated to methods of analysing risks and reducing risks for robots. A project that had been carried out for application in the space area will be used for demonstration.

Robots have the advantage that they can be very precise, repeat the same action without getting tired and work under hard conditions. On the other hand, robots do not have the same senses as a human being and are not able to react to unpredicted deviations from the normal process. From this constellation, risks can arise.

In the second section analyses are described that can be carried out in robotics to identify and reduce risks. Section three is dedicated to the example: a robot controlled by a virtual reality system. In the fourth section, the analyses are described that have been carried out in the example. Conclusions are given in the last section.

2. Safety Analyses for Robotics

Several analyses can be applied to identify risk and to reduce risk. Not all types of analyses described below need to be carried out for each project. They have to be tailored to the type of robot application. The analyses may help to show that the robot system is sufficiently safe.

- Failure Modes, Effects, and Criticality Analysis (FMECA)
  This type of analysis is well known and gives a bottom-up analysis of technical failures and their consequences.
- Fault tree analysis
  Combinations of failures and other events are studied. The occurrence of a dangerous event is computed.
- Hazard Analysis
  This analysis identifies hazards and builds accident scenarios. Risk estimates are provided and risk reduction can be facilitated with the help of this analysis.
- Operating Hazard analysis
  Hazards that arise from operation and interaction of the operators with the robot, e.g. during maintenance can be analysed.
- Consequence Tree Analysis
  A probabilistic analysis of event scenarios is carried out.

The analyses focus on special topics. That means, not each analysis is able to detect all hazards. For example, an FMECA detects all risks associated with failures of components, but will not be able to detect risks that arise from combinations of failures or from hazards as e.g. cutting edges. Therefore, the choice of a good combination of analyses is important to detect the existing risks and to reduce them.

On the other hand, tailoring is necessary to limit the expenses for analyses. Therefore, it is important to identify the boundaries of the robot system and give an abstract description in form of a functional block diagram, together with a functional description. Such an analysis should precede any other analysis.

3. The system VITAL

VITAL is a system that has been designed to conduct experiments onboard of a space station, operated from a ground station. Originally it was planned to install the system on the MIR space station but due to unfortunate circumstances the project has not been finalised.

VITAL consisted of two main parts: the virtual reality and the robot lab. Within the lab, an experiment is carried out that is similar to the so-called STATEX II-Experiment. In the experiment, the influence of gravitation on the development of gravity receptors in the inner ear of tadpoles and fish larvae is studied. One part of the larvae is kept as a reference group within a centrifuge that maintains 1 g gravity. The other larvae are kept under microgravity. The behaviour of the larvae is observed by videocameras. Before the end of the mission, the larvae are fixed. Therefore, a fixing agent is injected with a help of a fixing unit. All operations in the lab would be usually carried out by an astronaut. However, the astronaut is replaced by a robot. Application of a robot instead of an astronaut excludes many hazards, as e.g. heat radiation and convection or fluid jets.

The other part of the system consists of a virtual reality. A user is equipped with an eyephone and a dataglove. The eyephone shows a virtual reality of the lab with the help of two screens mounted in front of the eyes of the user. The user is able to open racks, take the fixing unit and inject the fixing agent. The motions of his hand are detected by the data glove and converted to commands for the robot. Sensors detect the position of the user inside the lab. The virtual reality brings specific hazard for the operator. For example, the operator might fall down, since he does not see the reality.

The project has been carried out by a consortium of different companies: CAE, MST, sitec, EFR, IRF, SDZ, VCS Engineering, graphikon and ASAP, a subsidiary of TÜV Rhineland.

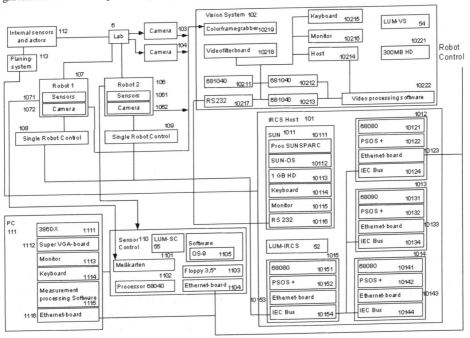

Figure 1 Excerpt of the Functional Block Diagram

4. Analyses for VITAL

A product assurance plan has been elaborated for the project and agreed within the project consortium.

4.1 Functional Analysis

As a basis, a functional analysis was elaborated. The functions of the hardware and software have been described. A number of about 90 functional blocks have been identified, described and predicted in a large functional block diagram. Figure 1 shows an excerpt of the diagram.

4.2 Failure Modes, Effects, and Criticality Analysis

Based on the Functional Analysis, a Failure Modes, Effects and Criticality Analysis (FMECA) has been elaborated. Each failure mode has been considered with its consequences. If necessary, design changes have been proposed. The following problems have been highlighted as important ones:
- Software has to be manufactured according to the safety requirements.
- Life time of hard disks must be sufficient.
- Virus protection has to be established.
- Different problems with the user in the virtual reality.

A user that works in a virtual reality will have to face several problems.
- In virtual reality, only visual information is supplied. Tactile information is supplied only in limited form.
- The user does not get feedback information about his own position. That might corrupt his sense of balance. Experiments in "swinging rooms" show that persons will get irritated about their position. As a consequence the user might fall.
- Using the eyephone can lead to problems, if the pictures are distorted or limited.
- Also, sensor failures can be critical since they might lead to uncontrolled reactions of the user within the virtual reality.

4.3 Hazard Analysis

In addition to the FMECA, a hazard analysis has been carried out. This approach is motivated by the fact that detection of hazards is only a by-product of the FMECA: Hazards are only detected by the FMECA, if they are controlled by a technical subsystem that is subject to the analysis and if a single point failure of this system will lead to an accident.

The hazard analysis was based on a standard from space industry, ESA-PSS 01-403 [1]. This standard provides the most systematic, rigorous and generic approach to hazard analysis.

Starting from hazards such as "high electric voltage", potential energy etc., hazardous conditions are defined, i.e. the particular manner in which the hazard is present in the system. Then, the initiating event is defined. The initiating event is necessary in order to start a chain of events that can finally lead to an accident. Also, this chain of events is elaborated. Figure 2 shows an example of a hazard tree scenario.

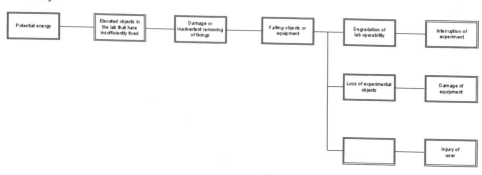

Figure 2 Example for Hazard Tree Scenario

The hazard analysis has been carried out in three steps. In the first step, a hazard analysis has been elaborated for the virtual reality. Special care has been dedicated to the man-machine interface. This analysis has been discussed inside the consortium. As a result of this discussion, a second report has been elaborated. The robotics lab has been subject to a separate analysis.

For the *virtual reality*, the main results can be summarised as follows.

There are some typical hazards for electrical installations as high temperature, high voltage and potential problems with electromagnetic compatibility. These problems can be resolved using standard equipment.

A more difficult problem is possible EMC interaction with the sensors that detect the position of the user inside the virtual reality. The sensors detect a magnetic field that is generated by magnets that are connected with the user. Shielding of the sensors is not applicable since this would reduce the magnetic field connected with the user as well. EMC problems could lead to small, unwanted virtual movements of the user in the virtual reality. In cases where EMC problems will cause faults of position sensing, there will be no consequence for the robotics lab, since both systems are decoupled. This is done by the following approach. First,. all actions are recorded in the virtual reality. Then, from the actions of the user in the virtual reality, commands are generated that are executed by the robot.

Falling down of the user during work has been excluded: The user is sitting on a chair during his work.

Finally, no serious problems have remained regarding the virtual reality.

For the *robotics lab*, the main findings are as follows. Mainly, they are introduced by the robot itself. There is pressurised air for operating the robot's arm, electrical voltage used in the control equipment and the motion of the robot. These problems are typical ones for industrial robots and have been covered by application of proven technology. Especially, methods of collision control have been applied.

Another hazard is the rotation of the centrifuge. For higher rotational velocities, the rack with the centrifuge would have to be secured by a separate lock.

4.4 Consequence Tree Analysis

In order to analyse the different accident scenarios and to find out which one would be the most probable, a Consequence Tree Analysis (CTA) was carried out. This kind of analysis has many commonalities with a Fault Tree Analysis. However, a combination of events does not yield another event instantaneously. The CTA is able to treat events that need a certain time to develop [2]. A typical example for such a kind of events are fires that emerge from the presence of hot objects and ignitable materials or events that are results of (rapid) ageing or wear of material.

Mostly, the following accidents will occur:
- injury of the user, especially cuts,
- destruction of one of the test objects,
- unintended contact with cutting edges.

These will be caused mainly by the following causes:
- wrong control of the robot caused by perception errors of the user,
- insufficient brightness in the virtual reality
- undefined direction of illumination, lack of shadows in the virtual reality.

The following three measures have been defined and partially implemented to reduce the risk:
- only introduced personnel is allowed to operate the system
- sufficient brightness and adequate selection of colours is necessary for the virtual reality
- for upcoming versions of the system, the objects in the virtual reality should have shadows.

5. Conclusions

A robotics system has been analysed by a set of general analyses. Results have been obtained and risk reduced, see [3].

The following main lessons can be learned from the project for robotics applications in general.

Safety analyses such as FMECA, Hazard Analysis and Consequence Tree Analysis are useful when new systems are developed for which only limited experience is available. These analyses are able to predict accident scenarios, thus giving the possibility to reduce risks already during system development.

In addition, the analyses can be used to show that the system has a sufficient level of safety.

A subset of safety analyses must be chosen to be effective. That means, the analyses should cover most new accident scenarios but should avoid in-depth analysis of already known accident scenarios.

When a virtual reality is used for tele-operation, a separation of the virtual reality and the robots are advantages. However, if the motions in the virtual reality have to be transferred into reality in real time, all sensing functions require special analysis to ensure their correctness and preciseness.

The methods described within this paper have also been applied in other areas of technology such as road vehicles, electronic control systems and railway systems. For railway systems, hazard analyses and risk reduction is described and requested by a specific standard, EN 50126 [4].

References

[1] ESA PSS-01- 403, Hazard Analysis, Issue 1 Draft 5, 1989 Sept. 10.
[2] Schäbe, H., A Stochastic Approach to Consequence Tree Analysis, Quality Rel. Engin. Internat. 10 (1994), 229-236.
[3] Application of Virtual Reality for Telerobotics, Main results of Product Assurance, Agency for Safety of Aerospace Products, 1994.
[4] EN 50126 EN 50126 Railway applications - The specification and demonstration of Reliability, Availability, Maintainability and Safety (RAMS), 1999.

Medical Navigation Based on Various Visualization Methods

R. R. Evbatyrov[1], G. G. Gubaidullin[1], M. Kunkel[2], T. E. Reichert[2]

[1] Dept. of Mathematics and Technology, RheinAhrCampus Remagen, University of Applied Sciences Koblenz, Suedallee 2, D-53424 Remagen

[2] Dept. of Oral and Maxillofacial Surgery, Johannes Gutenberg University Mainz, Augustusplatz 2, D-55131 Mainz

Abstract

Medical navigation has become a standard tool to support the surgeon with an artificial feedback during a surgical intervention. A navigation system has to be fast enough to reflect the current instrument location continuously. This paper evaluates and compares various visualization methods used in medical navigation: 2D projection, surface rendering and volume rendering. Problems of adequate and comprehensive patient model representation as well as visualization performance aspects are considered.

1 Introduction

Dictionaries usually define navigation as a science and practice of getting craft (ships, aircraft, spacecraft) from place to place by determining position, course, and distance traveled. Navigation is, thus, concerned with finding the way, avoiding collision, meeting schedules etc. Examples of navigation are airplanes flying in non-transparent fog by means of satellite-based navigation devices, spacecraft getting their orientation by star observation (celestial navigation) etc. We can see navigation examples in our everyday life: the Global Positioning System helps us to drive the car en-route on complex highway road junctions. Even the children playing "warmer-colder" games are "navigating" each other.

The navigation assistance can be based on acoustic, visual and/or haptic guidance. For example, a simple GPS car navigator doesn't display a road map but only indicates the direction with arrows and/or with voice prompts. More complex devices display a map (an image) related to the current car position. We can say here, that this navigation is visual or image guided.

Visualization is a representation of some data, idea, phenomenon or process in a visual form for the perception by human eye. A photo, a geographic map and a movie are examples of visualization.

The navigation process can be performed with or without visualization. For example, a sailor can steer a ship with the use of maps or GPS, or he can navigate just trusting his memory or intuition as thousands of years ago. It is clear that visualization can greatly improve the quality of navigation, make it more accurate, fast and reliable.

In the same way, surgeons hitherto bound to anatomic landmarks, use navigation means to guide medical instruments inside the patient's body now. The first devices to navigate inside the human brain, the so-called stereotactic frames, just provided an external frame to match (to register) the image space and physical space to each other and served as a mechanical platform for an aiming device [17]. Such systems have not been based on visualization. Later, the invention of 3D image acquisition technologies paved the way for image-guided navigation.

By medical navigation we will refer to a process of guidance of medical instruments by means of an artificial acoustic, visual and/or haptic feedback. In this paper, we will specifically consider visualization-supported medical navigation; moreover we will take into consideration only computer-based visualization.

2 Visualization Methods

The phenomenon of human vision is a product of co-operation work of the vision organs, the eyes, and the brain. The human eye is, due to its nature, unable to perceive 3D images directly; it sees 2D images projected on the retina only. The 3D mental image can only be reconstructed in our brain, **Fig. 1**. On the other hand, a computer screen can display 2D images only (at least now).

There are two major problems concerning visualization: 1) The visualization must adequately represent the 3D data on a 2D display. 2) The visualization has to be as comprehensive as possible, to help the surgeons build their mental 3D patient model easily.

There are a number of different visualization methods to represent a 3D patient image on a 2D computer screen. Each method or technique has different strengths for navigation purposes. Some methods are fast but deliver pictures where spatial relations of details are difficult to see. Other methods work very well in some applications but tend to loose important details in other ones. Finally, some methods look almost perfect but are very slow or require special hardware.

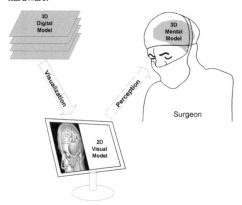

Fig. 1 Conversions of 2D and 3D patient data in visualization and perception

2.1 2D Projection

2D projection is a simple visualization method that represents one or several plane or specific surface sections of a volumetric dataset, **Fig. 2**. A section by plane can be parallel to the faces of the volume (orthogonal projection) or arbitrary. This visualization method is very fast, delivers the maximal available precision and is thus very common. The surgeon can easily define the specific points inside the patient model to measure extensions of tumors, lesions etc.

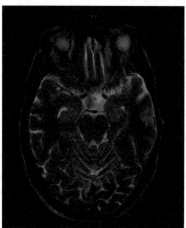

Fig. 2 2D orthogonal projection

The section of a volume dataset by plane in another direction than provided by CT or MR scanner (usually axial) is known as Multiplanar Reformat (MPR) technique. Beside the common sagittal and coronal orthogonal projections, any arbitrary (oblique) sections can be derived, **Fig. 3**. In some medical applications, such as endoscopy or vascular imaging, curved MPR can be very efficient.

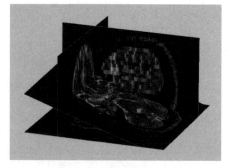

Fig. 3 Orthogonal and oblique projections

The main limitation of the 2D projection method is the inability to give a general view of a patient model. In other words, this method cannot provide a sense of depth, which challenges the surgeon's imagination. Other methods, also often referred to as 3D rendering techniques, are destined to overcome this limitation.

2.2 Surface Rendering

Surface rendering, also known as surface shading and indirect volume rendering, is a representation of a 3D object as a surface covering parts of the object with a specific voxel value (isosurface). The surface, typically a mesh of triangular cells, can be rendered on a computer screen by removing the hidden parts of the surface and by ignoring the z-coordinate. The perspective transformation can be applied to the image to emphasize the 3D effect, **Fig. 4**.

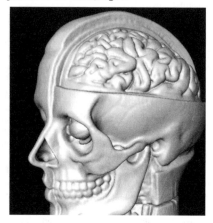

Fig. 4 An example of surface rendering

The creation of isosurfaces is known as surface extraction – and is usually a time-consuming process. The Marching Cubes algorithm [10] is the most popular surface extraction algorithm, but alternatives also exist [8, 9, 14]. The surface can contain hundreds of thousands of polygons. The more polygons constitute the surface, the more accurately it represents the real object, **Fig. 5**.

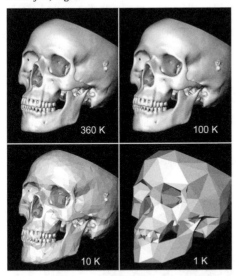

Fig. 5 Surfaces with different numbers of polygons: 360000, 100000, 10000 and 1000

Up-to-date computer systems usually support the hardware acceleration of polygon rendering that allows high-speed surface rendering visualization, even on a low-cost PC. Despite the high popularity of the surface rendering visualization, it has intrinsic limitations. Firstly, this method depicts the manner rather than the matter. It is impossible to see any details inside the isosurface, since all details are lost during the surface extraction. Secondly, the surface rendering technique is badly suitable for the visualization of amorphous cloud-like or fuzzy phenomena and smoothly varying flow fields. Thirdly, the method is geometry-sensitive, causing that the complexity of the extracted mesh can overwhelm the capabilities of the rendering system [11].

2.3 Volume Rendering

Volume rendering, also known as direct volume rendering, is a direct representation of a 3D volumetric dataset without intermediate conversion. The entire dataset is processed every time an image is displayed. Since all data are preserved, all the structures and internal details can be fully revealed, **Fig. 6**.

This technique conveys much more information than surface rendering, because each voxel contributes to the final image, **Fig. 7**. Another important feature of this method is its insensitivity to the complexity of the object. The volume rendering has become an invaluable visualization technique for a wide variety of applications [12].

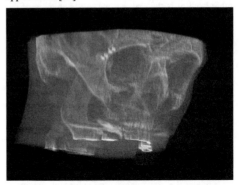

Fig. 6 Volume rendering visualization

Many physicians and radiologists have noted the principal advantages of the volume rendering technique over other methods in diverse medical areas, including musculoskeletal, thoracic, abdominal aortic applications [15, 7, 18], dental implantology [3] and others. However, the major drawback of this method is still the very high computational complexity.

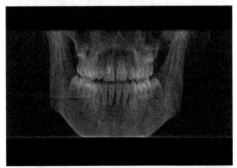

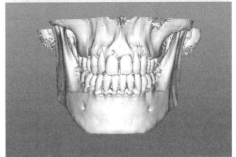

Fig. 7 Volume rendering (upper image) in comparison with surface rendering

There are a number of volume rendering algorithms; four of them have become the most popular ones: ray-casting, splatting, shear-warp and 3D texture mapping. The most common algorithm that delivers the best rendering quality is the ray-casting, but on the other hand, it is the slowest one. An evaluation of the popular volume rendering algorithms proved that none could be optimal for every case [11].

One should note, that an interesting simplification of the volume rendering method exists which is known as Maximal Intensity Projection (MIP). Here, not all the voxels along the ray cast through the volume, contribute to the final image, but the voxel with the maximal value only. Surprisingly, this method can provide contrast renderings of vascular structures, with no depth information, though.

The output of the volume rendering visualization significantly depends on the mapping defined by the color and transparency transfer functions. The mapping relates the voxel values into color and opacity. It is possible to highlight interesting parts of the patient model, e.g. bones, and make the unimportant details invisible. The creation of the transfer function, that provides optimal viewing, is not trivial and highly depends on the structures of interest, **Fig. 8**.

Medical navigation based on visualization has to be fast enough also to reflect the actual position of medical instruments. It is difficult to define a criterion for the real-time medical navigation, but such a criterion exists for the real-time visualization – an image on the screen should be updated faster than 24 frames per second [13]. The condition of real-time operating limits the use of slow visualization methods.

The 2D projection visualization method can be implemented for the real-time navigation on a low-cost modern PC without any special hardware. The application point of a medical instrument can be represented on a 2D projection as an intersection of thin lines related to the current instrument location. A single 2D projection viewport, i.e. visualization window, can be enough for navigation; however, the presence of more projections in different projecting directions, i.e. MPR is desirable, **Fig. 9**. The contents of the viewports should be constantly updated during the navigation to reflect the movement of the instrument in the direction of the projections. This looks like turning over the pages of a book.

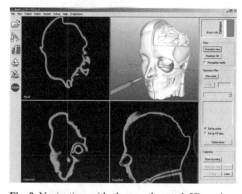

Fig. 9 Navigation with three orthogonal 2D projection viewports (axial, coronal and sagittal) and surface-rendering viewport

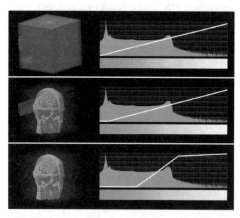

Fig. 8 Volume rendering visualization with different transfer functions

3 Visualization-Based Medical Navigation

One important aspect of the navigation is the providing of constantly updated position data. The position data must be acquired and displayed fast enough to perform the navigation. This condition is usually referred to as real-time operating. Real-time navigation stands for "fast enough" rather than simply "fast". For example, an aircraft location on a radar screen is updated every 5-10 seconds roughly, which is enough to navigate the airplane during the flight; but a landing radar should work much faster.

The surface rendering visualization method exploits a polygonal representation of the volume that requires a hardware acceleration to be rendered fast. Luckily, most modern computer video boards are readily equipped with inexpensive polygon-graphic accelerators, whose development has been spurred on by the computer games industry. Interactive frame rates can be achieved easily for the surface models with hundreds of thousands of polygons. So, the surface rendering representation is well suited for the navigation from the timing point of view. On the other hand, only surface data may be viewed here, that makes this visualization method almost useless for the navigation inside the patient body. Despite this limitation, the method can be used for some medical applications where surface information is important, such as traumatology, implantology, colonoscopy and others. Some researchers report that surgeons can prefer to view a portion of the skin surface as a landmark that can be useful for navigation [4].

A medical instrument can be displayed on a surface rendering viewport as a simple 3D model, e.g. as a cone or needle. Graphical libraries, such as OpenGL or Direct3D provide powerful features for the implementation of surface rendering.

The volume rendering visualization methods were initially considered not fast enough to implement real-time navigation; therefore, they were solely used for static renderings. Many efforts have been made to implement interactive volume rendering [2, 16], but it seems that high-quality rendering can be achieved only with hardware acceleration. The performance of computers is increasing fast, but the CT and MR scanners produce huge amounts of data that tend to increase, too. So, simplifications of the volume rendering, e.g. MIP visualization, can be useful in many applications. The implementation of MIP algorithms can achieve the interactive frame rate.

The high demand of fast volume rendering has forced the development of specific hardware at reasonable prices [12]. These devices perform the volume rendering in near-real-time for the volume 512x512x512; that is sufficient for the implementation of navigation.

Another aspect of the visualization interactivity is also important. When the visualization is interactive, the physician can rotate the volume data during examination or navigation. Observing the volume during the rotation provides an excellent sense of depth, which is important for the navigation. This phenomenon is known as kinetic depth effect [6].

The current position of a medical instrument on a volume rendering viewport can be displayed with a simple model as on a surface rendering window, **Fig. 10**, or as a 3D line cursor. It should be noted that such an embedding of a polygonal instrument model into the volumetric patient model requires an intensive information exchange between these models.

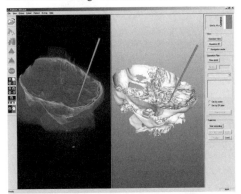

Fig. 10 Navigation with volume (left) and surface rendering

4 Conclusions

We can summarize the main features of the visualization methods mentioned above in one table to compare the advantages and drawbacks of each particular method, **Table 1**.

Table 1. A comparison of visualization methods

	2D Projection	Surface Rendering	Volume Rendering
Adequacy of visualization	good	poor	good
Navigation accuracy	good	mediocre	average
Sense of depth	poor	good	good
Measurements	good	difficult	difficult
Real-time implementation	yes	yes	with specific hardware only
Preprocessing is required	for MPR only	yes	no
Output is adjustable	yes	poorly	yes
Cost	low	low	high

As one can see, there is no ideal method that is optimal for every use. Currently, the volume rendering method has the widest future potential and will be developed and widespread further [1], whereas the surface rendering visualization seems to have achieved its maturity already.

Since each of the visualization methods discussed above has its own advantages and drawbacks, it is reasonable to integrate every visualization technique in one combined navigation system. This approach allows the exploitation of the advantages of every particular method and has proved to be efficient [4, 5].

5 Acknowledgement

The authors thank the Foundation Rhineland-Palatinate for Innovation for the sponsoring of the project "Robotic Manipulation under Augmented Navigation for Tooth-Implant Concepts" (ROMANTIC).

6 References

[1] Bartz, D.; Meissner, M.: Voxels versus Polygons: A Comparative Approach for Volume Graphics. Proceedings Volume Graphics '99, pp. 33-48, 1999

[2] Brady, M. L.; Jung, K. K.; Nguyen, H. T.; Nguen, T. P.: Interactive Volume Navigation. IEEE Transactions on Visualization and Computer Graphics, Vol. 4, No. 3, pp. 243-256, July-September 1998

[3] Cavalcanti, M. G.; Ruprecht, A.; Vannier, M. W.: 3D Volume Rendering Using Multislice CT for Dental Implants. Dentomaxillofacial Radiology, 31, pp. 218-223, 2002

[4] Gering, D. T.; Nabavi, A.; Kikinis R.; Hata N.; O'Donnel, L. J.; Grimson, W. E. L.; Jolesz, F. A.; Black, P. M.; Wells, W. M.: An Integrated Visualization System for Surgical Planning and Guidance Using Image Fusion and an Open MR. Journal of Magnetic Resonance Imaging, Vol. 13, pp. 967-975, 2001

[5] Gubaidullin, G. G.; Evbatyrov, R. R.: A Multi-Purpose Setup for Medical Robotic Experiments. Proceedings of the International Conference on Computing, Communications and Control Technologies CCCT '04, Austin, Texas, 2004

[6] Kaufman, A.; Dachille, F.; Chen, B.; Bitter, I.; Kreeger, K.; Zhang, N.; Tang, Q.: Real-Time Volume Rendering. International Journal of Imaging Systems and Technology, Vol.11, pp. 44-52, 2000

[7] Lawler, L. P.; Fishman, E. K.: Multi-Detector Row CT of Thoracic Disease with Emphasis on 3D Volume Rendering and CT Angiography. Radio Graphics 2001, Vol. 21, pp. 1257-1273

[8] Levoy, M.; Whitted, T.: The Use of Points as Display Primitives. Technical report, The University of North Carolina at Chapel Hill, Department of Computer Science, 1985

[9] Levoy, M.: Display of Surfaces from Volume Data. Computer Science Department, IEEE Computer Graphics and Applications, Vol. 8, No. 5, pp. 29-37, 1988

[10] Lorensen, W.; Cline, H.: Marching Cubes: A High Resolution 3D Surface Construction Algorithm. Computer Graphics, Vol. 21, No. 4, July, 1987, pp. 163-169

[11] Meissner, M.; Huang J.; Bartz, D.; Mueller, K.; Crawfis, R.: A Practical Evaluation of Popular Volume Rendering Algorithms. Proceedings of the 2000 IEEE Volume Visualization, pp. 81-90, 2000

[12] Pfister, H.; Hardenbergh, J.; Knittel, J.; Lauer, H.; Seiler, L.: The VolumePro Real-Time Ray-Casting System. Computer Graphics, SIGGRAPH 99 Proceedings, pp. 251-260. Los Angeles, CA, August 1999

[13] Pfister, H.: VolumePro: At the Frontier of Advanced 3D Graphics. Nikkei Science, 1999

[14] Poston, T.; Wong, T.-T.; Heng, P.-A.: Multiresolution Isosurface Extraction with Adaptive Skeleton Climbing. Computer Graphics Forum, Vol. 17, No. 3, pp. 137-148, September 1998

[15] Pretorius, E. S.; Fishman, E. K.: Volume-rendered Three-dimensional Spiral CT: Musculoskeletal Applications. Radio Graphics, 1999, Vol. 19, pp. 1143-1160

[16] Roettger, S.; Guthe, S.; Weiskopf, D.; Ertl, T.; Strasser, W.: Smart Hardware-Accelerated Volume Rendering. IEEE TCVG Symposium on Visualization VisSym '03, pp. 231-238, 2003

[17] Schiffbauer, H.: Neuronavigation in Brain Tumor Surgery. Academic Dissertation. Department of Neurosurgery at the Oulu University, Oulu, 1999

[18] Tam, R. C.; Healey, C. G.; Flak, B.; Cahoon, P.: Volume Rendering of Abdominal Aortic Aneurysms. Proceedings IEEE Visualization '97

Euler Angles and Quaternions in Robotics

Gail G. Gubaidullin

Dept. of Mathematics and Technology, RheinAhrCampus Remagen, University of Applied Sciences Koblenz, Suedallee 2, D-53424 Remagen, Germany

Abstract

In robotics Euler angles and quaternions are widely used for the description of the spatial orientation of rigid bodies. The orientation of the end-effectors of many industrial robots is described with the Euler angles. Diverse location sensors use the quaternions for the description of orientation, e. g. optical sensors in the field of medical robotics. In this paper, various Euler angle sets are discussed and universal algorithms for extracting the Euler angles from a rotation matrix and from a rotation quaternion are described.

1 Introduction

In robotics the position and orientation of various objects (the robot links, tools, workpieces etc.) in three-dimensional space are constantly considered. In order to describe the position and orientation of a rigid body object, a coordinate system, or frame, is rigidly attached to the object; then, the position and orientation of this frame relative to other frames are described. A number of different mathematical quantities are used for the description of position and orientation of the frames [1, 2].

In this paper, the description of orientation, especially with Euler angles and quaternions is considered. Various Euler angle sets are discussed and universal algorithms for extracting the Euler angles from a rotation matrix and from a rotation quaternion are described.

2 Rotation vectors, angles and matrices

The easiest way to represent a rotation is to give the rotation axis L with its unit vector $l = [l_x, l_y, l_z]^T$ and the rotation angle α. For a rotation of a frame $\{A\}$ about L by α we will write $Rot(A, L, \alpha)$. The orientation of $\{A\}$ relative to $\{B\}$ can be described with a rotation vector and angle ${}_B^A(l, \alpha)$ corresponding to a rotation through which $\{B\}$ would reach the orientation of $\{A\}$. The description of orientation with the rotation vectors and angles is clear and compact, but this method has no adequate algebra for the description of combined rotations and rotation sequences.

Another method, which has such an algebra, is the description of orientation with rotation matrices. Any rotation can be described with a 3x3 rotation matrix

$$\mathbf{R} = \begin{bmatrix} r_{11} & r_{12} & r_{13} \\ r_{21} & r_{22} & r_{23} \\ r_{31} & r_{32} & r_{33} \end{bmatrix}. \quad (1)$$

The rotation matrix corresponding to (l, α) is

$$\mathbf{R}_{L,\alpha} = \quad (2)$$

$$\begin{bmatrix} l_x l_x V\alpha + C\alpha & l_y l_x V\alpha - l_z S\alpha & l_z l_x V\alpha + l_y S\alpha \\ l_x l_y V\alpha + l_z S\alpha & l_y l_y V\alpha + C\alpha & l_z l_y V\alpha - l_x S\alpha \\ l_x l_z V\alpha - l_y S\alpha & l_y l_z V\alpha + l_x S\alpha & l_z l_z V\alpha + C\alpha \end{bmatrix}$$

with $S\alpha = \sin\alpha$, $C\alpha = \cos\alpha$, $V\alpha = 1 - C\alpha$. For the rotation matrix which describes the orientation of $\{A\}$ relative to $\{B\}$ we will write ${}_B^A\mathbf{R}$. If $\{A\}$, firstly coincident with $\{B\}$, makes a rotation $Rot(A, L, \alpha)$, then ${}_B^A\mathbf{R} = \mathbf{R}_{L,\alpha}$. Important special cases of (2) are the rotation matrices which describe rotations about the frame axes:

$$\mathbf{R}_{X,\alpha} = \begin{bmatrix} 1 & 0 & 0 \\ 0 & C\alpha & -S\alpha \\ 0 & S\alpha & C\alpha \end{bmatrix},$$

$$\mathbf{R}_{Y,\alpha} = \begin{bmatrix} C\alpha & 0 & S\alpha \\ 0 & 1 & 0 \\ -S\alpha & 0 & C\alpha \end{bmatrix}, \quad (3)$$

$$\mathbf{R}_{Z,\alpha} = \begin{bmatrix} C\alpha & -S\alpha & 0 \\ S\alpha & C\alpha & 0 \\ 0 & 0 & 1 \end{bmatrix}.$$

3 Rotation matrices and Euler angles

3.1 Euler angle set conventions

Any rotation matrix $\mathbf{R}$ can be represented as a product of three rotation matrices (3):

$$\mathbf{R} = \mathbf{R}_{IJK}(\varphi,\theta,\psi) = \mathbf{R}_{I,\varphi} \cdot \mathbf{R}_{J,\theta} \cdot \mathbf{R}_{K,\psi} \quad (4)$$

with $I, J, K \in \{X, Y, Z\}$, $I \neq J$, $J \neq K$. The three rotation angles φ, θ, ψ in (4) are called $I-J-K$ Euler angles and can be used for the description of the same rotation or orientation as $\mathbf{R}$.

There are six $I-J-K$ Euler angle sets with $I \neq J$, $J \neq K$, $K \neq I$:

$X-Y-Z$, $X-Z-Y$, $Y-X-Z$,
$Y-Z-X$, $Z-X-Y$, $Z-Y-X$.

In addition there are six $I-J-I$ Euler angle sets with $I \neq J$:

$X-Y-X$, $X-Z-X$, $Y-X-Y$,
$Y-Z-Y$, $Z-X-Z$, $Z-Y-Z$.

Any Euler angle set can be illustrated with various three-rotation sequences. For example, the $Z-Y-X$ Euler angle set can be illustrated with the help of two frames $\{A\}$ and $\{B\}$, which are firstly coincident, and then make one of the following rotation sequences:

1) $Rot(A, Z_A, \varphi), Rot(A, Y_A, \theta), Rot(A, X_A, \psi)$
2) $Rot(A, X_B, \psi), Rot(A, Y_B, \theta), Rot(A, Z_B, \varphi)$
3) $Rot(A, Y_A, \theta), Rot(A, X_A, \psi), Rot(A, Z_B, \varphi)$
4) $Rot(B, Z_B, -\varphi), Rot(A, Y_A, \theta), Rot(A, X_A, \psi)$

After any of these rotation sequences (and this list can be continued) we obtain the same orientation of $\{A\}$ relative to $\{B\}$:

$${}^A_B\mathbf{R} = \mathbf{R}_{ZYX}(\varphi,\theta,\psi) = \mathbf{R}_{Z,\varphi} \cdot \mathbf{R}_{Y,\theta} \cdot \mathbf{R}_{X,\psi} \quad (5)$$

3.2 Extracting the Euler angles from a rotation matrix

Given a rotation matrix $\mathbf{R}$, the equivalent $I-J-K$ Euler angles $(I \neq J, J \neq K)$ can be computed as follows, see [2]. Multiplying (4) from the left by the inverse rotation matrix $\mathbf{R}_{I,\varphi}^{-1}$, we obtain:

$$\mathbf{R}_{I,\varphi}^{-1} \cdot \mathbf{R} = \mathbf{R}_{J,\theta} \cdot \mathbf{R}_{K,\psi} \quad (6)$$

The matrix equation (6) consists of nine scalar equations with three unknown Euler angles φ, θ, ψ. One of these equations gives the value of $S\varphi/C\varphi$ and therefore two $Atan2$ formulas for the angle φ. Four other of these nine equations give $S\theta$, $C\theta$, $S\psi$ and $C\psi$ in dependency of φ and consequently the angles θ and ψ in the form of $\theta = Atan2(S\theta, C\theta)$, $\psi = Atan2(S\psi, C\psi)$.

Depending on the given rotation matrix $\mathbf{R}$ there can be either two solutions (*general case*) or an infinite set of solutions (*special cases*) for the unknown Euler angles φ, θ, ψ. An analysis of all twelve Euler angle sets shows that for any $I-J-K$ Euler angle set $(I \neq J, J \neq K)$ it can be decided whether this is the general or a special case with the help of the element r_{ik} of the matrix $\mathbf{R}$ where i is the number of the axis I (if $I = X$, then $i = 1$; if $I = Y$, then $i = 2$; if $I = Z$, then $i = 3$) and k is the number of the axis K in the name $I-J-K$ of the Euler angle set.

For any $I-J-K$ Euler angle set with $I \neq J$, $J \neq K$, $K \neq I$, it can be shown that:

a) if $|r_{ik}| \neq 1$, then this is the general case, i.e. there are two solutions $(\varphi_1, \theta_1, \psi_1)$ and $(\varphi_2, \theta_2, \psi_2)$ with $\varphi_2 = \varphi_1 \pm \pi$, $\theta_2 = \theta_1 \pm \pi$, $\psi_2 = \psi_1 \pm \pi$;

b) if $r_{ik} = 1$ or $r_{ik} = -1$, then this is a special case, i.e. there is an infinite set of solutions with

$$\theta = \frac{\pi}{2} \quad \text{or} \quad \theta = -\frac{\pi}{2}, \quad \text{and}$$

$$\varphi + \psi = const \quad \text{or} \quad \varphi - \psi = const.$$

For any $I-J-I$ Euler angle set with $I \neq J$, it can be shown that:

a) if $|r_{ii}| \neq 1$, then this is the general case, i.e. there are two solutions $(\varphi_1, \theta_1, \psi_1)$ and $(\varphi_2, \theta_2, \psi_2)$ with $\varphi_2 = \varphi_1 \pm \pi$, $\theta_2 = -\theta_1$, $\psi_2 = \psi_1 \pm \pi$;

b) if $r_{ii} = 1$ or $r_{ii} = -1$, then this is a special case, i.e. there is an infinite set of solutions with

$$\theta = 0 \quad \text{or} \quad \theta = \pi, \quad \text{and}$$

$$\varphi + \psi = const \quad \text{or} \quad \varphi - \psi = const.$$

As examples, we consider the extracting of the $Z-Y-X$ and $Z-Y-Z$ Euler angles from a rotation matrix.

3.3 Z-Y-X Euler angles

The $Z-Y-X$ Euler angles are related to the equivalent rotation matrix through

$$\mathbf{R} = \mathbf{R}_{ZYX}(\varphi,\theta,\psi) = \mathbf{R}_{Z,\varphi} \cdot \mathbf{R}_{Y,\theta} \cdot \mathbf{R}_{X,\psi}. \quad (7)$$

In order to extract the $Z-Y-X$ Euler angles from a given rotation matrix $\mathbf{R}$ with the method described above, we multiply (7) from the left by the inverse rotation matrix $\mathbf{R}_{Z,\varphi}^{-1}$:

$$\mathbf{R}_{Z,\varphi}^{-1} \cdot \mathbf{R} = \mathbf{R}_{Y,\theta} \cdot \mathbf{R}_{X,\psi},$$

$$\begin{bmatrix} C\varphi & S\varphi & 0 \\ -S\varphi & C\varphi & 0 \\ 0 & 0 & 1 \end{bmatrix} \cdot \begin{bmatrix} r_{11} & r_{12} & r_{13} \\ r_{21} & r_{22} & r_{23} \\ r_{31} & r_{32} & r_{33} \end{bmatrix} =$$

$$= \begin{bmatrix} C\theta & 0 & S\theta \\ 0 & 1 & 0 \\ -S\theta & 0 & C\theta \end{bmatrix} \cdot \begin{bmatrix} 1 & 0 & 0 \\ 0 & C\psi & -S\psi \\ 0 & S\psi & C\psi \end{bmatrix}. \quad (8)$$

From (8) we get nine equations with three unknowns: φ, θ, ψ. Five of these nine equations are:

$$-S\varphi \cdot r_{11} + C\varphi \cdot r_{21} = 0, \quad (9a)$$
$$r_{31} = -S\theta, \quad (9b)$$
$$C\varphi \cdot r_{11} + S\varphi \cdot r_{21} = C\theta, \quad (9c)$$
$$-S\varphi \cdot r_{13} + C\varphi \cdot r_{23} = -S\psi, \quad (9d)$$
$$-S\varphi \cdot r_{12} + C\varphi \cdot r_{22} = C\psi. \quad (9e)$$

The equation (9a) gives for the angle φ:

$$\varphi = Atan2(r_{21}, r_{11}), \quad (10a)$$
$$\varphi = Atan2(-r_{21}, -r_{11}). \quad (10b)$$

The equations (9b, c) give for the angle θ:

$$\theta = Atan2(-r_{31}, C\varphi \cdot r_{11} + S\varphi \cdot r_{21}). \quad (10c)$$

And the equations (9d, e) give for the angle ψ:

$$\psi = Atan2(S\varphi \cdot r_{13} - C\varphi \cdot r_{23},$$
$$-S\varphi \cdot r_{12} + C\varphi \cdot r_{22}). \quad (10d)$$

General case

If $|r_{31}| \neq 1$, then there are two solutions $(\varphi_1, \theta_1, \psi_1)$ and $(\varphi_2, \theta_2, \psi_2)$ with:

$$\varphi_1 = Atan2(r_{21}, r_{11});$$
$$\theta_1 = Atan2(-r_{31}, C\varphi_1 \cdot r_{11} + S\varphi_1 \cdot r_{21});$$
$$\psi_1 = Atan2(S\varphi_1 \cdot r_{13} - C\varphi_1 \cdot r_{23},$$
$$-S\varphi_1 \cdot r_{12} + C\varphi_1 \cdot r_{22}); \quad (11)$$

$$\varphi_2 = Atan2(-r_{21}, -r_{11});$$
$$\theta_2 = Atan2(-r_{31}, C\varphi_2 \cdot r_{11} + S\varphi_2 \cdot r_{21});$$

$$\psi_2 = Atan2(S\varphi_2 \cdot r_{13} - C\varphi_2 \cdot r_{23},$$
$$-S\varphi_2 \cdot r_{12} + C\varphi_2 \cdot r_{22}). \quad (12)$$

Special cases

1) If $r_{31} = 1$, then $r_{11} = r_{21} = 0$, both (10a) and (10b) give $\varphi = Atan2(0, 0)$, i.e. the angle φ can be chosen arbitrarily, and (10c) gives $\theta = -\pi/2$. By substituting the θ in (7) we obtain that

$$\mathbf{R} = \begin{bmatrix} 0 & -S(\varphi+\psi) & -C(\varphi+\psi) \\ 0 & C(\varphi+\psi) & -S(\varphi+\psi) \\ 1 & 0 & 0 \end{bmatrix},$$

and there is an infinite set of solutions with

$$\theta = -\pi/2, \quad \varphi+\psi = Atan2(-r_{12}, r_{22}). \quad (13)$$

2) If $r_{31} = -1$, then, analogously, $\theta = \pi/2$,

$$\mathbf{R} = \begin{bmatrix} 0 & -S(\varphi-\psi) & C(\varphi-\psi) \\ 0 & C(\varphi-\psi) & S(\varphi-\psi) \\ -1 & 0 & 0 \end{bmatrix},$$

and there is an infinite set of solutions with

$$\theta = \pi/2, \quad \varphi-\psi = Atan2(-r_{12}, r_{22}). \quad (14)$$

3.4 Z-Y-Z Euler angles

The $Z-Y-Z$ Euler angles are related to the equivalent rotation matrix through

$$\mathbf{R} = \mathbf{R}_{ZYZ}(\varphi,\theta,\psi) = \mathbf{R}_{Z,\varphi} \cdot \mathbf{R}_{Y,\theta} \cdot \mathbf{R}_{Z,\psi}. \quad (15)$$

The extracting of the $Z-Y-Z$ Euler angles from a given rotation matrix $\mathbf{R}$ is analogous to that of the $Z-Y-X$ Euler angles described above, so we give only the results:

General case

If $|r_{33}| \neq 1$, then there are two solutions $(\varphi_1, \theta_1, \psi_1)$ and $(\varphi_2, \theta_2, \psi_2)$ with:

$$\varphi_1 = Atan2(r_{23}, r_{13});$$
$$\theta_1 = Atan2(C\varphi_1 \cdot r_{13} + S\varphi_1 \cdot r_{23}, r_{33});$$
$$\psi_1 = Atan2(-S\varphi_1 \cdot r_{11} + C\varphi_1 \cdot r_{21},$$
$$-S\varphi_1 \cdot r_{12} + C\varphi_1 \cdot r_{22}); \quad (16)$$

$$\varphi_2 = Atan2(-r_{23}, -r_{13});$$
$$\theta_2 = Atan2(C\varphi_2 \cdot r_{13} + S\varphi_2 \cdot r_{23}, r_{33});$$
$$\psi_2 = Atan2(-S\varphi_2 \cdot r_{11} + C\varphi_2 \cdot r_{21},$$
$$-S\varphi_2 \cdot r_{12} + C\varphi_2 \cdot r_{22}). \quad (17)$$

Special cases

1) If $r_{33} = 1$, then $\theta = 0$,

$$\mathbf{R} = \begin{bmatrix} C(\varphi+\psi) & -S(\varphi+\psi) & 0 \\ S(\varphi+\psi) & C(\varphi+\psi) & 0 \\ 0 & 0 & 1 \end{bmatrix},$$

and there is an infinite set of solutions with

$$\theta = 0, \quad \varphi + \psi = Atan2(-r_{12}, r_{22}). \tag{18}$$

2) If $r_{33} = -1$, then $\theta = \pi$,

$$\mathbf{R} = \begin{bmatrix} -C(\varphi-\psi) & -S(\varphi-\psi) & 0 \\ -S(\varphi-\psi) & C(\varphi-\psi) & 0 \\ 0 & 0 & -1 \end{bmatrix},$$

and there is an infinite set of solutions with

$$\theta = \pi, \quad \varphi - \psi = Atan2(-r_{12}, r_{22}). \tag{19}$$

4 Euler angles and rotation quaternions

4.1 Rotation quaternions

Beside the matrix algebra, the algebra of four-dimensional numbers, or quaternions, is used for the description of orientation. A quaternion $\mathbf{Q}$ is

$$\mathbf{Q} = s + i \cdot a + j \cdot b + k \cdot c \tag{20}$$

with s, a, b, c – four real numbers; i, j, k – three different imaginary units. A rotation (l, α) can be described with the unit quaternion

$$\mathbf{Q}_{L,\alpha} = C\frac{\alpha}{2} + i \cdot l_x S\frac{\alpha}{2} + j \cdot l_y S\frac{\alpha}{2} + k \cdot l_z S\frac{\alpha}{2} \tag{21}$$

called rotation quaternion. For the rotation quaternion which describes the orientation of $\{A\}$ relative to $\{B\}$ we will write $^A_B\mathbf{Q}$. If $\{A\}$, firstly coincident with $\{B\}$, makes a rotation $Rot(A, L, \alpha)$, then $^A_B\mathbf{Q} = \mathbf{Q}_{L,\alpha}$. Special cases of (21) are the rotation quaternions describing rotations about the frame axes:

$$\mathbf{Q}_{X,\alpha} = C\frac{\alpha}{2} + i \cdot S\frac{\alpha}{2},$$

$$\mathbf{Q}_{Y,\alpha} = C\frac{\alpha}{2} + j \cdot S\frac{\alpha}{2}, \tag{22}$$

$$\mathbf{Q}_{Z,\alpha} = C\frac{\alpha}{2} + k \cdot S\frac{\alpha}{2}.$$

4.2 Extracting the Euler angles from a rotation quaternion

Corresponding to (4), any rotation quaternion $\mathbf{Q}$ can be represented as a product of three rotation quaternions (22):

$$\mathbf{Q} = \mathbf{Q}_{IJK}(\varphi, \theta, \psi) = \mathbf{Q}_{I,\varphi} \circ \mathbf{Q}_{J,\theta} \circ \mathbf{Q}_{K,\psi} \tag{23}$$

with $I, J, K \in \{X, Y, Z\}$, $I \neq J$, $J \neq K$.

Given a rotation quaternion $\mathbf{Q}$, the equivalent $I-J-K$ Euler angles ($I \neq J, J \neq K$) can be computed analogously to the extracting from a rotation matrix described above. Multiplying (23) from the left by the inverse rotation quaternion $\mathbf{Q}_{I,\varphi}^{-1}$, we obtain:

$$\mathbf{Q}_{I,\varphi}^{-1} \circ \mathbf{Q} = \mathbf{Q}_{J,\theta} \circ \mathbf{Q}_{K,\psi}. \tag{24}$$

The quaternion equation (24) consists of four scalar equations with three unknown Euler angles φ, θ, ψ. From these four equations and their squares the Euler angles φ, θ, ψ can be computed both in general and special cases.

Another way to compute the Euler angles from a given rotation quaternion $\mathbf{Q}$ is to calculate the corresponding rotation matrix

$$\mathbf{R} = \tag{25}$$

$$\begin{bmatrix} 2(a^2+s^2)-1 & 2(ba-cs) & 2(ca+bs) \\ 2(ab+cs) & 2(b^2+s^2)-1 & 2(cb-as) \\ 2(ac-bs) & 2(bc+as) & 2(c^2+s^2)-1 \end{bmatrix}$$

and then to extract the Euler angles from $\mathbf{R}$. In the general case, this algorithm gives the same results as the described quaternion method. But in the special cases, the quaternion method provides more computationally effective solutions.

As examples, we consider the extracting of the $Z-Y-X$ and $Z-Y-Z$ Euler angles from a rotation quaternion.

4.3 Z-Y-X Euler angles

The $Z-Y-X$ Euler angles are related to the equivalent rotation quaternion through

$$\mathbf{Q} = \mathbf{Q}_{ZYX}(\varphi, \theta, \psi) = \mathbf{Q}_{Z,\varphi} \circ \mathbf{Q}_{Y,\theta} \circ \mathbf{Q}_{X,\psi}. \tag{26}$$

In order to extract the $Z-Y-X$ Euler angles from a given rotation quaternion $\mathbf{Q}$ with the method described above, we multiply (26) from the left by the inverse rotation quaternion $\mathbf{Q}_{Z,\varphi}^{-1}$:

$$\mathbf{Q}_{Z,\varphi}^{-1} \circ \mathbf{Q} = \mathbf{Q}_{Y,\theta} \circ \mathbf{Q}_{X,\psi},$$

$$\left(C\frac{\varphi}{2} - kS\frac{\varphi}{2}\right) \circ (s + ia + jb + kc) =$$
$$= \left(C\frac{\theta}{2} + jS\frac{\theta}{2}\right) \circ \left(C\frac{\psi}{2} + iS\frac{\psi}{2}\right) \quad (27)$$

From (27) we obtain four equations with three unknowns: φ, θ, ψ:

$$C\frac{\varphi}{2} \cdot s + S\frac{\varphi}{2} \cdot c = C\frac{\theta}{2} \cdot C\frac{\psi}{2}, \quad (28a)$$

$$C\frac{\varphi}{2} \cdot a + S\frac{\varphi}{2} \cdot b = C\frac{\theta}{2} \cdot S\frac{\psi}{2}, \quad (28b)$$

$$C\frac{\varphi}{2} \cdot b - S\frac{\varphi}{2} \cdot a = S\frac{\theta}{2} \cdot C\frac{\psi}{2}, \quad (28c)$$

$$C\frac{\varphi}{2} \cdot c - S\frac{\varphi}{2} \cdot s = -S\frac{\theta}{2} \cdot S\frac{\psi}{2}. \quad (28d)$$

From (28a)·(28d) + (28b)·(28c) we get:

$$C\varphi \cdot (ab + cs) - S\varphi \cdot (a^2 + s^2 - 0.5) = 0, \quad (29a)$$

from (28a)·(28c) − (28b)·(28d):

$$bs - ac = 0.5 \cdot S\theta, \quad (29b)$$

from (28a)^2 + (28b)^2 − (28c)^2 − (28d)^2:

$$C\varphi \cdot (a^2 + s^2 - 0.5) + S\varphi \cdot (ab + cs) =$$
$$= 0.5 \cdot C\theta \quad (29c)$$

from (28a)·(28b) − (28c)·(28d):

$$S\varphi \cdot (ca + bs) - C\varphi \cdot (cb - as) = 0.5 \cdot S\psi, \quad (29d)$$

and from (28a)^2 − (28b)^2 + (28c)^2 − (28d)^2:

$$C\varphi \cdot (b^2 + s^2 - 0.5) - S\varphi \cdot (ba - cs) =$$
$$= 0.5 \cdot C\psi \quad (29e)$$

The equation (29a) gives for the angle φ:

$$\varphi = Atan2(ab + cs, a^2 + s^2 - 0.5), \quad (30a)$$

$$\varphi = Atan2(-ab - cs, -a^2 - s^2 + 0.5). \quad (30b)$$

The equations (29b, c) give for the angle θ:

$$\theta = Atan2[bs - ac,$$
$$C\varphi \cdot (a^2 + s^2 - 0.5) + S\varphi \cdot (ab + cs)]. \quad (30c)$$

And the equations (29d, e) give for the angle ψ:

$$\psi = Atan2[S\varphi \cdot (ca + bs) - C\varphi \cdot (cb - as),$$
$$C\varphi \cdot (b^2 + s^2 - 0.5) - S\varphi \cdot (ba - cs)]. \quad (30d)$$

General case

If $|ac - bs| \neq 0.5$, then there are two solutions $(\varphi_1, \theta_1, \psi_1)$ and $(\varphi_2, \theta_2, \psi_2)$ with:

$$\varphi_1 = Atan2(ab + cs, a^2 + s^2 - 0.5);$$
$$\theta_1 = Atan2[bs - ac,$$
$$C\varphi_1 \cdot (a^2 + s^2 - 0.5) + S\varphi_1 \cdot (ab + cs)];$$
$$\psi_1 = Atan2[S\varphi_1 \cdot (ca + bs) - C\varphi_1 \cdot (cb - as),$$
$$C\varphi_1 \cdot (b^2 + s^2 - 0.5) - S\varphi_1 \cdot (ba - cs)]; \quad (31)$$

$$\varphi_2 = Atan2(-ab - cs, -a^2 - s^2 + 0.5);$$
$$\theta_2 = Atan2[bs - ac,$$
$$C\varphi_2 \cdot (a^2 + s^2 - 0.5) + S\varphi_2 \cdot (ab + cs)];$$
$$\psi_2 = Atan2[S\varphi_2 \cdot (ca + bs) - C\varphi_2 \cdot (cb - as),$$
$$C\varphi_2 \cdot (b^2 + s^2 - 0.5) - S\varphi_2 \cdot (ba - cs)]. \quad (32)$$

Special cases

1) If $ac - bs = 0.5$, then $\theta = -\pi/2$. By substituting the θ in (26) we obtain that

$$\mathbf{Q} = \frac{1-j}{\sqrt{2}} \cdot C\left(\frac{\varphi + \psi}{2}\right) + \frac{i+k}{\sqrt{2}} \cdot S\left(\frac{\varphi + \psi}{2}\right),$$

and there is an infinite set of solutions with

$$\theta = -\pi/2, \quad \varphi + \psi = 2 \cdot Atan2(c, s). \quad (33)$$

2) If $ac - bs = -0.5$, then $\theta = \pi/2$,

$$\mathbf{Q} = \frac{1+j}{\sqrt{2}} \cdot C\left(\frac{\varphi - \psi}{2}\right) + \frac{k-i}{\sqrt{2}} \cdot S\left(\frac{\varphi - \psi}{2}\right),$$

and there is an infinite set of solutions with

$$\theta = \pi/2, \quad \varphi - \psi = 2 \cdot Atan2(c, s). \quad (34)$$

4.4   Z-Y-Z Euler angles

The $Z - Y - Z$ Euler angles are related to the equivalent rotation quaternion through

$$\mathbf{Q} = \mathbf{Q}_{ZYZ}(\varphi, \theta, \psi) = \mathbf{Q}_{Z,\varphi} \circ \mathbf{Q}_{Y,\theta} \circ \mathbf{Q}_{Z,\psi}. \quad (35)$$

The extracting of the $Z - Y - Z$ Euler angles from a given rotation quaternion $\mathbf{Q}$ is analogous to that of the $Z - Y - X$ Euler angles described above, so we give only the results:

General case

If $|2(c^2+s^2)-1| \neq 1$, then there are two solutions $(\varphi_1, \theta_1, \psi_1)$ and $(\varphi_2, \theta_2, \psi_2)$ with:

$\varphi_1 = Atan2(cb-as,\ ca+bs);$
$\theta_1 = Atan2[C\varphi_1 \cdot (ca+bs) + S\varphi_1 \cdot (cb-as),$
$\quad c^2+s^2-0.5];$
$\psi_1 = Atan2[-S\varphi_1 \cdot (a^2+s^2-0.5) +$
$\quad +C\varphi_1 \cdot (ab+cs),\ -S\varphi_1 \cdot (ba-cs) +$
$\quad +C\varphi_1 \cdot (b^2+s^2-0.5)];$ (36)

$\varphi_2 = Atan2(as-cb,\ -ca-bs);$
$\theta_2 = Atan2[C\varphi_2 \cdot (ca+bs) + S\varphi_2 \cdot (cb-as),$
$\quad c^2+s^2-0.5];$
$\psi_2 = Atan2[-S\varphi_2 \cdot (a^2+s^2-0.5) +$
$\quad +C\varphi_2 \cdot (ab+cs),\ -S\varphi_2 \cdot (ba-cs) +$
$\quad +C\varphi_2 \cdot (b^2+s^2-0.5)].$ (37)

Special cases

1) If $c^2+s^2 = 1$, then $\theta = 0$,

$$\mathbf{Q} = C\left(\frac{\varphi+\psi}{2}\right) + k \cdot S\left(\frac{\varphi+\psi}{2}\right),$$

and there is an infinite set of solutions with

$\theta = 0,\quad \varphi+\psi = 2 \cdot Atan2(c, s).$ (38)

2) If $c^2+s^2 = 0$, then $\theta = \pi$,

$$\mathbf{Q} = j \cdot C\left(\frac{\varphi-\psi}{2}\right) - i \cdot S\left(\frac{\varphi-\psi}{2}\right),$$

and there is an infinite set of solutions with

$\theta = \pi,\quad \varphi-\psi = 2 \cdot Atan2(-a, b).$ (39)

5 Example

Fig. 1 shows the links 4-6 of a six-jointed robot arm for three different sets of the joint angles $q_4 - q_6$. Frame $\{3\}$ with the origin O_3 and axes X_3, Y_3, Z_3 is rigidly attached to link 3. Link 4 is rotated relative to link 3 about joint axis 4 by q_4; link 5 is rotated relative to link 4 about joint axis 5 by q_5; link 6 is rotated relative to link 5 about joint axis 6 by q_6. Frame $\{6\}$ is rigidly attached to link 6 and the end-effector, which is, in this example, a gripper.

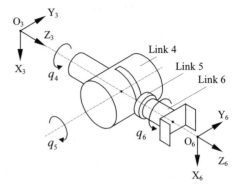

a) $q_4 = 0°,\ q_5 = 0°,\ q_6 = 0°$

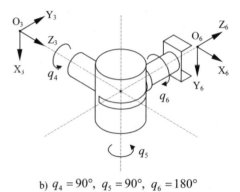

b) $q_4 = 90°,\ q_5 = 90°,\ q_6 = 180°$

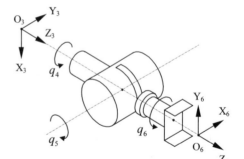

c) $q_4 = 0°,\ q_5 = 0°,\ q_6 = 90°$

Fig. 1 The three last links of a six-jointed robot arm

The orientation of $\{6\}$ relative to $\{3\}$ is considered. The position of the frames is not relevant here.

We consider the extracting of the $Z-Y-Z$ and $Z-Y-X$ Euler angles from the rotation matrix ${}^6_3\mathbf{R}_b$ and quaternion ${}^6_3\mathbf{Q}_b$ corresponding to Fig. 1b as well as from the rotation matrix ${}^6_3\mathbf{R}_c$ and quaternion ${}^6_3\mathbf{Q}_c$ corresponding to Fig. 1c.

The $Z-Y-Z$ Euler angles are interesting, because in this example for this Euler angle set

$$\varphi = q_4,\ \theta = q_5,\ \psi = q_6. \quad (40)$$

The $Z-Y-X$ Euler angles are interesting, because this Euler angle set is used for many robot arms with such kinematics, e. g. by Mitsubishi Electric, Reis Robotics etc., with the notation

$$\varphi = A,\ \theta = B,\ \psi = C. \quad (41)$$

In order to distinguish between both Euler angle sets we will use the notation (40) for $Z-Y-Z$ and the notation (41) for $Z-Y-X$.

Given the rotation matrix

$${}^6_3\mathbf{R}_b = \begin{bmatrix} 0 & 1 & 0 \\ 0 & 0 & 1 \\ 1 & 0 & 0 \end{bmatrix} \quad (42)$$

(see Fig. 1b), we obtain the following.

In $Z-Y-Z$ this is the general case, because $r_{33} = 0$. Thus, there are two solutions defined by (16) and (17):

1) $q_4 = 90°,\ q_5 = 90°,\ q_6 = 180°$; (43)
2) $q_4 = -90°,\ q_5 = -90°,\ q_6 = 0°$. (44)

In $Z-Y-X$ this is a special case, because $r_{31} = 1$. Thus, from (13) we get an infinite set of solutions with:

$$B = -90°,\ A+C = -90°. \quad (45)$$

Given instead of the rotation matrix (42) the corresponding rotation quaternion

$${}^6_3\mathbf{Q}_b = \frac{1}{2} - \frac{1}{2}i - \frac{1}{2}j - \frac{1}{2}k, \quad (46)$$

the same results (43), (44) and (45) can be obtained with the help of (36), (37) and (33).

Given the rotation matrix

$${}^6_3\mathbf{R}_c = \begin{bmatrix} 0 & -1 & 0 \\ 1 & 0 & 0 \\ 0 & 0 & 1 \end{bmatrix} \quad (47)$$

(see Fig. 1c), we obtain the following.

In $Z-Y-X$ this is the general case, because $r_{31} = 0$. Thus, there are two solutions defined by (11) and (12):

1) $A = 90°,\ B = 0°,\ C = 0°$; (48)
2) $A = -90°,\ B = 180°,\ C = 180°$. (49)

In $Z-Y-Z$ this is a special case, because $r_{33} = 1$. Thus, from (18) we get an infinite set of solutions with:

$$q_5 = 0°,\ q_4 + q_6 = 90°. \quad (50)$$

Given instead of the rotation matrix (47) the corresponding rotation quaternion

$${}^6_3\mathbf{Q}_c = \frac{\sqrt{2}}{2} + \frac{\sqrt{2}}{2}k, \quad (51)$$

the same results (48), (49) and (50) can be obtained with the help of (31), (32) and (38).

6 Conclusions

In this paper, various Euler angle sets were discussed and universal algorithms for extracting the Euler angles from a rotation matrix and from a rotation quaternion were described. These algorithms can be used in robotics and other application fields for all rotation matrices, rotation quaternions and Euler angle sets.

7 Acknowledgement

The author thanks the Foundation Rhineland-Palatinate for Innovation for the sponsoring of the project "Robotic Manipulation under Augmented Navigation for Tooth-Implant Concepts" (ROMANTIC).

8 References

[1] Craig, J.J.: Introduction to Robotics: Mechanics and Control. 2nd ed. Reading, Mass.: Addison-Wesley, 1997

[2] Fu, K.S.; Gonzalez, R.C.; Lee, C.S.G.: Robotics: Control, Sensing, Vision, and Intelligence. McGraw-Hill, 1987

Robot, Robotics, Medical Robotics

Gail G. Gubaidullin

Dept. of Mathematics and Technology, RheinAhrCampus Remagen, University of Applied Sciences Koblenz, Suedallee 2, D-53424 Remagen, Germany

Abstract

In science fiction literature and movies imaginary robots perform both physical and intellectual tasks like human beings. However, most of the real robots today are the robot arms which have been used since the 1960s in industry for relatively simple repetitive physical tasks like machine tending, welding etc. Since the 1980s such robot arms and other automated mechanisms as well as computer- and sensor-based visualization and navigation systems have been introduced in the medical field more and more frequently. In this contribution, this new branch and evolution phase of robotics is considered and a general scheme of a medical robot is presented.

1 Introduction

People have always dreamed of artificial beings which are able to perform diverse human physical and intellectual tasks on their own, and kept trying to design such beings. In 1921, the Czech dramatist Karel Čapek published his play "R. U. R."; there, the term "robot" was first used to refer to such artificial human beings. In the early 1940s the American science-fiction author Isaac Asimov invented the term "robotics" and the famous "Three Laws of Robotics" (both appeared firstly in his story "Runaround", 1942) [1]. In SF literature and movies the imaginary robots perform both physical and intellectual tasks like human beings. However, most of the real robots today are the robot arms which have been used since the 1960s in industry for relatively simple repetitive physical tasks like machine tending, welding etc. [2-3]. Since the 1980s such robot arms and other automated mechanisms as well as computer- and sensor-based visualization and navigation systems have been introduced in the medical field more and more frequently [4-9].

Over the last years, some companies, which initially planned a very fast breakthrough in the field of medical robotics, failed to succeed and went bankrupt. Some patients were dissatisfied with results of robot assisted medical operations, which led to court cases and negative statements in the mass media. Unfortunately, the inevitable disappointments of the initial stage of medical robotics are sometimes taken as a reason to call into question the importance and the future of the medical robotics in general.

In this contribution, medical robotics is considered as a new branch and logical evolution phase of medical engineering and robotics (**Fig. 1**) and a general scheme of a medical robot is presented.

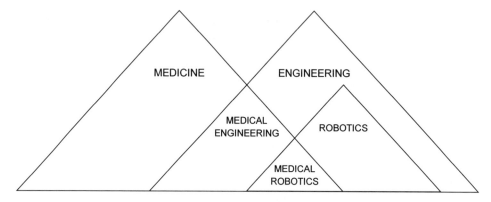

Fig. 1 Medical robotics – a new branch of medical engineering and robotics

2   Robot and Robotics

Any procedure which is carried out with the help of a robot (robot assisted procedure) can be analyzed, at a relatively high level of abstraction, with the general scheme of a human-robot interaction depicted in **Fig. 2**. Here, the term "robot" is used not only for "a reprogrammable, multifunctional manipulator ...", but in a broader sense: for a technical system based on computer- and sensor-controlled mechanisms, which performs some physical and intellectual tasks on its own [7]. We will not try here, to define the term "robot" more exactly; but we just briefly mention that, the more physical and intellectual tasks can be carried out by a robot, the more the robot resembles the human – and thus the more difficult it becomes to define this term.

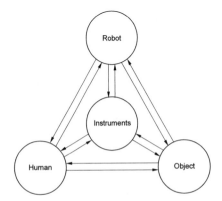

Fig. 2 Human-robot interaction on an object

Separate subtasks of the considered procedure can be carried out by the human alone (without or with special instruments), by cooperation between the human and the robot, or by the robot alone. Various natural and artificial feedforward and feedback channels and control loops are used for the carrying out of the separate subtasks and the whole procedure.

The basic physical and intellectual tasks, which have to be distributed between the human and the robot, are perception, execution, planning and control [7]. Correspondingly, a robot includes a perception subsystem (PCP) consisting of sensors and robot interface input devices, an execution subsystem (EXC) consisting of actuators and robot interface output devices, and a planning and control subsystem (PLC) consisting of computers and controllers, **Fig. 3**.

Robotics concerns firstly the construction of robots and has its foundations in several classical fields like control engineering, computer sciences, artificial intelligence, sensor engineering, mechanics, electronics, drive engineering etc. In addition, robotics deals with the interconnections between the robot on the one hand, and the human, object, instruments and the world (robotic term for all the things and persons outside the considered system) on the other hand; thus, robotics also builds up on human-robot interface engineering, the relevant areas of the object and task related sciences etc., Fig. 3.

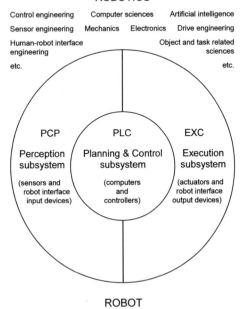

Fig. 3   Robot and robotics

Robotics is a relatively young science and engineering branch and is in a constant development process. The robots have become more and more accurate, fast, intelligent, safe and reliable; and they have been used in more and more new fields of application to work on more and more new objects.

In all these fields the famous "Three Laws of Robotics" by Isaac Asimov [1] are valid:

1) A robot may not injure a human being, or, through inaction allow a human being to come to harm.
2) A robot must obey the orders given it by human beings except where such orders would conflict with the First Law.
3) A robot must protect its own existence as long as such protection does not conflict with the First or Second Law.

It is also very important that there is no known law of nature that restricts the theoretical possibility of the development of robots imitating the physical and intellectual activities of human beings as exactly as desirable [10].

3 Medical Robotics

From the robotic point of view, a robot assisted medical procedure is only a special case of the general robot assisted procedure with a physician as the human subject and a patient as the object of the procedure (cf. Fig. 2 and **Fig. 4**).

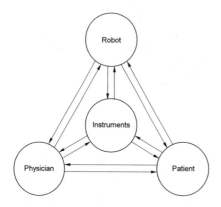

Fig. 4 Robot assisted medical procedure

A medical robot (R), which assists a physician (Ph) during a medical procedure on a patient (P) with medical instruments (I), interacts with Ph, P, I and the rest of the world (W) by means of sensors, actuators and special interfaces between R on the one hand, and Ph, P, I and W on the other hand, **Fig. 5**. In Fig. 5, the following notation is used:

Ph → R – physician-to-robot interface,
R → Ph – robot-to-physician interface,
I → R   – instruments-to-robot interface,
R → I   – robot-to-instruments interface,
P → R   – patient-to-robot interface,
R → P   – robot-to-patient interface,
W → R   – world-to-robot interface,
R → W   – robot-to-world interface.

Medical robots for various robot assisted medical procedures (diagnosis, treatment, rehabilitation, patient transportation, assistance to disabled and senior citizens, medical training etc.) with various technical means (patient visualization, instruments navigation, position/force robot control, diverse robot interface input and output devices etc.), see [4-9], can be considered as special cases of the general scheme depicted in Fig. 5. In [7, 8], some medical robotic systems were discussed from this point of view.

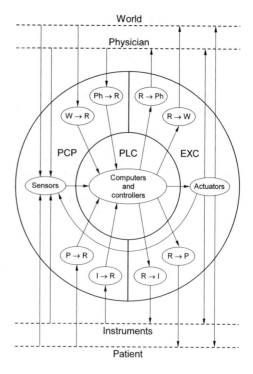

Fig. 5 General scheme of a medical robot

4 Conclusions

In this contribution, medical robotics was considered as a new branch and evolution phase of robotics. A general scheme of a medical robot was presented. The used system approach allows the comparative analysis of different medical robotic systems.

5 Acknowledgement

The author thanks the Foundation Rhineland-Palatinate for Innovation for the sponsoring of the project "Robotic Manipulation under Augmented Navigation for Tooth-Implant Concepts" (ROMANTIC).

6 References

[1] Asimov, I.: I, Robot. London, Grafton Books, 1986, 206 p.
[2] Craig, J.J.: Introduction to Robotics: Mechanics and Control. 2nd ed. Reading, Mass.: Addison-Wesley, 1997, 450 p.

[3] Fu, K.S.; Gonzalez, R.C.; Lee, C.S.G.: Robotics: Control, Sensing, Vision, and Intelligence. McGraw-Hill, 1987
[4] Cleary, K.; Nguyen, Ch.: State of the Art in Surgical Robotics: Clinical Applications and Technology Challenges. Computer Aided Surgery, Vol. 6, No. 6, Wiley-Liss, 2001, pp. 312-328
[5] Dario, P.; Guglielmelli, E.; Allotta, B.; Carrozza, M. C.: Robotics for Medical Applications. IEEE Robotics and Automation Magazine, 1996, pp. 44-55
[6] Davies, B. L.: A review of robotics in surgery. Proceedings of the Institution of Mechanical Engineers, Part H, Journal of Engineering in Medicine, 2000, Vol. 214 (H1): pp. 129-140
[7] Gubaidullin, G. G.: Medical Robotics – a New Field of Medical Engineering and Robotics. Proceedings of the 8th World Multiconference on Systemics, Cybernetics and Informatics SCI '04, Orlando, Florida, 2004
[8] Gubaidullin, G. G.; Evbatyrov, R. R.: A Multi-Purpose Setup for Medical Robotic Experiments. Proceedings of the International Conference on Computing, Communications and Control Technologies CCCT '04, Austin, Texas, 2004
[9] Howe, R. D.; Matsuoka, Y.: Robotics for Surgery. Annual Review of Biomedical Engineering, Vol. 1, Stanford University's HighWire Press, 1999, pp. 211-240
[10] Timofeev, A. V.: Robots and artificial intelligence. Moscow, Nauka, 1978, 192 p. (in Russian)

Biomechanik

Biomechanics

Biomechanische Untersuchung der Lage des Widerstandzentrums des oberen Frontzahnblocks

Susanne Reimann[1], Christoph Bourauel[1], Ludger Keilig[2] und Andreas Jäger[1]

[1]Poliklinik für Kieferorthopädie, Universitätsklinikum Bonn, Rhein. Friedrich-Wilhelms-Universität Bonn, Welschnonnenstraße 17, 53111 Bonn, Deutschland

[2]Abteilung für Zahnärztliche Propädeutik - Experimentelle Zahnheilkunde, Universitätsklinikum Bonn, Rhein. Friedrich-Wilhelms-Universität Bonn, Welschnonnenstraße 17, 53111 Bonn, Deutschland

Zusammenfassung

Die Lage des Widerstandszentrums (WZ) ist ein entscheidender Parameter bei der Planung kieferorthopädischer Zahnbewegungen. In der vorliegenden Untersuchung wurde das gemeinsame Widerstandszentrum der oberen Frontzähne mit Hilfe der Finite-Elemente-Methoden (FEM) numerisch untersucht. Auf Basis eines kommerziellen 3D-Datensatzes eines vollständig bezahnten Oberkiefers und bekannter Materialparameter wurden FE-Modelle der oberen Schneidezähne sowie des umgebenden Zahnhalteapparates generiert. Im FE-System wurde das Modell des Frontzahnblocks mit Drehmomenten von je 10 Nmm an den seitlichen Schneidezähnen belastet. Es zeigte sich, dass die einzelnen Zähne unabhängig voneinander bewegt wurden. Somit liegt für den Frontzahnblock kein gemeinsames Widerstandszentrum im klassischen Sinn vor und die Planung der kieferorthopädischen Zahnbewegungen sollte nicht mehr auf der Grundlage dieser Überlegungen erfolgen.

1 Einleitung

Die kieferorthopädische Biomechanik beschäftigt sich mit der Beschreibung der biologischen Reaktionen der dentalen Strukturen auf einwirkende mechanische Größen im Bereich der Kieferorthopädie. Hier sind insbesondere die Beschreibung von Zahnbewegungen mit numerischen Methoden, die kombinierte experimentelle und numerische Untersuchung des Materialverhaltens von Zahn, Parodontalligament (PDL) und Kieferknochen sowie die Analyse und das Design spezieller Behandlungselemente zu nennen [1 - 3].

Zur Behandlung von Zahnfehlstellungen mit Hilfe kieferorthopädischer Apparaturen werden gezielt Kraftsysteme auf die Zähne über sogenannte Brackets aufgebracht, die dann über Knochenumbauprozesse die Zahnbewegungen induzieren (siehe **Bild 1** und **2**). Dabei können Translationen bis zu 5 mm, in Einzelfällen auch mehr, und Rotationen deutlich über 10° erzielt werden. Kräfte und Drehmomente liegen in der Größenordnung von 1 N bzw. etwa 10 Nmm. Als ein entscheidender Faktor geht in die Planungen kieferorthopädischer Zahnbewegungen die Lage des Widerstandszentrums (WZ) ein. Mit Hilfe des WZ ermittelt der Kieferorthopäde das notwendige Kraftsystem, welches er für eine gewünschte Zahnbewegung auf die Zahnkrone aufbringen muss.

Physikalisch entspricht der Zahn einem gestützten, starren Körper, der im umgebenden Parodontalgewebe gelagert wird. Wird die Zahnkrone mit einem reinen Drehmoment belastet, so sollte der Zahn im Zahnhalteapparat um einen definierten Punkt, dem Widerstandszentrum, rotieren. Diese Eigenschaft wurde in einer größeren Zahl von experimentellen und numerischen Untersuchungen genutzt, um die Position des WZ bei einzelnen, teilweise idealisierten Zähnen zu bestimmen [z.B. 4 - 6].

Klinisch tritt häufig aber auch die Aufgabe auf, dass

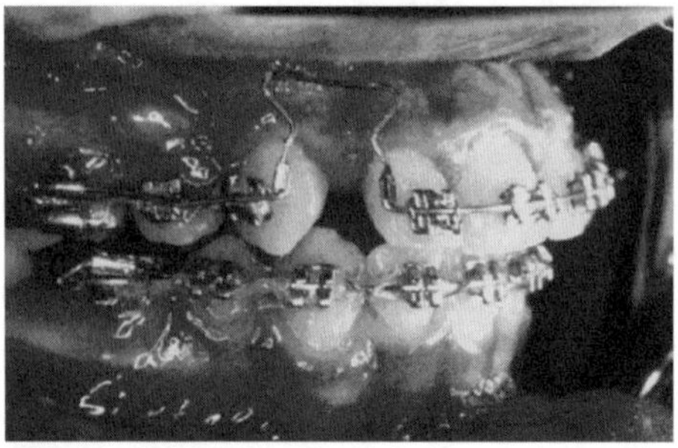

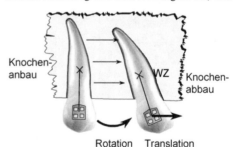

Bild 1 Retraktion der Oberkieferfront nach posterior mit einem superelastischen Retraktionsbogen.

Bild 2 Prinzip der orthodontischen Zahnbewegung durch Knochenmodellierung und -remodellierung.

das Segment, bestehend aus den vier zu einer Einheit verbundenen Frontzähnen, wie in Bild 1, gezeigt nach distal bewegt werden muss. Zur Lage des Widerstandszentrums des Frontzahnsegments wurden eingehende experimentelle und numerische Untersuchungen bislang nicht durchgeführt. Es existieren lediglich auf klinischen Beobachtungen beruhende Plausibilitätsüberlegungen oder Merkregeln. Diese basieren z.B. auf Herleitungen der Position des WZ des gesamten Frontzahnblocks mittels Hebelgesetzen aus der Überlagerung der WZ der einzelnen Zähne [6 - 8]. Werden auf der Grundlage dieser Überlegungen klinisch Frontretraktionen durchgeführt, so zeigen sich Widersprüche. In einer eigenen Studie wurde für die Frontretraktion ein exakt kalibriertes Behandlungselement eingesetzt, das für eine rein translatorische Bewegung das korrekte Kraftsystem erzeugen sollte. Ohne Korrekturen konnte die gewünschte Zahnbewegung jedoch klinisch nicht zufriedenstellend erzielt werden [10]. Die Lage des gemeinsamen Widerstandszentrums des oberen Frontzahnblocks scheint demnach bislang noch nicht eindeutig geklärt zu sein.

Ziel dieser Untersuchung war es daher, mit Hilfe der Finite-Elemente-Methoden das Last/Auslenkungsverhalten der oberen Frontzähne numerisch zu untersuchen und so das gemeinsame Widerstandszentrum des Frontzahnblocks zu bestimmen. Dabei sollte besonderes Augenmerk auf die anatomisch korrekte Darstellung der Morphologie der Zahnwurzeln und des Zahnhalteapparates gelegt werden. Diese haben entscheidenden Einfluss auf die Reaktion der Zähne auf Kraftsysteme und sie legen die Position des WZ fest [1].

2   Material und Methoden

2.1   Modellgenerierung

Auf der Basis von 3D-Datensätzen eines vollständig bezahnten Oberkiefers der Firma Viewpoint DataLabs® wurden FE-Modelle der oberen mittleren und seitlichen Schneidezähne sowie des umgebenden Zahnhalteapparates (PDL und Alveolarknochen) generiert. Bei dem verwendeten Datensatz handelte es sich um ein Oberflächenmodell, bestehend aus 4-Knotenelementen. Zahnwurzeln und Zahnkronen waren getrennt darstellbar, der Zahnhalteapparat jedoch nicht ausgeformt (vergl. **Bild 3** und **Bild 4**).

Daher wurde in einem ersten Schritt das Oberflächenmodell eines Kieferknochens mit Alveolen erstellt. Hierfür wurden die vorhandenen Zahnwurzeln so in das Kiefermodell eingepasst, dass sie anatomisch korrekte Alveolen ergaben (**Bild 5**). Die Modellzähne wurden anschließend um die Dicke des Parodontalligamentes (durchschnittlich 0,2 mm) verkleinert und in das Kiefermodell eingepasst. Aus den Oberflächenmodellen der Alveolen und der verklei-

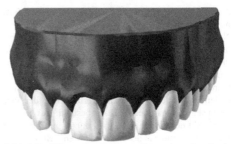

Bild 3 Dreidimensionale Darstellung eines Standardoberkiefermodells.

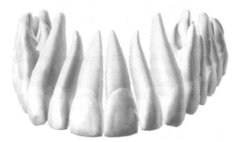

Bild 4 Korrekt vormodellierte Zähne, bei denen Wurzeln und Kronen getrennt darstellbar sind.

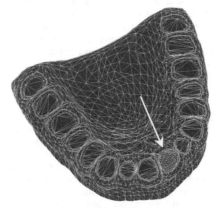

Bild 5 Modellierung einer Alveole an der Position eines rechten oberen Schneidezahns (Pfeil).

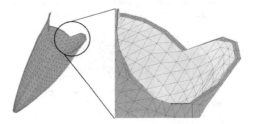

Bild 6 Erstellung der Geometrie und des FE-Netzes des PDL aus den Oberflächen der Zahnwurzeln und des Alveolarknochens.

nerten Zahnwurzeln wurde das PDL eines jeden Zahnes erstellt (**Bild 6**).

Nach Umwandlung der einzelnen Strukturen in Volumenmodelle wurden Alveolarknochen, Zähne sowie umgebendes PDL im FE-Programm knotenweise ineinandergesetzt und miteinander verknüpft. Sodann wurden auf den Zähnen idealisierte Brackets zur Lasteinleitung modelliert und die Verblockung der vier Frontzähne konstruiert. Diese bestand aus einem Stahldraht des Querschnitts 0,46 x 0,64 mm², der alle vier Zähne miteinander verband (siehe **Bild 7**). In einer weiteren Rechnung wurde der Querschnitt des Stahldrahtes auf 1,38 x 1,92 mm² erhöht, also verdreifacht. Dieser Querschnitt ist klinisch nicht zu erzielen, sollte aber den Einfluss der Verblockung demonstrieren.

Für Vernetzung und Berechnungen wurde das Programmsystem MSC.Marc/Mentat® verwendet. Die Netzgenerierung erfolgte mit dem isoparametrischen 10-Knoten-Tetraeder-Element, da auch nichtlineare Berechnungen mit Materialnichtlinearität sowie großen Verschiebungen und endlichen Rotationen durchgeführt werden sollten. Der Stahldraht zur Verblockung wurde mit Beamelementen generiert. Insgesamt bestand das Oberkiefermodell mit den vier Zähnen aus etwa 75.000 Knoten und 366.000 Elementen. Dies erforderte eine Reduktion der Elementzahl, um eine Berechnung in akzeptabler Zeit sicherzustellen. Daher wurde nur ein Teil des Knochens für die Berechnung herangezogen (dunkel markierter Bereich in Bild 7), so dass ein Modell mit etwa 150.000 Elementen entstand.

2.2 Modellberechnung

Zur Bestimmung des WZ wurden die Frontzähne über die seitlichen Schneidezahnbrackets mit Kräftepaaren von je 10 Nmm belastet, wobei der Knochen räumlich fixiert war (Bild 7). Die Materialparameter des Zahns und des Knochens wurden der Literatur entnommen [11], das nichtlineare Verhalten aus eigenen früheren experimentellen und numerischen Studien [12 - 13]. Sie sind in den **Tabellen 1** und **2** aufgeführt. Alle Rechnungen mussten unter der Annahme von isotropem und homogenem Verhalten durchgeführt werden, was gegenüber der Realität selbstverständlich eine

Material	Elastizitäts-modul (MPa)	Querkontraktionszahl µ
Zahn (Mittelwert)	20000	0,30
Knochen (Mittelwert)	2000	0,30
PDL	bilinear	0,30
Bracket (Stahl)	200000	0,30

Tabelle 1 Materialparameter von Zahn, Knochen und Bracket. Zahn und Knochen wurden nicht differenziert in Schmelz/Dentin bzw. Kortikalis/Spongiosa.

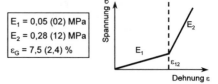

$E_1 = 0,05\ (02)$ MPa
$E_2 = 0,28\ (12)$ MPa
$\varepsilon_G = 7,5\ (2,4)$ %

Tabelle 2 Parameter zur Beschreibung des nichtlinearen Verhaltens des PDL. Mit dieser bilinearen Approximierung konnte das Verhalten des Zahnhalteapparats in experimentellen und numerischen Untersuchungen gut beschrieben werden.

Einschränkung bedeutet. Auf die Lage des WZ sollte dies jedoch keinen Einfluss haben. Zur Auswertung wurden für jeden Knoten und jedes Element die Spannungen, die Verzerrungen und die Verschiebungen berechnet bzw. gelistet.

3 Ergebnisse

Bild 8 zeigt die berechneten Auslenkungen der vier Frontzähne als Reaktion auf die angreifenden Drehmomente. Bereits in dieser Darstellung ist zu erkennen, dass die seitlichen Schneidezähne ganz offensichtlich deutlich weiter ausgelenkt werden als die mittleren. Die maximale Auslenkung an den Schneidekanten der seitlichen Schneidezähne liegt bei etwa 11 µm, bei den mittleren Schneidezähnen beträgt die-

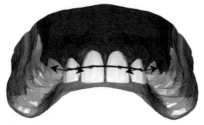

Bild 7 FE-Modell eines Oberkiefers mit den vier Frontzähnen. Die Pfeile deuten die Kräftepaare an.

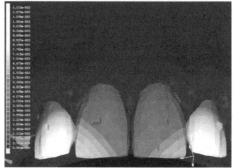

Bild 8 Darstellung der berechneten Auslenkungen der OK-Frontzähne. Die hellen Bereiche (Original: gelb) haben die größten Auslenkungen.

ser Wert etwa 5 µm. Demnach bewegen sich die vier Frontzähne trotz des Verblockungsdrahtes ganz offensichtlich nahezu unabhängig voneinander.

Betrachtet man die berechnete Bewegung ausschließlich der Zähne, also ohne Knochen und PDL, so kann man bei entsprechender farblicher Codierung sehr gut die Achsen identifizieren, um die die Zähne jeweils rotieren. Dies ist in **Bild 9** jeweils in der Okklusal- (oben) und der Frontalebene (unten) dargestellt. Nimmt man das Widerstandszentrum als den Punkt der Zahnwurzel mit der geringsten Auslenkung an, so kann man durch diese jeweils die momentanen Drehachsen legen.

In Bild 9 oben erkennt man zwar, dass in okklusaler Ansicht eine gemeinsame Drehachse für alle vier Frontzähne konstruiert werden kann, in dorsaler Ansicht erkennt man jedoch, dass sich die vier Zähne in der Frontalebene vollkommen unabhängig voneinander bewegen. Zudem liegen die Drehachsen der Einzelzähne schief im Koordinatensystem. Bezogen auf den einzelnen Zahn lagen die WZ in unseren Rechnungen 9 bzw. 12 mm apikal und 5 mm distal des Kraftangriffspunktes, also der seitlichen Schneidezahnbrackets. In Betrachtung des gesamten Zahnblockes ergeben sich so statt eines gemeinsamen WZ mehrere, jedoch in einer gemeinsamen Ebene liegende WZ. Diese liegen für die mittleren Schneidezähne 2-3 mm weiter in Richtung Wurzelspitze als in der Literatur angegeben, für die seitlichen Schneidezähne etwa auf dieser Höhe (vergl. **Bild 10**).

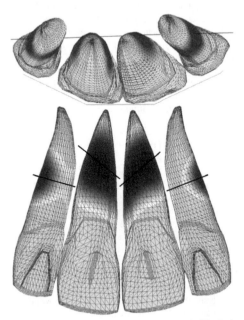

Bild 9 Position des WZ beim Frontzahnblock in okklusaler und dorsaler Ansicht. Die Drehachsen wurden jeweils durch die Bereiche mit der geringsten Auslenkung (dunkle Färbung, Original: blau) gelegt.

Durch eine Erhöhung der Steifigkeit des Verblockungsbogens sollte es möglich sein, den Effekt der unabhängigen Bewegung der Einzelzähne zu reduzieren. Dies ist in Bild 10 gezeigt. Trotz einer Erhöhung des Bogenquerschnitts auf den vierfachen Wert der stärksten in der Kieferorthopädie gebräuchlichen Bögen ist dies offensichtlich nicht zu erreichen. Die Widerstandszentren nähern sich zwar einander an, sie verlagern sich aber nicht so, dass eine gemeinsame Rotationsachse entsteht, die dem in der Literatur angegebenen Wert (9-10 mm apikal, 7 mm distal des seitlichen Schneidezahnbrackets) entsprechen würde.

Die Verteilungen der Verzerrungen und Spannungen sind in **Bild 11** dargestellt. Deutlich ist zu sehen, dass die verstärkte Auslenkung der seitlichen Schneidezähne auch zu erhöhten Spannungen und Verzerrungen um die Wurzeln der beiden seitlichen Zähne führt. Insbesondere bei den Verzerrungen zeigen sich die höchsten Werte im Bereich der Wurzelspitzen und des Alveolarkamms.

4 Diskussion

In der vorliegenden Untersuchung wurde das Last/Auslenkungsverhalten der oberen Frontzähne bei Anlegen eines Kräftepaares von 10 Nmm mit Hilfe der Finite-Elemente-Methoden untersucht. Im Gegensatz zu vorhergehenden Studien wurde besonderer Wert auf die gemeinsame Untersuchung der vier Zähne in einem Oberkiefermodell mit anatomisch korrektem Zahnhalteapparat und realistischer Verblockung der Einzelzähne mit einem konventionellen Stahlbogen gelegt.

Frühere Untersuchungen zur Bewegung und der Lage des Widerstandszentrums des Frontzahnblocks basierten teilweise auf Hebelgesetzen und der Annahme einer ideal-starren Verbindung der Einzelzähne. Da die klinische Anwendung im Rahmen einer Frontretraktion sich aber nicht auf den Einzelzahn konzentriert, sondern der gesamte Frontzahnblock als ein Segment bewegt werden soll, ist die Morphologie aller vier Zahnwurzeln und die Art der Verblockung von besonderer Wichtigkeit. Es ist anzunehmen, dass unterschiedliche Zahnwurzelformen die Biomechanik

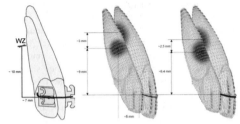

Bild 10 Bei einer Verstärkung der Verblockung (rechts: Drahtquerschnitt 1,38 x 1,92 mm²) nähern sich die einzelnen WZ einander an. Die in der Literatur angegebene Position (links) wird aber nicht erzielt.

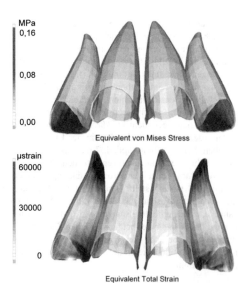

Bild 11 Spannungs- und Verzerrungsverteilungen im PDL um die bewegten Frontzähne. Die höchsten Spannungen (dunkel) konzentrieren sich auf das PDL der seitlichen Schneidezähne.

des Frontzahnblocks und seine Auslenkung im PDL ebenso entscheidend beeinflussen wie bei den Berechnungen der Bewegung von Einzelzähnen. Dies wird durch die vorliegenden Ergebnisse bestätigt:
Eine ideal-starre Verblockung der Frontzähne ist klinisch nicht zu erzielen. Da andererseits aber die Wurzeln der mittleren Schneidezähne deutlich größer sind, also einer Auslenkung im PDL einen höheren Widerstand entgegensetzen, ist von einer Torsion des Drahtbogens zwischen den seitlichen und mittleren Schneidezähnen auszugehen und die mittleren Schneidezähne werden mit deutlich geringeren Drehmomenten belastet als die seitlichen. Dies bedeutet im Umkehrschluss aber auch, dass die seitlichen Schneidezähne deutlich stärker mit dem aufgebrachten Kraftsystem belastet werden als vom Kieferorthopäden eigentlich geplant.
Klinisch ergeben sich hieraus zwei Konsequenzen:
1.) Zum einen wird die vom Kieferorthopäden geplante Bewegung nicht in der Art und Weise ablaufen, wie es gewünscht ist. Die veränderte Lage des WZ führt dazu, dass es verstärkt zu Kippungen, insbesondere der mittleren Zähne kommen wird. Zusätzlich kann angenommen werden, dass die seitlichen Schneidezähne durch eine erhöhte Knochenumbaurate stärker bewegt werden als die mittleren Zähne. Dies ist damit zu begründen, dass die Mehrzahl der auf diesem Gebiet arbeitenden Wissenschaftler davon ausgeht, dass das Knochenmodelling und -remodelling mit den Verzerrungen in Zusammenhang steht. Die Verzerrungen sind im PDL der seitlichen Schneidezähne durch die inhomogene Auslenkung aber signifikant höher (vergl. Bild 11). Dies könnte dazu führen, dass die seitlichen Schneidezähne den mittleren in gewissen Grenzen, die durch die Elastizität des Bogens gegeben ist, „vorlaufen" werden. Das hier beschriebene Verhalten deckt sich recht gut mit den Beobachtungen einer früheren eigenen klinischen Studie [10].
2.) Zum anderen birgt die große Auslenkung der seitlichen Schneidezähne aber auch die Gefahr, dass die Spannungen im PDL einen unphysiologisch hohen Wert erreichen. Vergleicht man die berechneten Maximalspannungen im PDL mit dem kapillären Blutdruck (0,0012 MPa), so sieht man, dass durch ein kieferorthopädisches Kraftsystem durchaus Spannungen erzeugt werden, die diesen Wert um das 100-fache übersteigen. Dies deckt sich auch mit den Ergebnissen anderer Untersuchungen.
Eine Merkregel besagt, dass die Spannungen im PDL durch Einsatz eines kieferorthopädischen Kraftsystems den kapillären Blutdruck nicht deutlich überschreiten sollten, da es sonst sowohl zu starken Schmerzen als auch zu Schädigungen am Parodontalligament und dem Wurzelzement kommen kann. Hier entstehen dann sogenannte Wurzelresorptionen, die teilweise vom Zahnhalteapparat selbst repariert, teilweise aber auch bis hin zu sichtbaren Wurzelverkürzungen führen können.
Die hier vorgestellten Untersuchungen sind ein interessantes Beispiel für den Einsatz theoretisch-biomechanischer Modelle mit unmittelbarem Bezug zum klinischen Alltag. Auch wenn durch die erforderliche Komplexität des Modells der Zeit- und Rechenaufwand sehr hoch war, hat sich gezeigt, dass die Ergebnisse diesen Aufwand rechtfertigen. Die weiteren Simulationen werden in jedem Falle ebenfalls in richtung komplexer, anatomisch korrekter Zahn/Kiefer-Modelle gehen.
Insgesamt muss die Empfehlung ausgesprochen werden, dass die Kieferorthopäden in Zukunft nicht mehr auf die einfache Merkregel der Position eines gemeinsamen Widerstandszentrums zurückgreifen sollten.

5 Literatur

[1] Burstone, C. J.: The biomechanics of tooth movement. in: Kraus BS, Riedel RA (Hrsg.): Vistas in orthodontics. S. 197-213, Lea & Fiebiger, Philadelphia 1962

[2] Bourauel, C., Nolte, L.-P., Drescher, D.: Numerische Untersuchung kieferorthopädischer Behandlungselemente aus pseudoelastischen NiTi-Legierungen. Biomed. Tech. 37 (1992), 46-53

[3] Hinterkausen, M., Bourauel, C., Siebers, G., Haase, A., Drescher, D., Nellen, B.: In vitro analysis of the initial tooth mobility in a novel optomechanical set-up. Med. Eng. Phys. 20 (1998), 40-49

[4] Burstone, C. J., Pryputniewicz, R. J.: Holographic determination of centers of rotation produ-

ced by orthodontic forces. Am. J. Orthod. 77 (1980), 396-409

[5] Dermaut, L. R., Kleutghen, J. P., De-Clerck, H. J.: Experimental determination of the center of resistance of the upper first molar in a macerated, dry human skull submitted to horizontal headgear traction. Am. J. Orthod. Dentofac. Orthop. 90 (1986), 29-36

[6] Vollmer, D., Bourauel, C., Jäger, A., Maier, K.: Determination of the centre of resistance in an upper human canine and idealized tooth model. Eur. J. Orthod. 21 (1999), 633-648

[7] Bauer, W., Diedrich, P., Wehrbein, H., Schneider, B.: Der Lückenschluß mit T-Loops (Burstone) – eine klinische Studie. Fortschr. Kieferorthop. 53 (1992), 192-202

[8] Vanden-Bulcke, M. M., Burstone, C. J., Sachdeva, R. C., Dermaut, L.R.: Location of the centers of resistance for anterior teeth during retraction using the laser reflection technique. Am. J. Orthod. Dentofac. Orthop. 91 (1987), 375-384

[9] Gjessing, P.: Controlled retraction of maxillary incisors. Am. J. Orthod. Dentofac. Orthop. 101 (1992), 120-131

[10] Bourauel, C., Drescher, D.: Retraktion der oberen Schneidezähne mit pseudoelastischen NiTi-T-Federn. Fortschr. Kieferorthop. 55 (1994), 36-44

[11] Abé, H., Hayashi, K., Sato, M. (Hrsg.): Data Book on Mechanical Properties of Living Cells, Tissues, and Organs. Springer Verlag, Tokyo, Berlin, Heidelberg, New York, 1996

[12] Vollmer, D., Haase, A., Bourauel, C.: Halbautomatische Generierung von Finite-Element-Netzen für zahnmedizinische Präparate. Biomed. Tech. 45 (2000), 62-69

[13] Poppe, M., Bourauel, C., Jäger, A.: Determination of the elasticity parameters of the human periodontal ligament and the location of the center of resistance of single-rooted teeth a study of autopsy specimens and their conversion into finite element models. J. Orofac. Orthop. 63 (2002), 358-370

[14] Kawarizadeh, A., Bourauel, C., Jäger A.: Experimental and numerical determination of initial tooth mobilities and material parameters of the periodontal ligament in rat molar specimens. Eur. J. Orthod. 25 (2003), 569-578

Anwendung der Biomechanik bei der ergonomischen Arbeitsgestaltung

Ellegast, Rolf, Berufsgenossenschaftliches Institut für Arbeitsschutz (BIA), 53754 Sankt Augustin, Deutschland

Kurzfassung

Biomechanische Modelle werden in der Ergonomie zur Abschätzung der Muskel-Skelett-Belastungen des arbeitenden Menschen herangezogen. Bezogen auf Wirbelsäulenbelastungen handelt es sich dabei um biomechanische Kettenmodelle, denen u.a. die Kinematik und die auf den Menschen wirkenden externen Kräfte als Eingabegrößen dienen. Diese in die Modelle einfließenden Daten werden üblicherweise mit Labor-Messsystemen erhoben. Dadurch ist eine Belastungsabschätzung der realen Arbeitssituation nur näherungsweise möglich.
Aus diesem Grund hat das Berufsgenossenschaftliche Institut für Arbeitsschutz (BIA) ein personengebundenes Messsystem entwickelt, welches eine kontinuierliche Erfassung und automatisierte Bewertung von biomechanischen Belastungsgrößen an der arbeitenden Person in deren realer Arbeitsumgebung erlaubt.

1 Einleitung

Erkrankungen des Muskel- und Skelettsystems stellen mit ca. 28% die häufigste Ursache von Arbeitsunfähigkeitstagen in Deutschland dar. Durch den damit verbundenen Produktionsausfall entstehen erhebliche volkswirtschaftliche Kosten. Die Bundesanstalt für Arbeitsschutz und Arbeitsmedizin (BAuA) schätzte diese für das Jahr 2002 auf ca. 12 Mrd € [1].
Zu den bereits bestehenden Berufskrankheiten im Bereich des Muskel- und Skelettsystems sind seit 1993 drei weitere hinzugekommen, welche sich auf bandscheibenbedingte Erkrankungen der Wirbelsäule beziehen. Die Erfassung und Bewertung von berufsbezogenen Muskel- und Skelettbelastungen sind somit im Sinne der Prävention ein wichtiges Aufgabengebiet der Berufsgenossenschaften.

2 Prinzipien biomechanischer Modelle zur Abschätzung von beruflichen Wirbelsäulenbelastungen

Zur Beurteilung der Gesundheitsgefährdungen für die Wirbelsäule gelten als Belastungsindikatoren die während der Ausführung einer Tätigkeit auf die Bandscheiben wirkenden Druckkräfte. Dabei insbesondere diese, welche auf die unterste Bandscheibe L5/S1 des lumbosakralen Übergangs wirken. Eine direkte in-vivo Messung des intradiskalen Drucks wurde bisher nur in wenigen Studien durchgeführt (siehe u.a. [2]) und verbietet sich neben dem hohen medizinischen Aufwand vor allem aus ethischen Gründen. Daher werden in der Regel Modellrechnungen zur Abschätzung der auf die Bandscheiben wirkenden Kompressionskräfte verwendet. Bei den dazu eingesetzten Modellen handelt es sich um biomechanische Kettenmodelle, in denen der menschliche Körper durch starre Kettenglieder, die über idealisierte Gelenke miteinander verbunden sind, modelliert wird. Ziel der Berechnungen ist zunächst die Ermittlung des äußeren Drehmomentes, welches im Zentrum der untersten Bandscheibe L5/S1 wirkt. Dazu können prinzipiell zwei Ansätze verwendet werden: der „top-down"- und der „bottom-up"-Ansatz (siehe z.B. [3]). Bei dem „top-down"-Modell werden alle Körperteile oberhalb einer Schnittebene durch den Mittelpunkt der lumbosakralen Bandscheibe L5/S1 berücksichtigt. Im stark vereinfachten, zweidimensionalen statischen Fall berechnet sich hier das am Ort des Bandscheibenzentrums wirkende Drehmoment $M_{L5/S1\_td}$ aus der Summe der Produkte von Körperteilgewichtskräften oberhalb der Schnittebene und deren horizontalen Abständen zum Bandscheibenmittelpunkt. Im Gegensatz dazu werden beim „bottom-up"-Modell alle Körperteile unterhalb der durch das Bandscheibenzentrum verlaufenden Schnittebene zur Bestimmung des resultierenden äußeren Drehmomentes $M_{L5/S1\_bu}$ berücksichtigt. Dieses berechnet sich aus dem Gesamtmoment, welches in den Boden geleitet wird, abzüglich aller Drehmomente der Körperteile unterhalb der Schnittebene durch L5/S1.
Bei beiden Modellansätzen erfolgt die Kompensation des äußeren Drehmomentes in erster Linie durch Muskelmomente, so dass sich das gesamte Drehmoment, welches auf das Bandscheibenzentrum wirkt, aufhebt. Die Kompressionskräfte auf die Bandscheibe L5/S1 berechnen sich dann aus der Summe der äußeren Kräfte und der über das oben beschriebene Momentengleichgewicht berechneten Muskelkräfte.

3 Messverfahren zur Belastungserhebung

Als Eingabegrößen für biomechanische Kettenmodelle werden neben den anthropometrischen Maßen, wie Länge der Körpersegmente, auch die Segmentgewichte, Lage der Segmentschwerpunkte, die Kinematik und die auf den Menschen wirkenden externen Kräfte benötigt. Die Kinematik, welche die Bewegung der Körpersegmente beschreibt, wird im Labor üblicherweise mit Infrarotkamerasystemen gemessen. Mehrere Kameras erfassen hierbei IR-Reflexionssignale von hochreflektierenden Markern, welche an den Körpersegmenten des Probanden befestigt sind. Bei vorheriger Kalibrierung des Systems mit einer bekannten Markeranordnung und der Kenntnis der Kamerastandorte sind somit eine Rekonstruktion der räumlichen Lage der Marker, und damit eine dreidimensionale Bewegungsanalyse der Segmente, möglich. Zur Bewegungserfassung an realen Arbeitsplätzen ist die Methode jedoch nicht geeignet, da sie nur unter Laborbedingungen in einem eingeschränkten Raumbereich anwendbar ist. Die auf den Probanden wirkenden Kräfte werden im Labor meist mittels stationärer Kraftmessplattformen gemessen.

Wegen der Beschränkung der oben beschriebenen Messverfahren auf den Laboreinsatz wurde am Berufsgenossenschaftlichen Institut für Arbeitsschutz (BIA) das CUELA-Messverfahren (Computerunterstützte Erfassung und Langzeitanalyse von Belastungen des Muskel- und Skelettsystems) entwickelt, mit welchem eine kontinuierliche Erfassung von Körperbewegungen und gehandhabten Lastgewichten direkt am Arbeitsplatz durchgeführt werden kann [4]. Das System besteht aus Sensoren zur Bewegungserfassung (Gyroskope, Inklinometer, Potentiometer) und einem tragbaren Miniaturcomputer, welche auf der Kleidung der Arbeitsperson angebracht werden können (siehe Bild 1).

Die Messdaten werden mit hoher Zeitauflösung (50 Hz) über eine Arbeitsschicht gemessen und auf einer Flash-Speicherkarte abgelegt. Die autarke Energieversorgung mit einer handelsüblichen Batterie ermöglicht den Feldeinsatz des Messsystems auch an nicht-stationären Arbeitsplätzen.

Mit dem Messsystem werden folgende Körper-/Gelenkbewegungen erfasst:
Oberkörperbewegungen in drei Dimensionen: Flexions- und Extensionsbewegungen im Lendenwirbelsäulen- und im oberen Brustwirbelsäulen-Bereich, der Verdrehungswinkel (Torsion) und der Seitneigungswinkel (Lateralflexion) des Oberkörpers, Bewegungen der Hüftgelenke in der Sagittalebene und Bewegungen der Kniegelenke in der Sagittalebene.

Eine Erweiterung des CUELA-Messsystems erfasst darüber hinaus die Bewegungen des Kopfes (Flexion/Extension), der Halswirbelsäule (Flexion/Extension), der Gelenke des Schulter-Arm-Hand-Systems (Schulterblatt, Schultergelenk, Ellenbogen, Unterarm und Handgelenk) und des Beckens (Neigung). Zeitgleich zur Körperwinkelbestimmung werden die Bodenreaktionskräfte durch ein Fußdruckmesssystem registriert.

Bild 1 Einsatz des CUELA-Messsytems im Bauwesen.

Die Messdaten des CUELA-Messsystems können als Eingabedaten für biomechanische Kettenmodelle zur Abschätzung der Kompressionskräfte auf die Bandscheibe L5/S1 genutzt werden [5]. Für die durch die Basisversion des Messsystems erzeugten Daten bietet sich ein „bottom-up"-Modell an. Mit der Erweiterungsvariante des CUELA-Messsystems, welche auch eine Bewegungserfassung der oberen Extremitäten beinhaltet, können Berechnungen auch mittels eines „top-down"-Modells erfolgen.

Die Messungen werden zusätzlich durch Videoaufnahmen dokumentiert. Durch eine einfache Synchronisation des Videofilms mit den Messdaten ist somit eine spätere Zuordnung der Belastungsmesswerte zu den Arbeitssituationen möglich.

Direkt nach Beendigung einer Messung können die Messdaten in eine speziell entwickelte Software eingelesen und dargestellt werden. Mit dieser Software ist es möglich, sich zu jedem beliebigen Zeitpunkt der Messung die Körperhaltung anhand einer dreidimensionalen Computerfigur sowie durch zeitabhängige Winkelgraphen der gemessenen Körper- und Gelenkwinkel anzeigen zu lassen. Gleichzeitig hierzu wird die zugehörige Arbeitssituation durch das Videobild automatisiert eingeblendet (siehe Bild 2).

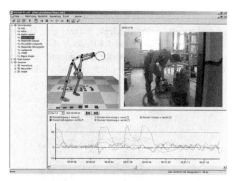

Bild 2 Darstellung der Messdaten mit der CUELA-Software

4 Anwendung in der ergonomischen Arbeitsgestaltung

Mit Hilfe des CUELA-Messverfahrens wurden in den letzten Jahren zahlreiche Arbeitsplätze in verschiedenen Branchen zur Ermittlung der Wirbelsäulenbelastungen analysiert. Ein Beispiel dafür ist eine 2004 abgeschlossene, umfassende Studie über die Belastung des Muskel- und Skelettsystems (insbesondere der Wirbelsäule), die für Flugbegleiterinnen beim Schieben und Ziehen von Servierwagen (Trolleys) entstehen [6].

Bild 3 Labormessaufbau zur Erfassung der Wirbelsäulenbelastungen von Flugbegleiterinnen beim Ziehen und Schieben von Trolleys

Auf Basis der ermittelten Belastungsdaten wurden konkrete Präventionsvorschläge sowohl im Hinblick auf die Arbeitsorganisation als auch in Bezug auf die Arbeitsmittelgestaltung möglich.

Bei der Analyse anderer Arbeitsplätze, z. B. von Näherinnen (vgl. [7]), war darüber hinaus auch eine Quantifizierung der Belastungssituation vor und nach den ergonomischen Interventionen und damit eine genaue Erfolgskontrolle der getroffenen Präventionsmaßnahmen realisierbar.

5 Literatur

[1] Bericht über Sicherheit und Gesundheit bei der Arbeit der Bundesregierung: Bundesanstalt für Arbeitsschutz und Arbeitsmedizin (BAuA), 2002
http://www.baua.de/info/statistik/stat_2002/kosten2002.pdf

[2] Nachemson, A. Morris J.M.: In Vivo Measurements of Intradiscal Pressure. Journal of Bone and Joint Surgery, 46-A Nr. 5, 1964

[3] Plamondon, A., Gagnon, M., Desjardin, P.: Validation of two 3-D segment models to calculate the net reaction forces and moments at the L5/S1 joint in lifting. Clinical Biomechanics 11, 1996, pp. 101-110

[4] Ellegast, R., Kupfer J.: Portable posture and motion measuring system for use in ergomomic field analysis. In: Landau Ergonomic Software Tools in Product and Workplace Design, Ergon Verlag Stuttgart, 2000, pp. 47-54

[5] Ellegast, R.: Personengebundenes Messsystem zur automatisierten Erfassung von Wirbelsäulenbelastungen bei beruflichen Tätigkeiten. BIA-Report 5/98, HVBG, Sankt Augustin, 1998

[6] Glitsch, U., Ottersbach H.-J., Ellegast, R., Hermanns I., Feldges, W., Schaub, K., Berg, K., Winter G., Sawatzki, K., Voß, J., Göllner R., Jäger, M.: Untersuchung der Belastung von Flugbegleiterinnen und Flugbegleitern beim Schieben und Ziehen von Trolleys in Flugzeugen. BIA-Report 5/2004, HVBG, Sankt Augustin, 2004
http://www.hvbg.de/d/bia/pub/rep/rep04/bia05 04.html

[7] Ellegast, R., Herda, C. Hoehne-Hückstädt, U., Lesser, W., Kraus, G. Schwan, W.: Ergonomie an Näharbeitsplätzen. BIA-Report 7/2004, HVBG, Sankt Augustin, 2004
http://www.hvbg.de/d/bia/pub/rep/rep04/bia07 04.html

Zur Biomechanik des Fahrradfahrens

T. Bildhauer, O. Schulyzk, U. Hartmann
RheinAhrCampus Remagen, Südallee 2, 53424 Remagen

Kurzfassung

Moderne Meßtechnik findet zunehmende Verbreitung im sportlichen Bereich. Ziel des Einsatzes der Meßtechnik ist die Optimierung des Sportgeräts im Hinblick auf seine biomechanische Verträglichkeit und eine Leistungssteigerung des Sportlers. In diesem Beitrag soll am Beispiel des Fahrradfahrens gezeigt werden, welche technischen Verfahren eingesetzt werden können. Neben Beschleunigungs- und Kraftmessungen am Fahrrad werden auch erste Ansätze zur biomechanischen Untersuchung des Fahrradfahrens vorgestellt. Die Integration der Ergebnisse aller Modalitäten kann in Zukunft zum Erstellen und Validieren eines Computermodells des Gesamtsystems Mensch - Sportgerät beitragen.

1 Einleitung

Moderne Meßtechnik hält seit geraumer Zeit Einzug in den Sport. Nicht nur im Bereich des Spitzensports dienen die Ergebnisse elektronischer Meßgeräte einem besseren Verständnis von komplexen Bewegungsabläufen und der permanenten Optimierung von Sportgeräten. Aufgrund der Miniaturisierung und der Erschwinglichkeit elektronischer Bauteile setzt sich der Einsatz von moderner Technik auch im Breitensport immer mehr durch. Das Interesse am eigenen Körper, d.h. an den eigenen sportlichen Leistungen aber auch den damit verbundenen Belastungen läßt sich am stetig steigenden Absatz von einfachen Messinstrumenten wie z.B. Pulsuhren ablesen. Es ist davon auszugehen, daß auch anspruchsvollere Meßtechnik z.B. zur Ortung per GPS oder zur Leistungsdiagnostik bei den sportlich Aktiven auf ein breites Interesse stoßen werden. Am Beispiel des Fahrrads (siehe Bild 1) sollen in diesem Beitrag meßtechnische Möglichkeiten zur Leistungs- und Belastungsmessung präsentiert werden. Allgemeine Betrachtungen zur Biomechanik des Fahrradfahrens findet man in [1, 2].

Bild 1   Fahrrad mit mobilem Data-Recording-System im Plexiglasgehäuse.

2 Methoden

2.1 Mobile Messwerterfassung mechanischer Belastungen am Fahrrad

Zur mobilen Messdatenerfassung wird zur Zeit ein mobiles Data-Recording-System für Fahrräder aufgebaut. Ziel ist es herauszufinden, welche Belastungen im Fahrbetrieb an einzelnen Komponenten, wie z.B. Rahmen, Gabel oder Lenker auftreten. Dazu wurden an einem Mountainbike mit Federgabel verschiedene Sensoren angebracht.

Ein Wegaufnehmer (Linearpotentiometer) an der Federgabel misst den zurückgelegten Federweg. Anhand dieser Messwerte lässt sich feststellen, ob der zur Verfügung stehende Federweg optimal genutzt wird. Interessant ist außerdem die Fragestellung, in wie weit sich unterschiedliche Abstimmungen der Federgabel (Federhärte, Dämpfung) auf die Federkennlinie auswirken.
Weiterhin erfassen zwei Dehnungsmessstreifen an Lenker und Vorbau die während der Fahrt auftretenden Kräfte. Ein Beschleunigungssensor misst zusätzlich noch die vertikale Rahmenbeschleunigung.
Um bei einer späteren Auswertung der Messdaten auch die Gegebenheiten des Untergrunds mit berücksichtigen zu können, werden zusätzlich über einen GPS-Empfängers die Koordinaten der Wegstrecke abgespeichert. Die so gewonnenen Wegdaten können dann in eine digitale Karte mit entsprechendem Koordinatengitter übertragen werden. Dadurch ist es bei der Auswertung der Messdaten möglich, Einflüsse des Untergrunds, z.B. Schotter, Waldweg oder Asphalt, auf das Messergebnis zu berücksichtigen.

Um möglichst viele Daten während einer Messfahrt aufzeichnen zu können, wird als Speichermedium eine handelsübliche Compact Flash Card verwendet. Dieses Speichermedium verfügt über eine standardisierte Schnittstelle und ist in Größen bis zu 1 GB erhältlich. Dadurch ist es auch bei einer Messfahrt über einen längeren Zeitraum möglich, die Daten zu speichern und anschließend über ein USB-Kartenlesegerät am PC auszulesen. Da die Messwerte auf der Compact Flash Card in Form einer Binärdatei gespeichert werden, ist die Auswertung bzw. Analyse der Daten am PC mit Softwarepaketen wie MATLAB oder LabView problemlos möglich.

Die Signalverarbeitung erfolgt mittels eines 16-bit Mikroprozessors der Firma Renesas. Neben der A/D-Wandlung der Sensordaten übernimmt er auch die Speicherung der Messwerte. Zur Ansteuerung der Compact Flash Card wird ein Teil der digitalen I/O-Ports als Daten- bzw. Steuerbus verwendet. Über ein LC-Display am Fahrrad lassen sich alle wichtigen Parameter der Messeinrichtung, z.B. Speicherauslastung oder aktuelle Abtastrate, abrufen.

Die gesamte Messelektronik befindet sich zusammen mit dem Akku spritzwassergeschützt in einem Plexiglasgehäuse, das im Rahmendreieck des Fahrrads befestigt wird.

Das Fahrrad verfügt weiterhin über ein Leistungsmessgerät der Firma Schoberer Rad Messtechnik. Mit diesem System lässt sich der direkte Krafteinsatz beim Rad fahren messen. Um die exakte Leistung bestimmen zu können, erfasst ein Drehmoment-Sensor in der Kurbel die vom Fahrer auf die Pedale ausgeübte Kraft. Aus dem Drehmoment und der Trittfrequenz berechnet ein Computer am Lenker die Leistung in Watt.

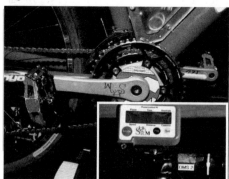

Bild 2 Drehmoment-Sensor in der Kurbel zur Leistungsmessung und Auswertecomputer am Lenker.

Mithilfe der verwendeten Messeinrichtungen lassen sich Untersuchungen in den folgenden Bereichen realisieren:

- Trainingssteuerung
- Leistungsdiagnostik
- Aerodynamik Tests
- Optimierung von Federwegen bei Mountainbikes
- Untersuchungen zur Kraftersparnis durch den Einsatz von Federungssysteme bei Fahrrädern

2.2 2D-Bewegungsanalyse

Bild 3 Eine Aufnahme der Hochgeschwindigkeitskamera: Die fünf reflektierendenMarker definieren anatomische Landmarken.

Um Berechnungen zur Biomechanik des Fahrradfahrens durchführen zu können, werden Videoaufnahmen einer Tretumdrehung auf einem räumlich fixierten Fahrrad aufgenommen. Um die Auswertung zu erleichtern wurde eine High-Speed-Kamera der Firma Vosskühler benutzt. Für unseren Versuch ist eine zeitliche Auflösung von 460 Bildern/Sekunde ausreichend um Verwischungseffekte zu vermeiden. Die räumliche Auflösung der Kamera beträgt 1024x1024 Pixel. Der Testfahrer wurde mit fünf retro-reflektiven Markern an anatomisch relevanten Stellen versehen:

1. Hüfte
2. Knie
3. Knöchel
4. Ferse
5. Vorderfuß

Da wir in unserem Versuch vorerst nur die unteren Extremitäten vermessen, enthält das Modell den Oberschenkel, den Unterschenkel, das Fersenbein und den Fuß. Diese Auswahl gibt ausreichend Information über die Korrektheit der Bewegung und die Angepaßtheit des Sportgeräts. Bild 3 zeigt als Beispiel eine Aufnahme der High-Speed Kamera für einen beliebig ausgewählten Zeitpunkt.

2.3 Pedobarographie

Die Pedobarographie ist ein Meßverfahren, das in der biomechanischen Ganganalyse eingesetzt wird, um den Druck auf die Fußsohle zu messen. Auf diese Weise kann die Qualität von Sportschuhen bestimmt oder Fehlbelastungen von Patienten ermittelt werden. In unserem Kontext wurde die Pedobarographie eingesetzt, um den Kraft- und Druckverlauf im Schuh des Fahrradfahrers verfolgen zu können.

Es werden drucksensitive Einlegesohlen der Firma PEDAR verwendet, die mit jeweils 99 kapazitiv arbeitenden Sensoren versehen sind. Die Meßfrequenz beträgt 50 Hz.

3 Ergebnisse

3.1 Bewegungsanalyse

Ein Tretzyklus dauert ungefähr 0.6 Sekunden. Bei einer Bildfrequenz von 460 fps (*frames per second*) entspricht das etwa 260 Bildern. In jedem dieser Bilder müssen die Positionen der Marker bestimmt werden, um anschließend weitere physikalische Größen wie Beschleunigung und Geschwindigkeit der Marker zu bestimmen.

Die Bewegungsanalyse wurde mit der SIMI Auswertesoftware [3] durchgeführt. Bei der Detektion der Marker handelt es sich um ein Bildverarbeitungsproblem, das nicht in jedem Fall leicht zu lösen ist. In unserem Beispiel führen bei automatischer Markererkennung Refexionen am Rahmen zu fälschlichen Detektion zusätzlicher Marker. Abhilfe schafft entweder die arbeitsaufwendige manuelle Lokalisierung der Marker durch Anklicken per Maus oder die für dieses Problem modifizierte Software, die in [4] beschrieben

Bild 4: Die Überlagerung der Beinstellungen zu verschiedenen Zeitpunkten des Tretzyklus liefert eine integrale Sicht auf den Bewegungsablauf.

ist. Nach dem Bearbeiten der 260 Einzelaufnahmen der Hochgeschwindigkeitskamera gelangt man zu einer abstrakten Strichdarstellung der Bewegung. Die Marker werden zu diesem Zwecke durch gerade Linien anatomisch sinnvoll verbunden.

In Bild 4 ist nur jedes zehnte Beinmodell in Strichdarstellung gezeichnet. Die Abbildung liefert eine anschauliche Darstellung der räumlichen Bereiche, die von den einzelnen Körpersegmenten während eines Tretzyklus überstrichen werden. Schon an dieser Stelle könnte eine detaillierte Analyse der Paßgenauigkeit des Fahrrads für den individuellen Sportler ansetzen. Eine andere nicht weniger instruktive Darstellung sind die Trajektorien der Marker bzw. Körperpunkte (siehe Bild 5). Die drei geschlossenen Kurven im unteren Bereich der Abbildung sind die Bahnen des Vorderfusses, der Ferse und des Knöchels. An der Exzentrizität der Bahnkurven lassen sich fehlerhafte Bewegungsabläufe mathematisch

Bild 5 Die Trajektorien der Marker sind für Vorderfuß, Ferse und Knöchel geschlossene Kurven. Der Hüftmarker zeigt nur minimale Bewegung, das Knie bewegt sich auf einem Kreisausschnitt hin und her.

quantifizieren. Die Ergebnisse der Bewegungsanalyse beschränken sich jedoch nicht auf die geometrische Beschreibung. Da der zeitliche Abstand zwischen zwei Bildern bekannt ist, kann aus der Ortsänderung eines Markers seine Geschwindigkeit und daraus seine Beschleunigung berechnet werden (siehe Bild 6).

Beschleunigungen dienen ihrerseits zur Bestimmung der Kräfte und Drehmomente, die auf die Gelenke wirken. Ist die Kraft, die auf den Fuß von außen wirkt, bekannt (z.B. durch die im ersten Kapitel beschriebene Meßvorrichtung), stehen außerdem Massen und Trägheitsmomente der Körpersegmente zur Verfügung, dann kann unter Einbeziehung der Beschleunigungsdaten eine invers-dynamische Berechnung durchgeführt werden.

Darüber hinaus kann mit den erhobenen Daten ein biomechanisches Computermodell des Fahrradfahrers erstellt werden.

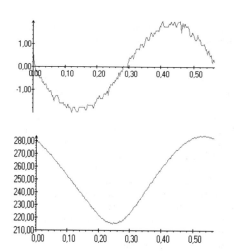

4 Literatur

[1] Gressmann, M., Fahrradphysik und Biomechanik, ISBN: 3895950238, 6. Auflage 2003
2] Pawlik, R., Biomechanik des Radfahrens, Dissertation U Wien 1995
[3] SIMI Motion Software, Manual Version 5, www.simi.com
[4] Richarz, S., Preißler, S., Hartmann, U. Ein Verfahren zur Vermessung der Pronation beim Laufen, Beitrag für die 2. Remagener Physiktage, siehe nächste Seite

Bild 6 oben: Zeitlicher Geschwindigkeitsverlauf des Vorderfußmarkers. Unten: Zeitlicher Verlauf des Kniewinkels.

3.2 Pedobarographie

Bild 7 stellt die Ergebnisse der Druckverteilungsmessung im Schuh des Radfahrers dar. Die gemittelte Druckverteilung im rechten Bild zeigt die erwarteten Spitzendrücke im Vorderfussbereich. Die zeitliche Kraft- und Druckentwicklung für acht Tretzyklen ist im linken Bild zu finden. Man erkennt einen fast linearen Anstieg, der von einem abrupten Druckabfall gefolgt wird

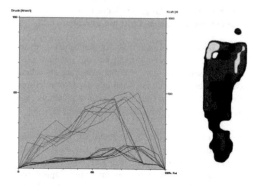

Bild 7: Ergebnisse der pedobarographischen Untersuchung des Drucks im Schuh des Fahrradfahrers.

Ein Verfahren zur Vermessung der Pronation beim Laufen

S. Richarz, S. Preissler, U. Hartmann
RheinAhrCampus Remagen, Südallee 2, 53424 Remagen

Kurzfassung

Die Wahl des optimalen Sportschuhs ist abhängig vom Laufstil und der individuellen Anatomie. Beim Sportschuhdesign ist ein wichtiges Ziel, eine starke Auftrittsdämpfung zu erreichen, um die mechanische Belastung der Gelenke zu minimieren. Ein zweiter wichtiger Aspekt bei der Konzeption und dem Erwerb eines Laufschuhs ist das Maß an Pronation, das der jeweilige Schuh zuläßt. Überpronation führt mit hoher Wahrscheinlichkeit nach einiger Zeit zu Beschwerden am Bewegungsapparat. Um die Auswahl des optimalen Sportschuhs im individuellen Fall zu erleichtern, wurde ein robustes Verfahren entwickelt, das auf der Basis von Filmaufnahmen des Laufens den Pronationswinkel bestimmt.

1 Einleitung

Pronation ist eine natürliche Bewegung beim Abrollen des Fußes (Bild 1), allein ihr Ausmaß entscheidet darüber, ob sie zu krankhaften Veränderungen führt.

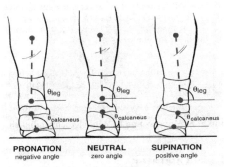

Bild 1 Zur Erklärung des Begriffs der Pronation am Beispiel des rechten Beins.

Bei der Pronation rotiert der Unterschenkel nach innen. Von außen betrachtet scheint das obere Sprunggelenk mit Knöchel nach innen zu auszuweichen [1]. Übermäßige Pronation (Knickfuß) führt mittelfristig zu einem Plattfuß. Verstärkt wird die Tendenz zur Pronation, wenn man sich in jeden Schritt mit gestrecktem Knie und der Ferse voran hineinfallen läßt. Bei Überpronation werden die Innenbänder des Sprunggelenks überdehnt und können ihre stabilisierende Wirkung nicht mehr entfalten. Durch die Verwendung von festen, stabilisierenden Elementen auf der Innenseite des Schuhs kann man den mit der Überpronation einhergehenden Symptomen sinnvoll entgegenwirken. Einlagen sind hier angeraten, da der Fuß nur durch den Schuh häufig nicht ausreichend zu stabilisieren ist. Ein Maß für die Pronation ist der Pronationswinkel, der aus der Differenz des Beinwinkels und des Calcaneuswinkels ergibt (siehe Bild 1). Von Überpronation spricht man, wenn der Pronationswinkel, größer als 4 Grad ist. Die möglichen körperlichen Folgen der Überpronation sind vielgestaltig:

- Knieprobleme
- Überlastung der Gelenke
- Muskelermüdung
- Probleme mit der Achillessehne.

Diese Vielzahl von Problemen macht deutlich, wie wichtig eine verläßliche Messung des Pronationswinkels für die Auswahl des passenden Sportschuhs ist [2, 3]. Die technische Lösung sollte jedoch nicht zu aufwändig sein, um die Erschwinglichkeit des Systems nicht zu gefährden und den Aufwand für die technische Betreuung gering halten zu können. Im folgenden Kapitel soll ein Prototyp eines solchen Systems beschrieben werden.

2 Methoden

Auf einer Platine sind zwei Leuchtdioden in einem definierten Abstand angebracht. Diese aktiven Marker werden auf der Rückseite des zu testenden Schuhs befestigt. Der Proband läuft auf einem Laufband und wird dabei von einer Hochgeschwindigkeitskamera mit 460 Bildern/s gefilmt. Durch die Verwendung einer Hochgeschwindigkeitskamera vermeidet man störende Verwischungseffekte der Marker (siehe Bild 2). Dies erleichtert die automatische Lokalisation mittels digitaler Bildverarbeitung. Die Genauigkeit des Verfahrens ist hauptsächlich durch die Auflösung der Kamera (1024x1024 Pixel) gegeben. Zur Lokalisation der Marker in den (bis zu) 512 Einzelbildern der Hochgeschwindigkeitsaufnahme wurde unter MATLAB® eine Auswertesoftware

Bild 2: Aufnahme der aktiven Marker am Schuh mit der Hochgeschwindigkeitskamera.

entwickelt, die automatisch den zeitlichen Verlauf des Pronationswinkels ermittelt. Bild 3 zeigt den Verlauf der Markerkoordinaten. Die Verbindungslinien zwischen zwei zueinander passenden Markern sind nur für das Anfangs- und Endbild eingezeichnet. Zusammen mit den gestrichelten Hilfslinien definieren sie den aktuellen Pronations- bzw. Supinationswinkel.

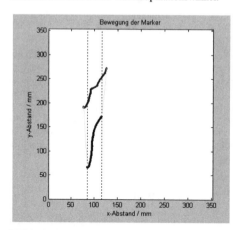

Bild 3: Gesamtübersicht der Markerkoordinaten.

3. Ergebnisse

Aus den Koordinaten der Marker wird der zeitliche Verlauf des Supinations- und Pronationswinkels berechnet (Bild 4). Man erkennt, dass kurz nach Aufsetzen des Fusses der Winkel sein Maximum erreicht. Der Winkel sinkt dann auf Null Grad ab, um danach eine Pronation von ca. 2 Grad zu beschreiben. Dieser Wert ist als unkritisch zu bezeichnen.

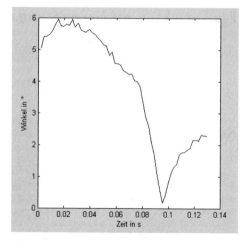

Bild 4: Zeitlicher Verlauf des Pronationswinkels.

Biomechanisch verwertbar sind des weiteren die Geschwindigkeiten der Marker (Bild 5). Insbesondere die Winkelgeschwindigkeit (5b) kann bei der Schuhentwicklung oder der Schuhauswahl als Maß für die Güte eines Sportschuhs betrachtet werden.

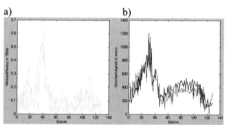

Bild 5: a) Marker- und b) Winkelgeschwindigkeit.

4 Literatur

[1] Nigg, B. Biomechanics of running shoes, Human Kinetics 1986, ISBN 0-87322-002-1
[2] Zatsiorsky, V. Biomechanics in sport, Blackwell Science 2000, ISBN 0-632-05392-5 , 161-184
[3] International Society of Biomechanics Technical Group on Footwear Biomechanics. www.uni-essen.de /%7Eqpd800/FWISB/sneakers.html

Neue Messverfahren in der Golfschwunganalyse

J. Bongartz[1], G. Laschinski[2] und U. Hartmann[1]
[1]RheinAhrCampus Remagen, Südallee 2, 53424 Remagen,
[2]B&L Meßsysteme, Melissenstrasse 21, 41466 Neuss

Kurzfassung

Es werden neue Messverfahren zur Golfschwunganalyse vorgestellt. Mit Hilfe von Hochgeschwindigkeitskameras (230 Bilder/sec.) ist es möglich, die Schwungbewegung eines Golfspielers sehr exakt zu analysieren. Eine speziell entwickelte Software erlaubt es, verschiedene Hochgeschwindigkeits-Aufnahmen zu überlagern, so dass die Bewegung unterschiedlicher Spieler (z.B. Golfschüler und –lehrer) räumlich und im zeitlichen Verlauf miteinander verglichen werden können.
Zusätzlich wird eine Methode vorgestellt, die mit einem am Golfschläger befestigten Messmodul die auftretende Beschleunigung in radialer Richtung aufzeichnet und per Bluetooth an einen Pocket-Computer überträgt. Es können anhand des Verlaufs der Beschleunigung einzelne Schwungphasen eindeutig identifiziert werden, so dass neben der Ermittlung absoluter Beschleunigungswerte auch die Dauer der einzelnen Schwungphasen bestimmt werden kann.

1. Einleitung

Der Golfschwung ist einer der komplexesten Bewegungsabläufe im Sport. Der Schwungrhythmus und das Timing im Treffmoment entscheiden darüber, ob die Bewegung optimal in Form eines dynamischen Schlages umgesetzt werden kann. Aus diesem Grund werden immer neue Trainingsmethoden entwickelt, mit denen der Golfspieler seine Schwungdynamik und seinen Schwungrhythmus trainieren und überprüfen kann, um den Bewegungsablauf zu stabilisieren. Am weitesten verbreitet ist bisher die Analyse der Schlagbewegung mittels konventioneller Videotechnik. Dabei wird der Spieler mit einer Videokamera nacheinander aus der rückwärtigen und der seitlichen Perspektive aufgezeichnet. Nachteilig bei dieser Methode ist die geringe zeitliche Auflösung der Videotechnik von ca. 25 Bilder/s und die Verwendung von nur einer Kamera, so dass bei einem einzelnen Schlag nicht beide Perspektiven gleichzeitig aufgezeichnet werden können. Im Folgenden soll der Einsatz von Hochgeschwindigkeitskameras und eines lokal am Schläger befestigten Beschleunigungs-Messmoduls zur Golfschwunganalyse vorgestellt werden.

2. Der Golfschwung

Der Golfschwung kann prinzipiell in 7 Abschnitte unterteilt werden. Beim Ansprechen (1) richtet der Spieler den Schläger und seinen Körper zum Ball aus. Daran schließt sich die Aufschwungphase (2) des Schlägers an, bis der Umkehrpunkt (3) der Ausholbewegung erreicht ist. In der Abschwungphase (4) beschleunigt der Schlägerkopf, um im Treffpunkt (5) eine maximale Impulsübertragung auf den Golfball zu erzielen. Nach dem Treffpunkt erfolgt der Durchschwung (6) des Schlägers. Die Endposition der Schwungbewegung wird als Finish (7) bezeichnet.

Ansprechen   Aufschwung   Umkehrpunkt

Abschwung   Treffpunkt   Durchschwung

Abb. 1:

Die einzelnen Phasen eines Golfschwungs. Die Koordination der einzelnen Bewegungsabläufe entscheidet über die Effizienz des Schlages.

3. Videoanalyse

Ein Golfschwung dauert ca. 2 Sekunden. Dabei treten Schlägerkopfgeschwindigkeiten von bis zu 200 km/h auf, die bei optimaler Impulsübertragung Ballgeschwindigkeiten von bis zu 300 km/h ermöglichen. Die Impulsübertragung auf den Ball dauert wenige Millisekunden.

Eine Aufnahme dieser Bewegung mit konventioneller Videotechnik, die eine zeitliche Auflösung von 25 Bildern/s aufweist, zeigt deutliche Einschränkungen. Bei hohen Schlägergeschwindigkeiten sind die Abstände der Schlägerpositionen in den Einzelbildern sehr groß, so dass man bei der Analyse keine stetige Bewegung mehr nachvollziehen kann. Aufgrund relativ langer Belichtungszeiten kommt es zudem zu einem starken Verwischungseffekt, so dass eine genaue Lokalisierung des Schlägers oftmals nicht möglich ist. Der Einsatz von Hochgeschwindigkeitskameras vom Typ Vosskühler HCC-1000 [2] gestattet eine deutliche Erhöhung der zeitlichen Auflösung der Bewegungsanalyse. Diese Kameras erreichen bei einer räumlichen Auflösung von 1024x1024 Pixeln eine Bildrate von bis zu 462 Bildern/s. Da der lokale Bildspeicher der Kamera aber bei dieser Geschwindigkeit nur für eine Aufnahmedauer von 1 Sekunde ausreicht, wurde die Bildrate auf 231 Bilder/s reduziert, um den gesamten Golfschwung aufzuzeichnen. Dies entspricht aber immer noch einer zehnmal höheren zeitlichen Auflösung im Vergleich zur konventionellen Videotechnik.

Durch den gleichzeitigen Einsatz von 2 Kameras ist es zudem möglich, einen identischen Schwung gleichzeitig aus der rückwärtigen und seitlichen Perspektive aufzuzeichnen.

Die speziell für diesen Anwendungsbereich entwickelte Analyse-Software (*GolfLAB®-Viewer*) ermöglicht eine bildgenaue und synchrone Auswertung der einzelnen Aufnahmesequenzen. Zudem ist eine Überlagerung von Bildsequenzen bzw. Einzelbildern möglich, so dass verschiedene Spieler (Abb. 3a) oder verschiedene Zeitpunkte einer Bewegung (Abb. 3b) miteinander verglichen werden können. Die Überlagerungsfunktion verschiedener Spieler ermöglicht es z.B., die Bewegung eines Golfschülers mit der des Golflehrers zu überlagern, was die Fehleranalyse des Golflehrers durch den direkten Vergleich deutlich vereinfacht.

a) b)

Abb. 3: a) Bildüberlagerung von zwei verschiedenen Spielern b) Bildüberlagerung verschiedener Zeitpunkte einer Bewegung.

4. Beschleunigungs-Messmodul

Durch Fortschritte im Bereich der Mikrotechnik sind seit einigen Jahren miniaturisierte Halbleiter-Beschleunigungssensoren verfügbar. Aufbauend auf dem Beschleunigungssensor ADXL250 der Firma Analog Devices [1] wurde ein kompaktes Messmodul aufgebaut, das direkt an einem Golfschläger befestigt werden kann. Der ADXL250 ermöglicht die Messung von Beschleunigungen von maximal 50g mit einer Messrate von bis zu 1 kHz.

Mit dem in Abb. 4 gezeigten Messmodul werden Beschleunigungsmessungen während eines Golfschwungs durchgeführt und mit den Daten einer gleichzeitigen Videoanalyse verglichen. Untersucht wird, ob aus dem zeitlichen Beschleunigungsverlauf Aussagen über den Golfschwung abgeleitet werden können.

Abb. 2: Zwei Hochgeschwindigkeitskameras ermöglichen die gleichzeitige Aufzeichnung verschiedener Perspektiven.

Abb. 4:

Messeinheit, die während eines Golfschwungs die auftretenden Beschleunigungen aufzeichnet.

Das Messmodul enthält neben dem ADXL250 einen Mikrocontroller vom Typ Atmel AVR Mega32 [3], der für die zentrale Steuerung des Moduls zuständig ist. Die vom Sensor ermittelten Beschleunigungswerte werden mit dem im Mikrocontroller integrierten AD-Wandler mit einer Frequenz von 230 Hz digitalisiert und im internen Speicher abgelegt. Nach Abschluss des Golfschwungs übermittelt das Messmodul die digitalen Messwerte drahtlos per Bluetooth, mit Hilfe eines integrierten Bluetooth-Modul BlueRS+I [4] der Firma Stollmann, an einen Pocket-Computer vom Typ Palm Tungsten T3. Auf dem Palm T3 erlaubt eine mit NSBasic 4.3 entwickelte Software die Weiterverarbeitung der Daten. Neben einer graphischen Darstellung ist auch die Speicherung der Daten und eine Übertragung an einen PC per Docking-Station möglich.

gangs Messdaten vor und nach dem Triggersignal im lokalen Digitalspeicher vorhanden und somit ist ein vollständiger Golfschwung gespeichert. Dieser Messdatensatz wird anschließend per Bluetooth übertragen.

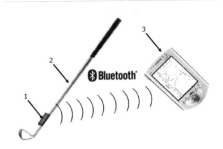

Abb. 5: Die Daten des Messmoduls werden drahtlos per Bluetooth an einen Pocket-Computer übertragen. (1:Messmodul; 2:Golfschläger; 3:Pocket-Computer)

Um einen Golfschwung sinnvoll zu erfassen, muss eine geeignete Triggerung des Messvorgangs erfolgen. Ein Golfschwung dauert ca. 2 Sekunden, wobei im Treffpunkt, der zeitlich gesehen ungefähr in der Mitte der Schlagbewegung liegt, die höchsten Beschleunigungen auftreten. Die Triggerung erfolgt auf das eigentliche Beschleunigungs-Messsignal. Dazu wird kontinuierlich die Beschleunigung gemessen und die digitalen Messwerte werden in einem Ringspeicher abgelegt, der die Messdaten eines bestimmten Zeitraums speichern kann. Die notwendige Speicherkapazität des Ringspeichers ergibt sich aus der Abtastfrequenz des Beschleunigungssignals und der Periodendauer, die der Ringspeicher abdecken soll. Überschreitet zu einem bestimmten Zeitpunkt der Beschleunigungs-Messwert eine vorgegebene Schwelle, so ist dies ein Zeichen für einen durchgeführten Durchschwung und ein Triggersignal wird ausgelöst. Das Triggersignal bewirkt, dass die Messelektronik den Messvorgang nach einer einstellbaren Zeitspanne beendet. Die einstellbare „Posttrigger-Zeitdauer" entspricht in der Regel der halben Periodendauer des Ringspeichers. Aufgrund der Ringspeicheranordnung sind nach Beendigung des Messvor-

Abb. 6: Golfschwung mit einem Golfschläger mit integriertem Messmodul. Die gemessene radiale Beschleunigungskomponente ist mit einem Pfeil gekennzeichnet.

In einer ersten Versuchsreihe hat ein PGA-Golflehrer bei zusätzlicher Videoanalyse einige Golfschwünge (ohne Ball) mit einem Golfschläger mit integriertem Messmodul durchgeführt. Mit dem Messmodul wurde dabei die radiale Beschleunigungskomponente aufgezeichnet, die in Abb. 6a-d jeweils mit einem Pfeil gekennzeichnet ist.

5. Ergebnisse

Ein typischer zeitlicher Verlauf der radialen Beschleunigungskomponente während eines Golfschwungs ist in Abb. 7 dargestellt. Es lassen sich im Kurvenverlauf der Beschleunigung einige Extremwerte identifizieren, die im Diagramm mit 1-5 gekennzeichnet sind. Durch den Vergleich mit der parallel durchgeführten Videoanalyse ist es möglich, den einzelnen Abschnitten einzelne Phasen des Golfschwungs eindeutig zuzuordnen. Nach der Ansprechphase beginnt im Punkt 1 der Aufschwung, die Radialbeschleunigung steigt und erreicht im Punkt 2 ein lokales Maximum von ca. 4,5 g. Die Aufschwungphase endet beim Umkehrpunkt, der im

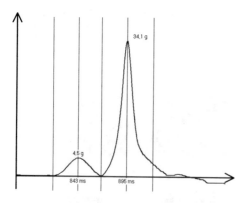

Abb. 7: Zeitlicher Verlauf der radialen Beschleunigungskomponente während eines Golfschwungs.

Diagramm dem Punkt 3 entspricht, hier ist die Beschleunigung wieder auf Null abgefallen. Beim Abschwung steigt die Beschleunigung stark an, um im Treffpunkt einen maximalen Wert von ca. 34 g zu erreichen (Punkt 4). Nach dem Treffpunkt fällt die Beschleunigung während des Durchschwungs wieder ab und läuft gegen Null, da im Finish der Golfschläger zu Ruhe kommt.

Neben der Ermittlung der absolut auftretenden Beschleunigungswerte ermöglicht die Messung, auch die Dauer der einzelnen Schwungphasen relativ exakt zu messen. So dauert der Aufschwung ca. 843 ms und die Abschwung- und Durchschwungphase 895 ms.

6. Diskussion

Der Einsatz moderner Messtechnik bietet vielfältige Vorteile bei der Golfschwunganalyse. Durch Hochgeschwindigkeitskameras mit ihrer besseren zeitlichen Auflösung sind im Vergleich zur konventionellen Videotechnik deutlich verbesserte Schwunganalysen möglich. Der parallele Betrieb von zwei Kameras erlaubt zudem eine gleichzeitige Beobachtung aus verschiedenen Perspektiven.

Mit Hilfe einer neu entwickelten Software lassen sich die aufgezeichneten Bewegungssequenzen überlagern, so dass verschiedene Spieler bzw. verschiedene Schwünge eines Spielers sowohl räumlich als auch zeitlich miteinander verglichen werden können.

Mit einem neu entwickelten Messmodul konnte die Radialkomponente der Beschleunigung während eines Golfschwungs aufgezeichnet werden. Aus dieser Kurve konnten nicht nur die absoluten Beschleunigungswerte ermittelt werden, sondern es konnten auch eindeutig die einzelnen Schwungphasen identifiziert werden. Dies ermöglicht, die Dauer der verschiedenen Schwungphasen zu bestimmen.

Die Verwendung eines Pocket-Computers erlaubt den Einsatz der Beschleunigungsmessung mit relativ geringem Aufwand. Zur Kontrolle des Schwungrhythmus, d.h. der Dauer der einzelnen Schwungphasen, benötigt man ein am Schläger befestigtes Messmodul und einen Pocket-Computer, der die Messdaten auswertet und anzeigt.

7. Literatur

1) Datenblatt ADXL250, Analog Devices, www.analog.de
2) Datenblatt HCC-1000, VDS Vosskühler, www.vdevossk.de
3) Datenblatt AVR Mega32, Atmel Inc., www.atmel.com
4) Datenblatt BlueRS+I, Stollmann Entwicklungs- und Vertriebs-GmbH www.stollmann.de

Biosignalverarbeitung

Bio-Signal Processing

Messung der Komplexität in Herzfrequenz- und Blutdruckvariabilitätszeitreihen mittels Kompressionsentropie

Andreas Voss, Mathias Baumert
Fachhochschule Jena, FB Medizintechnik, Carl-Zeiss-Promenade 2, D-07745 Jena, Deutschland

Kurzfassung

Die kardiovaskuläre Regulation erfolgt mittels eines komplexen Systems von verschiedenen interagierenden Kontrollmechanismen. Klinische Studien haben gezeigt, dass eine kardiovaskuläre Dysfunktion mit dem Grad der Komplexität der Herzfrequenzvariabilität korreliert. Mit Hilfe eines modifizierten LZ77 Algorithmus untersuchten wir die Komplexität von Herzfrequenz- und Blutdruckzeitreihen anhand ihrer Komprimierbarkeit. Bei der Früherkennung von Präeklampsien in der Schwangerschaft konnte die Kompressionsentropie signifikante Unterschiede zwischen 80 normotensiven Schwangeren und 44 Patientinnen mit Präeklampsie aufzeigen. Auch bei der Vorhersage von Tachyarrhythmien in implantierbaren Defibrillatoren war die Kompressionsentropie allen Standardverfahren signifikant überlegen. Zusammenfassend ermöglicht die Komplexitätsanalyse mittels Kompressionsentropie eine bessere Charakterisierung kardiovaskulärer Dysfunktionen.

1 Einleitung

Die kardiovaskuläre Regulation erfolgt mittels eines komplexen Systems von verschiedenen interagierenden Kontrollmechanismen. Klinische Studien haben gezeigt, dass eine kardiovaskuläre Dysfunktion mit dem Grad der Komplexität der Herzfrequenzvariabilität korreliert [1,2]. Ein Verlust von autonomer Regulation führt vermutlich zu einem weniger komplexen Systemverhalten [3].
Um die Komplexität zu beurteilen, wurden verschiedene Ansätze benutzt, unter anderem aus dem Bereich der nichtlinearen Dynamik. Ein einfacher Weg, mit dem die Komplexität eines Systems geschätzt werden kann, kommt aus der Informationstheorie und basiert auf Entropiemaßen. Um die Frage: "Wie schwer ist das System zu beschreiben?" zu beantworten, gibt die Entropie an, wie viel Information nötig ist, um den nächsten Wert mit einer bestimmten Genauigkeit vorherzusagen. Nach Shannon ist die Entropie einer gegebenen Sequenz x begrenzt [4]. Der kleinste Algorithmus, der x generiert, entspricht der Entropie von x (Algorithmische Entropie; Kolmogorov-Komplexität) [5]. Obwohl ein solcher Algorithmus theoretisch unmöglich ist, stellen verlustfreie Datenkompressionsalgorithmen eine gute Näherung dar.
Mit Hilfe eines modifizierten LZ77-Algorithmus untersuchten wir die Komplexität von Herzfrequenz- und Blutdruckzeitreihen anhand ihrer Komprimierbarkeit. In einer klinischen Pilotstudie wurde untersucht, inwieweit sich die Entropie in Herzfrequenz- und Blutdruckzeitreihen von Schwangeren mit Präeklampsie, d.h. Patientinnen mit Hypertonie und gleichzeitiger Proteinurie, von normotensiven Schwangeren unterscheidet, und ob damit eine Charakterisierung der kardiovaskulären Veränderungen möglich ist [6, 7].
In einer zweiten klinischen Studie wurde untersucht, ob mit Hilfe der Kompressionsentropie eine Vorhersage von ventrikulären Tachyarrhythmien in implantierbaren Kardioverter-Defibrillatoren möglich ist.

2 Methodik

2.1 Daten

2.1.1 Normotensive Schwangere und Schwangere mit Präeklampsie

Es wurden die kontinuierlich nichtinvasiv gemessenen Blutdrucksignale (Portapres M2®, TNO Biomedical Instrumentation) von 80 normotensiven Schwangeren (CON) und 44 Patientinnen mit Präeklampsie (PE) ausgewertet. Das maternale Alter (CON 28 (24-31) Jahre vs. PE 27 (22-31) Jahre) und die mittlere Schwangerschaftswoche (CON 35. (32.-37.) Woche vs. PE 32. (30.-36.) Woche) waren in beiden Gruppen vergleichbar. Aus den 30-minütigen Aufzeichnungen, die jeweils vormittags unter Ruhebedingung in der Universitätsfrauenklinik Leipzig stattfanden, wurden die Zeitreihen der Herzschlag-Intervalle (BBI) und der systolischen Blutdruckwerte SP extrahiert und anschließend auf ihre Komplexität hin untersucht.

2.1.2 Patienten mit implantierbaren Kardioverter-Defibrillatoren

An der Franz-Volhard-Klinik Berlin wurden Herzfrequenzdaten von 50 Patienten mit schwerer Herzinsuffizienz akquiriert. Alle Patienten hatten einen Kardioverter-Defibrillator vom Typ PCD 7220/ 7221 Medtronic implantiert, welcher in der Lage ist, die letzten 1024 Schlag-zu-Schlag-Intervalle vor dem Einsetzen einer ventrikulären Tachyarrhythmie zu speichern (siehe Abbildung 1). Basierend auf diesen Daten wurde eine HRV-Analyse durchgeführt und mit der von individuell, während Follow-up Untersuchungen akquirierten arrhythmiefreien Kontrollzeitreihen verglichen. Kein Patient erhielt vor der Studie Antiarrhythmika der Klasse I oder III.

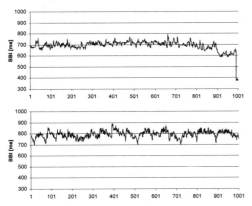

Bild 1: Beispiele von im implantierbaren Kardioverter-Defibrillator gespeicherten Herzfrequenzzeitreihen. Oben: Herzfrequenzzeitreihe unmittelbar vor dem Auftreten einer ventrikulären Tachyarrhythmie. Unten: Kontrollzeitreihe des gleichen Patienten.

2.2 Die Kompressionsentropie

Lempel und Ziv entwickelten 1977 einen universellen Algorithmus zur verlustfreien Datenkompression, welcher auf Sequenz-Matching in einem ‚Sliding Window' basiert [8]. Der Algorithmus ist in Abb. 2 dargestellt und sei hier kurz beschrieben:
Eine Sequenz von Symbolen $x = x_1, x_2, \ldots$ der Länge L eines gegebenen Alphabets Θ der Größe $\Phi=|\Theta|$ soll komprimiert werden. Subsequenzen $(x_m, x_{m+1}, \ldots, x_o)$ von x seien definiert durch x_m^o.
Der Algorithmus behält die zuletzt kodierten Symbole im ‚Window' der Größe w. Die noch nicht kodierte Sequenz von Symbolen ist im ‚Lookahead Buffer' der Größe b gespeichert. Der Kodierer, positioniert an der Stelle p, sucht nach der längsten Übereinstimmung zwischen der noch nicht kodierten Sequenz im ‚Lookahead Buffer' und der bereits kodierten Sequenz im ‚Window', beginnend an der Stelle v. Die übereinstimmende Sequenz von n Symbolen kann somit einfach durch die Werte n und v kodiert werden, d.h. einem Zeiger of das vorige Auftreten der Sequenz im ‚Window' und der Läge der Übereinstimmung.
Der Algorithmus beinhaltet somit folgende Schritte:
1) Kodiere die ersten w Symbole ohne Komprimierung
2) Setze den Zeiger p=w+1
3) Finde für ein v im Bereich $1 \leq v \leq w$ das größte n im Bereich $1 \leq n \leq b$, so dass $x_p^{p+n-1} = x_{p-w+v}^{p-w+v+n-1}$
4) Kodiere n und v im Binärcode und das Symbol $x_{p+n} \in \Theta$ ohne Komprimierung
5) Setze den Zeiger p auf p=n+1 und gehe zu Schritt 3 (iteriere)

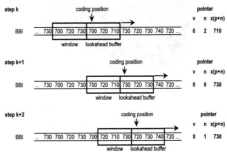

Bild 2: Schema des Kompressionsalgorithmus. ‚Window' (w=3) und ‚Lookahead Buffer' (b=3) werden durch die Sequenz von BBI geschoben. Der Zeiger speichert die Position v, die Länge der übereinstimmenden Sequenz n, und das nachfolgende BBI x(p+n).

Als Maß für die Entropie wurde das Verhältnis der Länge der komprimierten Sequenz zur Länge der unkomprimierten Sequenz gebildet [9]. Hohe Entropiewerte entsprechen somit einer geringen Komprimierbarkeit, geringe Entropiewerte einer hohen Komprimierbarkeit.

2.2.1 Optimierung des Kompressionsalgorithmus

Innerhalb des Kompressionsalgorithmus kann neben der Länge des Windows auch die Länge des Lookahead Buffers frei gewählt werden. Anhand der Daten aus 2.1.2 wurden beide Größen schrittweise im Bereich von 1-20 für das Window bzw. 1-10 für den Lookahead Buffer variiert und die Signifikanz der Gruppentrennung zwischen Messungen vor dem Auftreten von ventrikulären Tachyarrhythmien und Kontrollzeitreihen geprüft. In Abbildung 3 ist die Kompressionsentropie innerhalb der Kontrollgruppe in Abhängigkeit von Window und Lookahead Buffer dargestellt. Die beste Gruppentrennung ergab sich bei einem Window der Größe 7 und einem Lookahead Buffer der Größe 3.

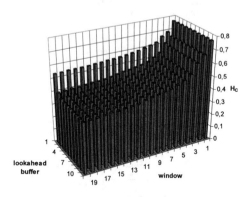

Bild 3: Mittlere Kompressionsentropie der Kontrollzeitreihen in Abhängigkeit von Window- und Lookahead Buffer-Größe.

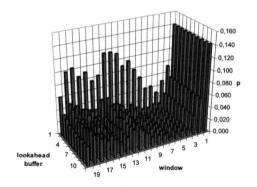

Bild 4: Wilcoxon-Testergebnisse von Kontroll- und Tachyarrhythmiezeitreihen in Abhängigkeit von Window- und Lookahead Buffer-Größe.

2.2.2 Test auf Nichtlinearität

Um zu prüfen, inwieweit die Kompressionsentropie auch nichtlineare Eigenschaften der zu komprimierenden Zeitreihen erfasst, wurde auf das von Theiler et al. vorgeschlagene Verfahren der Surrogat-Analyse zurückgegriffen [10]. Das Prinzip der Surrogat-Analyse besteht darin, aus gegebenen Originalzeitreihen Ersatzdaten (Surrogate) zu erzeugen, welche sich hinsichtlich der linearen Eigenschaften nicht von den Originaldaten unterscheiden.

Anschließend wurde verglichen, ob der zu prüfende Parameter zwischen Original- und Surrogat-Daten unterscheidet. Treten signifikante Unterschiede auf, so ist davon auszugehen, dass der Parameter auch nichtlineare Eigenschaften misst.

Der Test wurde anhand von simulierten Herzfrequenzzeitreihen durchgeführt. Es wurden 20 Zeitreihen mit einer Dauer von 30 Minuten generiert, bei denen die mittlere Herzfrequenz und gleichzeitig die Varianz variiert wurden. In den Daten sind die für die HRV typischen VLF-, LF- und HF-Schwankungen simuliert.

Zur Generierung des nichtlinearen Signalanteils wurde ein Rössler-Attraktor-System benutzt.

Die Ergebnisse sind in Tabelle 2 zusammengefasst. Während die linearen Standardparameter der HRV meanNN, sdNN, rmssd, LFn (Definition sieh Tabelle 1) sich erwartungsgemäß nicht zwischen Originalzeitreihen und deren Surrogate unterschieden, zeigte die Kompressionsentropie der Surrogate signifikante Erhöhungen. Die erhöhte Entropie in den Surrogaten deutet auf eine Zerstörung der nichtlinearen Struktur bei der Erzeugung der Surrogate hin.

Parameter	Definition
meanNN	Mittelwert aller normaler RR-Intervalle; in ms
sdNN	Standardabweichung der RR-Intervalle; in ms
rmssd	Root mean square aller aufeinanderfolgender RR-Intervalldifferenzen; in ms
LFn	Normierte LF-Leistung der RR-Intervalle

Tabelle 1: Standardparameter der Herzfrequenzvariabilität.

Parameter	ORG		SUR		t-Test
	Mean	SD	Mean	SD	p
meanNN	782	154	782	154	1,00
sdNN	38	13	39	13	0,99
rmssd	22	7	19	6	0,26
LF	0,83	0,17	0,82	0,05	0,85
H_c	0,49	0,01	0,61	0,06	<0,0001

Tabelle 2: Ergebnisse der Surrogatanalyse. Mittelwert und Standardabweichung sowie t-Testergebnisse. ORG-Originalzeitreihen; SUR-Surrogate

3 Ergebnisse

3.1 Normotensive Schwangere und Schwangere mit Präeklampsie

Die Ergebnisse der Variabilitätsanalysen sind in Tabelle 3 und 4 dargestellt. Weder die Standardparameter der Herzfrequenzvariabilität noch die Komprimierbarkeit der Herzfrequenzzeitreihen war bei den Patientinnen mit Präeklampsie signifikant verändert (SW: 0,58 (0,52-0,63) vs. PE: 0,58 (0,53-0,65); p=0,7). Im Gegensatz dazu zeigten sich in der Kompressionsentropie der Blutdruckzeitreihen (SW: 0,63 (0,59-0,65) vs. PE: 0,67 (0,63-0,71); p<0,001) sowie in einigen weiteren Standardparametern der Blutdruckvariabilitätsanalyse signifikante Unterschiede (Tabelle 4).
Der Korrelationskoeffizient zwischen der Kompressionsentropie der Blutdruckzeitreihen und dem mittleren systolischen Blutdruck (CON: 117 (111-130) vs. PE: 154 (131-168); p<0,001) lag bei r=0,4.

	SW		PE		
Parameter	Median	IQA	Median	IQA	p
MeanNN	673	612-670	727	628-796	n.s.
sdNN	44	31-53	43	29-51	n.s.
rmssd	16	10-24	18	12-28	n.s.
LFn	0,72	0,63-0,78	0,70	0,59-0,84	n.s.
H_c	0,58	0,52-0,63	0,58	0,53-0,65	n.s.

Tabelle 3: Median und Interquartilabstand der Herzfrequenzvariabilitätsparameter bei normotensiven Schwangeren (SW) und Schwangeren mit Präeklampsie (PE) sowie u-Testergebnisse: n.s. – nicht signifikant.

	SW		PE		
Parameter	Median	IQA	Median	IQA	p
MeanNN	117	111-130	154	131-168	*
sdNN	8	7-9	9	8-10	*
rmssd	2,5	2,1-2,8	3,1	2,8-3,6	*
LFn	0,86	0,81-0,89	0,87	0,80-0,91	n.s.
H_c	0,63	0,59-0,65	0,67	0,63-0,71	*

Tabelle 4: Median und Interquartilabstand der Blutdruckvariabilitätsparameter bei normotensiven Schwangeren (SW) und Schwangeren mit Präeklampsie (PE) sowie u-Testergebnisse. n.s. – nicht signifikant; * - p<0.01

3.2 Patienten mit implantierbaren Kardioverter-Defibrillatoren

Die Kompressionsentropie der Herzfrequenzzeitreihen unmittelbar vor dem Auftreten von ventrikulären Tachyarrhythmien war verglichen mit den Kontrollzeitreihen signifikant verringert (0,48 (0,41-0,61) vs. 0,49 (0,41-0,53); p=0.007). Kein Standardparameter der Herzfrequenzvariabilität zeigte ansonsten signifikante Unterschiede (siehe Tabelle 5).

	CON		VT		
Parameter	Median	IQA	Median	IQA	p
meanNN	745	656-853	699	585-802	n.s.
sdNN	50	30-66	38	23-56	n.s.
rmssd	17	11-28	15	12-25	n.s.
LFn	0,73	0,62-0,80	0,67	0,51-0,76	n.s.
H_c	0,48	0,42-0,61	0,49	0,41-0,53	*

Tabelle 5: Median und Interquartilabstand der Herzfrequenzvariabilitätsparameter in Kontrollzeitreihen (CON) und vor dem Auftreten von ventrikulären Tachyarrhythmien (VT) sowie Wilcoxon-Testergebnisse: n.s. – nicht signifikant; * - p<0.01

4 Diskussion

Es wurde ein Verfahren vorgestellt, mit Hilfe dessen die Entropie von kardiovaskulären Zeitreihen anhand ihrer Komprimierbarkeit geschätzt werden kann. Die Methodik wurde angewendet, um die Entropie von Herzfrequenz- und Blutdruckzeitreihen bei Patientinnen mit Präeklampsie mit der von normotensiven Schwangeren zu vergleichen und mögliche Hinweise auf eine veränderte kardiovaskuläre Regulation zu erhalten. Des weiteren wurde die aus implantierbaren Kardioverter-Defibrillatoren ausgelesene Herzfrequenzvariabilität vor dem Auftreten von ventrikulären Tachyarrhythmien untersucht, um eine Vorhersage zu ermöglichen.
Die Analysen zeigten, dass die Entropie in den Zeitreihen des systolischen Blutdrucks bei Präeklampsie signifikant vergrößert ist. Offenbar kommt es aufgrund der endothelialen Dysfunktion und dem damit alterierten arteriellen Gefäßzustand, welcher nicht an die Schwangerschaft angepasst ist [11], bei Präeklampsie zu einer reduzierten Redundanz in den Blutdruckschwankungen. Diese Veränderungen stehen jedoch nicht bedingt mit dem Bluthochdruck selbst in Zusammenhang, da die Korrelation zwischen mittlerem systolischen Blutdruck und Kompressionsentropie mit r=0,4 relativ gering war. Folglich lassen sich mit der Kompressionsentropie Informationen über das

kardiovaskuläre Regulationsverhalten bei Präeklampsie gewinnen.

Die Auswertung der Herzfrequenzdaten vor dem Auftreten vor ventrikulären Tachyarrhythmien zeigte eine signifikant verringerte Entropie im Vergleich zu Kontrollzeitreihen, welche auf eine reduzierte Komplexität und damit reduzierte kardiovaskuläre Regulation hindeutet. Dies entspricht auch den Beobachtungen von anderen Autoren, welche eine reduzierte Herzfrequenzvariabilität vor ventrikulären Tachyarrhythmien beschreiben und als Verschiebung der sympathovagalen Balance in Richtung des Sympathikus interpretieren [12, 13, 14, 15].

Ein wesentlicher Vorteil der Kompressionsentropie gegenüber klassischen Variabilitätsmaßen aus der Zeitbereichs- und Frequenzbereichsanalyse besteht neben dem Erfassen von nichtlinearen Eigenschaften in dessen Stabilität und Robustheit. Da zur Komprimierung nur die aktuell im Window bzw. Lookahead Buffer vorliegenden Sequenzen betrachtet werden, haben Artefakte und Extrasystolen nur einen lokal begrenzten Einfluss.

5 Danksagung

Diese Studie wurde mit Mitteln der Deutschen Forschungsgemeinschaft (Projekt Vo505/4-2) und der AiF (KF0318502KLF3) gefördert.

6 Literatur

[1] Ho KK, Moody GB, Peng CK, Mietus JE, LARSon MG, Levy D, Goldberger AL: Predicting survival in heart failure case and control subjects by use of fully automated methods for deriving nonlinear and conventional indices of heart rate dynamics., Circulation., 96, pp. 842-8, 1997

[2] Voss A, Kurths J, Kleiner HJ, Witt A, Wessel N, Saparin P, Osterziel KJ, Schurath R, Dietz R: The application of methods of non-linear dynamics for the improved and predictive recognition of patients threatened by sudden cardiac death., Cardiovasc Res., 31, pp. 419-433, 1996

[3] Signorini MG, Cerutti S, Guzzetti S, Parola R: Non-linear dynamics of cardiovascular variability signals., Methods Inf Med., 33, pp. 81-4, 1994

[4] Shannon CE: A mathematical model of communication., The Bell System Technical J., 27, 379-423 and 623-656, 1948

[5] Li M, Vitany P: An introduction to Kolmogorov complexity and its applications., Springer, 1997

[6] Voss A, Malberg H, Schumann A, Wessel N, Walther T, Stepan H, Faber R: Baroreflex sensitivity, heart rate, and blood pressure variability in normal pregnancy. Am J Hypertens., pp. 1218-25, 2000

[7] Faber R, Baumert M, Stepan H, Wessel N, Voss A, Walther T: Baroreflex sensitivity, heart rate, and blood pressure variability in hypertensive pregnancy disorders. J Hum Hypertens., pp. 707-12, 2004

[8] Lempel A, Ziv J: Universal Algorithm for Sequential Data-Compression. IEEE Trans Inf Th., pp. 337-343, 1977

[9] Baumert M, Baier V, Haueisen J, Wessel N, Meyerfeldt U, Schirdewan A, Voss A: Forecasting of life threatening arrhythmias using the compression entropy of heart rate. Methods Inf Med., pp. 202-6, 2004

[10] Theiler J: Two tools to test time series data for evidence of chaos and/or nonlinearity., Integr Physiol Behav Sci., 29, pp. 211-6, 1994

[11] VanWijk MJ, Kublickiene K, Boer K, VanBavel E: Vascular function in preeclampsia., Cardiovasc. Res., 47, pp. 38-48, 2000

[12] Pruvot E, Thonet G, Vesin JM, Van-Melle G, Seidl K, Schmidinger H, Brachmann J, Jung W, Hoffmann E, Tavernier R, Block M, Podczeck A, Frommer M: Heart rate dynamics at the onset of ventricular tachyarrhythmias as retrieved from implantable cardioverter-defibrillators in patients with coronary artery disease., Circulation., 23, pp. 2398-404, 2000

[13] Wessel N, Ziehmann C, Kurths J, Meyerfeldt U, Schirdewan A, Voss A: Short-term forecasting of life-threatening cardiac arrhythmias based on symbolic dynamics and finite-time growth rates., Phys Rev E Stat Phys Plasmas Fluids Relat Interdiscip Topics., 61, pp. 733-9, 2000

[14] Mani V, Wu X, Wood MA, Ellenbogen KA, Hsia PW: Variation of spectral power immediately prior to spontaneous onset of ventricular tachycardia/ventricular fibrillation in implantable cardioverter defibrillator patients., J Cardiovasc Electrophysiol., 10, pp. 1586-96, 1999

[15] Lombardi F, Porta A, Marzegalli M, Favalle S, Santini M, Vincetti A, DE Rosa A: Heart rate variability patterns before ventricular tachycardia onset in patients with an implantable cardioverter defibrillator. Participating Investigators of ICD-HRV Italian Study Group., Am J Cardiol., 86, pp. 959-63, 2000

Die Erholungsfunktion des Hörnerven und ihr Einfluss auf die Anpassung von Cochlear-Implants

Andre Morsnowski[1,2], Gerd Pfister[1], Joachim Müller-Deile[2],
[1] Institut für Experimentelle und Angewandte Physik, Christian-Albrechts-Universität Kiel
[2] Klinik für Hals-, Nasen-, Ohrenheilkunde, Kopf- und Halschirurgie, Phoniatrie und Pädaudiologie,
Dir: Prof. Dr. P. Ambrosch, Christian-Albrechts-Universität Kiel

Kurzfassung

Ein Cochlear-Implant kann bei taub geborenen Kindern sowie ertaubten und resthörigen Kindern und Erwachsenen die ausgefallene Innenohrfunktion ersetzen. Hierbei werden Sprache und Umgebungsgeräusche in einem Sprachprozessor entsprechend ihrer Frequenzen zerlegt und auf einen im Innenohr implantierten Elektrodenstrang übertragen. Die elektrische Stimulation erzeugt im Hörnerv ein Summenaktionspotential, welches dann im Gehirn einen Höreindruck erzeugt. Die Messung dieser Potentiale kann zur Feinabstimmung der Übertragungscharakteristik der Sprachsignale verwendet werden. Hierzu wird vor allem die Nachweisschwelle elektrischer Stimulation benutzt. Diese ist durch die Refraktärzeit des Hörnervens und dessen Erholungsverhalten beeinflußt, was durch Optimierung der Meßparameter minimiert werden kann.

1 Einleitung

1.1 Die Frequenz-Ort-Abbildung in der Cochlea

Cochlear-Implants sind Hörhilfen, die direkt am Hörnerv ansetzen und diesen elektrisch reizen. Zielgruppe sind postlingual ertaubte Erwachsene und gehörlose Kinder mit angeborener oder erworbener Taubheit mit erhaltener Hörnervfunktion [1]. Mittlerweile werden auch resthörige bisher mit Hörgeräten Versorgte implantiert, da sich gezeigt hat, daß auch diese stark von Cochlear-Implants profitieren können.

Das Cochlear-Implant stellt die natürliche Schallverarbeitung des Ohres nach: beim gesunden Ohr wird die mechanische Energie des Schalls in neuronale Impulse in der Hörschnecke, der Cochlea, umgewandelt. Diese besteht aus 2 ½ Windungen zweier übereinanderliegender Gänge. Die Schwingungen des Steigbügels auf das ovale Fenster am Eingang der Cochlea erzeugen eine Druckwelle, die sich im oberen Gang bis zur Spitze ausbreitet und dann den unteren Gang wieder herunterläuft. Dabei versetzt die Druckwelle die dazwischen liegende Basilarmembran in Schwingungen. Diese übertragen sich auf die Haarzellen im Cortischen Organ, welche ihre Auslenkung in neuronale Impulse umsetzt.

Da die Basilarmembran in ihrer Breite von unten zur Spitze hin zunimmt und die Steifigkeit sich ändert, entsteht eine Wanderwelle, deren Amplitudenmaximum sich je nach der Frequenz des Tones an einer anderen Stelle der Basilarmembran ausbildet. Es liegt also eine Frequenz-Ort-Abbildung, Tonotopie genannt, entlang der Basilarmembran vor.

Fallen die Haarzellen durch Krankheit oder Unfall aus, so kann man durch Implantation eines Elektrodenstrang mit bis zu 22 Elektroden den Hörnerv direkt elektrisch reizen.

1.2 Das Cochlear-Implant

Das System Cochlear-Implant teilt sich in das eigentliche Implantat, welches vollständig unter der Haut des Trägers verborgen ist, und in einen äußeren Teil, welches der Patient am Körper trägt (**Bild 1**).

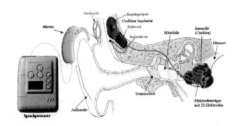

Bild 1 Querschnitt durch ein mit einem Cochlear-Implant versorgten Ohr

Der Patient trägt ein Mikrophon über dem Ohr. Die Signale werden zu einem Sprachprozessor übertragen, der als Filterbank die Signale zu Stimulationsmustern für die einzelnen Elektroden umsetzt. Diese werden über eine Senderspule per Funk durch die Haut an das Implantat übertragen, welches dann die einzelnen (bis zu 22) Elektroden in der Cochlea steuert. Der Elektrodenstrang selbst wird im unteren Gang der Cochlea, also unterhalb des Cortischen Organs, in Lymphe schwimmend plaziert.

Dabei entspricht die Frequenzzuordnung der Wahl einer Stimulationselektrode, während die Lautheit durch die Höhe des Stimulationsstroms repräsentiert wird. Die Abstimmung dieser Parameter erfolgt möglichst individuell für die Elektroden. Der Sprachprozessor führt eine gefensterte Spektralanalyse des aufgenommenen akustischen Signals durch und wählt einzelne zu stimulierende Elektroden aus. Dabei gilt das Ortsprinzip: tiefe Frequenzbänder sind Elektroden in der Spitze der Cochlea zugeordnet, während hohe Frequenzbänder auf Elektroden am Eingang der Cochlea abgebildet werden.

1.3 Sprachprozessor-Anpassung

Weiter muß die Stimulationsstromstärke und damit die wahrgenommene Lautstärke für jede Elektrode einzeln (wie an einem Mischpult) angepasst werden: Es wird die Stromstärke bestimmt, ab der man Geräusche wahrnimmt. Diese wird Threshold oder T-Wert genannt. Weiter wird der „Comfort"- oder C-Wert ermittelt, also die Schwelle, bis zu der noch gerade angenehm laut gehört wird. Geräusche unterhalb des T-Wertes werden im Sprachprozessor nicht verarbeitet. Signale oberhalb des C-Wertes werden nur mit der Stromstärke des C-Wertes selbst übertragen.

Die Versorgung von taubgeborenen bzw. ertaubten Kleinkindern ist besonders dringend, da die Strukturen zum Hören im Gehirn stimuliert werden müssen, damit der Spracherwerb nicht beeinträchtigt wird bzw. überhaupt erfolgen. Diese sollten möglichst in den ersten beiden Lebensjahren versorgt sein. Da diese aber sich nicht mitteilen können, ist es notwendig, diese Schwellen mittels eines objektiven Verfahrens zu schätzen.

1.4 Neural Response Telemetry / NRT-Schwelle

Die Neural Response Telemetry™ ist ein objektives Verfahren, bei dem das Implantat ein Aktionspotential im Hörnerv auslöst, indem von einer Elektrode des Elektrodenstrangs zu einer Elektrode auf dem Implantat ein elektrisches Feld aufgebaut wird. Das Implantat zeichnet über eine Nachbarelektrode dieses Potential auf und sendet die Messwerte zum Sprachprozessor zurück [2].

Abhängig von der Stimulushöhe, also der Stromstärke, wird die Amplitude des Aktionspotentials gemessen. Senkt man die Stimulushöhe von Messung zu Messung, lässt sich die Schwelle der Elektrostimula-

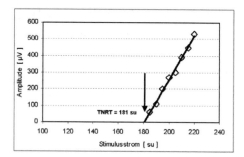

Bild 2 Die Schwelle elektrisch evozierter neuraler Antwort, T-NRT, ist die lineare Extrapolation der Amplitudenwachstumsfunktion.

tion extrapolieren. Diese wird NRT-Schwelle (T-NRT) genannt und für alle Elektroden einzeln bestimmt, was in nur 4 Minuten Messzeit am Patienten durchführbar ist (**Bild 2**). Eine Schätzung für die Hörschwelle, die T-Werte, erhält man, indem der T-Wert einer einzelnen Elektrode mit einem subjektiven Verfahren bestimmt und die anderen T-Werte durch konstante Verschiebung des T-NRT-Profils gewonnen werden [3]. Obwohl dieses Verfahren in der Praxis gute Ergebnisse liefert, gibt es im Einzelfall systematische Abweichungen zwischen psychoakustisch bestimmten Threshold-Werten und den NRT-Schwellen bei postlingual ertaubten Erwachsenen (**Bild 3**).

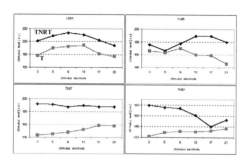

Bild 3 Die T-NRT- und T-Profile von vier postlingual ertaubte Erwachsenen

2 Methode

2.1 Artefaktunterdrückung

Die Erholungsfunktion wird gemessen, indem man zwei Stimuli (Masker und Probe) mit unterschiedlichem Abstand, dem Masker-Probe-Intervall MPI, auf den Hörnerven gibt und die Antwort des Hörnerven aufzeichnet (**Bild 4**). Da die Stimuli im Vergleich zur neuralen Antwort ca. 1:1000 größer sind, nutzt man aus, daß der Hörnerv auf den zweiten von zwei sehr kurz nacheinander gegebenen Stimuli mit dem Abstand refMPI nicht reagieren kann, da die Zellmembran erst ihr ursprüngliches Potential wieder aufbauen muß. Zieht man die Antwort auf eine solche Stimulation von der ursprünglichen Antwort ab, so heben sich die Stimulationsartefakte auf, die neurale Antwort der ersten Spur bleibt erhalten. Weitere Messungen werden benötigt, um die Artefakte und die neuralen Antworten der Masker zu entfernen.

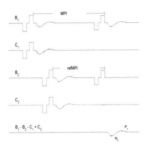

Bild 4 Die elektrisch evozierten Potentiale (rund, stark vergrößert) werden zur Unterdrückung von Stimulationsartefakten als Differenz von maskierten Stimulationen (eckig) gemessen.

2.2 Die Erholungsfunktion

Die Erholungsfunktion wird gemessen durch Variation des MPI. Bei sehr kurzen MPI kann der Hörnerv auf den Probestimulus nicht mit einem Aktionspotential antworten. Erst mit größerem zeitlichen Abstand der Stimuli antwortet der Hörnerv zunächst nur schwach und dann immer stärker bis zur vollen Antworthöhe. Diese Erholungsfunktion beschreiben wir durch ein exponentiell gedämpftes Anwachsen mit einer reziproken Zeitkonstanten α gegen eine Sättigung A. Der Startwert T_0 ist ein Schätzer für die absolute Refraktärzeit (**Bild 5**).

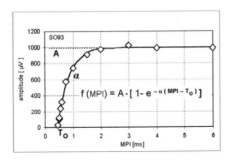

Bild 5 Die Erholungsfunktion mit Parametrisierung

2.3 Schätzung von T_0

In einer Vorstudie [4] konnten wir an 71 Elektroden von 14 Patienten zeigen, daß dieser Wert T_0 in 60 Elektroden kleiner als 500 µs ist (**Bild 6**). Die Amplitudenwachstumsfunktion wird aber traditionell mit einem refMPI von 500 µs gemessen, was bedeutet, daß die Erholungsfunktion einen direkten Einfluß auf die Artefaktreduktion nimmt und damit die T-NRT-Schwellen verschieben sollte.

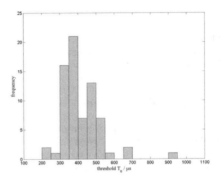

Bild 6 Geschätzte Schwellen T_0 von 71 Elektroden

2.4 Hypothese

Die Erholungsfunktion sollte also den Abstand zwischen T-NRT und T-Wert beeinflussen. Dies wurde in einer mehrzentrischen Studie [5] untersucht, indem bei 17 postlingual ertaubten CI-Trägern an sechs Elektroden mittels NRT Amplitudenwachstumsfunktion und Erholungsfunktion gemessen und psychoakustisch die Schwellen des angenehmen Hörens (C) und die Hörschwelle (T) bestimmt wurden.

3 Ergebnisse

3.1 Innerhalb eines Patienten

An den jeweils sechs Elektroden pro Patient wird der Abstand zwischen T-NRT und T-Wert berechnet und über die Rangreihenkorrelation nach Spearman mit der reziproken Zeitkonstante α der Erholungsfunktion verglichen. Die Rangreihenkorrelation hat den Vorteil, daß sie gegenüber Ausreißern unempfindlich ist. In der Regel erhalten wir eine negative Rangkorrelation. Allerdings ist die lineare Korrelation nicht immer derart ausgeprägt wie im angegebenen Beispiel (**Bild 7**).

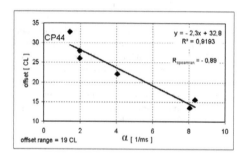

Bild 7 Der Abstand (offset) T-NRT zu T ist gegen die reziproke Zeitkonstante der Erholungsfunktion für eine Patientin aufgetragen.

3.2 Über alle Patienten

Da Korrelationen nur bei geeignet großer Variation der Daten sinnvoll bestimmt werden können, haben wir für 12 der 17 Patienten mit einer genügenden Variation des Abstand zwischen T-NRT und T-Wert die Spearman-Korrelation dargestellt. Die genügende Variation wird ausgedrückt durch den range, also die Differenz von Maximum zu Minimum des Abstandes zwischen T-NRT und T-Wert entlang der ausgewählten sechs Elektroden, von wenigstens 15 Stimuluseinheiten. Dabei ergibt sich, daß alle bis auf eine Korrelation negativ sind. Darüber hinaus sind 5 der 12 Korrelationen sogar auf dem 5% Niveau signifikant negativ (**Bild 8**).

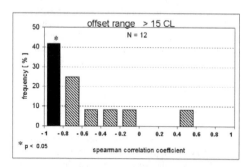

Bild 8 Abstand T-NRT zu T vs. reziproke Zeitkonstante der Erholungsfunktion für eine Patientin

4 Schlußfolgerung und Ausblick

Der Einfluß der Erholungsfunktion auf den Abstand zwischen T-NRT und T-Wert kann über ihren Einfluß bei der Artefaktreduktion erklärt werden und wurde über die Korrelation des obigen Abstandes zur reziproken Zeitkonstante α der Erholungsfunktion nachgewiesen. Derzeit werden die Amplitudenwachstumsfunktionen mit einem verminderten refMPI vermessen, um die Reduzierung des Einflusses der Erholungsfunktion nachzuweisen und so die Hypothese abschließend zu entscheiden.

5 Danksagung

Ein herzlicher Dank geht an die Kollegen in Lyon, Prof. L. Collet, B. Charasse, H. Thai-Van, sowie in Mechelen, M. Killian, mit denen die hier beschriebene Studie gemeinsam durchgeführt wurde, sowie an die Cochlear AG für die finanzielle Unterstützung.

6 Literatur

[1] Lehnhart, E.: Praxis der Audiometrie, 7. Aufl., Thieme, Stuttgart, 1996

[2] Lai, WK.: An NRT Cookbook. Guidelines for making NRT recordings, Schweiz, ISBN 3-9521853-0-2, 1999

[3] Brown, CJ.; Hughes, ML.; Luk, B.; Abbas, PJ.; Wolaver, A.; Gervais, J.: The relationship between EAP and EABR thresholds and levels to program the Nucleus 24 Speech Processor: data from adults, Ear and Hearing, 21 (2) 2000, 151-163

[4] Morsnowski, A.; Charasse, B.; Collet, L.; Killian, M.; Müller-Deile, J.: Reference Masker-Probe-Interval to estimate the recovery function of the auditory nerve using Neural Response Telemetry, 4th International Symposium on Electronic Implants in Otology & Conventional Hearing Aids, Toulouse, June 5-6-7, 2003

[5] Müller-Deile, J.; Morsnowski, A.; Charasse, B.; Thai-Van, H.; Killian, M.; von Wallenberg, E.; Collet, L.: Can auditory nerve refractoriness explain the offset between auditory nerve response thresholds and psychophysical thresholds in cochlear implant recipients?, 4th International Symposium on Electronic Implants in Otology & Conventional Hearing Aids, Toulouse, June 5-6-7, 2003

Neural Response Telemetry™ ist eingetragenes Warenzeichen der Cochlear Limited.

Sound Investigation for Electronic Artificial Larynx*

Thorsten M. Buzug[#] and Michael Strothjohann
Department of Mathematics and Technology, RheinAhrCampus Remagen, Suedallee 2, 53424 Remagen

Abstract

After total laryngectomy the normal voice can be replaced by an electronic artificial larynx (AL). However, the results of the surrogate voice are not overall satisfying due to a robotic clattered sound of that prosthesis. In this paper the results of a sound investigation for an AL are presented. This work is part of our AL research project of laryngectomee's speech enhancement. As a result an autoregressive model of the hearing process is derived that can directly be used for speech enhancement.

1 Introduction

Total laryngectomy, i.e. resection of larynx, is the standard therapy for advanced laryngeal cancer. Within this intervention the connection between the lungs and the mouth and nose is eliminated (see Fig. 1).

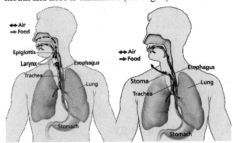

Figure 1: Normal anatomy (left). Anatomy after total laryngectomy showing a so called stoma (right) [1].

As a consequence, patients lost their normal voice. This has extreme social consequences for the patients having undergone the surgery. For those who are not able to learn the so-called oesophageal speech, the electronic voice aid known as artificial larynges (AL) is an option. Figure 2 shows today's ALs of different providers.

Figure 2: Today's product range of speech aids for laryngectomees.

However, today's ALs produce a sound that can be described as monotonous and 'robotic' giving an overall unpleasant clattered voice and unintelligible speech. The main goal of our project is the generation of a more natural AL voice.

In this paper the sound quality of a SMT Servox AL is investigated. Fig. 3a shows the neck-type electronic-voice aid in its charger station. A laryngectomee uses this prosthesis by pressing the device on the neck (see Figure 3b).

The signal is transferred to the mouth cavity as a tone replacement.

This surrogate tone turns to speech by the normal articulation, i.e. interaction of tongue, mouth cavity, teeth and lips. The basic principle of the SMT Servox AL is the electromagnetic driven oscillation of a plunger (Fig. 3d) hitting a hard membrane (see Fig. 3c).

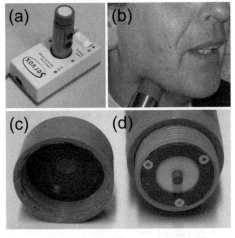

Figure 3: Electronic artificial larynx (AL) and its automatic charger station (a). The AL is pressed on the neck for signal transfer into the mouth (b). Inner parts: Hard membrane (c) and electromagnetic driven plunger (d).

---

* This work has been supported by the BMBF via *Arbeitsgemeinschaft industrieller Forschungseinrichtungen* (AiF) *Otto von Guericke e.V.* Grant 1705600.
[#] Contact: buzug@rheinahrcampus.de, phone: +49 (0) 26 42 / 932 – 318, fax: +49 (0) 26 42 / 932 – 301.

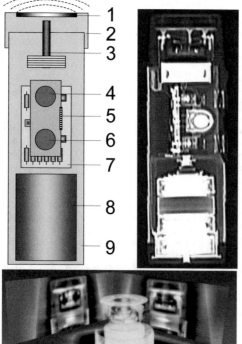

The well known speech production model (Fant [2]) uses a two-state excitation (i.e. impulse-train for voiced and random noise for unvoiced speech), a composition of a vocal tract filter and a lip radiation filter (see Fig. 5a). Speech generated by the laryngectomee using the electronic AL is different. A pitch-controlled impulse-train is applied to the neck and used for voiced and unvoiced speech (see Fig. 3b). The periodically excited hard membrane is producing equally spaced spectral lines at frequencies

$$f_k = k f_0, \quad k = 1, 2, \ldots, \quad f_0 = 70\text{-}100 \text{ Hz}. \qquad (1)$$

The acoustic impulse response of the hard membrane and the corresponding spectrum is shown in Fig. 3.

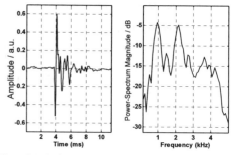

Figure 6: A single acoustic pulse from a common artificial larynx (left) and its spectrum (right). Note the two spectral lines at $f = 0.95$ kHz and $f = 2.2$ kHz are the "finger print" of the mechanical properties of the artificial larynx.

Figure 4: Top left: Principle sketch of artificial larynx: Hard membrane-1, top-unit-2, plunger-3, basis-tone button-4, volume-5, second-tone (pitch) button-6, controller pcb-7, battery-8, titanium body-9. Top right: X-ray fluoroscopy of the AL. Bottom: CT 3D reconstruction of the electric plunger unit.

Figure 4 shows the principle sketch of the SMT AL as well as a X-ray fluoroscopy and the respective 3D CT reconstruction of the electro-mechanical plunger unit. All artificial larynges shown in Fig. 2 are basically working on the same principle.

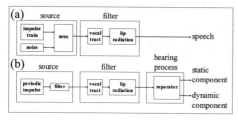

Figure 5: a) Source-filter speech production model. b) Source-filter and hearing-process model for the electronic artificial larynx.

2 Methodology

2.1 Hearing Process

We start with a model of the hearing process to distinguish the artificial from the natural sound. An overview of the simplified behavioral model for this process is shown in Fig. 7.

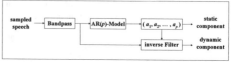

Figure 7: Overview of the hearing-process model used in this paper. Note that the separation of the two components are modeled by an AR(p) model.

The purpose of this model is to distinguish the static speech component from the dynamic component. This is done by a separation of the N-dimensional signal space V^N into two subspaces

$$V^N = V^p_{static} \oplus V^{N-p}_{dynamic}, \qquad (2)$$

where the static speech component is characterized by a p-dimensional vector in the space V^p_{static}.

Please note we are not modeling the speech-signal generation process but the speech-signal hearing process. In our framework this process is not memoryless. Consequently, it is based on an autocorrelation leading straightforwardly to an AR-model. The associated sub-space projection

$$P_N^p : V^N \to V_{static}^p \qquad (3)$$

is sensitive to static speech components and insensitive to dynamic speech components. Therefore, it is predicted to be low dimensional as will be shown below.

2.2 Auto-Regressive (AR)-Model

The hearing process predicts each speech sample s_n by a linear combination of the last p samples (compare Fig. 8):

$$\hat{s}_n = \sum_{i=1}^{p} a_i s_{n-i} \qquad (4)$$

with the prediction error

$$e_n = \sum_{i=1}^{p} a_i s_{n-i} - s_n. \qquad (5)$$

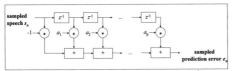

Figure 8: Signal flow for the AR(p) model. The delays z^{-1} are used as storage elements for the past speech samples.

The p coefficients a_i are obtained by minimizing the prediction error energy (Marple [3])

$$\varepsilon_p = \frac{1}{N-p} \sum_{n=p+1}^{N} e_n^2. \qquad (6)$$

2.3 Model Dimension

For each model dimension p the coefficients a_i are determined using the fast *Levinson-Durbin* Algorithm (Ljung [4]), based on the symmetry and *Toepliz* property of the autocorrelation matrix. The elements of this matrix are obtained using the estimator

$$r_k = \frac{1}{N-m} \sum_{n=1}^{N-k} s_{n+k} s_n, \qquad (7)$$

where $k = 0, ..., p-1$. For each model dimension p the associated inverse filter – compare Figure 7 – is build and the energy $E(p)$ for the resulting dynamic component is calculated using *Parseval's* theorem. In our framework the relative model energy

$$R(p) = \frac{E(p)}{E(0)} \qquad (8)$$

is used for the hearing process to find the "best" model dimension. Here, $E(0)$ is the corresponding energy for unprocessed speech.

3. Results

3.1 Relative Model Energy

We analyze the speech of an experienced artificial larynx user (exemplarily shown for a timescale of 4.5 seconds) in the way proposed in the section above. The result is shown in Fig. 9 where the relative model energy $R(p)$ is plotted versus the model dimension p. A sharp energy edge can be found at $p_c = 40$. From that we conclude that the AR(40)-filter is the projection described by Eq. (3) and, furthermore, the corresponding inverse filter is the appropriate transformation for our speech enhancement unit.

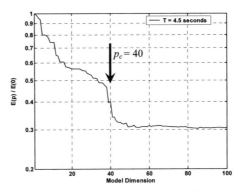

Figure 9: Relative energy $R(p)$ as a function of model dimension p. Note the sharp edge at $p_c = 40$ and the constant energy $R(p) = 0.31$ for $p > 50$.

3.2 Speech-Enhancement Unit

The speech-enhancement unit described here is designed as close as possible to our p_c-dimensional hearing model. The decorrelating (i.e. whitening) inverse filter is realized by a commercial digital signal processor (DSP) that is composed out of 8 parallel processing units and controlled by a Very-Long-

Instruction-Word (VLIW). The processing units are used as data-path units, routed by software and acting on the pre-filtered and sampled data from the analog-digital converter (ADC). Our development system prototype includes a trace buffer and supports for tracing the program flow, executing coverage and profiling the code.

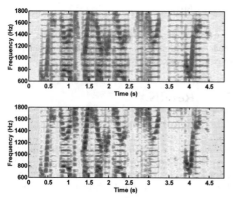

Figure 10: Spectrogram of the German phrase "*Guten Tag. Meine Damen und Herren Sehr schön wie sie wohnen*" spoken by an experienced user of the SMT Servox electronic artificial larynx. Top: Unprocessed artificial speech. Note the constant spectral lines produced by the artificial larynx. Bottom: Dynamic speech component. The ratio of the interline energy to the line energy is enhanced reducing the unpleasant clatter.

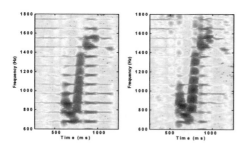

Figure 11: (Enlargement of Fig. 10) Note the fast changing formant frequency producing a high dynamic speech component between the equal spaced artificial-larynx lines (left). This jet-like structure is not seen by our AR(p)-Model (right).

3.3 Dynamic Speech Component

The spectrograms of the unprocessed and processed signals, respectively, are shown for an entire phrase in Figure 10. The constant spectral lines are weaker in the dynamic speech component, thus reducing the unpleasant clatter.

Obviously, the speech production mechanism for the artificial larynx is different from normal speech. It is not an amplitude modulation of the equally spaced spectral lines, but a quasi-adiabatic chirp, produced by a rapidly changing formant (see Fig. 11).

3.4 Hadamard Training-Unit

We design a training unit for the novice user of the artificial larynx measuring the relative interline energy. The unit is adaptable to different sequences or phrases. For performance reasons we use a different system of basis vectors adapted to the equally spaced spectral lines produced by the electronic artificial larynx.

3.4.1 Hadamard-Transform

Any $M \times M$ matrix $\mathbf{H}_M$ with elements $+1$ and -1 satisfying

$$\mathbf{H}_M \; \mathbf{H}_M^T = M \, \mathbf{I}_M \tag{9}$$

is a *Hadamard*-Matrix, where $\mathbf{I}_M$ is the identity of order M. The rows of $\mathbf{H}_M$ form a basis in any M-dimensional space V^M such that

$$\mathbf{x}'_M = \mathbf{H}_M \; \mathbf{x}_M \tag{10}$$

defining a *Hadamard*-Block-Transformation. When M is chosen to be a power of 2, the *Hadamard*-Matrices $\mathbf{H}_M$ are given by factorization (Lee and Kaveh [5]).

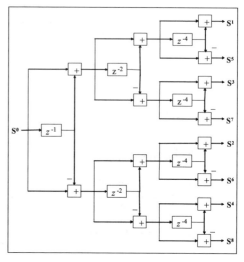

Figure 12: Signal flow for the filter bank F_8. Note the delay lines z^{-k} storing the pre-calculated values.

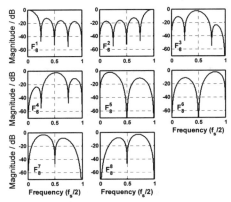

Figure 13: Normalized frequency response for the *Hadamard* FIR-filters $F^i{}_8$. f_s is the sample frequency.

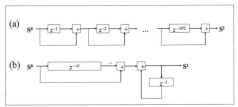

Figure 14. (a) Signal flow for the FIR-filter $F^1{}_M$. (b) Signal flow for the equivalent MA-AR-filter, reducing the arithmetic complexity to two operations per sample.

With

$$\mathbf{H}_1 = \mathbf{I}_1 = [+1], \tag{11a}$$

and

$$\mathbf{H}_M = \begin{bmatrix} +\mathbf{H}_{M/2}, & +\mathbf{H}_{M/2} \\ +\mathbf{H}_{M/2}, & -\mathbf{H}_{M/2} \end{bmatrix} \tag{11b}$$

the fast *Hadamard*-transform (FHT) can be defined similar to the FFT idea.

3.4.2 Hadamard-Filterbank

We interpret the M rows of the *Hadamard*-matrices $\mathbf{H}_M$

$$\mathbf{F}^i_M = \begin{pmatrix} h_{i,1} & h_{i,2} & \cdots & h_{i,M} \end{pmatrix} \tag{12}$$

with $M = 2^k$ and $i = 1, ..., M$ as M finite-impulse response (FIR) filters acting on the speech signal $\mathbf{s}^0$

$$\mathbf{s}^i = \mathbf{F}^i_M \otimes \mathbf{s}^0, \tag{13}$$

where $i = 1, 2, ..., M$. The signal flow and frequency response for the $\mathbf{F}^i{}_8$-*Hadamard* filter-bank are shown in Fig. 12 and Fig. 13.

Depending on the spoken sequence a linear combination of the band-energies for the different channels $\mathbf{s}^i$ is a first measure for the performance of the user. In its simplest form – as shown in Fig. 14 – this linear combination is reduced to one channel $\mathbf{s}^1$.

No floating-point operation is needed for the described orthogonal *Hadamard* filter-bank. Therefore, we are able to transfer it to a common embedded micro-controller reducing the cost and power requirements of the final speech-training unit.

4 Nonlinear Time-Series Analysis

Unfortunately, the efficiency of the linear methods described above is limited, if one wants to give quantitative characterizations of time series obeying nonlinear dynamics. Better results are obtained with more powerful methods which are briefly described in this section.

4.1 Phase-Space Reconstruction

To classify the time series one first reconstruct the phase space (or better embedding space) of the nonlinear dynamical system. This is usually done with Takens' delay time coordinates [6], where a vector in the embedding space is given by

$$\mathbf{x}(t_i) = \left(s(t_i), s(t_i + \tau), ..., s(t_i + \tau(dim_E - 1))\right)^T \tag{14}$$

with $i = 1, ..., N_{dat} - \tau/T_a(dim_E - 1)$, dim_E is the embedding dimension, N_{dat} the number of sampled data points, τ the delay time and T_a the sampling time. To find optimal embedding parameters i.e. the proper delay time τ and a sufficiently large embedding dimension dim_E, one has to calculate the fill-factor $f_{dim_E}(\tau)$ which is a measure of the utilization of the embedding space in any embedding dimension. The fill-factor is defined by

$$f_{dim_E}(\tau) := \log\left(\frac{\sum_{k=1}^{N_{ref}} V_{dim_E,k}(\tau)}{N_{ref}\langle V_{dim_E}\rangle}\right) \tag{15}$$

where $V_{dim_E,k}(\tau)$ is the k-th volume of a parallelepiped defined by $(dim_E + 1)$ corner points which are arbitrarily distributed on the attractor, $\langle V_{dim_E}\rangle$ is a normalization by the volume which covers the attractor in each embedding dimension dim_E and N_{ref} is the number of reference points. The first maxima of the fill-factor, corresponding to a maximum spanned attractor in the embedding space, provide proper delay times. A sufficiently large embedding dimension can be obtained by the convergence of the qualitative structure of the fill-factor for successively increasing embedding dimension. A detailed description of this method can be found in [7].

4.2 Measures of Complexity

To indicate a low dimensional strange attractor one usually calculates the dynamical variables like fractal dimensions, Lyapunov spectra or entropies for increasing embedding dimensions looking for a convergence of these values.

4.2.1 Correlation Dimension

To estimate the fractal dimension of the reconstructed strange attractors in phase space we calculate the correlation dimension D_2 [8]

$$C(R) \propto R^{D_2} \Rightarrow D_2 = \lim_{R \to 0} \frac{\log_{10}(C(R))}{\log_{10}(R)} \qquad (16)$$

R is the scaling radius and C(R) is the correlation integral

$$C(R) \approx \frac{1}{N_{ref}} \sum_{j=1}^{N_{ref}} \frac{1}{N_{dat}} \sum_{i=1}^{N_{dat}} \sigma(R - \|x_i - x_j\|), \qquad (17)$$

where σ is the Heaviside function, N_{dat} is the number of points in phase space and N_{ref} a sufficiently large number of reference points. In Fig. 15 the attractors of normal and artificial voice are embedded in a reconstructed phase spaces via delay-time coordinates – eq. (14). On the attractor of the artificial voice (Fig. 15 right) the construction principle of the correlation integral eq. (17) is visualized for a single reference point.

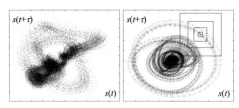

Figure 15. 2D projection of reconstructed phase space of normal (left) and artificial voice (right) embedded via delay-time coordinates.

An inherent stretching, shrinking and folding process leads to a very complicated self-similar structure of the strange attractor which we measured with the correlation dimension.

4.2.2 Lyapunov Exponents

To measure the dynamical behaviour, i.e. the averaged time constants for the stretching and shrinking process on the attractor, one estimates the so called Lyapunov exponents. This must be done by an approximation of the linear flow map T_j using a least squares fit. From the flow map the spectrum of Lyapunov exponents are obtained via

$$\lambda_k = \lim_{t_{ev} \to \infty} \frac{1}{t_{ev}} \log \|T_{j \cdot t_{ev}} e_j^k\| \qquad (18)$$

where e_j^k ($k=1,...,dim_E$) is an orthonormal base. Details of this linear fit are described in [9].

4.2.3 Kaplan-York Dimension and Entropy

From the spectrum of Lyapunov exponents the so called Kaplan-Yorke-Dimension D_{KY} using the conjecture

$$D_{KY} := j + \frac{\sum_{k=1}^{j} \lambda_k}{|\lambda_j + 1|}, \qquad (19)$$

where $\sum_{k=1}^{dim_E} \lambda_k < 0$, $\sum_{k=1}^{j} \lambda_k \geq 0$ and $\sum_{k=1}^{j+1} \lambda_k < 0$.

An essential measure of the "chaoticity" of a systems state is the entropy, which can be estimated from the Lyapunov exponents via

$$h = \sum_k \lambda_k^+. \qquad (20)$$

λ^+ denotes the positive exponents.

5 Conclusions

We proposed a model of the hearing process leading to an enhancement of the electronic artificial larynx speech. The model may also be useful in determining a better exiting function for a new larynx generation, based on piezoelectric actors. In this framework a training unit is presented measuring the training success for novice laryngectomees.

6 Acknowledgement

The authors acknowledge the granting of the project „Klangharmonisierung bei elektronischen Sprechhilfen für Kehlkopflose" [10,11] by BMBF, W 29025 via Arbeitsgemeinschaft industrieller Forschungseinrichtungen „Otto von Guericke" e.V. Köln).

References

[1] National Cancer Institute, http://www.nci.nih.gov/cancertopics/wyntk/larynx
[2] G. Fant, *Acoustic Theory of Speech Production*, Gravenhage, Netherlands: Mounton and Co., 1960.
[3] S. L. Marple, *Digital Spectral Analysis with Applications*, Englewood Cliffs, NJ: Prentice-Hall, 1987, ch. 8.
[4] L. Ljung, *System Identification: Theory for the User*, NJ: Prentice-Hall, 1987, pp. 278–280.
[5] M. H. Lee and M. Kaveh, *IEEE Trans. Acoust., Speech, Signal Processing*, vol. 34, no. 6, pp. 1666–1667, 1986.
[6] F. Takens, Lecture Notes in Math. **898** (Springer, 1980) 230.
[7] Th. M. Buzug and G. Pfister, Physica D (1992) 127.
[8] P. Grassberger and I. Procaccia, Phys. Rev. Lett. **50** (1983) 346.
[9] Th. M. Buzug, *Analyse chaotischer Systeme*, BI-Wissenschaftsverlag, Mannheim, 1994.
[10] M. Strothjohann, *Klangharmonisierung bei elektronischen Sprechhilfen für Kehlkopflose*, Projektbericht, RheinAhrCampus Remagen, 2002.
[11] M. Strothjohann and Th. M. Buzug, *Speech Enhancement for an Artificial Larynx Using a Low-Dimensional Model of the Hearing Process*, Proc. of IEEE EMBS, San Francisco, 2004, pp. 707-710.

Novel Modulating Deep Brain Stimulation Techniques based on Real Time Model in the Loop Concepts

C. Silex [1], M. Schiek [1], N. Hermes [1], H. Rongen [1],
U. B. Barnikol [2], C. Hauptmann [2], H.J. Freund [2], V. Sturm [3], P. A. Tass [2,3]
(1) Central Institute for Electronics, Research Centre Jülich, 52425 Jülich, Gemany
(2) Institute of Medicine, Research Centre Jülich, 52425 Jülich, Gemany
(3) Departemnt of Stereotaxic and Functional Neurosurgery, University Hospital, 50924 Cologne, Germany

Abstract

Synchronization of neuronal activity appears to be the hallmark of several neurological diseases like Parkinson's disease and essential tremor. In patients who do not respond to medication, permanent electrical high frequency (greater than 100 Hz) deep brain stimulation (DBS) turned out to be the therapeutic gold standard. This standard DBS basically mimics the effect of tissue lesoning. However, its mechanism is still not sufficiently understood.
To improve DBS we have studied stimulation induced processes in target areas for DBS in mathematical models. We have implemented these neuronal networks in real time models. This enables us to simulate stimulation induced processes in patients under real time conditions. Our approach makes it possible to overcome typical real time problems, e.g. time delay in information extraction caused by filtering, and thus to evaluate and optimize software and hardware realizing of our novel desynchronizing DBS techniques (so-called model in the loop concept).

1 Graphical programming of real time models

We used a graphical programming language (SIMULINK by Mathworks) to code the mathematical models for real time simulations. This language has several advantages:
- modular programming
- easy reusability of model parts
- transparent and fast modification of model parameters
- easy switching between different types of models (e.g. between instantaneous and delayed coupling)
- online variation of model parameters
- real time simulation on standard PC components

2 Model description

We implemented a model of globally coupled phase oscillators[1]:

$$\frac{d\Psi_j}{dt} = \omega_j - K\left[\sum_{i \neq j}\sin(\Psi_j - \Psi_i)\right] + X(t)S_j(\Psi_j) + F_j(t)$$

The model of 400 phase oscillators was calculated on a time grid of 0.5 ms. The firing density of the whole population and of 4 equally sized sub clusters with additive noise of modifiable amplitude have been converted to analog signals using the National Instruments PCI-6713 D/A card. All four sub clusters could be stimulated independently via the National Instruments PCI-6071 A/D card (12 bit resolution).

$X(t) = \{1 \sim \text{Stimulatin on}; 0 \sim \text{Stimulatin off}\}$
Noise: $\langle F_j(t)\rangle = 0$ and $\langle F_j(t)F_i(t')\rangle = D\delta_{ji}\delta(t-t')$
Stimuli: $S(\Psi_i) = I\cos(\Psi_i)$

Almost all model parameters like coupling strength K, mean and standard variation of the phase oscillator frequencies or noise amplitude D could be modified online during the simulations.

3 Real time evaluation of novel DBS techniques

We tested our stimulation hardware together with a DBS technique based on a coordinated reset of neural subpopulations [2].
The basic idea of this method is to excite a cluster state, i.e. to split the pathological synchronized neurons into several synchronized but phase shifted sub populations (clusters). Within one cluster all neurons are synchronized in phase but the phase difference between the clusters from $2\pi/n$ to $2\pi(n-1)/n$, n being the number of clusters.
In the real time simulations we applied low frequency stimulations to all four sub clusters, with equal stimulation frequency but delayed stimulation onset as shown in **figure 1**. Repetitively the stimulation was applied over 2 cycles followed by a pause of up to 7 cycles.

Due to the so called ‚slaving principle' the model passed into a desynchronized state during the stimulation pause and resynchronized towards the end of the pause (**figure 1**). The desynchronizing effect proved to be robust against significant variations of the model parameters. First intraoperative tests have been encouraging.

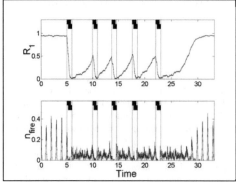

Figure 1: Pattern and desynchronizing effect of phase shifted pulse trains applied via two different electrodes. R denotes the order parameter (synchronization) and n denotes the firing density off all neurons.

3 Literature

[1] Tass, P. A. : Phase Resetting in Medicine and Biology (Springer, Berlin,1999)
[2] Tass, P. A. : Biol Cybern 2003 Aug;89(2):81-8

Zeitaufgelöste diffuse Nahinfrarot-Reflektometrie zur Bestimmung der Hirndurchblutung von Schlaganfallpatienten am Krankenbett

M. Möller[1], A. Liebert[1,2], H. Wabnitz[1], J. Steinbrink[3], R. Macdonald[1], H. Obrig[3]
[1] Physikalisch-Technische Bundesanstalt, Abbestr. 2-12, 10587 Berlin
[2] Institute of Biocybernetics and Biomedical Engineering, Trojdena 4, 02-109 Warschau, Polen
[3] Klinik für Neurologie, Charité, Humboldt-Universität, Schumannstr. 20/21, 10117 Berlin

Kurzfassung

Die zeitaufgelöste diffuse Nahinfrarot-Reflektometrie am Kopf ist ein minimal-invasives Verfahren zur Bestimmung der Durchblutung des Gehirns, das am Krankenbett eingesetzt und in kurzen Abständen wiederholt werden kann. Nach intravenöser Injektion eines Bolus des Farbstoffes Indocyaningrün (ICG) wird dessen Durchgang durch das intra- und extrazerebrale Gewebe simultan in beiden Hemisphären verfolgt. Wir habe diese Methode erfolgreich bei zwei Patienten mit Durchblutungsstörungen im Gehirn angewendet und dabei Unterschiede in der Transitzeit des Bolus zwischen beiden Hirnhälften beobachtet. Unsere Ergebnisse zeigen die Notwendigkeit eines Verfahrens mit sub-Nanosekunden-Zeitauflösung zum Erreichen der benötigten Tiefenselektivität.

1 Einleitung

Die Bestimmung der Durchblutung des Gehirns ist entscheidend für das Überwachung von Patienten, die an zerebrovaskulären Krankheiten leiden, z.B. nach einem Schlaganfall. Alle bisher entwickelten Verfahren, die es erlauben, den zerebralen Blutfluss oder das zerebrale Blutvolumen abzuschätzen, können allerdings nicht kontinuierlich bzw. am Krankenbett eingesetzt werden. Die Kernspintomographie (MRT), gestützt auf die Analyse der Kinetik des Anflutens und des Auswaschens eines intravenös injizierten Kontrastmittels (Gd-DTPA) im Hirngewebe, hat sich zwar zum Standardverfahren entwickelt, um die zerebrale Durchblutung bei einem akuten Schlaganfall zu bestimmen. Allerdings verlangt dieses Verfahren den Transport des Patienten zur Tomographieanlage, wodurch außerdem klinische Risiken erhöht werden. Andere Verfahren erfordern die Anwendung von Radioisotopen oder das Einführen eines arteriellen Katheters. Damit ist eine einfache Methode für nicht-invasive oder minimal-invasive quasi-kontinuierliche Bestimmung der Gehirndurchblutung am Krankenbett immer noch ein Forschungsziel.

Es sind bisher bereits einige Versuche unternommen worden, diffuse Nahinfrarot-Reflektometrie (NIRR) bei mehreren Wellenlängen, kombiniert mit der intravenösen Injektion optischer Kontrastmittel, zur Bestimmung des zerebralen Blutflusses anzuwenden. Allerdings wurde in fast allen diesen Studien mit kontinuierlichem Licht und nur einem einzigen Abstand zwischen Lichtquelle und Detektor gemessen; was dazu führt, dass Veränderungen der Absorption in extrazerebralen Schichten nicht von intrazerebralen Absorptionsänderungen unterschieden werden konnten.

Die wesentliche Quelle von Unsicherheiten bei solchen nicht-invasiven NIRR-Messungen an Erwachsenen ist der große und sehr variable Beitrag aus dem extrazerebralen Gewebe zum Messsignal. Hier können Veränderungen in der systemischen Hämodynamik die Messung verfälschen und so eine verlässliche Bestimmung der fraglichen zerebralen Perfusion vereiteln.

Vor kurzem haben Kohl-Bareis et al. [1] NIRR im Frequenzbereich eingesetzt, um die der Kontrastmittelinjektion folgenden Absorptionsänderungen getrennt in zwei Schichten zu messen. Sie analysierten Änderungen in der Abschwächung und Laufzeit des Lichtes, wobei sich jedoch selbst bei gesunden Probanden eine recht große Variabilität der Änderungen zeigte.

Unseren Untersuchungen zufolge erlaubt die zeitaufgelöste NIRR eine tiefenaufgelöste Analyse von Absorptionsänderungen [2]. Wir werden zeigen, daß mit unserer Technik, die auf dem Nachweis des Durchgangs eines ICG-Bolus – in Analogie zum Gd-DTPA-Bolus bei der Kernspintomographie – beruht, eine verläßliche Überwachung der zerebralen Hämodynamik möglich ist und dass die Beiträge der extrazerebralen Perfusion von dem zu untersuchenden zerebralen Signal getrennt werden können. [3]

2 Theoretischer Hintergrund

Bei der zeitaufgelösten diffusen Nahinfrarot-Reflektometrie werden Pikosekunden-Laserpulse in den Kopf eingestrahlt und die diffus reflektierten Photonen mittels zeitkorrelierter Einzelphotonenzählung, d.h. unter Messung der Laufzeit zwischen Einstrahlung und Detektion, nachgewiesen. Das Messergebnis sind *Laufzeitverteilungen* (LZV) der Photonen, bestehend aus M Zeitkanälen, mit jeweils N_k gezählten Photonen im k-ten Kanal, d.h. mit Laufzeiten $t_k \leq t < t_{k+1}$.

Aus den LZV berechnen wir die folgendermaßen definierten Momente:

$$N_{tot} = \sum_{k=1}^{M} N_k ,$$
$$\langle t \rangle = \sum_{k=1}^{M} t_k N_k \Big/ N_{tot} , \qquad (1)$$
$$V = \left(\sum_{k=1}^{M} t_k^2 N_k \Big/ N_{tot} \right) - \langle t \rangle^2 ,$$

dabei ist N_{tot} die Gesamtzahl der gezählten Photonen (Integral), $<t>$ ist die mittlere Laufzeit, und V die Varianz, also das nullte, das erste und das zweite zentrale Moment der gemessenen LZV. Die mittlere Laufzeit gibt die mittlere Weglänge der Photonen von der Quelle zum Detektor wieder, wogegen die Varianz ein Maß für die Streuung der verschiedenen durchlaufenen Weglängen darstellt.

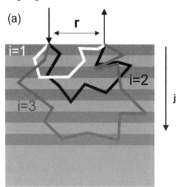

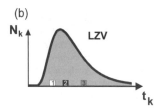

Bild 1: (a) Modell des Kopfes bestehend aus 9 homogenen Schichten (Index j) der gleichen Dicke über einer unteren, semi-infiniten Schicht ($j = 10$). Außerdem sind schematische Trajektorien dreier Photonenbündel (Index i) gezeigt.
(b) Die dazugehörige Photonen-Laufzeitverteilung, Die Laufzeiten der drei in (a) betrachteten Photonenbündel sind eingezeichnet.

Für homogene Medien können Absorptions- und reduzierte Streukoeffizienten leicht aus diesen Momenten bestimmt werden [4]. Allerdings ist der Kopf eines Erwachsenen keineswegs homogen, vielmehr müssen die detektierten Photonen Kopfhaut, Schädel und Gehirnflüssigkeit durchlaufen haben, bevor sie das zu untersuchende Gewebe, d.h. die Großhirnrinde, erreichen. Daher sind Schichtmodelle sehr viel angemessener als die gemeinhin benutzten homogenen Gewebemodelle. Wir haben kürzlich gezeigt, wie Absorptionsänderungen in unterschiedlichen Schichten eines diffus streuenden, mehrschichtigen Modells des Kopfes aus Änderungen in den Momenten abgeleitet werden können [2].

Wir betrachten daher die Ausbreitung von Photonen durch ein semi-infinites, diffus streuendes, vielschichtiges Medium (**Bild 1a**). Es besteht aus j_{max}-1 homogenen Schichten gleicher Dicke, die die tiefstliegende, unendliche dicke Schicht j_{max} bedecken. Weiterhin betrachten wir Bündel von Photonen, die in der obersten Schicht ($j = 1$) in das Gewebe eingestrahlt werden und nach der Propagation – mit zahlreichen Streuvorgängen – durch einige oder alle der betrachteten Schichten die oberste Schicht im Abstand r vom Ort der Einstrahlung wieder verlassen. In **Bild 1a** sind exemplarisch drei solche Propagationswege eingezeichnet.

Unterschiede in der Abschwächung ΔA, in der mittleren Laufzeit $\Delta <t>$ und Varianz ΔV sind folgendermaßen definiert:

$$\Delta A = -\log(N_{tot} / N_{tot,0}) ,$$
$$\Delta \langle t \rangle = \langle t \rangle - \langle t \rangle_0 , \qquad (2)$$
$$\Delta V = V - V_0 ,$$

wobei Größen mit Index 0 die Momente bezeichnen, die vor der Änderung der Absorption infolge der Ankunft des Bolus gemessen wurden.

Die Empfindlichkeiten der verschiedenen Momente für Änderungen der Absorption sind in unterschiedlicher Weise tiefenabhängig. Dies wird beschrieben durch die folgenden *Empfindlichkeitsfaktoren*, bezeichnet als „mean partial pathlength", *MPP*, „mean time of flight sensitivity factor", *MTSF* und „variance sensitivity factor", *VSF*:

$$\frac{\partial A}{\partial \mu_{a,j}} = MPP_j ,$$
$$\frac{\partial \langle t \rangle}{\partial \mu_{a,j}} = MTSF_j , \qquad (3)$$
$$\frac{\partial V}{\partial \mu_{a,j}} = VSF_j ,$$

dabei ist $\mu_{a,j}$ der Absorptionskoeffizient in der j-ten Schicht. Diese Faktoren hängen mit den Weglängen der Photonen in den verschiedenen Schichten zusammen und können durch Monte-Carlo-Simulationen berechnet werden.

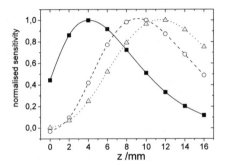

Bild 2: Normierte Empfindlichkeitsfaktoren: *MPP* (Quadrate), *MTSF* (Kreise) und *VSF* (Dreiecke) als Funktion der Tiefe z der Schicht, berechnet für einen Quelle-Detektor-Abstand von 30 mm. Daten aus Monte-Carlo-Simulationen unter der Annahme eines mehrschichtigen Gewebemodells bestehend aus 10 Schichten mit identischen optischen Eigenschaften ($\mu_a = 0.1$ cm^{-1}, $\mu_s = 10$ cm^{-1}, Brechungsindex $n = 1.4$). Eingezeichnete Linien sind nur zur Verdeutlichung.

Zur Illustration zeigt **Bild 2** den Betrag von *MPP*, *MTSF* und *VSF*, jeweils auf das Maximum normiert, als Funktion der Tiefenposition z der jeweiligen Schicht, berechnet für einen festen Quelle-Detektor-Abstand von $r = 30$ mm. Wie man sieht, treten bei festem Quelle-Detektor-Abstand die Maxima von *MPP*, *MTSF* und *VSF* bei immer größeren z auf, d.h., die Varianz ist empfindlicher für Änderungen in tieferen Schichten, verglichen mit den anderen Momenten, wohingegen die Abschwächung am empfindlichsten für Änderungen in oberflächlichen Schichten ist. Die höhere Empfindlichkeit der Varianz für Absorptionsänderungen in tieferen Schichten wird verursacht durch den Faktor t^2 in Gleichung (3): lange Trajektorien von Photonen tragen besonders stark zur Varianz bei.

Veränderungen in der Varianz sollten also am besten geeignet sein, Veränderungen des Absorptionskoeffizienten in tieferen Gewebeschichten – und damit den Durchgang des ICG durch das Hirngewebe – zuverlässig zu verfolgen.

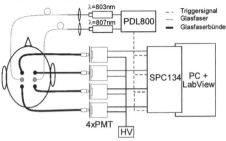

Bild 3: Experimenteller Aufbau (schematisch): PMT = Photomultiplier, PDL800 = Diodenlasernetzgerät, SPC134 = Elektronik zur zeitkorrelierten Einzelphotonenzählung, HV = Hochspannungsversorgung.

3 Methoden

3.1 Experimenteller Aufbau

Das zeitaufgelöste Vierkanalsystem (siehe **Bild 3**), das in der vorliegenden Studie benutzt wurde, misst die diffuse Reflexion des NIR-Laserlichts simultan an vier Positionen auf dem Kopf des Probanden bzw. Patienten. Zwei Pikosekunden-Diodenlaser (PDL-800; PicoQuant, Berlin) emittieren Laserpulse mit einer Breite von etwa 100 ps bei einer Repetitionsrate von 60 MHz bei annähernd gleichen Wellenlängen (803 bzw. 807 nm). Die Wellenlängen wurden so gewählt, daß sie mit dem Absorptionsmaximum von ICG bei ungefähr 800 nm übereinstimmen. Das Licht wurde durch optische Fasern (Stufenindex, Durchmesser 630 µm, numerische Apertur 0.38, Länge 1,5 m) zu einander entsprechenden Positionen auf den beiden Hemisphären geleitet. Das diffus reflektierte Licht wurde an zwei Stellen auf jeder Hemisphäre (jeweils 3 cm frontal und caudal von der Sendefaser) durch Glasfaserbündel (Durchmesser 4 mm, numerische Apertur 0.54, Länge 1,5 m; Loptek Glasfasertechnik, Berlin) gesammelt und zu Photomultiplier-Röhren (PMT, Typ R7400U-02; Hamamatsu Photonics, Japan) geleitet. Die PMTs wurden bei einer Hochspannung von 900 V betrieben. Die Einzelphotonenpulse der vier Detektoren wurden durch vier separate Vorverstärker (Becker & Hickl, Berlin) verstärkt und an ein vierkanaliges System zur zeitkorrelierten Einzelphotonenzählung (TCSPC; SPC 134; Becker & Hickl, Berlin) weitergeleitet.

Mit dieser Anordnung konnten in Intervallen von 50 ms jeweils vier Laufzeitverteilungen mit jeweils 1024 Zeitkanälen (Breite je 12.2 ps) simultan gemessen werden. Die Breite der Apparatefunktion betrug etwa 500 ps.

Für die Messungen an Patienten wurde ein Halter für die Optoden (**Bild 4**) entwickelt, der während der Messung von Hand auf dem Kopf des Patienten festgehalten wurde. Durch den Verzicht auf eine Befestigung mittels Bandagen konnte so die Messung schneller und mit weniger Unannehmlichkeiten für die Patienten durchgeführt werden.

Bild 4: Handhalter für optische Sensoren

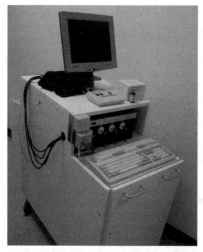

Bild 5: Meßaufbau zur Aufstellung am Krankenbett.

Der Gesamtaufbau bildet ein kompaktes, fahrbares Gerät (Länge 100 cm, Breite 60 cm, Höhe 120 cm, **Bild 5**) das zur Messung an das Krankenbett gefahren werden kann.

3.2 Probanden und Patienten

Zwei gesunde männliche Probanden (zwei der Autoren) im Alter von 33 bzw. 38 Jahren wurden untersucht. Sie hatten keine Vorgeschichte irgendwelcher neurologischer oder sonstiger Erkrankungen, insbesondere keine chronischen Gefäßerkrankungen.
Patientin 1 (57 Jahre) wurde aufgenommen mit einem ischämischen Schlaganfall. Drei Stunden nach dem Auftreten der Symptome wurde eine Thrombolysebehandlung begonnen. Eine Stunde danach wurde sie zum ersten Mal nach unserem Protokoll untersucht. Die Doppler-Sonographie zeigte einen anfänglichen Verschluss der rechten *Arteria cerebri media*. Diese Okklusion wurde im Verlauf der folgenden Tage infolge von Rekanalisation aufgelöst, wie eine erneute Untersuchung zeigte.
Patient 2 war ein 60 Jahre alter Mann, der an einem permanenten Verschluss der rechten und einer etwa 50-prozentigen Stenose der linken *Arteria carotis interna* litt.
Eine Einverständniserklärung wurde von beiden Patienten erhalten. Studie und Studienprotokoll wurden von der Ethikkommission der Charité gebilligt.

3.3 Meßprotokoll und ICG-Bolus

Alle Probanden bzw. Patienten wurden in Rückenlage untersucht. Bei den gesunden Probanden wurden die Optoden über den linken und rechten Zentralregionen mittels Bandagen sicher am Kopf befestigt. Bei den Patienten wurde ein Paar Sensoren über der durch die Computertomografie definierten ischämischen Läsion positioniert, während das andere Paar über dem homologen Gebiet der nicht betroffenen Hemisphäre positioniert wurde.
Indocyaningrün ist ein gut etablierter Farbstoff, der in der Ophthalmoskopie oder zur Untersuchung der Leberfunktion verwendet wird. Bei den Patienten und Probanden wurden insgesamt drei aufeinanderfolgende Boli von 5 mg ICG in 3 ml Kochsalzlösung intravenös injiziert. Die Messung begann 30 s vor der Verabreichung eines Bolus und wurde 3 min lang fortgesetzt. Die Injektionszeit betrug etwa 1 s und die Injektionen wurden erst nach 5 min wiederholt, um den Einfluß der Absorption des restlichen ICG der vorherigen Injektion zu minimieren.

3.4 Datenanalyse

Die gemessenen Laufzeitverteilungen wurden zunächst vorverarbeitet, indem die Hintergrundphotonenzahlen subtrahiert wurden und die differentielle Nichtlinearität der TCSPC-Elektronik korrigiert wurde. Aus den korrigierten LZV wurden das Integral, die mittlere Laufzeit und die Varianz berechnet.
Zur Analyse der Änderungen der Momente als Funktion der Zeit (Abtastfrequenz 20 Hz) wurde zunächst eine laufende Mittelung mit einem Zeitfenster von 10 s (d.h. 200 Werte) durchgeführt. Um den Vergleich zwischen den Änderungen der unterschiedlichen Momente an den einzelnen Messpositionen auf dem Kopf zu vereinfachen, wurden dann die Änderungen dieser Signale auf den Bereich [0, 1] skaliert. Die resultierenden Signale sind in **Bild 6** gezeigt.

4. Ergebnisse

4.1 Probanden

Die beiden oberen Reihen in **Bild 6** zeigen Änderungen in der Abschwächung ΔA, in der mittleren Laufzeit $\Delta <t>$ und in der Varianz ΔV von LZV, die nach der Verabreichung von ICG bei den beiden gesunden Probanden gemessen wurden. Der Bolus verursachte einen schnellen Anstieg in der Abschwächung (A) und einen Abfall in den anderen beiden Momenten ($<t>$, V). Ihre zeitlichen Profile unterscheiden sich, was auf unterschiedliche Anflutungs- und Auswaschungskinetik des Kontrastmittels in verschiedenen Schichten des Gewebes hindeutet.
Bei dem ersten Probanden (oberste Reihe in **Bild 6**) erreicht die an verschiedenen Optoden gemessene ICG-induzierte Änderung ΔA der Abschwächung ihren Scheitelwert nach unterschiedlichen Zeiten, und zeigt unterschiedliche Steigungen der Vorderflanke. Tatsächlich unterscheiden sich sogar die Latenzzeiten der ΔA-Signale innerhalb einer Hemisphäre deutlich zwischen den beiden Optoden. Im Gegensatz dazu

zeigen $\Delta<t>$ und ΔV keine solche Streuung.
Bei dem anderen Probanden zeigen sowohl ΔA als auch $\Delta<t>$ eine ortsabhängige Streuung der Latenzzeiten des Scheitelpunktes. Hier zeigen nur die Änderungen in ΔV ein simultanes Minimum, unabhängig von der Position auf der jeweiligen Hemisphäre.

Um die Reproduzierbarkeit der Änderungen der Varianz ΔV zu überprüfen, wurden Proband 1 drei aufeinanderfolgende Boli injiziert. Es ergaben sich sehr ähnliche Latenzzeiten und Kurvenformen von ΔV nach allen drei Boli.

Berücksichtigt man, daß die Tiefe, in der die mittlere Laufzeit am empfindlichsten auf Änderungen der Absorption ist, zwischen den Tiefen der größten Empfindlichkeit für Abschwächung und Varianz liegt (siehe **Bild 2**), und daß die Dicke von Haut und Schädel wie auch deren optische Eigenschaften sich zwischen den Probanden unterscheiden, so legen diese Ergebnisse nahe, daß zur Änderung der mittleren Laufzeit auch das extrazerebrale Gewebe – in unterschiedlichem Maße – beiträgt.

Die einzelnen Beobachtungen können dadurch erklärt werden, daß die Perfusion der Hirnrinde relativ schnell und homogen ist – verglichen mit der Perfusion des extrazerebralen Gewebes, die infolge des inhomogenen Blutflusses in der Haut sehr variabel ist.

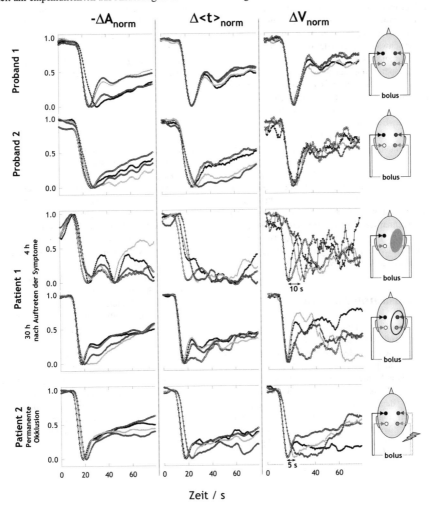

Bild 6: Normierte Änderungen der Momente der Photonen-Laufzeitverteilungen (ΔA = Abschwächung, $\Delta<t>$ = mittlere Laufzeit und ΔV = Varianz) nach der Verabreichung eines ICG-Bolus (Injektion zur Zeit $t = 0$). Zur Normierung: siehe Text. Erste und zweite Reihe: Ergebnisse von gesunden Probanden. Dritte und vierte Reihe: Ergebnisse von Patient 1 (Schlaganfall rechtsseitig). Dritte Reihe: Messung 4 Stunden nach Beginn der Symptome. Vierte Reihe: Messung 30 Stunden nach Beginn der Symptome. Fünfte Reihe: Patient 2 (permanenter Verschluss der rechten und 50%-Stenose der linken *Arteria carotis interna*)

4.2 Patienten

Die dritte und vierte Reihe in **Bild 6** zeigen die Ergebnisse von Patientin 1, die an einer akuten zerebralen Ischämie durch Verschluss der rechten *Arteria cerebri media* litt. An der Patientin wurde zweimal Messungen vorgenommen:
Die erste Messung fand 4 Stunden nach dem Auftreten der Symptome statt. In dieser akuten Phase konnte im zeitlichen Verlauf der Änderung der Varianz ein klarer Unterschied zwischen den beiden Hemisphären beobachtet werden: über der betroffenen Hemisphäre wird das Minimum später erreicht als über der nicht betroffenen Hemisphäre. Dieser Zeitunterschied ist bei der Abschwächung und der mittleren Laufzeit viel weniger klar zu sehen.
Einen Tag später war der aus ΔV bestimmte Unterschied in der Latenzzeit zwischen den Hemisphären, verschwunden. Dies könnte anzeigen, daß eine Rekanalisation des Verschlusses der Arterie durch die Thrombolyse-Behandlung erreicht worden war.
Wie schon angesprochen, litt der andere Patient an einem vollständigen Verschluss der rechten und einer 50-prozentigen Stenose der linken *Arteria carotis interna*. Die zeitlichen Profile, die zur rechten Hemisphäre gehören (untere Reihe in **Bild 6**) ähnelten den Ergebnissen bei der Patientin 1 in der akuten Phase des Schlaganfalls. Änderungen in der Varianz sind über der rechten Hemisphäre deutlich verzögert. Wiederum ist der Unterschied zwischen den Hemisphären bei den zeitlichen Änderungen der Abschwächung und der mittleren Laufzeit weniger ausgeprägt.

5. Zusammenfassung

Wir haben ein nicht-invasives Verfahren zur Verfolgung der Hirndurchblutung entwickelt, das am Krankenbett eingesetzt werden kann. Es basiert auf mehrkanaliger zeitaufgelöster diffuser Reflektometrie, wobei der Durchgang eines intravenös injizierten Kontrastmittelbolus durch das Gehirn verfolgt wird.
Die aus den gemessenen Photonen-Laufzeitverteilungen abgeleitete Varianz ist ein Maß für die Streuung der Weglängen des Lichtes durch den Kopf und ein geeigneter Parameter, um Absorptionsänderungen in der Großhirnrinde, getrennt von denen in Haut oder Schädel, zu verfolgen. Die durch die Auswertung der Varianz gegebene Tiefenselektivität ist insbesondere bei erwachsenen Patienten unverzichtbar, womit die Varianz klar als Parameter zur Verfolgung des Durchgangs des Kontrastmittels durch das *intrazerebrale* Gewebe favorisiert ist.
Das Kontrastmittel Indocyaningrün gilt nach bestem Wissen als sicher, was wiederholte Messungen in kurzen Zeitabständen erlaubt.
Wir haben unser Verfahren erfolgreich an zwei Patienten mit einseitig beeinträchtigter Perfusion eingesetzt. Insbesondere konnte bei einem Patienten mit akutem Schlaganfall ein signifikanter Unterschied zwischen beiden Hemisphären beobachtet werden, der, nach Behandlung, nach einem Tag verschwand.
Der Nutzen des Verfahrens zur Überwachung der Hirndurchblutung muss im Rahmen einer klinischen Untersuchung an einer größeren Zahl von Patienten validiert werden.

8 Literatur

[1] Kohl-Bareis M., Obrig H., Steinbrink J., Malak J., Uludag K., Villringer A., *Noninvasive monitoring of cerebral blood flow by a dye bolus method: Separation of brain from skin and skull signals*, J. Biomed. Optics, 7, 464-470 (2002).

[2] Liebert A., Wabnitz H., Steinbrink J., Obrig H., Möller M., Macdonald R., Villringer A., Rinneberg H., *Time-resolved multi-distance NIR spectroscopy of the adult head: intra- and extracerebral absorption changes from moments of distributions of times of flight of photons*, Appl. Opt., 43, 3037-3047 (2004)

[3] Liebert A., Wabnitz H., Steinbrink J., Möller M., Macdonald R., Rinneberg H., Villringer A., Obrig H., *Bed-side assessment of cerebral perfusion in stroke patients based on optical monitoring of a dye bolus by time-reolved diffuse reflectance*, NeuroImage (im Druck)

[4] Liebert A., Wabnitz H., Grosenick D., Möller M., Macdonald R., Rinneberg H., *Evaluation of optical properties of highly scattering media using moments of distributions of times of flight of photons*, Appl. Opt., 42, 5786-5792 (2003).

Muscle oxygenation and blood content of muscle at rest and during exercise assessed by near-infrared spectroscopy

Oliver Rohm[1], Christiane Andre[1], Roland Gürtler[1], Patrick Neary[2], Matthias Kohl-Bareis[1]
1 RheinAhrCampus Remagen, University of Applied Sciences Koblenz, Südallee 2, D-53424 Remagen
2 Faculty of Kinesiology, University of New Brunswick, Fredericton, Canada

Abstract

The local haemoglobin status and blood flow is an important parameter for the assessment of muscle physiology, particularly during exercise in both health and disease. Here we investigate the prospect of a non-invasive, local measurement of oxygen saturation of haemoglobin and myoglobin using optical spectroscopy with the aim of i) development and validation of algorithms, and ii) supplying a tool for applications in muscle physiology and sports medicine. The methods applied encompass spatially resolved and time–resolved optical spectroscopy in the NIR wavelength range with diffusion approximation as a description of the transport of photons in tissue. The paradigms include arterial and venous cuff occlusion on the arm, and muscle exercise during dynamic incremental cycling.

1 Introduction

Currently, the main emphasis in near infrared (NIR) monitoring of tissue haemoglobin is on changes in the oxygenated and deoxygenated components (oxy-Hb and deoxy-Hb). However, in many applications the knowledge of the tissue haemoglobin oxygen saturation SO_2, i.e. the fraction of oxy-Hb with respect to the total haemoglobin concentration is needed to understand the factors that affect oxygen delivery and consumption to the myocyte.

A relatively simple approach has been suggested to obtain SO_2 from measurements of the slope of intensity (attenuation) with respect to the distance of light source and detector on the tissue [1, 2]. This method is called spatially resolved spectroscopy (SRS) and is implemented by a commercial instrument, the NIRO-300 from Hamamatsu Photonics (Japan), which uses four wavelengths in the NIR range and two detectors separated by a few millimetres. This approach has been validated theoretically and in tissue simulation models, and has already been used in physiological and clinical trials [3, 4]. However, there are a number of spectroscopic questions which have not been thoroughly scrutinized and remain unanswered.

Figure 1 Experimental set-up for the analysis of muscle oxygenation during exercise

The emphasis of this paper is twofold. First, the aim is the validation of the SRS approach. Therefore, the experimental set-up is built to closely adopt the measurement conditions of the NIRO-300, but with a number of important differences. On the one hand, the number of wavelengths is not limited to 4 but a full spectrum (550-1000nm) approach is followed here which allows a truly spectroscopic inspection of the oxygenation data. Second, six detectors are used instead of 4 for an assessment of errors due to the inhomogeneities of the tissue, especially close to the skin surface. In addition, a coregistration and comparison with time-resolved spectroscopy (TRS) is included. The second rationale of this paper is the application of these methods for the non-invasive monitoring of muscle oxygenation during rest and dynamic exercise. NIR spectroscopy has already been used for an investigation of muscle physiology [5 - 9], however, mainly with systems that are "trend-indicating" and can only measure changes in haemoglobin concentration rather absolute values and saturation. Thus, methodological questions, as raised here, need to be answered to allow a broader application of this optical technology in muscle and sport physiology.

2 Method

2.1 Experimental Setup

The spectroscopy system is based upon a Peltier-cooled slow-scan CCD-camera (Spec-10, Roper Scientific Germany) with a 1340 x 400-detector array and 16 bit dynamic range. The detector is coupled to a spectrometer (SP-150, Acton Research) with a diffraction grating of 300 grooves / mm and a blaze wavelength of 750 nm resulting in an overall detection range between 550 and 1000 nm. On the skin six light detecting fiber bundles (Loptek GmbH) of 50 μm core fibers and a diameter of 1 mm each were arranged in a line separated by $\Delta\rho = 2.5$ mm at a mean distance $\rho = 35$ mm from the light delivering bundle (5 mm diameter; 50 μm core fibers). The detection bundles were connected to the input slit of the spectrometer with a separation of 0.5 mm to clearly separate the six independent spectra recorded. The output of a standard 50 W halogen light source (at 12 V) was focused with a condenser optic into the light delivering fiber bundle. A mechanical shutter (Prontor Magnetic, Prontor GmbH Germany) synchronized with the acquisition cycle of the CCD-camera blocked the light input during CCD-readout time thus minimizing the cross-talk between the different fiber signals. An integrating sphere of 8 inch diameter (IS-080-SF, Labsphere Inc.) served as a normalisation standard for the relative light intensities of the six detector channels and for correction of the wavelength dependence of the whole optical system. To acquire the 6 reference spectra ($I_{0,1-6}(\lambda)$) the fiber probe was moved to a two inch aperture of the integrating sphere with the illuminating fiber irradiating through a ½ inch aperture. The acquisition rate was between 0.5 and 2 Hz including a readout time of the CCD-chip of approximately 16 ms. No pixel binning was set for the horizontal (wavelength) axis of the CCD while for its vertical axis binning was selected over the pixels illuminated by each fiber. Including the broadening due to the entrance slit of the spectrometer the wavelength resolution was approximately 8 nm. Literature haemoglobin extinction spectra were smoothed accordingly. Count rates were typically in the range of 3000 s^{-1} pixel^{-1} for each fiber. Data acquisition and on-line analysis was programmed in Labview 6.0 (National Instruments Inc.) based on the driver package SITK (Roper Scientific) giving full control of all camera functions as well as an on-line calculation of absorption coefficient $\mu_a(\lambda)$, saturation SO_2 and chromophore concentrations. All raw spectra were stored for additional and independent off-line data evaluation in Matlab (Version R12; Mathworks Inc., USA). A second set-up based on time-correlated-single-photon counting technique consisted of three diode lasers at wavelengths 686, 783 and 836 nm (Sepia PicoQuant GmbH, Germany) with a pulse length of about >100 ps and up to 80 MHz repetition rate, four photomultiplier tubes (H7422-P50; Hamamatsu Inc. Japan) for photon counting and detector electronics based on 4 PCI-cards (SPC-134; Becker & Hickl GmbH, Germany). A measurement chamber with defined fiber distance and variable attenuation from neutral density filters served as absolute reference for the reflectance time with respect to the laser pulses and for measurement of the instrument response function.

2.2 Paradigm

In the first paradigm, muscle oxygenation was measured in 8 volunteers with a mean age of 29.3 years during cuff occlusion. A pneumatic cuff was placed around the left and right arm above the elbow and was inflated up to 260 mm Hg for 8 min in order to induce an arterial and venous occlusion. The light guiding fibers were placed on the main body of the brachioradialis. The SRS - and TRS - system was placed on the left and right arm, respectively.

In the second paradigm, muscle oxygenation of the left vastus lateralis muscle was recorded in 12 volunteers with the SRS – instruments during an ergometer cycling exercise. The exercise was incremental in steps of 2 min duration (power 50 – 350 W) on a racing mountain bike mounted on a stationary training system. Power, cadence as well as heart rate were recorded and stored together with the optical data.

2.3 Data Analysis

SRS - method

Intensities at the 6 detectors ($I_{1-6}(\lambda)$) were converted into attenuation changes ($A_{1-6}(\lambda) = \log_{10}(I_{0,1-6}(\lambda)/I_{1-6}(\lambda))$, where $I_{0,1-6}(\lambda)$ are the reference intensities measured in the integrating sphere. Following Patterson et al. (1989) [5] and Matcher et al. (1993) [1] $\mu_a(\lambda)$ can be derived from the slope of attenuation with respect to ρ :

$$\mu_a(\lambda) = \frac{1}{\mu_s'(\lambda)} \cdot \frac{1}{3} \cdot \left(\ln(10) \frac{\partial A(\lambda)}{\partial \rho} - \frac{2}{\rho} \right)^2. \quad (1)$$

Here the exact knowledge of $\mu_s'(\lambda)$ is required, which is generally not known. For most tissues $\mu_s'(\lambda)$ is a simple function of wavelength and can approximately be described by a function $\mu_s'(\lambda) = \mu_{s,0}' \cdot (1 - s \cdot \lambda)$. Matcher et al. (1993) [11] measured values of the intercept $\mu_{s,0}'$ and the slope s for leg, arm and head and found values $\mu_{s,0}' = 1.55$ mm^{-1} and $s = 5.9 \cdot 10^{-4}$ mm$^{-1} \cdot$nm^{-1} for the arm. While from the $\mu_a(\lambda)$-spectra absolute haemoglobin concentrations can only be calculated when absolute $\mu_s'(\lambda)$ are assumed, haemoglobin oxygen saturation values SO_2 = oxy-Hb/(oxy-Hb + deoxy-Hb) can be derived when contributions from other tissue chromophores are neglected or subtracted.

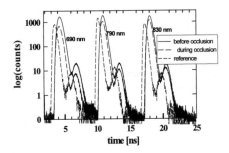

Figure 2 Time – resolved reflectance spectra measured for three wavelengths. For each wavelength the reference spectrum as well as the tissue reflectance spectra before and during cuff occlusion is shown.

TRS - method
Time-resolved reflectance data (see Fig. 2) were fitted to analytical model solutions for a homogeneous half-space taking into account the finite instrument response with the transport scattering μ_s' and absorption coefficients μ_a as free parameters. According to Patterson et al. [10], the reflectance can be written as a function of time t and distance d between source and detector as:

$$R(d,t) = (4\pi D)^{-3/2} z_o t^{-5/2} \exp(-\mu_a ct) \cdot \exp\left(-\frac{d^2 + z_o^2}{4Dt}\right)$$

where $z_o = 1/\mu_s'$ and $D = \dfrac{c}{3(\mu_a + \mu_s')}$ and the velocity of light c. Again, from the absolute absorption coefficients at the tree wavelengths values for SO_2 were obtained.

Test of SRS - analysis
The analysis of the collected data allowed for an assessment of the SRS - method with respect to the following criteria: i) influence of the wavelength range used for data fitting, ii) noise in the slope $\partial A/\partial \rho$, iii) effect of uncertainties in the slope of $\mu_s'(\lambda)$ with respect to λ and iv) uncertainties in the contributions of other chromophores like water and fat.

3 Results

In Figure 3 an example of the attenuation spectra $A_{1-6}(\lambda)$ is shown before and during arm cuff occlusion with the prominent feature close to 760 nm as well as the increase towards shorter wavelengths mainly due to deoxy-Hb and the strong water absorption peaking at 970 nm. The change in the spectra resulting from the change in the oxygenation is apparent. From the wavelength dependent of the slope $\partial A/\partial \rho$ μ_a – spectra were calculated from Eq. 1. From Fig. 4 it is apparent that $\mu_a(\lambda)$ changes rather dramati-

cally during occlusion. Fitting the haemoglobin extinction spectra gives a good description of the experimental data though the residual depends on assumed water and fat concentration and absorption background as well as the wavelength dependence of μ_s'.

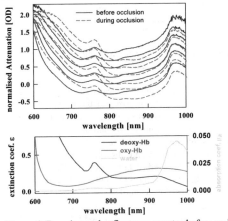

Figure 3 Experimental reflectance spectra before and during cuff occlusion measured on the forearm for six source – detector spacings between xxx and xxx mm. In the lower figure the extinction coefficients of the haemoglobin components as well as water is shown.

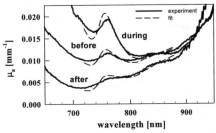

Figure 4 Spectra of tissue absorption coefficient μ_a derived from SRS obtained before, during and after cuff occlusion of the forearm. The spectra obtained by fitting the haemoglobin spectra are shown in dotted lines.

The time course of SO_2 values during arm cuff occlusion is shown in Fig. 5 for a single volunteer. There is an overall agreement between the two independent measurements with a gradual decrease in SO_2 down to values of about 10%. After pressure release higher blood flow results in values higher than at the beginning. Similar time courses were obtained in all volunteers.

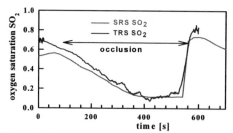

Figure 5 Time course of SO2 during cuff occlusion measured with SRS and TRS systems.

Though the agreement indicated by Fig. 5 is rather assuring, the question arises of the uncertainties and errors in the conversion of μ_a - spectra into haemoglobin and SO_2 values. To analyse this points the influence of wavelength range used for the fitting and of assumed background absorption due to water and lipid was calculated (data not shown). The conclusion is that the broad band SRS approach has the inherent advantage to allow a detailed spectroscopic analysis.

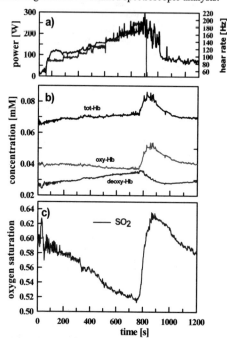

Figure 6 Time course of a) power output during a cycling ergometer exercise and heart rate, b) haemoglobin/myoglobin concentration based on the SRS approach and c) muscle SO_2.

In the second paradigm the SRS approach was tested for a registration of muscle oxygenation during a cycling exercise. In Fig. 6 and 7 typical data are shown for a volunteer with the closely correlated increase in heart rate with power. A higher power is linked with a higher oxygen extraction resulting in the overall decrease in oxygenation. The rise in total haemoglobin shown in Fig. 6b signifies an increase in vessel diameter. After ending the exercise, SO_2 overshoots which is consistent with an increased blood flow when compared with rest conditions.

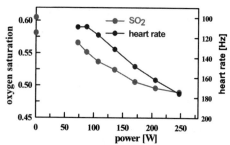

Figure 7 Correlation of power and oxygen saturation and heart rate for the data during a cycling exercise shown in Fig. 6.

Similar data were obtained in all volunteers. However, comparable to the data during cuff occlusion the calculated SO_2 values and its decrease during exercise was dependent on the assumptions made for the tissue composition.

4 Discussion

The measurement of muscle oxygenation based on optical methods has the clear advantage of providing in-vivo and localised values of physiological data.
A measurement based on the different absorption coefficients of haemoglobin and myoglobin seems obvious and straightforward, however, difficulties arise from the diffuse propagation of photons in tissue, the inhomogeneity and the unknown optical properties of the tissue. As there is no other non-invasive tool to compare with, the approach followed here is to register apparently redundant information, i.e. broad band spectra at 6 detector positions for the SRS in addition to time-resolved data. This permits a crosscheck and estimation of errors for all the calculated data.

The data analysis performed for the cuff occlusion experiments point at the following conclusions: i) There is an overall good agreement between saturation values obtained from SRS– and TRS- systems (Fig. 4). However, there is an uncertainty of between 5 to 10 % in the SO_2 values due to the choice in wavelengths used for the fitting (data not shown here) with larger errors for low SO_2 values. This might explain the discrepancies in similar experiments when two commercial instruments were compared [3]; ii)

There is no indication that the TRS method has an inherent advantage over cw methods. On the contrary, the main error is probably due to unknown tissue background absorption which can only analyse with multi-wavelengths data. In this light the broad band SRS approach seems to be the most prominent. Currently, we are undertaking further analyse by developing techniques to separate tissue components for different tissue layers (i.e., skin, fat, muscle).

The analysis of muscle performance based on optical spectroscopy seems promising. In a future study an extension is planned to investigate the influence of oxygen supply on muscle performance and oxygenation.

The ultimate measurement system for non-invasive muscle oxygenation would combine haemoglobin/myoglobin measurement based on reflectance spectroscopy and blood flow or perfusion registration based on laser-Doppler flowmetry in a portable unit. Though ambitious, this does not seem unrealistic in the light of recent technological developments.

5 References

[1] Matcher S J, Kirkpatrick P, Nahid K, Delpy D T (1993) Absolute quantification method in tissue near infrared spectroscopy *SPIE 2389*, 486 – 495

[2] Suzuki S., Takasaki S., Ozaki T., Kobayashi Y. (1999) A tissue Oxygenation monitor using NIR spatially resolved spectroscopy. *SPIE 3597*, 582 – 592.

[3] Komiyama T, Quaresima V, Shigematsu H, Ferrari M (2001) Comparison of two spatially resolved nearinfrared photometers in the detection of tissue oxygen saturation: poor reliability at very low oxygen saturation *Clinical Science* **101**, 715–718

[4] Al-Rawi P., Smielewski P., Kirkpatrick P. J. (2001) Evaluation of a Near-Infrared Spectrometer (NIRO 300) for the Detection of Intracranial Oxygenation Changes in the Adult Head *Stroke*. 32, 2492 - 2500.

[5] Neary, J.P., McKenzie, D.C., Bhambhani, Y.N. (2002). Effects of short-term endurance training on muscle deoxygenation trends using NIRS. Med Sci Sports Exerc 34:1725-1732.

[6] Bhambhani, Y.N., Buckley, S., Susaki, T. (1997). Detection of ventilatory threshold using near infrared spectroscopy in men and women. Med Sci Sports Exerc 29(3):402-409.

[7] Bhambhani, Y.N. Maikala, R., Esmail, S. (2001). Oxygenation trends in vastus lateralis muscle during incremental and intense anaerobic cycle exercise in young men and women. Eur. J. Appl. Physiol. 84:547-556.

[8] Miura, H., McCully, K., Chance, B. (2003). Application of multiple NIRS imaging device to the exercising muscle metabolism. Spectroscopy. 17:549-558.

[9] Ferrari, M., Binzoni, T., Quaresima, V. (1997). Oxidative metabolism in muscle. Philos. Trans. R. Soc. Lond. B. Biol. Sci. 352:677-683.

[10] Patterson M., Chance B., Wilson C. (1989) Time resolved reflectance and transmittance for the noninvasive measurement of tissue optical properties, *Applied Optics* **28**, 2331–2336

[11] Matcher S J, Cope M, Delpy D T (1997) In vivo measurements of the wavelength dependence of tissue scattering coefficients between 760 and 900 nm measured with time resolved spectroscopy, *Applied Optics* **36**, 386- 396

System for the measurement of blood flow and oxygenation in tissue applied to neurovascular coupling in brain

Matthias Kohl-Bareis[1], Roland Gürtler[1], Ute Lindauer[2], Christoph Leithner[2], Heike Sellien[2], Georg Royl[2], Ulrich Dirnagl[2]

1 RheinAhrCampus Remagen, University of Applied Sciences Koblenz, Südallee 2, D-53424 Remagen
2 Experimentelle Neurologie, Charité, Humboldt Univ. Berlin, Schumann Str. 20/21, 10098 Berlin

Abstract

We designed a system incorporating the independent measurement of blood flow and oxygenation of haemoglobin. This is based on laser-Doppler spectroscopy with NIR wavelengths which gives a measure for changes in blood flow or tissue perfusion as well as reflectance spectroscopy in the VIS wavelength range for the calculation of the oxygenated and deoxygenated haemoglobin components. The co-registration of these parameters allows the neurovascular coupling of brain to be investigated. This is demonstrated by recording functional activity of the rat brain during electrical forepaw stimulation.

1 Introduction

The determination of absolute haemoglobin concentrations (oxy-Hb, deoxy-Hb) and oxygen saturation SO_2 = oxy-Hb/(oxy-Hb + deoxy-Hb) in tissue is still subject of a number of recent improvements in experimental methods and instrumentation. The basic spectroscopic task of calculation of chromophore concentrations is obstructed by the effect of the scattering of photons in tissue. In any non-scattering medium the standard approach of the Lambert-Beer law exploits the linear relationship between absorption coefficient μ_a ($\mu_a = c \cdot \varepsilon$, where c is the concentration and ε the extinction coefficient of the chromophore) and light attenuation A of light: $A = \mu_a \cdot x = c \cdot \varepsilon \cdot x$, where x is the optical pathlength. However, in scattering media the pathlength itself is a function depending on the absorption as well as the scattering properties and hence no simple relationship exists to recover absorption coefficients or chromophore concentrations from attenuation measurements.

There have been different methods proposed to calculate haemoglobin oxygen saturation of tissue. The focus has been on the NIR range for its larger tissue penetration with techniques varying from spatial resolved spectroscopy to time and frequency domain [1 – 4]. All these techniques aim at probing tissue samples in the volume range of a few cm³ with applications ranging from muscle and brain physiology to tumour monitoring. In these cases the diffusion approximation gives a reasonable description of the transport of photons in tissue and absorption coefficients, and subsequent SO_2, can be obtained as well as the scattering properties.

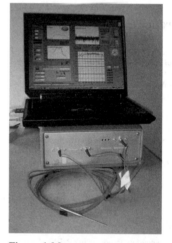

Figure 1 Measurement prototype incorporating a laser – Doppler channel for blood flow and perfusion measurement and a VIS spectrometer for quantification of haemoglobin. The fiber tip incorporates four separate optical fibers for both methods.

When the tissue under investigation is smaller the use of visible wavelengths is advantageous since the haemoglobin absorption is stronger and the effect of an unknown absorption background smaller. For volumes ≤ 1 mm³, time and frequency domain methods are ruled out as the light path is very short and measurement of the time spread would require a subpicosecond resolution. Obviously multi sensor approaches as e.g. established by Bevilaqua et al. [5] probe larger volumes and do not work in these circumstances. Therefore the recovery of absorption coefficients and SO_2 values must be base purely on spectroscopic methods with measurements of reflectance intensity as a function of wavelength at a certain distance between light source and detector.

It has been suggested before [6] that the effect of scattering results in an offset in the measured light attenuation and that this offset can simply be subtracted from the spectra when it is assumed to be wavelength independent. A second approach is to measure in-vitro reference spectra of haemoglobin at various oxygen saturations in a tissue simulating, light scattering environment like milk or intralipid and take these scattering corrected spectra as a basis or look-up table for the conversion of experimental spectra into haemoglobin concentrations. However, these simplifications are flawed.

In a physical model, the reflected intensity has to be described as a function of the basic tissue optical properties, i.e. absorption coefficient μ_a, scattering coefficient μ_s, the scattering phase function expressed by the anisotropy function g and the refractive index n. The non-linear relationship between these quantities and the reflectance makes it notoriously difficult to recover μ_a and subsequently chromophore concentrations.

In earlier publications it has been shown that an accurate model of the light transport in tissue has a severe effect on the calculation of changes in haemoglobin concentration following cortical activation when certain assumptions about absolute values are made [7 – 9].

In the instrument designed, the measurement of SO_2 and haemoglobin concentrations was combined with simultaneous recording of the blood flow via laser-Doppler flowmetry (LDF) in virtually the same tissue volume. This was applied to the assessment of cortical signals in a rat model following electrical activation. To induce a large variation in tissue oxygenation, hyperbaric hyperoxygenation (HBHO) was induced by exposure to a pressure of 3 atmospheres of pure oxygen.

2 Method

2.1 Animal Model and Paradigm

Male Wistar rats (240-360g; n = 7 control group; n = 9 hyperbaric hyperoxygenation group) were anesthetized with isoflurane (70 % N_2O, 30 % O_2) and artificially ventilated. Body temperature was kept constant using a heating pad. The left femoral artery and vein were cannulated for arterial blood gas analysis, arterial blood pressure monitoring and infusion of physiological saline solution. The animals were placed in a stereotactic frame and the skull was thinned to translucency over the right somatosensory cortex. Somatosensory evoked potentials were recorded via an epidurally placed Ag/AgCl ball electrode using a commercial EEG recording unit. After surgery, anesthesia was switched to α-chloralose / urethane (40 mg/kg body weight α-chloralose with 400 mg/kg body weight urethane as i.v. bolus injection, followed by continuous i.v. infusion as necessary). Neuronal activation was induced by electrical stimulation of the left forepaw (stimulation period 10 s at 3 Hz; single stimulus duration 0.3 ms; intensity 1.0 – 1.4 mA; interstimulation interval 60 s). Changes of rCBF (all animals) and changes of rCBO (HBHO group n = 3 out of 9) were measured through the thinned skull by Laser Doppler flowmetry (LDF) and by microfiber optical haemoglobin-spectroscopy, respectively.

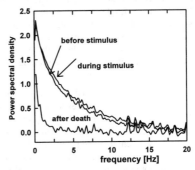

Figure 2 Laser – Doppler (LDF) signal: Power spectral density of reflected light before and during somatosensory stimulation measured on the rat brain. After sacrificing the animal there is a strong drop in the LDF signal.

Absorption spectra and LDF data were recorded continuously with a time resolution of about 0.3 s. HBHO was induced in a hyperbaric chamber flushed with 100% O_2 and pressurized to 3 atmospheres. Within the hyperbaric chamber the animals breathed spontaneously and were disconnected from i.v. anesthesia supplementation. Following disconnection, arterial pCO_2 slightly increased from 39 ± 7 to 42 ± 8 mmHg (control group) and from 40 ± 3 to 47 ± 5 mmHg (hyperbaric hyperoxygenation group), respectively. All other physiological parameters were within normal ranges. In the control group disconnection from anesthesia infusion and respiration did not alter SEP amplitudes or rCBF responses to functional activation under normobaric normoxia. The rCBF and rCBO responses under hyperbaric hyperoxia were compared with stimulations performed in each animal during control conditions under normobaric normoxia before hyperbaric hyperoxia in the same group.

2.2 Experimental Setup

The instrument (see Figure 1) combines two channels for measurement of blood flow based on the laser Doppler-flowmetry (LDF) and haemoglobin concentration and oxygen saturation based on reflection spectroscopy.

In the spectroscopy unit of the instrument a white light emitting LED (LXHL-LW5C; Luxeon, USA) emitting in the wavelength band between 450 and 650 nm is coupled to a single optical fiber with 200 µm core diameter resulting in an optical power of about 2 mW at the tip of the fiber. The detector fiber (200 µm core) is positioned at a distance of 120 µm, resulting in a center – center distance of 320 µm. Reflected light was detected with an USB-based spectrometer (USB2000; Ocean Optics, USA) with a grating of 600 grooves / mm blazed at a wavelength of 500 nm. A colour filter (UG 40, Schott GmbH, Germany) suppressed the signal of the laser of the LDF channel in the spectrometer. The effective wavelength range was between 400 and 700 nm.

A $BaSO_4$ reflectance probe served to collect a reference spectrum $I_{ref}(\lambda)$ prior to each experiment. Due to its small wavelength dependent reflection properties, dividing all spectra $I(\lambda)$ collected on tissue by $I_{ref}(\lambda)$ eliminated the wavelength dependence of the light source, fiber transmission and detector sensitivity. These corrected intensity spectra were converted into haemoglobin concentrations (oxy-Hb and deoxy-Hb) based on a specially designed algorithm taking into account the optical properties of the tissue and its wavelength dependence. Details will be presented elsewhere. From the haemoglobin spectra the oxygen saturation was calculated:

$$SO_2 = oxy-Hb / (oxy-Hb + deoxy-Hb)$$

The radiation of a continuous wave laser diode (780 nm, 3 mW) was coupled to a 50 µm optical fiber by a lens and transported to the tissue. A fraction of the scattered and back reflected light was collected by a second 50 µm core fiber and subsequently detected by an avalanche photo diode (APD) module (C5460-01; Hamamatsu Photonics, Japan). The voltage output of this module was read by a PCMCIA analog/digital card (PC-Card DAS16/16; Measurement Computing Inc., USA) at 100 kHz with 16 bit resolution. The frequency of the light scattered by moving blood cells is Doppler shifted and interferes with light scattered from tissue at rest. This results in an intensity modulation at the difference (beat) frequency which is in the range up to a few kHz depending on the velocity of the blood cells. According to Shepherd and Öberg (1989) blood flow perfusion F (i.e. regional cerebral blood flow rCBF) can be calculated from the power spectral density of this interference signal as

$$F = \int_{v_{low}}^{v_{high}} v \cdot PSD(v) dv . \qquad \text{Eq. 1}$$

PSD is the power spectral density of the detector signal as function of the frequency ν and v_{low} and v_{high} are cut-off frequencies to suppress instrument noise which were set to 0.15 and 12 kHz, respectively. The measurement of rCBF gives only relative values, i.e. only changes can be detected.

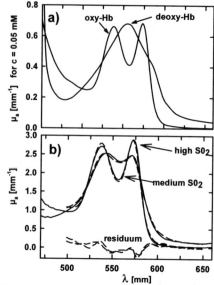

Figure 3 a) Spectra of absorption coefficient μ_a of oxy-Hb and deoxy-Hb for concentrations c = 0.05 mM. b) Experimental absorption spectra for brain tissue at medium and high oxygen saturation corresponding to t = 10 s and 2500 s in Fig. 4 (solid lines). The dashed lines show the fitted spectra for a wavelength range between 500 and 650 nm. The residua in the lower part of the figure are scaled by a factor 2. The saturation values for low and high obtained were SO_2 = 44 % and 97 % when fitting $\mu_a(\lambda)$.

Data collection and online-evaluation of blood flow, haemoglobin concentrations and oxygen saturation was programmed in Labview (Version 6.1; National Instruments, USA). All analysed data as well as all raw data (reflection spectra, power spectral density of the LDF signal) were stored for detailed off-line examination (Matlab R12, Mathworks, USA). Collection frequency can be set up to 10 Hz and was typically 3 – 4 Hz.

The four optical fibers were arranged in x-form with the two fibers for the LDF and the spectroscopy channel opposite to each other (distance 320 µm), thus probing approximately the same tissue volume when

when the different tissue penetration depths due to differences in wavelength are ignored. All four fibers of the sensor were bonded in a metal tube of 1.5 mm diameter for easy positioning on tissue.

3 Results

An example of the data obtained during somatosensory activation measured on cortical tissue is shown in Fig. 2 – 5. In Fig. 2 the laser – Doppler power spectra are shown before and during cortical activation. The shift of the spectra to higher frequencies corresponds to an increased blood perfusion by about 15 % (compare with Fig. 4). Furthermore, the LDF signal is shown after sacrificing the animal with a much lower intensity in the power spectrum reflecting a negligible movement of the red blood cells.

In Fig. 3 the reflectance spectra are shown for cortical tissues which clearly show the typical pattern of the haemoglobin components. The shape of the spectra is distinctly different when the oxygen supply is altered, with the two prominent peaks of the oxy-Hb component much stronger when the oxygen level is high.

Fitting the textbook absorption spectra of haemoglobin based on a model for the propagation of light in tissue gives a good description of the experimental data with the residua shown in the lower part of Fig. 3b. The haemoglobin concentrations obtained from these fits for all spectra are shown in Fig. 4 together with the time courses of oxygen saturation SO_2 and blood flow. During normoxia (t < 900 s) there is a distinct increase in oxy-Hb for each stimulus concurrent with a decrease with deoxy-Hb. The average of these values over the first ten stimuli during normoxia is shown in Fig. 5. This time course is in agreement with the basic concept of neurovascular coupling [11], where the higher oxygen consumption during cortical activation is compensated by an elevated blood flow. Increasing the oxygen supply during hyperbaric hyperoxygenation is mirrored by an increase in oxygen saturation to values between 95 – 97 % when the physical model for the analysis of the reflectance spectra is used. Without this model, i.e. when the textbook haemoglobin spectra are directly used for the fitting of the reflectance spectra, values for SO_2 are much lower. During HBHO the response of haemoglobin and flow to the stimulus is altered with much smaller amplitudes in the haemoglobin changes.

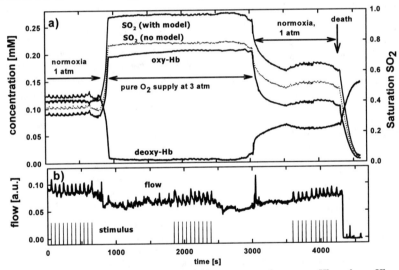

Figure 4 a) Time course of haemoglobin concentrations oxy-Hb, deoxy-Hb and oxygen saturation SO_2 recorded on the brain of a rat during normoxia (1 atm with 21% oxygen) and hyperoxia (3 atm at pure O_2). At t = 4300 s the animal was sacrificed. SO_2 was calculated with and without applying the model explained in the text. b) Recorded blood flow in arbitrary units and stimulus marker signals for cortical activation. For each stimulus there is a definite change in oxygenation as well as blood flow. This neurovascular coupling is altered by the oxygen supply.

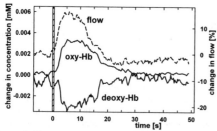

Figure 5 Changes in haemoglobin concentration and blood flow following electrical forepaw stimulation at t = 0 s. Data represent the average over 10 stimulation periods (t < 750 s in Fig. 4). The concentrations at the beginning are set to zero. Data are smoothed over 4 points.

In the last phase of the experiment the animal was sacrificed. This is reflected by a drop in SO_2 and blood flow to values close to zero

4 Discussion and Summary

We have shown that the measurement system designed is able to simultaneously monitor haemoglobin changes, oxygen saturation values and blood flow (rCBF). The validity of the haemoglobin concentrations is very much dependent on the model used for their calculations. The correctness of the data and algorithms was checked by variations of the wavelength range used for the fitting (data not shown), of the assumed underlying tissue properties like water and of the scattering parameters.

The method proposed here is based on a physical model of the light transport in tissue correlating the chromophore concentrations with the experimental assessable quantity attenuation. Compared with an analysis scheme based on pre-recorded reference spectra at different SO_2 values this holds advantages: First, recorded reference spectra at various SO_2 – settings are prone to experimental errors and, second, have to be obtained for each single source – detector distance used. Furthermore, errors due to varying haemoglobin concentrations and different scattering properties can not be estimated easily. The physical model used here does overcome these limitations.

However, there are a few limitations have to be pointed out. The basic assumption here is that the tissue absorption coefficient is dominated by haemoglobin with negligible contributions from other chromophores. It is difficult to estimate the errors by disregarding e.g. the cytochromes. Furthermore, the model assumes that the sample probed is homogeneous. There are no definite means to judge the validity of this assumption. In future work we will extend the analysis proposed here with multi-distance measurements.

The haemoglobin and oxygen saturation values during normal conditions were found to be in the range 0.4 – 0.5 which is in agreement with literature values for rodent [12]. Furthermore, variations of the physiological conditions as done between normal and hyperbaric hyperoxygenation are the best means to challenge the underlying analysis. The results obtained underline the validity of the method.

The physiological question of the influence of alterations in the neurovascular coupling and its dependence on oxygen supply will be discussed in a separate paper.

5 Reference

[1] S. J. Matcher, P. Kirkpatrick, K. Nahid, D. T. Delpy, "Absolute quantification method in tissue near infrared spectroscopy," *SPIE 2389*, 486 – 495, 1993.

[2] M. Patterson, B. Chance, and C. Wilson, "Time resolved reflectance and transmittance for the noninvasive measurement of tissue optical properties," *Applied Optics* **28**, 2331–2336, 1999.

[3] S. Suzuki, S. Takasaki, T. Ozaki, and Y. Kobayashi, "A tissue Oxygenation monitor using NIR spatially resolved spectroscopy, " *SPIE 3597*, 582 – 592, 1999.

[4] S. Fantini, M. A. Franceschini, J. S. Maier, S. A. Walker, B. Barbieri, and E. Gratton, "Frequency-domain multichannel optical detector for non-invasive tissue spectroscopy and oximetry," *Opt. Eng.* **34**, 32–42, 1995.

[5] F. Bevilacqua, D. Piguet, P. Marquet, J. Gross, B. Tromberg, and C. Depeursinge, "*In vivo* local determination of tissue optical properties: applications to human brain," *Appl. Opt.* **38**, 4939–4950, 1999.

[6] D. Malonek, A. Grinvald, "Interactions between electrical activity and cortical microcirculation revealed by imaging spectroscopy: implications for functional brain mapping," *Science* **272(5261)** 551-4, 1996.

[7] J. Mayhew, D. Johnston, J. Berwick, M. Jones, P. Coffey, Y. Zheng, "Spectroscopic Analysis of Neural Activity in Brain: Increased Oxygen Consumption Following Activation of Barrel Cortex," *NeuroImage* **12**, 664–675, 2000.

[8] M. Kohl, U. Lindauer, G. Royl, M. Kühl, L. Gold, A. Villringer, and U. Dirnagl, "Physical model for the spectroscopic analysis of cortical intrinsic optical signals," *Phys. Med. Biol.* 45, 3749–3764, 2000.

[9] U. Lindauer, G. Royl, C. Leithner, M. Kühl, L. Gold, J. Gethmann, M. Kohl-Bareis, A. Villringer, and U. Dirnagl," No Evidence for Early Decrease in Blood Oxygenation in Rat Whisker Cortex in Response to Functional Activation," *NeuroImage* **13,** 988–1001, 2001.

[10] P. Shepherd, P. A. Öberg, "Laser-doppler blood flowmetry," Kluwer Academic Publishers, 1989.

[11] Villringer A., Dirnagl U., Coupling of brain activity and cerebral blood flow: basis of functional neuroimaging. Cerebrovasc Brain Metab Rev. 1995; 7(3):240-76.

[12] H. Miyake, S. Nioka, A. Zaman, and B. Chance, "The detection of cytochrome oxidase heme iron and copper absorption in the blood-perfused and blood free brain in normoxia and hypoxia," *Anal. Biochem.* **192:** 149–155, 1991.

Forensik

Forensics

Forensic Facial Reconstruction

Thorsten M. Buzug[#], Dirk Thomsen and Jan Müller
Department of Mathematics and Technology, RheinAhrCampus Remagen, Suedallee 2, 53424 Remagen

Abstract

A new method for forensic facial soft-tissue reconstruction is presented. It is based on a nonlinear warping technique using radial basis functions known as thin-plate splines extended to 3D space. As for the manual facial reconstruction procedure the forensic expert has to attach soft tissue on a skull find. However, since the conventional, manual 4-step approach of i) *examination of the skull*, ii) *development of a reconstruction plan*, iii) *practical sculpturing* and iv) *mask design* is very time consuming, multi-modality elastic matching of 3D MRI soft tissue onto the 3D CT image of a skull find is proposed.

1 Introduction

In this paper a multi-modality approach to forensic facial reconstruction is proposed. We deal with an application of the 3D thin-plate spline method to the problem of identifying unknown persons. It is based on a 3D computed tomography image of a skull find and a 3D magnetic resonance image providing the facial template. Computed tomography (CT) is used, because it yields a 3D scan of forensic skull finds that is free of any geometrical distortion [1].

Generally, there are two possible alternatives to proceed with the 3D CT image stack. As one alternative, which will be described in this article, it is of course conceivable to use the 3D CT dataset as the basis of a virtual soft tissue reconstruction, especially since it is stored in the computer as a representation that can be directly used for further work.

The other alternative uses the CT data as the basis for the so-called rapid prototyping technique that produces plastic copies of the skull in its true size; to these the anthropologist can add the soft tissues with traditional plasticine without putting the archaeological find at risk.

Following the first alternative of a virtual soft tissue reconstruction, the method for the elastic matching of a 3D soft tissue template with a 3D reconstructed CT skull will briefly be described. According to literature, so-called B-splines are often attached to the 3D skull [2]; the depth of the soft tissue which varies over the skull is pre-defined by standard distances at characteristic anatomical landmarks (Helmer et al. [3]). Based on these distances, the B-splines so to speak form a soft tissue tent over the cranial bone. In our case an alternative method is chosen which uses the images created by another 3D modality.

2 Tomography and Visualisation

2.1 Computed Tomography

Today, clinical practice can no longer be imagined without computed tomography. Computed tomography was the first method to produce images showing the inside of the human body without superimposition, and without having to cut the body open. In the seventies of the last century this was an enormous step within the range of diagnostic possibilities in medicine. Today, there are a number of rival methods, the most important being nuclear magnetic resonance imaging. However, due to its ease of use and clear physical diagnostic result, as well as the progress in detector technology, reconstruction mathematics, and reduction of radiation exposure, computed tomography will retain and expand its established position in the field of radiology.

Recently interesting technical, anthroposophical, forensic and archaeological and palaeontological applications of computed tomography have been developed which further strengthen the method's position as a general diagnostic tool of non-destructive material testing and three-dimensional visualisation beyond its medical uses. Nuclear magnetic resonance imaging fails whenever the object to be examined is practically dehydrated, so that computed tomography is the three-dimensional imaging method to be used. These cases include the examination of archaeological skull finds.

The central problem of computed tomography can be easily described: Reconstruct an object from the shadows – or, more precise, projections – of this object. An x-ray source penetrates the object to be examined, i.e. the skull find, with a fan beam. Depending on the path of rays, the x-rays are attenuated to varying extents when penetrating the object, and local

---

[#] Contact: buzug@rheinahrcampus.de, phone: +49 (0) 26 42 / 932 – 318, fax: +49 (0) 26 42 / 932 – 301.

absorption is measured by a detector array. Of course, the shadow cast in only one direction is not an adequate basis for the determination of the spatial structure of the three-dimensional object. But in order to determine this structure, it is necessary to transmit rays through the object from all directions. Figure 1 schematically illustrates this principle. If we plot the different attenuation or absorption profiles over all angles of rotation of the x-ray tube we obtain a sinusoidal arrangement of the absorption coefficients of the skull.

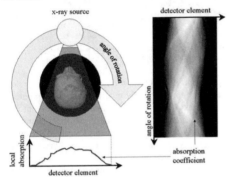

Figure 1: The x-ray source rotates around the object to be examined, penetrating it with a fan beam. Absorption is measured by an x-ray detector and saved for each angle. Therefore, the absorption data – plotted over all angles (right) – take a sinusoidal course.

Figure 2 (left) illustrates the methodology of data acquisition for modern computer tomographs. The measurements of the skull used in our example were taken with a Philips Secura scanner at the laboratory for computed tomography of RheinAhrCampus Remagen. This tomograph offers very high spatial resolution and the so-called spiral capability. During the spiral measurement process the x-ray tube is continuously rotated, at the same time the examination table is moved into the measuring field. This scan process produces data which take a spiral course (resembling a cut radish).

 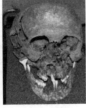

Figure 2: Left: Laboratory for computed tomography at RheinAhr-Campus Remagen: Modern computer tomographs use the so-called spiral scanning method. In this process the examination table is moved through the measuring field while the x-ray tube rotates continuously. In this way, the data are given a spiral structure that resembles a cut radish. Right: The original skull find in the tomograph. The skull is placed on a foam mat in order to facilitate subsequent image processing.

This method offers the possibility of computing any number of layers at a later time, so that we can expect a very accurate three-dimensional rendering.

From a mathematical point of view image reconstruction by computed tomography is the task of computing the spatial structure of an object that casts shadows by using these shadows. This is a so-called inverse problem. In 1961 the solution to this problem was for the first time applied to a sequence of x-ray projections for which an anatomical object had been measured from different directions. *Allen MacLeod Cormack* (1924 - 1998) and *Sir Geofrey Hounsfield* (born in 1919) are pioneers of medical computed tomography who in 1979 received the Nobel Prize in Medicine for their epochal work during the sixties and seventies. Figure 2 (right) shows the positioning of the archaeological skull find in the Secura computer tomograph. It is not necessary to fix motionless objects in position. The skull is placed on a foam mat the x-ray absorption of which is only very slight. This considerably facilitates the post-processing of the images for visualisation.

2.2 Visualisation

Figure 3 shows a comparison of the results of the projection x-ray technique and computer tomography for the skull find.

Figure 3: Comparison of different representations. Left: Simple x-ray projection from the three main directions. Centre: Sequence of layers produced by the spiral scanning method, and right: Translucent volume rendering.

The left picture shows the simple projections from the three main directions; for better orientation they have been projected onto the inner sides of a cube. Although representations like these show for example internal cracks, layer by layer measurement cannot be dispensed with, since the contrast in CT images, and hence the possibility of detecting minor damages, is substantially better. The picture in the middle of figure 3 shows a sequence of five layers through the skull which can be computed from the spiral or helix which is computer added in figure 2 (left).

In practice we have not only reconstructed 5, but 150 layers which – stacked at intervals of 1 mm above each other – result in a data stack which subsequently must be adequately displayed. The right picture of figure 3 shows a so-called volume rendering. This method assigns a physical light reflection and scattering value to each spatial pixel, the so-called voxel (*volume x element* as an extension of *picture x element*). The computer is then used to light this "data fog" with a virtual light source and to compute the optical impression a real reflection and scattering of light would create. The result of this visualisation technique is shown above.

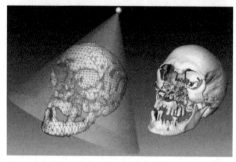

Figure 4: Surface rendering: A method of visualisation of CT data.

An alternative to volume rendering is the so-called surface rendering. The individual shades of grey of the layers in the data stack represent the degree of physical attenuation of the x-ray beam. In a clinical context deviations from the normal distribution of these values may give clues to pathological changes in the patient. Figure 4a shows a layer from the stack. For better orientation a representation of the skull created by using the volume rendering technique is also shown.

During visualisation we can decide to visualise certain ranges of values, and to selectively suppress others. If the viewer chooses a constant grey value, all spatial points with this value are displayed in space as a so-called iso-surface. In figure 4b a single iso-line is drawn within the chosen layer. Figure 4c shows that the stack of iso-line images forms the basis for the visualisation, because these lines are used in the triangle-based graphical model of the grey value iso-surface. Figure 4d shows the result of this so-called triangulation.

Then the mosaic of triangles is again lighted and displayed by the virtual method described above. Figure 4e shows this process schematically. The greater the chosen number of mosaic pieces for the reconstruction of the surface, the more lifelike the result. In figure 4e approximately 10,000 triangles have been computed, in figure 4f 1,000,000. The left picture in figure 5 shows a front view of the skull which was produced in this way. The side view (right picture of figure 5) additionally shows the inside and the outside of the skull in a semi-translucent way; this conveys an impression of the thickness of the skull.

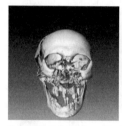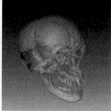

Figure 5: Left: Front view of the skull as a surface rendering. Right: Translucent side view of the surface rendering of the skull.

Apart from its ability to produce an image of the skull that can be interactively modified via the computer – e. g. the skull can be rotated, and virtual tours into the skull are possible, so that the three-dimensional impression is enhanced – the central merit of computed tomography lies in the fact that we obtain precise three-dimensional measurements of the skull on the basis of which we can carry out further analyses and especially, reconstructions. In our example the data shown above form the basis for the rapid prototyping technique that produces plastic copies of the skull in its true size; to these the anthropologist can add the soft tissues with traditional plasticine without putting the original archaeological find at risk. Figure 6 shows a plastic skull reproduced in this way.

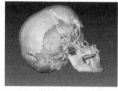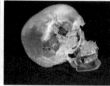

Figure 6: Left: Translucent side view of the surface rendering of the skull. Right: Reproduced plastic model to which plasticine can be added for soft tissue reconstruction. (Produced by C. Tille and H. Seitz, caesar Rapid Prototyping group [9]).

In addition, our team is investigating virtual methods of adding the facial soft tissues on the basis of computed tomography by means of computer simulation. The following chapter will deal with this issue.

3 Reconstruction Methodology

We can indeed discover a solution to the central problem of soft tissue depth, if we look at three-dimensional images produced by magnetic resonance tomographs (MRT), because MRTs are especially designed to show the soft tissues (see fig. 7). Actually, it is particularly easy to segment the MRI (magnetic resonance imaging) fat signal of the skin, so that in this way the facial soft tissue depth of an individual can be made directly available. Apart from magnetic resonance imaging, ultra-fast holography according to Hering et al. [4] is another modality for three-dimensional capturing of the facial surface – i.e. a soft tissue template. Unlike the MRI soft tissue template a holographic template has the advantage that it measures the individual facial expression of a subject in a sitting position. The results of the different approaches are compared in our working group.

The so-called thin-plate spline method will be used to adjust the MRI soft tissue template to the CT cranial bone. This landmark-based method is known from registration studies in medicine; in our context we will show the 3D enhanced implementation of Booksteins thin-plate spline method [5]. It is essentially a transformation of coordinates using radial basis functions. To avoid errors occurring through misplaced landmarks a regularized version of this algorithm is used. In general, the entire facial reconstruction method follows four basic steps:

A. Computed tomography of the skull find
This step includes the taking of skull measurements. The method presented in this article uses above all anatomical landmarks, which first have to be selected and entered into the computer. Digitalisation by the computer tomograph offers the advantage that further, and even very complex marks of the skull form, e.g. so-called "crest lines" or other structural features defined by differential geometry, may be determined later.

B. Matching of a MRI template from a database
The development of a reconstruction plan aims at identifying the "correct" soft tissue template in a database with the help of additional information. This additional information which should originate from the results of the forensic and anthropological examination and CID (*criminal investigation division*) finds at the place where the body was discovered (e.g. hair fragments) is still indispensable and is also used in preparation of step *D*. Here again, the matching of anatomical landmarks of the MRI soft tissue template with the anatomical landmarks of the skull find has to be reviewed.

C. Warping of the Template to the CT skull find
In a first step the computer carries out the global elastic warping of the MRI template onto the skull find using the landmarks and, if necessary, different interpolation methods. Usually, subsequent interactive corrections of individual parts of the face are necessary.

D. Texture mapping
Texture mapping includes the application of patterns, shading and colours to surfaces. Today, computers are very well able to carry out this task. The real problem is rather the scope for artistic design. As for the manual method the decisions in this field are still left to the medico-legal expert and the anthropologist. But decision-making can be supported by the computer, and wrong decisions can be easily corrected.

These basic steps aim at supplementing and above all at accelerating traditional procedures. In this contribution we will only deal with step *A* to *C*.

4 Point-Based 3D Elastic Matching

This chapter will describe an inherently three-dimensional method of facial soft tissue reconstruction. In our case an elastic image warping will be exclusively based on points, in contrast to the "crest line" based method by Quatrehomme et al. [6].

The central prerequisite for the point-based registration is that in two corresponding images the positions and, if necessary, further parameters, such as curvatures which characterise the anatomical landmarks, are known. On the basis of such homologous set of landmarks from the pair of images to be registered elastic transformation is possible, which essentially corresponds to the estimate of a non-linear mapping between the coordinates of the two images.

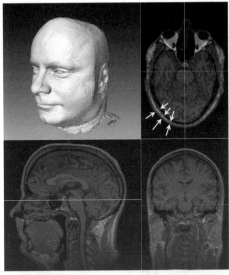

Figure 7: Surface visualisation of the skin from an MRI dataset. Beside and below an axial, sagittal and coronal section through the head is shown. The fatty tissue of the skin can be identified very well in the MRI sections. The arrows exemplarily mark the tissue depth of the soft tissue layer on the skull.

The so-called thin-plate spline method provides a good starting point for an elastic image transformation in which topology is preserved. This method is an interpolation which must be extended into an approximation by methods of regularization, in order to achieve acceptable results even with a set of deficient landmarks. Before an image can be mapped elastically onto another image, matching features of both images must be identified. Geometrical objects, such as points, lines or planes of the corresponding anatomical objects, can be used for this purpose, points – if necessary together with geometrical information about their neighborhood – being the easiest to determine. In this article these anatomical point landmarks will be selected interactively.

For the results at hand Bookstein's *principal warps* (1989) [5] extended to 3D have been implemented; they represent the application of an interpolation within the scope of elastic registration on the basis of base points which are irregularly scattered in space. Fig. 8 shows the interaction scheme by which homologous landmarks are identified in the corresponding datasets.

In this example, $Q_k = (x_k, y_k, z_k)$ is the set of points in the target image (i.e. in the CT image of the forensic skull find), and $P_k = (x'_k, y'_k, z'_k)$ is the set of corresponding points in the MRI soft tissue template which is to be adjusted to the target image by elastic transformation; $k = 1,...,N$. N is the number of anatomical landmarks.

The registration approach falls into the category of global basis functions. We will first select a set of N basis functions $U_k(x',y',z')$ for the set of the N anatomical landmarks P_k. Then, the coefficients w_k of the basis functions will be computed in a way in which the linear combination $\Sigma w_k U_k$ interpolates the data.

In literature (Franke [7]), a variety of basis functions have been compared with respect to accuracy, smoothness, sensitivity to variations of parameters, and ease of implementation. Franke reports that Hardy's so called multiquadric methods can be expected to provide the best results with respect to smoothness and accuracy. With respect to smoothness and accuracy the results of the similar method which uses surface splines and which was developed in order to model the physical bend of a thin plate which is warped at several points, are comparable to the results of Hardy's multiquadrics. For the mathematical details of the warping method used in this paper see [8].

5 Application of the Method to Forensic Problems

This last chapter will present a first preliminary result of the interpolation method described in chapter 4. It is a facial soft tissue reconstruction of an unknown person on the basis of the skull find. The traditional plasticine reconstruction produced by R. P. Helmer is used as the basis for comparison. Fig. 9 first shows the acquisition of the data basis via computed tomography. The plasticine head which contains the original skull is placed in the iso-centre of a Philips Secura CT scanner on a plastic dish (Fig. 9, a and b).

By way of example, 10 axial and sagittal sections each of the reconstructed section data are shown in Fig. 10. It is very easy to see the difference between the original skull and the plasticine layer. The metal fixations for the skull used for preparation purposes do not substantially interfere with CT image acquisition.

Figure 9: a) Positioning of the plastic skull reconstruction in the CT scanner on a plastic dish and b) positioning of the head in the iso-centre of the CT scanner.

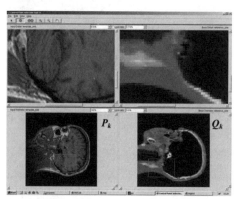

Figure 8: Interaction scheme for the selection of homologous anatomical landmarks which are used as a computational basis for the elastic transformation of coordinates.

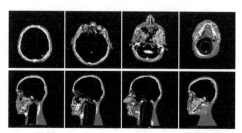

Figure 10: Representation of several skull sections of the head reconstruction. Upper row axial sections and lower row sagittal sections.

6 Results and Discussion

On the basis of interactively defined anatomical landmarks the MRI soft tissue template is mapped onto the skull of the person to be identified. Fig. 11 allows a visual comparison of the results of the matching process.

In Fig. 11a we first see a photo of the plasticine reconstruction in profile, Fig. 11b shows the surface of the MRI facial soft tissue template which is to be matched to the skull find. Fig. 11c shows the anatomical 3D landmarks which were defined with the interactive tool (see Fig. 8).

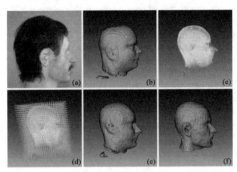

Figure 11: Visual comparison of the result of an elastic mapping of the MRI template onto the skull bone of the person to be identified. The plasticine reconstruction of the face produced by Helmer et al. [3] forms the basis for comparison.

On the basis of this landmark-based transformation the space is subjected to distortions which can be seen in Fig. 11d. Fig. 11e shows the transformation result. For comparison the CT surface reconstruction of the plasticine head is shown in Fig. 11f. There are certain similarities which are very encouraging. A quantitative comparison of the two reconstructions does not make sense, since Fig. 11f is also speculative.

7 Summary and Conclusions

We have presented results of a 3D thin-plate spline method for forensic facial soft tissue reconstruction which are based on a CT scan of a skull find and an arbitrarily chosen MRI facial soft tissue template. In comparison with a traditional plasticine reconstruction the results are encouraging.
The choice of an MRI soft tissue template is certainly a critical step. Not all the warping which may be necessary can be done by 3D thin-plate splines. In particular, only those transformations are possible which preserve topology.
We therefore plan to create an MRI soft tissue database, and to make an adequate choice for each individual case. A disadvantage of the MRI modality is the acquisition situation in which the individuals for the MRI database are positioned in the scanner. The head is moved into the measuring field of the MRI scanner in a lying position and slightly fixed in position. Therefore, pulsed holography will be used for facial measurement in a further step [4]. In the process, the facial template is captured very fast while the individuals are in a normal sitting position, so that a more natural facial expression is generally available for the matching process. We are currently working on a new method for 3D interaction based on navigation principles of image-guided surgery.

References

[1] Th. M. Buzug, P. Hering and R. P. Helmer, *3D Tomography as a Basis for Anthropological and Forensic Facial Reconstruction*, Proc. 1. Int. Conf. on Reconstruction of Soft-Facial Parts, Potsdam, p. 91.

[2] D. W. Bullock, *Computer Assisted 3D craniofacial reconstruction*, Master Thesis, University of British Columbia, 1999.

[3] R. P. Helmer, S. Rohricht, D. Petersen and F. Mohr, *Forensic Analysis of the Skull*, Chapter 17: Assessment of the Reliability of Facial Reconstruction, Wiley-Liss, 1993, p. 229.

[4] P. Hering, Th. M. Buzug und R. P. Helmer, *Ultra-fast Three-Dimensional Facial Profile Measurement Through Pulsed Holography for Forensic Facial Reconstruction*, Proc. 1. Int. Conf. on Reconstruction of Soft-Facial Parts, Potsdam, p. 109.

[5] F. L. Bookstein, *Principal warps: Thin-plate splines and the decomposition of deformations*, IEEE Trans. PAMI **11** (1989) 567.

[6] G. Quatrehomme, S. Cotin, G. Subsol, H. Delingette, Y. Garidel, G. Grevin, M. Fidrich, P. Bailet and A. Ollier, *A Fully Three-Dimensional Method for Facial Reconstruction Based on Deformable Models*, J. Forensic Sci **42** (1997) 649.

[7] R. Franke, *Scattered data interpolation: Tests of some methods*, Math. of Computation **38** (1982) 181.

[8] J. Müller, A. Mang, D. Thomsen and Th. M. Buzug, *Regularized 3D Thin-Plate Splines for Soft-Tissue Reconstruction*, Biomedizinische Technik **49**, 2 (2004) 134-135.

[9] C. Tille and H. Seitz, caesar Rapid Prototyping group, private correspondence 2003.

3D Warping for Forensic Soft-Tissue Reconstruction

Jan Müller, Andreas Mang and Thorsten M. Buzug[#]

Department of Mathematics and Technology, RheinAhrCampus Remagen, Suedallee 2, 53424 Remagen, Germany

Abstract

We deal with an application of the 3D thin-plate spline method to the problem of identifying unknown persons. It is based on a 3D computed tomography image of the skull find and a 3D magnetic resonance image providing the facial template. Computed tomography (CT) is used, because it yields a 3D scan of forensic skull finds that is free of any geometrical distortion. Generally, there are two possible alternatives to proceed with the 3D CT image stack. As one alternative, which will be described in this article, it is of course conceivable to use the 3D CT dataset as the basis of a virtual soft tissue reconstruction, especially since it is stored in the computer as a representation that can be directly used for further work. The other alternative uses the CT data as the basis for the so-called rapid prototyping technique that produces plastic copies of the skull in its true size; to these the anthropologist can add the soft tissues with traditional clay without putting the archaeological find at risk. Following the first alternative of a virtual soft tissue reconstruction, the method for the elastic matching of a 3D soft tissue template with a 3D reconstructed CT skull will briefly be described. According to literature, so-called B-splines are often attached to the 3D skull; the depth of the soft tissue which varies over the skull is pre-defined by standard distances at characteristic anatomical landmarks. Based on these distances, the B-splines so to speak form a soft tissue tent over the cranial bone. In our case an alternative method is chosen which uses the images created by another 3D modality.

1 Introduction

In this paper a new multi modality approach for facial reconstruction is proposed. It is based on the elastic matching of an MRI soft tissue template onto a CT scan of a skull find. Regularized Thin Plate Splines (TPS) are used for this topology preserving transformation. Facial reconstruction is important in several scientific areas, especially in forensic science and archaeology. In both areas the base of all work is a skull find of a dead person which should be reconstructed. This helps with the identification of a skeleton from an open case of death or the comparison of facial features between modern and ancient human beings.

During the last years several methods have been developed to deal with this task. The oldest one is the manual approach using a copy of the skull find as a base and clay as soft tissue imitation. The skull can today be copied with the help of rapid prototyping where a CT scan of the skull is printed out with a 3D printer. Onto this copy dowels are attached at standardized positions as an orientation for the process of reconstruction. The individual height of each dowel must be determined from the soft tissue thickness data of similar persons which is taken from living individuals via ultrasound. After the adjustment of the dowels the face can be modelled with clay. Figure 1 shows such a clay-based traditional reconstruction which is used later as a comparison with our new method. The problem with this technique is that it is time consuming and one needs deep anatomical and forensic knowledge to create such a reconstruction.

When it comes to applying the soft tissue thickness of one person to the skull of another, the dowel height data has to be chosen carefully. The person from which the data is taken has to have similar attributes with respect to gender, age, weight, ethnical belonging, etc. Otherwise, a badly chosen thickness dataset would lead to an inaccurate reconstruction.

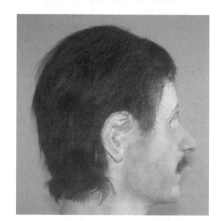

Fig. 1: The clay reconstruction produced by Helmer et al. which is used for comparison with our results.

[#] Contact: buzug@rheinahrcampus.de

With the development of computer tomography several new computer based methods for craniofacial reconstruction were created. They use a CT scan of the skull as the base for the reconstruction work because this non invasive process yields a distortion free 3D image of the skull [8]. Using this as a point to start from there are several ways of tissue reconstruction.

One method is very similar to the manual approach although it is completely computer based [1]. The dowels are directly integrated into the CT image via software and can be edited easily. After the dowel placement the software tries to replicate the soft tissue with the help of B-Splines attached to all virtual dowels. This could be compared with a tent or a foil which is stretched over the different dowels. By adjusting the height of the dowels the appearance of the face can be changed according to the personal attributes of the person. This method results in a head which looks somewhat similar to the dead person but also quite unnatural. The reason for this is the fact that only a few dozen points with height information are used for the reconstruction, i.e. the largest parts of the face are build upon interpolated height information. To decrease this inaccuracy more dowels could be attached but the result would still lack of details.

2 Methods

2.1 Thin-Plate Splines

Starting with a CT scan of the skull a new method for soft tissue reconstruction is presented in this paper. In contrast to the approach described above the soft tissue thickness is not interpolated through a few single points but all data contained in an MRI scan of a living individual is used for reconstruction. Not being forced to interpolate large areas of the face but being able to use the real thickness of soft tissue should lead to more natural looking reconstructions.

There are several good reasons for combining CT and MRI scans. As described the CT provides a copy of the skull free of geometrical distortion without putting it to risk damage. This method is chosen for the reference dataset because it makes the bones of the skull clearly visible which is necessary for the following work. MRI imaging is chosen for the template datasets of the living individuals because this type of scan is specialized in displaying soft tissue.

The idea behind the new method is not to directly extract the soft tissue thickness from the MR image but to transform this data in a way that it afterwards perfectly fits onto the CT scan of the skull. The soft tissue data can be thought as some kind of rubber glove which is wrapped around the skull.

To accomplish this some sort of mathematical transformation algorithm is necessary. The known methods can be divided into two groups, the rigid and nonrigid transformation. The rigid transformations only allow simple geometrical transformations like translation, rotation and shearing, whereas the nonrigid ones allow more complex deformations.

A rigid transformation alone is insufficient for our application because the two datasets are too different to get them aligned properly. But nevertheless this kind of transformation is important because it is used for the overall global alignment. To achieve better results the ability for local deformation is needed in addition to the simple rigid transformation. From the family of nonrigid transformations the elastic transformations are a good choice because on one hand they offer the ability for both local and global deformation and on the other hand they deliver a smooth and continuous result. Several of these elastic transformations are based upon two sets of corresponding landmarks (e.g. points, lines, surfaces and volumes) to influence the degree and the kind of transformation. The process of landmark localization can be done either automatically by software (e.g. edge detection) or manually by the user.

In contrast to Quatrehomme et al. who use a crest line based algorithm [9] we have chosen to implement an exclusively point based method called Thin Plate Splines (TPS). It was first applied by Bookstein [1] for medical image application and is based on two sets of corresponding point landmarks which are set manually in this application. It was decided to employ this non-rigid algorithm as it provides smooth global and local deformations.

The basic version of TPS does an interpolation between the two landmark sets which means that every landmark from one dataset is mapped exactly onto its corresponding counterpart. With perfectly set landmarks this behaviour is desired but in real world applications errors in landmark localization are very likely to occur. Without any counter measure this would lead to insufficient results. Hence an extended TPS version which is more an approximation than an interpolation is presented later on.

2.2 Mathematics

The description which follows afterwards is just a general view onto the algebra of TPS, for a comprehensive coverage we refer to Bookstein [2].

The task is to find a function f which maps all the points from the target volume data into their corresponding positions in the source volume data:

$$f(p^t) = p^s \qquad (1)$$

The labels 'target' and 'source' denote from the role of both volume data sets in the actual implementation.

The algorithm walks through the entire target data calculating the corresponding point in the source data for each voxel and then copies the colour information found at this positions back into the target image.

One of the main concepts behind TPS is the use of a radial basis function (RBF) for the elastic part of the transformation. These functions are called radial because they only depend on the Euclidian Distance of their associated data point from their origin [2]. Since we are dealing with a 3D implementation the RBF is as follows:

$$U(r) = f(x,y,z) = \|r\|, \quad (2)$$

where

$$\|r\| = \sqrt{x^2 + y^2 + z^2}. \quad (3)$$

Radial basis functions for other dimensionalities can be found in [1,2]. The condition that a function really is an RBF is that is solves the biharmonic equation [4]

$$\Delta^2 f = 0, \quad (4)$$

i.e. $U(r)$ must satisfy equation

$$\Delta^2 U(r) = \left(\frac{\partial^4}{\partial x^4} + \frac{\partial^4}{\partial y^4} + \frac{\partial^4}{\partial z^4} + \ldots \right. \quad (5)$$
$$\left. \ldots + 2\frac{\partial^4}{\partial x^2 \partial y^2} + 2\frac{\partial^4}{\partial x^2 \partial z^2} + 2\frac{\partial^4}{\partial y^2 \partial z^2}\right) \cdot U(r) \propto \delta(x,y,z)$$

where Δ^2 is the two times iterated bilaplacian operator and $\delta(x,y,z)$ is the Dirac distribution.

TPS can be compared in a physical manner with a thin metal plate which is deformed by external forces. The more the plate is bend the higher is the bending energy. Considering only small deviations and ignoring gravitation such a plate will bend in such a way that the physical bending energy is minimal [5].

This behaviour can be expressed with the following term, it is one of Duchons' semi-norms [2][3], more precisely the one of second order for three dimensions.

$$J(f) = \int_{\mathbb{R}^3} \left(\frac{\partial^2 f}{\partial x^2}\right)^2 + \left(\frac{\partial^2 f}{\partial y^2}\right)^2 + \left(\frac{\partial^2 f}{\partial z^2}\right)^2 \ldots \quad (6)$$
$$\ldots + 2\left(\frac{\partial^2 f}{\partial x \partial y}\right)^2 + 2\left(\frac{\partial^2 f}{\partial x \partial z}\right)^2 + 2\left(\frac{\partial^2 f}{\partial y \partial z}\right)^2 dxdydz$$

In order to minimize this term an appropriate interpolant $f(x,y,z)$ needs to be chosen.

As stated before this function f describes the correspondence between the two sets of landmarks and can be split up into 3 functions, one for each dimension:

$$f(p_i') = \begin{bmatrix} f_x(p_i') \\ f_y(p_i') \\ f_z(p_i') \end{bmatrix} = p_i^s \quad i \in \{1,\ldots,n\} \quad (7)$$

where n donates the number of landmarks. The interpolation function must be continuously defined and will map each landmark from the target image exactly onto its corresponding landmark in the source image. According to [1] and [2]

$$f_\bullet(x,y,z) = a_{0\bullet} + a_{1\bullet}x + a_{2\bullet}y + a_{3\bullet}z + \sum_{k=1}^{n} w_{\bullet k} U(\|P_k - P\|) \quad (8)$$

minimizes $J(f)$. $P=(x,y,z)$ and P_k denotes the x, y and z coordinates of the target points. The point denotes the index for the x-, y- or z-coordinates. The function consists of 2 parts, where the first is the 3D version of the normal affine transformation. The second part takes care of the local elastic deformation and contains the previously mentioned Radial Basis function $U(r)$. The way the volume data is deformed is controlled by the still unknown parameters a and w.

Further investigations on eq. 8 show that the elastic part of the function needs boundary conditions to ensure that local deformations do not affect the whole area:

$$\sum_{k=1}^{n} w_{k\bullet} = 0 \quad (9)$$

$$\sum_{k=1}^{n} w_{k\bullet} x_k = 0 \quad (10)$$

$$\sum_{k=1}^{n} w_{k\bullet} y_k = 0 \quad (11)$$

$$\sum_{k=1}^{n} w_{k\bullet} z_k = 0 \quad (12)$$

N denotes the number of landmarks and $w_{k\bullet}$ is the elastic parameter of eq. 8. According to Bookstein [1] these terms limit the translation and rotation rate of the elastic part. They ensure that terms with a more than linear increase far away from the landmarks will not be considered.

The different coefficients a of the affine and w of the elastic part of eq. 8 can be determined by solving the following system of linear equations built from eq. 8.

$$Kw + Pa = Y$$
$$P^T w = 0 \quad (13)$$

This can also be written as

$$L^{-1} Y = (W | a_{0\bullet}, a_{1\bullet}, a_{2\bullet}, a_{3\bullet}) \quad (14)$$

Where Y is a matrix containing the source landmarks

$$Y = \begin{bmatrix} x_1^s & y_1^s & z_1^s \\ x_2^s & y_2^s & z_2^s \\ \vdots & \vdots & \vdots \\ x_n^s & y_n^s & z_n^s \end{bmatrix} \text{ and } L = \begin{bmatrix} K & P \\ P^T & 0 \end{bmatrix} \quad (15,16)$$

K is a matrix containing the values of the Radial Basis function between each target landmark

$$K = \begin{bmatrix} 0 & U(r_{12}) & \ldots & U(r_{1n}) \\ U(r_{21}) & 0 & \ldots & U(r_{2n}) \\ \vdots & \vdots & \ddots & \vdots \\ U(r_{n1}) & U(r_{n2}) & \ldots & U(r_{nn}) \end{bmatrix} \quad (17)$$

and P is a matrix containing the target landmarks

$$P = \begin{bmatrix} 1 & x_1^t & y_2^t & z_1^t \\ 1 & x_2^t & y_2^t & z_2^t \\ \vdots & \vdots & \vdots & \vdots \\ 1 & x_n^t & y_n^t & z_n^t \end{bmatrix}. \quad (18)$$

2.3 Regularization

As mentioned before the problem with this algorithm is that both landmark sets are exactly mapped onto each other. With properly located landmarks this isn't a problem but in practical application there are always inaccuracies in the landmark placement. These errors come either from the user who gives wrong input or from the general uncertainty while setting landmarks. In image data sets of bad quality it is often difficult to decide where to exactly put a landmark so a method is needed to weaken these errors.

To minimize the amount of errors a regularized TPS version was implemented [2,7]. Where eq. 1 requires an exact one to one correspondence the extended version allows some degree of freedom denoted as ε [2].

$$\sum_{i=1}^{n} \left\| f(p_i^t) - p_i^s \right\|^2 \le \varepsilon \quad (19)$$

The only change that is needed is the addition of a regularization matrix to eq. 14

$$(K + \lambda I)w + Pa = Y$$
$$P^T w = 0 \quad (20)$$

where I is an identity matrix of appropriate size and λ is the regularization parameter. It determines the relative weight between the approximation behaviour and the smoothness of the transformation. For the case $\lambda=0$ one can see that eq. 20 equals eq. 13, i.e. the transformation is not regularized. The bigger λ grows the weaker the elastic transformation gets. For large λ there is only an affine transformation left. In eq. 21 the whole equation system for transformation parameter can be seen, it is the detailed written form of eq. 20. For $\lambda=0$ it is equivalent to eq. 13.

$$\begin{pmatrix} x_1^s & y_1^s & z_1^s \\ \vdots & \vdots & \vdots \\ x_n^s & y_n^s & z_n^s \\ 0 & 0 & 0 \\ 0 & 0 & 0 \\ 0 & 0 & 0 \\ 0 & 0 & 0 \end{pmatrix} = \begin{pmatrix} \lambda & \cdots & U(r_{1n}) & 1 & x_1^t & y_1^t & z_1^t \\ \vdots & \ddots & \vdots & \vdots & \vdots & \vdots & \vdots \\ U(r_{n1}) & \cdots & \lambda & 1 & x_n^t & y_n^t & z_n^t \\ 1 & \cdots & 1 & 0 & 0 & 0 & 0 \\ x_1^t & \cdots & x_n^t & 0 & 0 & 0 & 0 \\ y_1^t & \cdots & y_n^t & 0 & 0 & 0 & 0 \\ z_1^t & \cdots & z_n^t & 0 & 0 & 0 & 0 \end{pmatrix} \begin{pmatrix} w_1^x & w_1^y & w_1^z \\ \vdots & \vdots & \vdots \\ w_n^x & w_n^y & w_n^z \\ a_0^x & a_0^y & a_0^z \\ a_1^x & a_1^y & a_1^z \\ a_2^x & a_2^y & a_2^z \\ a_3^x & a_3^y & a_3^z \end{pmatrix} \quad (21)$$

3 Implementation

For actual use one needs to implement eq. 8 for the transformation process and eq. 14 for determination of the transformation coefficients a and w. The CT scan of the skull is used as the target volume data and the MRI template is used as the source data.

To assure smooth results a trilinear interpolation has been implemented to suppress artefacts which are likely to occur when using a simple nearest neighbour interpolation.

Our current implementation is written in Matlab 6.5 and it takes about 10 minutes to morph a typical volume dataset (resolution: 256*256*256, 30 landmarks) on a 2.2 GHz PC.

4 Results

We have presented the 3D thin-plate spline method for craniofacial reconstruction based on a CT scan of a skull find and an arbitrarily chosen MRI soft-tissue template. On the basis of this landmark-based transformation the space is subjected to distortions which can be seen in Fig. 2.

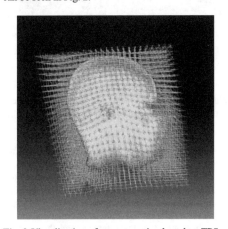

Fig. 2: Visualization of space warping based on TPS.

The first results received by this method are encouraging compared to traditional clay-based reconstructions. Fig. 3 and 4 show a case where an arbitrary MRI template was morphed onto the CT scan of a skull find from an unsolved case of murder. In Fig. 3 it can clearly be seen that after the transformation the template fits much better on the template than before. The region around the nose is a good example for the smoothness of the transformation.

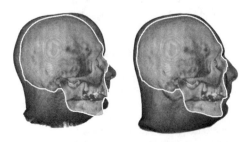

Fig. 3: A skull find is overlaid over an arbitrary template (left) and the result of the transformation (right).

In Fig. 4 the template can be compared to the result of the transformation. The two most deformed regions are the forehead and the nose which both looks way different than in the template. It can bee seen that the nose needs some adjustment because it looks somewhat unnatural.

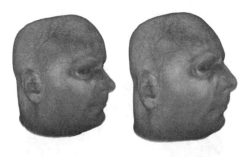

Fig. 4: Comparison between an MRI template (left) and the result of a transformation (right).

A first comparison between our new method and the manual approach is presented in Fig. 5 and Fig. 6. In figure 5 an overlay of the facial reconstruction done by Helmer et al. over a second template and the result of the transformation respectively.
As shown before in Fig. 3 the transformation leads to a much better fit of the reference and the template data. The displacement of the neck can be explained with the fact that the landmarks are only set upon the skull surface, a few additional landmarks are necessary to correct this error.

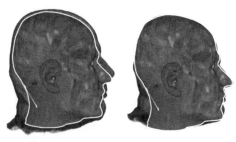

Fig. 5: The work of Helmer et al. is overlaid over an arbitrary template (left) and the result of the transformation (right).

A comparison in 3D can be seen in Fig. 6. Once again the region around the nose is the one with the most obvious deformation. These images also show that the selection of the template is important because obviously both persons do not exactly have the same weight. Fig. 6 does not allow a qualitative comparison between both methods because Helmer's work is hypothetical as well.
As seen in Fig. 6, TPS are not capable of all the warping which is potentially needed because only those transformations which preserve topology are possible. We therefore plan to create an MRI soft-tissue database containing different templates according to ethnic groups, gender, age, weight, physiognomy etc. where appropriate initial templates can be chosen from.

5   Future Work

As said before the quality of the reconstructed faces is highly influenced by the choice of an appropriate template MRI. To simplify this task it is planned to build up a template database from which the user can easily select a adequate template based on the age, the gender and other criteria. These templates could have preset landmarks so that after the landmark setting on the reference dataset every template from the database could be morphed without any additional work.
The disadvantage of the data acquisition using MRI modality is the lying position and the fixation of the individuals. Hence no natural facial expression is available for the matching process.
Another point of research is the 3D interaction during the process of setting the landmarks. We are currently

working on a new method based on navigation principles of image-guided surgery to simplify this task.

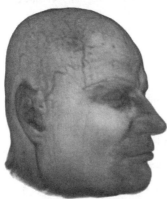

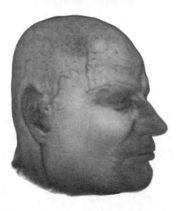

Fig. 6: Comparison between the head produced by Helmer et al. (first), the result of the transformation (second) and the template (third).

6 Conclusion

We have presented a new computer based method for facial reconstruction. It is based on regularized Thin Plate Splines and leads to encouraging results in the first step. For actual application further development is necessary to overcome the problems mentioned in section 5.

7 References

[1] Bullock, D. W.: *Computer Assisted 3D craniofacial reconstruction*, Master Thesis, University of British Columbia, 1999.

[2] Bookstein, F. L.: *Principal Warps: Thin Plate Splines and the Decomposition of Deformations*. IEEE Trans. Pattern Anal. Mach. Intell. 11, June 1989.

[3] Kybic J.: *Elastic Image Registration using Parametric Deformation Models*, Ph.D. dissertation. Ecole Polytechnique Fédérale de Lausanne, Lausanne, Switzerland. 2001.

[4] Duchon, J.: *Interpolation des fonctions de deux variables suivant le principe de la flexion des plaques minces*. RAIRO Analyse Numérique 10, 5-12, 1976.

[5] Weisstein, E. W.: *Biharmonic Equation*. From MathWorld - A Wolfram Web Resource. http://mathworld.wolfram.com/BiharmonicEquation.html.

[6] Weisstein, E. W. et al.: *Thin Plate Spline*. From MathWorld - A Wolfram Web Resource. http://mathworld.wolfram.com/ThinPlateSpline.html.

[7] Rohr K., Stiehl H. S., Sprengel R., Buzug Th. M., Weese J. and Kuhn M. H.: *Landmark-based elastic registration using approximating thin-plate splines*, IEEE Trans. Medical Imaging 20 (2001) 526.

[8] Buzug Th. M., Hering P. and Helmer R. P.: *3D Tomography as a Basis for Anthropological and Forensic Facial Reconstruction*, Proc. 1. Int. Conf. on Reconstruction of Soft-Facial Parts, Potsdam, p. 91.

[9] Quatrehomme G., Cotin S., Subsol G., Delingette H., Garidel Y., Grevin G., Fidrich M., Bailet P. and Ollier A.: *A fully three-dimensional method for facial reconstruction based on deformable models*. J. Forensic Sci., 42(4):649–652, 199.

Gradient Vector Flow Based Active Contours for Facial Reconstruction

Andreas Mang, Jan Müller and Thorsten M. Buzug[#]
Department of Mathematics and Technology, RheinAhrCampus Remagen, Suedallee 2, 32524 Remagen

Abstract

A new method for virtual forensic facial reconstruction is proposed. The reconstruction of the face of dead individuals based on the shape of their skull is for interest in forensic, archaeology as well as in anthropology. Several methods have been developed during the last decade.
The aim of the presented method is to extract a facial template from magnetic resonance images (MRI) and fit this template to the computed tomography (CT) of the skull. Currently we are working on an automatic segmentation strategy for MRI fat signal to obtain the template. An active contour or snake algorithm, respectively, is proposed. This approach is well known from image processing applications to locate boundary objects. It was first proposed by M. Kass et al. [5]. Further the gradient vector flow based version of active contours is presented. The flow field leads to a contour which is capable of entering concave regions. As a side effect it will help to avoid the rotation of the snake, as it is intended to adjust the template to the CT data according to the same algorithm.

1 Introduction

Post mortem identification of human remains is a challenging task in forensic science. The main target of facial reconstruction is to find the appropriate look of the deceased person corresponding to the forensic findings. A recommended traditional method for the reconstruction of the face of dead individuals is using clay or plasticine.
The aim of computer based reconstruction is, to depend less on artistic skills and experience than the mentioned traditional method. The method presented is based on a 3D computed tomography image of a skull and a 3D magnetic resonance image providing the facial template (see Fig. 1). As MRI is especially designed to show soft tissue data we do not have to refer to measurements of the soft tissue thickness using e.g. ultrasound. Collecting the data with computed tomography ensures that the virtual skull is free of any geometrical distortion.
Other computerized procedures known from literature use e.g. B-splines [2] or Booksteins Thin-Plate Splines [8] for virtual facial reconstruction. The first method employs "dowels" representing the soft tissue thickness and the second multimodality approach for facial reconstruction deals with a landmark based morphing algorithm. Corresponding landmarks are set on a CT of the skull and an MRI representing the facial template. Afterwards the facial template is morphed onto the CT scan.
In order to reduce the interaction of the user we propose a new method, based on active contours, so called "snakes" [5], known from image processing applications, to locate boundaries. In our context we deal with the classical [5] and the gradient vector flow [3,4,6,10,11,12,13,14,15,17] enhanced versions of this algorithm. Due to its efficiency the snake algorithm offers several applications including edge detection, shape modelling, segmentation and motion tracking [15].
The idea is to collect the soft tissue thickness from MR images taken from living individuals and adopt this facial template to the computed tomography image of the skull. We use the snake algorithm to extract the soft tissue thickness of the MR image (see marks in Fig. 1).
In our opinion the collected facial template could be morphed onto the CT image of the skull using the same algorithm. In 3D we plan to employ balloons, an extension to the snake algorithm, to adjust the facial template. Currently we focus on the segmentation of the soft tissue data.

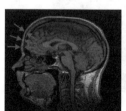 

Fig. 1: Surface visualisation of the skin from an MRI dataset and the sagittal section through the head. The arrows exemplarly mark the tissue depth of the facial template.

[#] Contact: buzug@rheinahrcampus.de

2 Material and Methods

In this section, we deal with the conventional snake algorithm and the gradient vector flow enhanced version. The mathematical description of snakes and the gradient vector flow are presented.

2.1 Snake Algorithm – Classical Formulation

In literature there are two different versions of active contours: the geometric and the parametric [15]. We focus on the parametric ones. In 2D a snake is a dynamic, elastic curve C defined within an image domain

$$C = \{v(s,t) = (x(s,t); y(s,t))\ /\ s \in [a;b]\}. \quad (1)$$

Usually the movement of these active contours is influenced by potential forces.

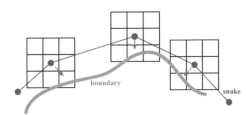

Fig. 2: Snake movement schematically illustrated.

The snake moves through the image domain to the desired features, usually object boundaries, under the influence of external forces computed from the image data and internal forces, which are geometrical properties the snake contains itself (see eq. 5). The movement of the snake is exemplarily shown in Fig. 2.

The energy functional, which is minimized and controls the behaviour of the snake, is associated with the curve. The functional consists of two terms, the external and the internal energy:

$$E_{snake} = \int E_{int}v(s) + E_{ext}v(s)ds \overset{!}{=} \min \quad (2)$$

The first term represents the internal energy, dominated by the geometrical properties of the active contour. As mentioned before, the second term, E_{ext}, is computed from the image. A snake minimizing the mentioned energy functional (eq. 2) has to satisfy the following Euler Lagrange equation [16]:

$$\frac{\partial}{\partial s}\left(\alpha\frac{\partial v(s)}{\partial s}\right) - \frac{\partial^2}{\partial s^2}\left(\beta\frac{\partial^2 v(s)}{\partial s^2}\right) - \nabla E_{ext}(v(s)) = 0 \quad (3)$$

Equation 3 simply represents the balance between external and internal forces. The external energy E_{ext} is defined in eq. 6. In order to solve eq. 3 the term needs to be treated as a time dependent function [14]:

$$\frac{\partial v(s,t)}{\partial t} = \alpha\frac{\partial v(s,t)}{\partial s} - \beta\frac{\partial^2 v(s,t)}{\partial s^2} - \nabla E_{ext}(v(s,t)) \quad (4)$$

When the snake reaches a steady state the solution of eq. 3 is found and the expression (eq. 4) disappears, meaning it is zero.

2.1.2 The Internal Energy

The internal energy is composed of two terms, which are dominated by the geometrical properties of the active contour:

$$E_{int} = \int_a^b \alpha\left\|\frac{\partial v(s,t)}{\partial s}\right\| + \beta\left\|\frac{\partial^2 v(s,t)}{\partial s^2}\right\| ds \quad (5)$$

The internal energy tends to preserve the shape of the snake, as it is dependent on the shape of the contour. It controls the physical behaviour of the snake. Thus it could be seen as internal spline energy caused by stretching and bending. Assuming that there is a large gap between the successive points of the snake the first derivative minimizes the total length of the contour, with respect to the minimization of the energy term. This part of the formula denotes the curvature. The second derivative represents the rigidity and keeps the snake from bending too much. It is responsible for the stiffness of the active contour. α and β are non-negative weighting parameters. Commonly these parameters are constant. A high α (controlling the elasticity) tends to force the active contour to shrink and a high β (controlling the tension) prevents sharp bending, as mentioned before.

2.1.1 The External Energy

The external energy pushes the snake towards the object boundaries and is computed from the image data defined as the potential energy functional of the image $I(x,y)$:

$$E_{ext} = f(x,y) = -\omega\|\nabla I(x,y)\|^2,\quad \omega > 0. \quad (6)$$

∇ represents the gradient operator

$$\nabla \equiv \left(\frac{\partial}{\partial x},\frac{\partial}{\partial y}\right) \quad (7)$$

and ω (see eq. 6) is a non-negative weighting factor for the external energy.

For a grey level image the most commonly used calculation of the external energy is given by the Laplacian of Gaussian (LOG) of $I(x,y)$ (see Fig. 3):

$$E_{ext} = -\omega \|\nabla[G_\sigma(x,y) \otimes I(x,y)]\|^2, \quad \omega > 0, \quad (8)$$

where $I(x,y)$ is the image intensity and $G_\sigma(x,y)$ is a 2D Gaussian function with standard deviation σ. The vectors determined by $\nabla I(x,y)$ point towards the edges of the object and have large magnitudes near the boundaries of the object. In homogenous regions the edge map is zero, as $I(x,y)$ is constant; consequently, the gradient is zero.

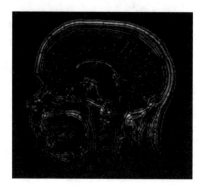

Fig. 3: Edge map of the magnetic resonance image using the LOG with a 9x9 kernel, which represents the external energy in the conventional snake algorithm.

To increase the capture range of external energy, the kernel size of the LOG operator is increased. This leads to blurry edges (see Fig. 3) in comparison to the result derived by the Sobel operator (see Fig. 7).

2.2 Gradient Vector Flow (GVF)

Several problems occur using the traditional formulation of the snake algorithm.
First of all the snake needs to be initialized close to the object boundaries, as the gradient of the image has a restricted influence range. Another difficulty is that the snake does not move into concave regions, as the internal energy tends to straighten the snake.
Moreover, the traditional snake is very sensitive to parameters (see eq. 5, eq. 6). To solve these problems a new external force, the so called gradient vector flow field, presented by Xu and Prince [13,12,11], is employed. The result is a new kind of snake, which can be initialised far away from object boundaries, but still tends to move towards the object and is capable of boundary concavities. Furthermore, the snake is less influenced by the present noise in the image.

Different from the traditional snake algorithm the GVF provides not only local information about the edges. The gradient information derived from the image (see eq. 6) of the vector flow field is diffused across the whole image. These acquired vectors point directly to the edges of the image data (see Fig. 4).
The GVF-field represents a vector field minimizing the following energy functional [14]:

$$E = \iint \mu\left(\left(\frac{\partial u}{\partial x}\right)^2 + \left(\frac{\partial u}{\partial y}\right)^2 + \left(\frac{\partial v}{\partial x}\right)^2 + \left(\frac{\partial v}{\partial y}\right)^2\right) + \ldots$$
$$\ldots + \|\nabla f\|^2 \|V - \nabla f\|^2 \, dxdy \quad (9)$$

where μ is a regularization parameter set according to the amount of noise present in the image. $V(x,y) = [u(x,y), v(x,y)]$ denotes the GVF-field; it is also a potential force field, just like the classical external energy. In this context f is the edge map, which, in our case, is computed with the Sobel filter:

$$f(x,y) = \|\nabla I(x,y)\|^2 \quad (10)$$

It also might be computed with other edge-emphasizing filters. Analogous to eq. 3 the force balance functional of the active contour is given by [6]

$$\alpha \frac{\partial^2 v(s,t)}{\partial s^2} + \beta \frac{\partial^3 v(s,t)}{\partial s^3} + \gamma V = 0, \quad (11)$$

where γ denotes a proportional coefficient.
The sum of squares of the partial derivates of eq. 9 makes the resulting vector flow $V(x,y)$ varying smoothly. This term is known from the optical flow by Horn and Shunk [4]. The second term represents the difference between the vector flow and its initial status. Hence, if ∇f is small - in homogeneous regions - this formula (eq. 9) is dominated by the partial derivatives of the vector flow field [6]. This yields to a slowly varying field. When ∇f is large the functional is dominated by the second term. The initial value of $V(x,y)$ is determined by the gradient of the edge map of $I(x,y)$:

$$V(x,y) = [u(x,y), v(x,y)] = \nabla f(x,y) = \ldots$$
$$\ldots = \nabla(\nabla I(x,y)) \quad (12)$$

Using calculus of variation it can be shown that the GVF $V(x,y)=[u(x,y), v(x,y)]$ is found by solving the following Euler equations [3,11]:

$$\mu \nabla^2 u - \left(u - \frac{\partial f}{\partial x}\right)\left(\left(\frac{\partial f}{\partial x}\right)^2 + \left(\frac{\partial f}{\partial y}\right)^2\right) = 0 \quad (13)$$

$$\mu \nabla^2 v - \left(v - \frac{\partial f}{\partial y}\right)\left(\left(\frac{\partial f}{\partial x}\right)^2 + \left(\frac{\partial f}{\partial y}\right)^2\right) = 0, \quad (14)$$

where $u(x,y)$ and $v(x,y)$ denote the partial derivates and ∇^2 the Laplacian operator. In homogeneous regions, the second term of the equations above is zero and for that reason $u(x,y)$ and $v(x,y)$ are determined by the Laplacian operation. Therefore, the gradient information is diffused into the homogeneous regions [11] (see Fig 4, Fig. 5). These two formulations (see eq. 13, eq. 14) can be solved separately by treating $u(x,y)$ and $v(x,y)$ as a function of time [9,13]:

$$u_t(x,y,t) = \mu \nabla^2 u(x,y,t) - \ldots \\ \ldots - (u(x,y,t) - f_x(x,y)) \cdot S_{xy} \quad (15)$$

$$v_t(x,y,t) = \mu \nabla^2 v(x,y,t) - \ldots \\ \ldots - (v(x,y,t) - f_x(x,y)) \cdot S_{xy} \quad (16)$$

where $S_{xy} = (f_x(x,y)^2 + f_y(x,y)^2)$, . This leads, with respect to the number of iterations, to a vector flow over the entire image. The resulting vectors point to the object edges from the image obtained by eq. 10.

3 Results and Discussion

Detecting the inner boundary of the fat signal in MRI data is a challenging task with many problems. Applying the GVF-enhanced version to the magnetic resonance image (Fig. 1) does not lead to the diffused vector flow with a large capture range, as it is shown in Fig. 4, as the MRI contains a lot of textures. In these areas with many textures being present, the vectors point into many different directions. This yields to problems during the approach of the snake to the inner edge of the fat signal. Hence, we currently use the traditional snake algorithm for the segmentation of the fat signal. To avoid this problem we think about using CT instead of MRI to get the fat signal information, as CT also contains the desired soft tissue information. For our purpose this might be difficult as it is quite hard to get different CT of the complete skull in order to build up a database of different facial templates.

As we know that the snake tends to rotate around the detected edges, which might become a serious difficulty for the shrinking of the facial template, we plan to use the vector information obtained with the GVF, to overcome this problem. This might not lead to the same problems compared to the MRI data, since the CT of the skull has less complex textures.

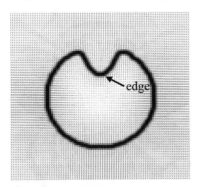

Fig. 4: Gradient vector flow field. This picture shows the diffusion of the gradient information over the whole image. The vectors are pointing to the edge of the object (see Fig. 5).

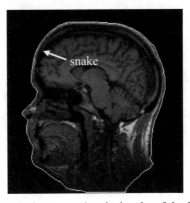

Fig. 6: Snake representing the boarder of the facial template.

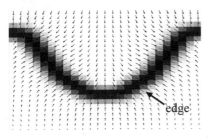

Fig. 5: Close-up of Fig. 4. It clearly can be seen, that the vectors directly point to the edge of the shape.

The idea is to locate the first edge of the image according to the snake algorithm and to delete it afterwards. In our implementation α is set to 1.0, β to 0.25 and γ to -188 in order to locate the outer boundary. The gradient information derived from the image is normalized to values between 0.0 and 1.0. The located outer boundary is deleted in the Sobel filtered image.

The second edge represents the inner boundary of the fat signal (see Fig. 7). Therefore, the snake is attracted a second time but now by the inner boundary. This leads to a positive result in the upper areas of the MRI, but around the nose it is quite hard to get good results, as there are several textures (see Fig. 6, Fig. 8).

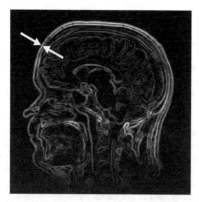

Fig. 7: Sobel edge map of MRI. The arrows point to first and second edge, referring to exterior and interior boundary of the facial soft tissue.

Hence, we will have to optimise our algorithm in order to get adequate results and further researches should be done. In Fig. 8 first result of the detection of the inner boundary of the fat signal could be seen.

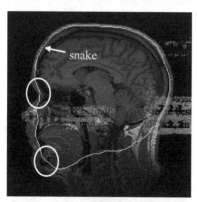

Fig. 8: First results of the location of the inner soft tissue boundary.

This figure exemplarily shows that there still are problems in the segmentation which have to be solved in future. Mainly in the front parts of the face (exemplarily shown with red circles) it is a challenging problem to locate the inner boundary of the fat signal automatically, as there are many textures, resulting in diffused boundaries (see Fig. 7). Another major problem is that the traditional snake algorithm is very sensitive to parameters, which additionally complicates the segmentation of the facial soft tissue template.

4 Conclusion and Future Work

In this paper an approach for segmenting the facial soft tissue thickness was presented. After segmenting the facial template the aim is to morph the extracted soft tissue data (see Fig. 6) according to the snake algorithm in 2D onto the gradient vector flow data of the CT data of the skull. Hence, the snake needs to have – in addition to elasticity and rigidity – one more property i.e. the soft tissue thickness (see Fig. 9).

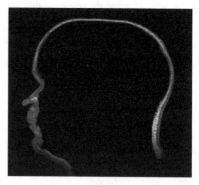

Fig. 9: This figure exemplarily shows the extracted soft tissue data, which we plan to employ as a kind of snake.

The results will show if it is possible to use the snake algorithm not only for segmentation, but also for warping. If we achieve encouraging results, we plan to use the volumetric MR image (see Fig. 1) and will warp this onto the 3D CT (see Fig. 10) data of the skull according to the 3D version of the active contours, the balloons.

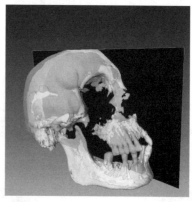

Fig. 10: 3D-CT scan of a historical skull find (Duke of Saxony, Widukind, late 8 century).

Balloons are a kind of extension concept to the described snake algorithm. It mainly follows the same idea as the 2D active contours, as it is also moved through the image domain by an energy minimizing process. The easiest way is to initialise it as 2D active contours slice by slice of the volumetric data, but the results are often insufficient. Hence, we plan to im-

plement forces operating in 3D - thus the points of the active contour are able to move in every direction of the three dimensional space.
The internal energy of the surface has got to be determined in a different way. Therefore, the Euler Lagrange equation, which has got to be minimized by calculating the according values of $v(r,s)$, (eq. 3) must be modified [1]:

$$\nabla E_{ext}(v, f) + \alpha_s \frac{d^2 v(r,s)}{ds^2} + \alpha_r \frac{d^2 v(r,s)}{dr^2}$$
$$\ldots - \beta_s \frac{d^4 v(r,s)}{ds^4} - \beta_r \frac{d^4 v(r,s)}{dr^4} - \ldots \quad (17)$$
$$\ldots - \beta_{sr} \frac{d^4 v(r,s)}{ds^2 dr^2} = 0$$

Just like in 2D this equation denotes that the internal forces of the active contour shall balance the forces extracted from the image data. The formulation for the internal energy is given by [7]:

$$E_{int} = \iint \alpha_s \left| \frac{dv(r,s)}{ds} \right|^2 + \alpha_r \left| \frac{dv(r,s)}{dr} \right|^2 + \ldots$$
$$\ldots + \beta_s \frac{d^2 v(r,s)}{ds^2} + \beta_r \frac{d^2 v(r,s)}{dr^2} + \ldots \quad (18)$$
$$\ldots + \beta_{sr} \frac{d^2 v(r,s)}{ds\, dr} ds\, dr$$

where the parameter (eq. 17, eq. 18) α_s and α_r define the weighting of the elasticity and the parameter β_s and β_r represent the loading of the rigidity along the corresponding s- and the r- axis. The last term of each equation (eq. 17, eq. 18) is controlled by β_{sr}, defining the inhibition of the rotation.

5 References

[1] Ahlberg, J.: Active Contours in three Dimensions, Department of Electrical Engineering, Computer Visions Laboratory, Linköpings Universitet, M. Sc Thesis No. LiTH-ISY-EX-1708, (1996).

[2] Bullock, D. W., Computer Assisted 3D facial reconstruction. Master Thesis, University of British Columbia, (1999).

[3] Hang, X., Greenberg, N. L., Thomas, J. D.: A Geometric Deformable Model for Echocardiographic Image Segmentation, Computers in cardiology, vol. **29**, (2002), pp. 77-80.

[4] Horn, B. K. B., Schunk, B. G.: Determining Optical Flow, Artificial Intelligence, pp. 185-203 (1981).

[5] Kass, M., Witkin, A., Terzopoulus, D.: Snakes: active contour models, Int. Journal of Computer Vision (1987), pp. 321-331.

[6] Luo, S., Li, R., Ourselin S.: A new deformable model Using dynamic Gradient Vector flow and Adaptive balloon forces, APRS Workshop on Digital Image Computing, Brisbane, Australia, (2003).

[7] Mansard, C. D., et. al.: Qantification of multi-contrast vascular MR images with NLSnake, an Active Contour Model: In Vitro Validation and In Vito Evaluation, magnetic resonance in medicine, vol. **51** (2), (2004), pp. 370-379.

[8] Müller, J., Mang, A., Thomsen, D., Buzug, T. M.: Regularized 3D Thin-Plate Splines for Soft-Tissue Reconstruction, Biomedizinische Technik, vol. **49**, 2 (2004) 134-135.

[9] Ntalianis, K. S., Doulamis, N. D., Doulamis, A. D., Kollias D. S.: Multiresolution Gradient Vector Flow Field, A fast Implementation Towards Video Object Plane Segmentation, IEEE, (2001)

[10] Paragios, N., Mellina-Gottardo, O., Ramesh, V.: Gradient Vector Flow Fast Geometric Active Contours, IEEE Transaction on Pattern Analysis and Machine Intelligence, vol. **6**, (2004).

[11] Santarelli, M. F., Positano, V., Michelassi, C., Lombardi, M., Landini, L.: Automated cardiac MR image segmentation: theory and measurement evaluation, Med. Eng. Phys., (2003), pp. 149-159.

[12] Wei, M., Zhou, Y., Wan, M.: A fast snake model based on non-linear diffusion for medical image segmentation, Computerized medical imaging and graphics, vol. **28** (3), (2004), pp. 109-117.

[13] Xu. C., Prince, J. L.: Gradient Vector Flow: A New External Force, CVPR, (1997), pp. 66-71.

[14] Xu. C., Prince, J. L.: Generalized gradient vector flow external forces for active contours, Signal Processing, vol. **71** (1998), pp. 131-139.

[15] Xu, C., Prince, J. L.: Snakes, shapes and gradient vector flow, IEEE Transactions on Image Processing, (1998), pp. 359-369.

[16] Xu, C., Pham, D. L., Prince, J. L.: Image Segmentation using Deformable Models, Handbook of Medical Imaging, vol. **2**, chapter 3. SPIE Press, (2000).

[17] Yu, Z., Bajaj, C.: Normalized Gradient Vector Diffusion and Image Segmentation, Proc. Eur. Conf. Comput. Vision (2002), pp. 517-530.

Forensische Schussrekonstruktion mit 3-dimensionalen individuell angepassten Opfer- und Tätermodellen im CAD-generierten 3D Tatort

Jörg Subke, Fachhochschule Gießen-Friedberg, Biomechanik-Labor, Gießen, Hessen

Kurzfassung

Das Ziel dieser Arbeit ist die Einführung digitaler 3D CAD Techniken in die forensische Rekonstruktion von Schussereignissen. Am Beispiel eines Falles werden die einzelnen Verfahrensschritte der digitalen Rekonstruktion gezeigt. Die Eckpunkte des Verfahrens sind die Erfassung und Dokumentation der Falldaten, die Generierung der digitalen Modelle der Opfer, des Täters, der Waffe und des Tatorts und schließlich die Rekonstruktion. In der Rekonstruktion wird eine der wichtigen Fragestellungen des Gerichtsverfahrens, wie das Opfer erschossen worden ist, diskutiert.

1 Einführung in den Fall

Um 3 Uhr morgens bricht ein Mann mittleren Alters gewaltsam in das Haus seines Schwagers, in einem Dorf in Süddeutschland, ein. Der Mann kennt sich von früheren Besuchen sehr gut aus und schleicht sich in das Schlafzimmer ohne Licht zu machen.
Dort angekommen, feuert er mit seiner Pistole insgesamt achtmal. Dabei tötet er die Frau seines Schwagers und verletzt seinen Schwager im Bereich des Halses und der Hüfte. Der Täter flüchtet, kann aber wenig später von der Polizei festgenommen werden.
Die Rekonstruktion wird sich im folgenden auf die Frau konzentrieren, die tot auf dem Bett vorgefunden wurde (**Bild 1**), da die Todesumstände eine entscheidende Rolle bei der Urteilsfindung hatten.

Bild 1 Tatort: Schlafzimmer des Ehepaares

Die Frau stirbt an den Folgen eines Kopfschusses, der in unmittelbarer Nähe auf sie abgegeben worden ist. Die Nähe der Schussabgabe lässt sich anhand der Schmauchspuren am Wundrand bestimmen (**Bild 2**). Ein Anzeichen für einen sofortigen Tod ist die gelöste Körperhaltung, in der die Frau vorgefunden wurde. Ebenso lassen sich Abwehrhandlungen der Frau ausschließen.
Zudem lässt sich aus der geradlinig verlaufenden Blutablaufspur an der Eingangswunde folgern, dass die Frau sich kurz nach dem Schuss nicht mehr bewegt hat bzw. nicht mehr bewegt worden ist.

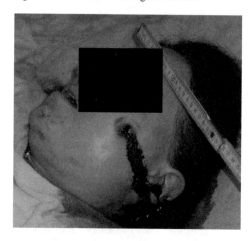

Bild 2 Tödlicher Nahschuss auf die Frau

Die Obduktion ergibt, dass die Kugel die Frau von der linken Wange durch das Rückenmark kurz unterhalb des Schädels nach hinten rechts durchdrungen hat. Durch diesen Schuss wurde das Rückenmark fast vollständig durchtrennt, so dass der Tod sofort eintrat.

2 Dokumentation und Generierung der 3-dimensionalen Modelle

Die Grundlage für die Generierung der digitalen Modelle der Opfer, des Täters und des Tatorts ist eine 3-dimensionale forensische Dokumentation. Bei der anschließenden Modellgenerierung werden die für die Rekonstruktion wichtigen Details herausgearbeitet.

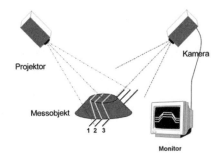

Bild 3 Prinzip der Streifenlichttopometrie SLT

Bild 4 Mobile Streifenlichttopometrie-Anlage

Für die forensische 3-dimensionale Dokumentation der Frau, des Ehemannes und der Spuren wird die Methode der Streifenlichttopometrie (SLT) eingesetzt [1] [2] [3]. Das Messprinzip basiert auf der Projektion von einer Serie von Streifenmustern auf das zu messende Objekt, die mit einer Videokamera aufgenommen werden (**Bild 3**). Aus der Verkrümmung der Streifen werden computergestützt die 3D Koordinaten der Oberfläche berechnet und die Farbinformation hinzugefügt.

Für die Dokumentation der Frau und der Spuren am Tatort wird eine mobile SLT-Anlage eingesetzt, die mit einem Messkopf ausgestattet ist, der aus einer Kamera und einem Projektor besteht (**Bild 4**).

Nach den 3D-Aufnahmen der Frau im Obduktionssaal wird anhand der Messdaten ein individuelles, maßgetreues digitales Modell der Frau erzeugt (**Bild 5**) [4]. Dabei werden bei der Modellgenerierung verschiedene Techniken eingesetzt.

Der Kopf wird in hoher Auflösung dargestellt, um die wichtigen Details der Verletzungen und der Spuren erkennen zu können.
Die Körperbereiche, die unverletzt sind und keine Spuren tragen, werden auf die geometrischen Umrisse reduziert, da keine weitere Information benötigt wird.

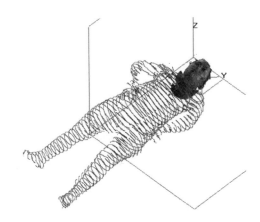

Bild 5 3D-Modell der Frau

Die Punkte der Körperoberfläche werden ausgedünnt, indem mit einer CAD-Funktion Profilschnitte durchgeführt werden. Auf diese Weise entsteht ein Konturenmodell, dessen Handhabung einfacher wird und das bei der grafischen Darstellung schneller bewegt werden kann.
Eine ähnliche Technik wird bei der Darstellung des Schusskanals angewendet. Mit einer CAD-Funktion kann die Anzahl der dargestellten Oberflächenpunkte variiert werden, um eine geschlossene Oberfläche transparent zu machen. Auf diese Weise kann der Ver-

lauf des Schusskanals in Bezug zur Körperoberfläche gezeigt werden (**Bild 6**).

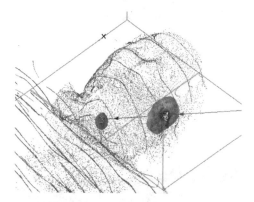

Bild 6 Transparentes 3D-Kopfmodell der Frau

Ein ähnliches 3D Modell wird für das zweite Opfer, den verletzten Ehemann, erzeugt. Es finden sich an ihm insgesamt vier Schussspuren; drei am Hals und eine an der Hüfte (**Bild 7**).

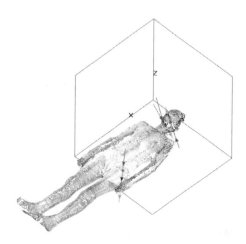

Bild 7 3D-Modell des verletzten Ehemanns

Für den Täter muss bei der Modellerstellung ein anders Vorgehen gewählt werden, da er nicht bereit ist, sich mit der Streifenlichttopometrie-Anlage dreidimensional vermessen zu lassen.
Seine Körpermaße nimmt ein Kriminaltechniker mit Hilfe eines Maßbands und eines Meterstabs im Untersuchungsgefängnis auf. Zudem wird festgestellt, dass der Täter Rechtshänder ist und über eine normale Beweglichkeit verfügt.

Diese Ergebnisse werden verwendet, um mit der Software Poser, der Firma Curious Labs, ein angepasstes Menschmodell zu erzeugen [5], das die Körperlängen, die Beweglichkeit sowie die Informationen über die Körpermassen und Massenschwerpunkte des Täters in erster Näherung wiedergibt (**Bild 8**).

Als Waffe verwendet der Täter eine Pistole Typ Browning (7,65 mm Kaliber), die ebenfalls als CAD-Modell 3-dimensional generiert wird.

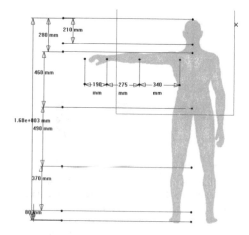

Bild 8 3D-Modell des Täters

Im nächsten Schritt wird ein digitales Modell des Tatorts, d.h. des Schlafzimmers des Ehepaares generiert. Zunächst werden das Schlafzimmer und die Möbel mit einem Meterstab vermessen und mit Hilfe eines CAD-Systems 3-dimensional digital rekonstruiert.

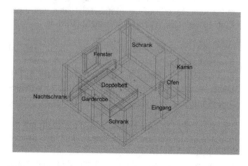

Bild 9 3D-Modell des Schlafzimmers

Das Schlafzimmer hat eine Grundfläche von insgesamt 15 Quadratmetern. Im Raum befinden sich neben einem Doppelbett zwei Nachtschränkchen, ein Ofen, ein Kamin sowie zwei große Kleiderschränke.

Eine Garderobe neben dem Bett der Ehefrau den nimmt zusätzlich Raum ein.

Die Kriminaltechniker haben insgesamt fünf Einschüsse im Raum festgestellt. Die Einschüsse befinden sich bei der Garderobe, in der Decke, im Bett, im Fensterrahmen und in der Wand unterhalb des Fensters. Zur Dokumentation der Position der Einschußstellen werden die Einschüsse 3-dimensional mit der mobilen Streifenlichttopometrie-Anlage aufgenommen und in das CAD-Modell des Schlafzimmers integriert.

Bild 10 Zwei Einschüsse auf der Fensterseite des Schlafzimmers

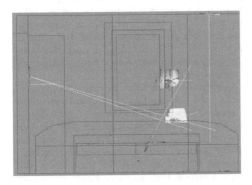

Bild 11 SLT-Aufnahmen der beiden Einschüsse auf der Fensterseite im CAD-Modell des Schlafzimmers

Der Vorteil der streifenlichttopometrischen Aufnahmen ist, dass die Schussverläufe beziehungsweise die Schussrichtungen in den 3-dimensionalen Daten der Einschüsse enthalten sind (siehe Übersicht in **Bild 12**).

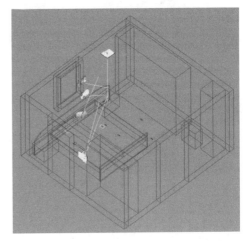

Bild 12 SLT-Aufnahmen der fünf Einschüsse im CAD-Modell des Schlafzimmers (isometrische Ansicht)

3 Forensische Rekonstruktion

Für die Rekonstruktion werden die digitalen Modelle in das CAD-Tatortmodell integriert. Dabei werden ausgehend von den Schussrichtungen im Raum die möglichen Positionen des Täters analysiert und die Position im Moment als der tödliche Schuss auf die Frau abgegeben wurde bestimmt.

Im ersten Schritt werden die Schussrichtungen anhand der 3D dokumentierten Einschüsse mit Hilfe von CAD-Funktionen bearbeitet, so dass die Einschussformen geometrisch durch Linien verlängert werden. Unsicherheiten im Verlauf werden als Varianten durch weitere Schussgeraden dargestellt (siehe Garderobenseite in **Bild 11** und **Bild 12**).

Der nächste Schritt ist die Integration des Modells der Frau und des Schussverlaufs am Kopf in das 3D-CAD-Modell des Tatorts (**Bild 13**).

Hierbei werden mögliche Kopfbewegungen der Frau berücksichtigt, die sich durch eine Reaktion der Muskulatur kurz nach Zerstörung des Rückenmarks ergeben haben könnten. Es wird von Rotationen in der horizontalen und der frontalen Ebene von jeweils ±10 Grad ausgegangen. Das Minimum und Maximum in der jeweiligen Ebene wird durch eine Linie markiert und es entsteht ein pyramidenförmiger Bereich, innerhalb dessen alle möglichen Schussrichtungen liegen.

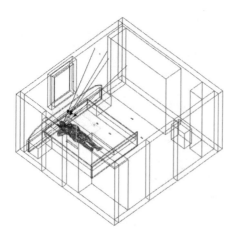

Bild 13 Integration des 3D-Modell der Frau in den CAD-Tatort; pyramidenförmiger Variationsbereich der Einschussrichtungen am Kopf des Modells

In der 3-dimensionalen Übersicht, in der alle Schussrichtungen dargestellt werden, ist zu erkennen, dass die Projektilbahnen von den Einschusslöchern zur rechten Bettseite weisen (**Bild 14**). Geht man davon aus, dass der Täter in zeitlich kurzen Abständen schoss, so muss er sich während der gesamten Zeit der Schussabgaben an der rechten Bettseite aufgehalten haben.

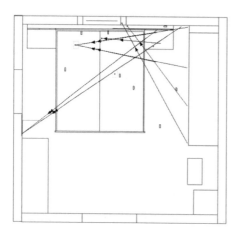

Bild 14 Rekonstruktion der Täterposition über die Schussrichtungen im CAD-Tatort (Aufsicht)

Nach Positionierung des Modells der Frau wird das Tätermodell in den CAD-Tatort eingebracht. Der aktuelle Rekonstruktionsvorgang besteht in der Variation möglicher Schusspositionen und dem Vergleich mit den Beweismitteln.

Eine offensichtliche Annahme ist, dass der Täter in einer vorgebeugten Haltung von der rechten Bettseite aus geschossen hat. Aber wie die Rekonstruktion zeigt, ist es für den Täter nicht möglich einen Nahschuss auf die Frau durch einfaches Vorbeugen ohne zusätzliche Stützen abzufeuern. Biomechanisch gesehen liegt hier der Schwerpunkt des Täters sehr weit vor der Standfläche, d.h. die Haltung des Täters ist im höchsten Maß instabil. Zudem ist die Entfernung Pistole zur Wange zu groß, um eine Schmauchspur, wie sie vorgefunden wurde zu erzeugen.

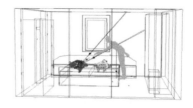

Bild 15 Rekonstruktion der Täterposition; stehende nach vorne gebeugte Haltung

Um einen fast aufgesetzten Schuss abgeben zu können, muss der Täter sich in eine Position nahe der Frau bringen. Zudem muss diese Position biomechanisch stabil sein. Dies ist nur möglich, wenn er sich auf dem Bett des Ehemanns abstützt. Das bedeutet, dass das Modell des Ehemanns in das CAD-Modell einbezogen werden muss, um die Rekonstruktion vervollständigen zu können.
Da die Position des Ehemanns nicht eindeutig festgelegt werden kann, wird die für den Täter ungünstigste Lage des Ehemanns, die Rückenlage angenommen.
Es wird davon ausgegangen, dass der Täter ein Knie auf dem Bett des Ehemannes abgestützt hat und das andere gestreckt hielt, um ein schnelles Zurückweichen zu ermöglichen. Zusätzlich muss er die linke Hand aufsetzen, um eine stabile Haltung einnehmen zu können. Der Täter muss auf diese Weise mit der Pistole, die er in der rechten Hand hält, die Einschusswunde an der Wange der Frau erreichen und die verschiedenen Einschusswinkel innerhalb des durch die Schusspyramide vorgegebenen Bereichs einstellen können.

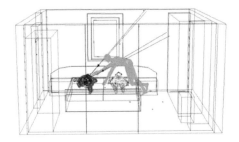

Bild 16 Rekonstruktionsvariante der Täterposition; über den Ehemann gebeugt.

Das Resultat dieser Rekonstruktion wird in der 3D-Darstellung des CAD-Systems deutlich, die die einzige Möglichkeit des Täters zeigt, den tödlichen Nahschuss auf die Frau abzugeben.

Wenn man mögliche Abwehrhandlungen der zweiten Person und eine schnelle Schussfolge berücksichtigt, kann daraus geschlossen werden, dass dieser Nahschuss der erste Schuss gewesen ist, den der Täter abgegeben hat. Damit ist die Rekonstruktion der Schussabgabe auf die Frau abgeschlossen.

4 Zusammenfassung

Die forensische Rekonstruktion, die sich auf die getötete Frau konzentriert, zeigt die wesentlichen Vorteile der computergestützten Vorgehensweise.

Dadurch dass der komplette Tatort als 3-dimensionales CAD-Modell aufgebaut ist, können die Entfernungen und die Winkel zwischen den dargestellten Objekten (Opfer, Täter etc) jederzeit exakt bestimmt werden. Ebenso lassen sich die verschiedenen Perspektiven einstellen, um den Tatort aus der Sicht der Opfer oder des Täters zu betrachten.

Die Opfer und der Täter werden als individuelle Modelle, inklusive der Verletzungen und der Schmauchspuren in das 3D CAD-Modell des Tatorts eingebracht und lassen eine maßgetreue 3-dimensionale Rekonstruktion zu.

Darüber hinaus enthalten die 3D-Dokumentationen der Einschüsse Informationen über die Kinematik der Projektile und damit Informationen über die Schussrichtungen und die Projektilverläufe, die in das 3D-Modell des Tatorts mit eingebunden werden.

Die Simulation der Schussereignisse lässt sich in einer relativ kurzen Zeit durchführen. Die Varianten der Körperhaltungen und Positionen von Opfer und Täter können zu jeder Zeit mit allen verfügbaren Beweismitteln geprüft werden.

Ebenso erlauben die Modelle die Bestimmung von Unsicherheiten, die als Variantionssbereich 3-dimensional dargestellt werden (z.B. die Kopfbewegungen des Opfers, Schusspyramide).

Die Menschmodelle können für eine bessere Darstellung der Schussereignisse animiert werden.

Jede Information, die sich als geometrisches Objekt darstellen lässt, kann für die Rekonstruktion als weiteres Modell in den CAD-Tatort integriert werden.

Letztlich die Möglichkeit, verschiedene Daten in ein 3-dimensionales Gesamtmodell zu speichern sowie die Vollständigkeit und die Genauigkeit der Daten machen die Anwendung der 3D CAD-Techniken im Bereich der forensischen Rekonstruktion zu einem wertvollen Werkzeug.

Danksagung

Wir bedanken uns für die freundliche Unterstützung bei den Firmen ABW GmbH, Frickenhausen und NTSI New Technology Solution GmbH, Kirchentellinsfurt.

5 Literatur

[1] Wolf, H.: Structured lighting for Upgrading 2D Vision Systems to 3D, International Symposium on Lasers, Optics and Vision for Productivity in Manufacturing I, Micropolis, Besancon, France, 10.-14. Juni 1996

[2] Wolf, H.: Getting 3D-Shape by Coded Light Approach in Combination with Phase Shifting. Numerisation 3D / Human Modelling, Salon et Congress sur la Numerisation 3D, Paris, 20.1.-21.1.1998

[3] Subke, J., Wehner H.D., Wehner F. and Wolf H.: Wundtopographie mittels Streifenlichttopometrie. Rechtsmedizin Supplement I zu Band 8, 1998

[4] Subke, J., Wehner, H.D., Roeser, P., Wolf, H.: Streifenlichttopometrie (SLT) in Legal Medicine. From Forensic Documentation to Individual Biomechanical 3-D Models for the Reconstruction of Traumatic Events. In: Computer Methods in Biomechanics & Biomedical Engineering - 3 (eds.) J. Middleton, M.L. Jones and G.N. Pande, Gordon and Breach Science Publisher, 2001

[5] Subke J., Haase S., Wehner H.D., Wehner F.: Computer aided shot reconstructions by means of individualized animated 3-dimensional victim models. Forensic Sci Int 125 (2002) 245-249

Zerstörungsfreie Materialprüfung

Non-Destructive Testing

Micro-CT in der Archäologie –
Untersuchung eines Neandertalerzahnes

Thorsten M. Buzug[1] und Axel von Berg[2]
1) Department of Mathematics and Technology, RheinAhrCampus Remagen, Südallee 2, 53424 Remagen
2) Archäologische Denkmalpflege, Amt Koblenz, Festung Ehrenbreitstein, 56077 Koblenz

Abstract

Seit einiger Zeit wird in der Materialprüfung mit so genannten Micro-CTs gearbeitet, die im Wesentlichen einer miniaturisierten Form moderner Kegelstrahl-CTs entsprechen und zur zerstörungsfreien, dreidimensionalen Mikroskopie genutzt werden. Das durchstrahlte Messfeld ist mit typischerweise 2 cm^3 so klein, dass medizinische Anwendungen auszuscheiden scheinen. Tatsächlich werden diese Geräte eher in der Materialprüfung und Materialanalyse verwendet, aber auch medizinische Anwendungen rücken zunehmend in das Zentrum des Interesses. Humanmedizinische Fragestellungen sind zum Beispiel Untersuchungen ganzer Zahnpräparate oder die Analyse der Trabekularstruktur von Knochen. In diesem Beitrag werden die Ergebnisse der Vermessungen eines archäologischen Zahnpräparates – eines Neandertalerfundstückes – präsentiert und das verbesserte Auflösungsvermögen des Micro-CTs gegenüber den klinischen CTs dargestellt. Der Fundort des Zahnes ist die Kratermulde des sog. „Schweinskopfes", Teil eines mittelpleistozänen Schlackenvulkankomplexes, der vor 200.000 Jahren im Rahmen des Osteifelvulkanismusses aktiv war. Die heutigen Auffüllschichten der Kratermulde beinhalten kaltzeitliche Sedimente des späten Mittelpleistozäns und des Spätglazials. Bei Grabungen der Archäologischen Denkmalpflege Koblenz wurde im Jahr 2001 in Siedlungsschichten aus der Zeit der frühen Neanderthaler zwei Hominidenzähne entdeckt, die nach vorläufigen Datierungen ein Alter von 180 – 200.000 Jahre haben. Bei dem im Micro–CT untersuchten Zahn handelt es sich um einen Prämolar, der durch Hitzeinwirkung leicht deformiert ist.

1 Einleitung

In den siebziger Jahren des letzten Jahrhunderts war es innerhalb der diagnostischen Möglichkeiten der Medizin ein enormer Schritt, dass mit der Computertomographie überlagerungsfreie Bilder aus dem menschlichen Körper gemessen werden konnten. Heute gibt es einige Konkurrenzverfahren dieses Röntgenverfahrens, allen voran die Kernspintomographie. Aufgrund der einfachen Handhabung und der klaren physikalisch-diagnostischen Aussage sowie der Fortschritte in der Detektortechnologie, Rekonstruktionsmathematik und der Reduktion der Strahlenbelastung wird aber die Computertomographie ihren festen Platz im radiologischen Umfeld behalten und weiter ausbauen.

Neuerdings kommen interessante technische, anthroposophische und forensische sowie archäologische und paläontologische Anwendungen der Computertomographie hinzu, die die Stellung des Verfahrens über die Verwendung in der Medizin hinaus als allgemein-diagnostisches Werkzeug der zerstörungsfreien Materialprüfung und dreidimensionalen Visualisierung weiter stärken. Immer dann, wenn das zu untersuchende Objekt praktisch kein Wasser mehr enthält, versagt die Kernspintomographie, so dass die Computertomographie als dreidimensionales Abbildungsverfahren zu verwenden ist. Im Rahmen der Untersuchung archäologischer Funde ist dies in der Regel der Fall.

Das zentrale Problem der Computertomographie ist leicht formuliert: Man rekonstruiere ein Objekt aus den Schattenwürfen – oder präziser Projektionen – dieses Objektes. Eine Röntgenquelle durchstrahlt das zu untersuchende Objekt, also den Zahnfund, kegelförmig. Die Röntgenstrahlung wird beim Durchgang durch das Objekt je nach Strahlenweg unterschiedlich geschwächt und die lokale Absorption mit einem Flächendetektor gemessen.

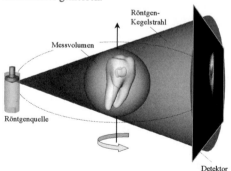

Abb 1: Bei Micro-CT-Scannern ist die Abtastanordnung aus Röntgenquelle und Flächendetektor fixiert und das Untersuchungsobjekt dreht sich in dem Röntgenkegelstrahl. Das Messvolumen ist etwa 2 cm^3 groß.

Nun kann man aus dem Schattenwurf unter einer einzigen Richtung natürlich nicht die räumliche Struktur des dreidimensionalen Objektes ermitteln. Dazu ist es vielmehr nötig, dass jeder Punkt des Objektes über 180° hinweg durchleuchtet wird. Abbildung 1 zeigt dieses Prinzip schematisch.

Wenn man das zu untersuchende Objekt dreht, kann man aus der Folge der Projektionsbilder das Bild des Objektes rekonstruieren. Eine Einführung in die Rekonstruktionsprinzipien findet man in (Th. M. Buzug 2004).

Mathematisch handelt es sich bei der Bildrekonstruktion der Computertomographie um die Aufgabe, wie man aus Schattenwürfen die räumliche Struktur des schattenwerfenden Objektes berechnet. Diese Fragestellung ist ein so genanntes inverses Problem. 1961 wurde die Lösung dieses Problems zum ersten Mal auf eine Sequenz von Röntgenprojektionen angewandt, die aus unterschiedlichen Richtungen von einem anatomischen Objekt gemessen wurden. Pioniere der medizinischen Computertomographie sind *Allan MacLeod Cormack* (1924-1998) und *Sir Geoffrey Hounsfield* (geb. 1919), die für ihre bahnbrechenden Arbeiten in den sechziger und siebziger Jahren 1979 den Nobelpreis für Medizin erhielten.

2 Geologie und Vulkanismus an Mittelrhein und Mosel

Die Besiedlung und früheste Geschichte einer Region ist vorrangig von ihrer Entstehung und damit von der geologischen und klimatologischen Entwicklung abhängig. Das nördliche Mittelrheingebiet liegt geographisch im Bereich der Mittelgebirgszone und wird durch die zwei großen Flussläufe Rhein und Mosel bestimmt. Während der Rhein die Hauptverbindungsachse zwischen Süden und Norden darstellt, erschließt die Mosel die westlichen Landschaftsteile bis Ostfrankreich. Rhein, Mosel und im Osten die Lahn teilen diese Mittelgebirgszone in vier Abschnitte, in Eifel, Hunsrück, Westerwald und Taunus, die als devonisches Grundgebirge oft beträchtliche Höhen erreichen.

Zentrale Landschaftseinheit ist das Neuwieder Becken, das zusammen mit den westlich anschließenden Landschaftsteilen Pellenz und Maifeld eine etwa 20 x 30 km große Beckenlandschaft bildet. Die Entwicklung der Region an Mittelrhein und Mosel und damit auch die Relikte frühen menschlichen Lebens sind im Wesentlichen durch zwei Faktoren eng miteinander verknüpft. Zum einen bestimmt die Flussgeschichte, zum anderen der Vulkanismus Landschaft und Menschen des Mittelrheingebietes bis in unsere heutige Zeit.

Die Region an Mittelrhein und Mosel wird durch ein quartäres Vulkanfeld bestimmt, das als Osteifelvulkanfeld bezeichnet wird. Das Zentrum der vulkanischen Tätigkeit ist das Gebiet zwischen Koblenz und Mayen (H. U. Schmincke 1988). Die vulkanischen Aktivitäten des Vulkanfeldes reichen weit bis in das Altpleistozän zurück und haben bislang mit der Eruption des Laacher See Vulkans vor etwa 13.000 Jahren vorläufig ihren Abschluss gefunden (W. Meyer 1986, 365; M. Street, M. Baales u. B. Weninger 1994). Anhand der Chronologie der Vulkanausbrüche kann eine räumliche Entwicklung festgestellt werden, die westlich des Laacher Sees beginnt und sich Richtung Osten bis in das Rheintal und in das Neuwieder Becken fortsetzt (W. Meyer 1986).

Der geologisch junge Vulkanismus der Osteifel hat seinen Beginn vor ca. 700.000 Jahren. Die Datierung der Vulkane ist durch die quartären Schichten der Kraterfüllungen möglich, in denen vulkanische Ablagerungen jüngerer Vulkane eingebettet sind. Einzelne Vulkanbauten können so mit naturwissenschaftlichen Verfahren zeitlich genau eingeordnet werden. Das Vulkanfeld der Osteifel gehört geologisch zu den kontinentalen sog. „Intraplattenvulkanen" und liegt auf einer sich hebenden alten Scholle. Der Vulkanismus steht im Zusammenhang mit der Aufwölbung des Rheinischen Schildes und dem gleichzeitigen Absenken der Niederrheinischen Bucht.

Etwa 150 Vulkane prägen die Landschaft zwischen Rhein und Mosel, wobei meist mächtige Schlackenkegel das Geländerelief gliedern. Sie entstanden durch die Förderung von mehr als 1000° heißem Magma, das sich bei der Eruption um den Schlot herum anhäufte. Der Basisdurchmesser dieser Schlackenkegel beträgt zwischen 300 und 1000 m und ihre Höhe zwischen 50 und 200 m. Oft stieg Lava aus mehreren dicht benachbarten Kratern auf, so dass einzelne Vulkane in Gruppen beisammen liegen. Von diesen Schlackenkegeln gehen Lavaströme aus, die der Geländeneigung folgend zum Rhein hin abflossen (Schmincke 1988).

3 Kratermulden der Osteifel

Das Eiszeitalter, auch Quartär genannt, ist bestimmt durch zyklisch auftretende, teils kräftige Klimaschwankungen, die in den Warm- und Kaltzeiten ihren Ausdruck finden. Insgesamt 9 Vereisungskomplexe sind bisher am Mittelrhein bekannt. Hierbei ist die Warmzeit meist von kürzerer Dauer, während die Kaltzeit den längeren Abschnitt umfasst. In der Regel bedeutet die Abfolge einer Warm- und einer Kaltzeit einen Zeitraum von etwa 100.000 Jahren. Sedimente vom Beginn des Eiszeitalters aus der Zeit vor über einer Million Jahren sind an Mittelrhein und Mosel bisher nur in Ansätzen nachweisbar. Die Aufschlüsse in dieser Region umfassen die Gliederung des mittleren und späten Abschnittes des Quartärs, aus der Zeit vor 800.000 bis 10.000 Jahren (A. von Berg u. H. H. Wegner 2001).

Die Landschaft an Mittelrhein und Mosel war während der langen Zeiträume der Kaltzeiten nie von Gletschern überzogen. Im ersten Teil einer Kaltzeit herrschte meist ein extrem trockenes Klima ohne größere Kälteschwankungen. Im zweiten und auch kälteren Teil einer Eiszeit wurde im Mittelrheingebiet Löß angeweht, der als feiner gelber Staub die Oberfläche versiegelte. In dieser Zeit mit extremen Temperaturunterschieden zwischen Sommer und Winter breitete sich eine lockere Steppenlandschaft aus. Während der darauf folgenden Warmzeiten herrschte an Mittelrhein und Mosel meist ein feucht gemäßigtes Klima, ähnlich dem heutigen, in dem sich ein dichter Laubwald ausbildete. Die Abfolge von Böden, Humuszonen und Löß lassen geologisch den Wechsel von Klima und Umwelt erkennen. Gerade im Neuwieder Becken sind alle diese Erscheinungen durch ständige Überlagerung mit neuen Sedimenten besonders in den Kratermulden erhalten geblieben. Zusätzlich sind diese Profile durch zahlreiche vulkanische Ablagerungen unterschiedlichster Eruptionen differenziert aufgegliedert (A. von Berg u. H. H. Wegner 2001)

Die meisten Vulkanbauten der Osteifel werden derzeit bergindustriell genutzt und sind schon weitgehend zerstört. Die jeweilige Sedimentfüllung im Schlot und auch in den Nebenkratern ist für die Industrie wertlos und wird mit Großgeräten bis auf die anstehende Vulkanschlacke abgeräumt und auf großen Abraumhalden gelagert. Fast alle Schlackenkegel mit ihren tief ausgeprägten Kratertrichtern haben für die Altsteinzeitforschung und die Quartärgeologie eine herausragende Bedeutung (Baales 2002). Die Kratermulden bilden effektive Sedimentfallen in denen sich pleistozäne Sedimente, Faunenreste und prähistorische Besiedlungsspuren vorzüglich erhalten haben (Bosinski et al. 1986). Nach dem Vulkanausbruch befand sich auf dem Berg ein schroffer Kratertrichter der von steilen Kraterwällen umgeben war. In der nachfolgenden Zeit rutschten die Ränder durch die schnell verwitternde Lava nach und der Trichter wurde sehr schnell mit Lavabrocken und Lößstaub ausgeglichen und abgedichtet. Gerade in den ersten Auffüllungsphasen boten die Mulden mit zeitweiliger Wasserfüllung günstige Aufenthaltsmöglichkeiten und Jagdplätze für den frühen Menschen, die die jeweiligen Orte mehrfach aufgesucht haben. Da die Sedimentation in einer solchen geschlossenen Hohlform recht schnell erfolgte, sind die vorhandenen Reste in der Regel gut erhalten. Die bisher bekannten und auch untersuchten Kraterfüllungen sind dabei immer nach gleichem Schema aufgefüllt. Die Abfolgen zeigen in der Regel Sedimente der vorletzten und letzten Kaltzeit. Die Ablagerungen sind dabei unter unterschiedlichen Bildungsbedingungen wie Windverfrachtung, Solifluktion, Wassertransport entstanden und durch Bodenbildungen gegliedert. Eingebettet in den Sedimenten der Kratermulden lassen sich dann zahlreiche vulkanische Ablagerungen jüngerer Osteifelvulkane feststellen, die für die geochronologische Einordnung der Profile von größter Bedeutung sind. Diese Ablagerungen liefern einzigartige Zeitmarken für die Datierung der jeweiligen Füllungen (P. v. d. Bogaard u. H. U. Schmicke 1990). So ergibt sich aus der Kombination der verschiedenen Sedimente eine detaillierte Gliederung der letzten 700.000 Jahre mit Veränderungen des Klimas, der Umwelt, der Vegetation, der Tierwelt und des Menschen. In den Kratern der Osteifel sind bisher zahlreiche altsteinzeitliche Rast- und Siedlungsplätze hauptsächlich aus der Zeit der Neandertaler untersucht und dokumentiert worden. Die Mulden boten den Menschen der Altsteinzeit hervorragende Siedlungsbedingungen. Die hohen Kraterwälle schützten vor Witterungseinflüssen, gerade während der Kaltzeiten. Der weite Ausblick von den Kraterwällen herab ermöglichte gerade für die Jagd auf Großwild beste Voraussetzungen. Die meisten Fundplätze in den Kratermulden gehören in die offene Steppenlandschaft der Kaltzeiten. In diesem extremen Biotop war die jagdstrategische Lage der Vulkanberge sehr vorteilhaft. Die im jahreszeitlichen Rhythmus durchziehenden Tierherden konnten von den hohen Kraterwällen aus von weitem ausgemacht werden.

Alle bisher untersuchten Siedlungsplätze weisen anhand der überlieferten Knochen eine gemischte Jagdbeute (Baales 2002) auf. Zur Beute des frühen Menschen gehörte nach den Grabungsfunden Nashorn, Elefant, Wildesel, Pferd, Hirsch und Rind. Eine solche gemischte Beute ist gerade für die Zeit der Neandertaler an Mittelrhein und Mosel charakteristisch. Art und Zustand der Jagdreste weisen darauf hin, dass die Jagd vorwiegend unterhalb der Vulkankegel in den freien Steppenlandschaften stattfand. Beweis hierfür sind immer nur ausgewählte Stücke, die auf die Vulkankegel zur Verwertung gebracht wurden. Der Mensch kam mehrfach und vermutlich im jahreszeitlichen Wechsel zu diesen Plätzen (Bosinski 1992). Kleine Menschengruppen durchstreiften das Gelände und hielten sich in den siedlungsgünstigen Mulden auf. Hinweise auf Wanderbewegungen liefern die vor Ort entdeckten Steinwerkzeuge, die mitgeführt wurden und die immer wieder an den jeweiligen Fundorten zurückgelassen sind. Das Rohmaterial erlaubt in der Herkunftsbestimmung die Wanderung der Menschen bis in die Region der Maas in einem Radius von etwa 100 km zu rekonstruieren (N. J. Conard 2001). Einen ersten Hinweis auf den frühen Menschen an Mittelrhein und Mosel selbst lieferte eine 1997 entdeckte Schädelkalotte eines frühen Neandertalers mit einem Alter von mind. 160.000 Jahren aus einer Muldenfüllung der Vulkangruppe „Wannenköpfe" in der Gem. Ochtendung (A. von Berg, S. Condemi, M. Frechen 2000)

4 Der „Schweinskopf" bei Bassenheim

Im nördlichen Gemarkungsbereich von Ochtendung, unmittelbar westlich der Autobahn A48 liegt die Karmelenberg - Vulkangruppe, deren westlicher Ausläufer als „Schweinskopf" bezeichnet wird (siehe Abb. 2).

Abb. 2: Bassenheim, Kratermulde „Schweinskopf". Gesamtansicht der noch vorhandenen Lössfüllung mit Grabungsfläche.

Der Karmelenberg in der Gemarkung Ochtendung ist mit 372,4 m Höhe über NN einer der am höchsten gelegenen Vulkane der Osteifel. Diesem heute unter Naturschutz stehenden Vulkanbau ist unmittelbar am nordöstlichen Fuß ein kleinerer Vulkankegel der sog. „Schweinskopf" vorgelagert, der heute der Gemarkung Bassenheim zugeordnet ist. Zusammen mit mehreren anderen kleinen Ausbruchstellen im Umfeld gehört der Schweinskopf zur Karmelenberg - Vulkangruppe des Osteifel Feldes. Sämtliche Schlackenkegel an dieser Stelle werden seit Jahren abgebaut und die Lava der industriellen Nutzung zugeführt. Heute noch erhalten sind Teile der Lößfüllung der ehemaligen Kratermulde, die heute nach Abtrag der Kraterränder hoch aufragend als Sedimentberg die Landschaft prägt (Abb. 2). Der Schweinskopf ist ein sog. „Schlackenkegel", bei dem die ausgeworfene Lava am unmittelbaren Austrittsort zurückfällt und so einen Kegel bildet. Das heutige Erscheinungsbild ist vollkommen durch einen intensiven Abbau verändert. Die Vulkantätigkeit am „Schweinskopf" begann unter waldsteppenartigen Klimaverhältnissen am Ende der vorletzten Warmzeit vor etwa 213.000 Jahren (Schäfer 1990). Der Kegel hatte einen Durchmesser von ursprünglich etwa 100 m und eine Höhe von ca. 50 m. Der Kraterwall um die Muldenfüllung herum ist bis auf wenige Reste heute vollständig zerstört. Durch den intensiven Lavabbau an dieser Stelle lassen sich die verschiedenen Eruptionsphasen des Schlackenkegels deutlich ablesen.
Erste Grabungen wurden innerhalb der damals noch vollständig vorhandenen Muldenfüllung des Kraters in den Jahren 1983 – 1987 durchgeführt und hier erste Erkenntnisse zur Stratigraphie und Fundschichtenabfolge gewonnen. Mehrere Lagerplätze der Neandertaler konnten dabei untersucht werden, die beweisen, dass die Kratermulde des Schweinskopfes über einen längeren Zeitraum immer wieder durch den frühen Menschen aufgesucht wurde. Über das gesamte Mittelpaläolithikum hinweg sind aus der Schweinskopfmulde archäologische Funde wie Jagdbeutereste und Steinwerkzeuge bekannt (Schäfer 1987 und 1990).
Durch den fortschreitenden Lavaabbau wurde 1998 die Untersuchung der noch vorhandenen westlichen Sedimentfüllung der Kratermulde durch die Archäologische Denkmalpflege, Koblenz nötig. Im Zuge der Arbeiten konnte die gesamte westliche Sedimentfüllung bis auf die anstehende Basaltlava des Schlotes abgetragen und die jeweiligen Fundschichten und archäologischen Funde dokumentiert werden (Abb. 2). Hierbei wurde die Stratigraphie und Schichtfolge mit der Sedimentation neu festgelegt und mit den frühen Grabungen im heute nicht mehr vorhandenen östlichen Muldenabschnitt korreliert. Die Sedimentfüllung wurde dann jeweils bis auf die paläolithischen Fundschichten abgetragen. Insgesamt fünf paläolithische Fundkonzentrationen ehemaliger Rastplätze des Mittelpaläolithikums konnten festgestellt und die Befunde detailgenau dokumentiert werden.
Eine Sondage am Rand der Kratermulde zur Feststellung der Sedimentation im Sohlenbereich im Jahre 2001 führte zur Freilegung mehrerer Quadratmeter innerhalb der Fundschicht 1, die als älteste Fundkonzentration unmittelbar über der Basaltlava liegt (siehe Abb. 3).

Abb. 3: Bassenheim, Kratermulde „Schweinskopf". Fundschicht 1 an der Basis der Lößfüllung.

Das Alter der Fundschicht liegt geochronologisch und stratigraphisch am Beginn der vorletzten Kaltzeit vor etwa 200.000 Jahren und datiert in die Zeit unmittelbar nach dem Vulkanausbruch. Die Funde lagen in einer stark mit Vulkanasche durchsetzten Lößschicht, die zwischen den teils mächtigen von den Randbereichen der Kratermulde eingerollten Basaltlavabrocken abgelagert wurde und den Beginn der Auffüllung der Mulde kennzeichnet (Abb. 3).

Auf einer Fläche von 10 Quadratmetern wurde Faunenreste der Jagdbeute und einige Steinartefakte aus Quarz dokumentiert. Bei den Jagdbeuteresten handelte es sich um Röhrenknochen des postcranialen Skelettes von Nashorn (*Coelodonta antiquititatis*) und Pferd (*Equus caballus*), die durch Sedimentdruck zwischen den Basaltlavabrocken oft kleinteilig zerbrochen waren. Unmittelbar neben einer etwa einen Quadratmeter großen ovalen Fläche mit leicht durch Hitze verziegelter Unterkante, die als Rest einer Feuerstelle gedeutet werden kann, fanden sich drei Quarzartefakte, zwei retuschierte Abschläge, ein Schaber aus Gangquarz. Nach den Verziegelungsspuren an der Unterkante war die Feuerstelle leicht konisch zwischen einigen größeren Basaltlavageröllen angelegt. Einen weiteren Hinweis auf die Feuerstelle lieferten mehrere kalzinierte Knochenreste von Großsäugern, die unmittelbar von der Oberkante der Verziegelungsspur stammten. Hier lagen auch die beiden Hominidenzähne, die ebenfalls durch Hitzeeinwirkung verändert waren.

5 Micro-CT-Messungen

Seit einiger Zeit wird mit so genannten Micro-CTs gearbeitet, die im Wesentlichen einer miniaturisierten Form moderner CTs entsprechen und zur zerstörungsfreien, dreidimensionalen Mikroskopie genutzt werden. Das durchstrahlte Messfeld ist mit typischerweise 2 cm³ so klein, dass medizinische Anwendungen auszuscheiden scheinen. Tatsächlich werden diese Geräte eher in der Materialprüfung und -analyse verwendet, aber auch medizinische Anwendungen rücken zunehmend in das Zentrum des Interesses.

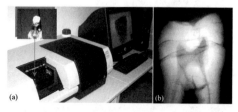

Abb. 4: Micro-CT-Tischgerät SkyScan 1072 (a) am RheinAhr-Campus Remagen. Auf einem Drehteller in der Messkammer wird das zu untersuchende Objekt platziert. Das Messvolumen ist etwa 2 cm³ groß. Schon das einfache Durchleuchtungsergebnis eines archäologischen Zahnfundes zeigt eine beeindruckende Vielfalt an Strukturen. Neben dem Wurzelkanal sind auch feine Risse zu erkennen, die sich im Laufe von Jahrtausenden bei diesem Zahn eines Neandertalers gebildet haben. Der kleine Pfeil markiert die Höhe, in der eine rekonstruierte Schicht näher betrachtet werden soll. Die planare, 1024 x 1024 Pixel große CCD-Matrix ist gekühlt und besitzt eine Detektorgröße $b_D < 10$ µm.

Humanmedizinische Fragestellungen sind zum Beispiel Untersuchungen der Trabekularstruktur von Knochen. Micro-CTs sind darüber hinaus ideale Geräte, um radiologische Diagnostik an Kleintieren zu betreiben (N. M. De Clerck, D. van Dyck and A. A. Postnov, 2003).

Abb. 4 zeigt ein Micro-CT-System der belgischen Firma SkyScan im Labor für Computertomographie des RheinAhrCampus Remagen. Es ist als Tischgerät ausgelegt und besitzt eine Messkammer, die mit Bleiwänden gegen nach außen dringende Röntgenstrahlung vollständig abgeschirmt ist, so dass keine weiteren Schutzmassnahmen ergriffen werden müssen. Das zu untersuchende Objekt wird auf einem Drehteller platziert, der von einem Schrittmotor gesteuert wird.

In Abbildung 2a ist die Vorbereitung einer Untersuchung eines Neandertalerzahnes zu sehen. Dieser Zahn ist 200.000 Jahre alt, so dass man sich heute für die Veränderungen des Zahninneren interessiert. Abbildung 2b zeigt eine Kegelstrahldurchleuchtung des Zahnes. Neben dem Wurzelkanal bilden sich in der Projektion auch Risslinien sehr gut ab.

Die beiden entscheidenden Komponenten von Micro-CTs sind die Röntgenröhre und das Detektorarray. Hierbei sind es speziell die Fokusgröße und die Größe der Detektorelemente, die neben der mechanischen Genauigkeit der Drehbewegung das Auflösungsvermögen bestimmen. Ein Blick auf Abbildung 5 zeigt warum die Fokusgröße die Detaildarstellung beeinflusst.

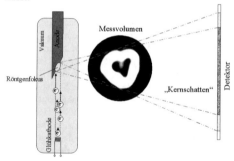

Abb. 5: Durch die Anschrägung der Anode kann man die Größe des optischen Brennflecks beeinflussen. Je größer der optische Brennfleck ist, desto unschärfer wird die Abbildung, da alle Objekte von einem Halbschattensaum umgeben sind.

Offenbar benötigt man eine so genannte Microfokusröhre. Dabei sind Röntgenfokusgrößen unterhalb von 10 µm wünschenswert. Natürlich kann bei einer so kleinen Elektronentargetfläche der Anodenstrom nicht sehr groß gewählt werden. Hier sind Ströme von $I < 100$ µA typisch. Da der Strom die Intensität des Röntgenspektrums steuert, unterliegt man in Bezug auf die zu untersuchenden Materialien natürlich gewissen Einschränkungen. Als Detektor wird ein gekühlter 12 Bit Röntgen-CCD-Chip mit einer Pixelmatrix von 1024 x 1024 genutzt, der über eine Fiberoptik an einen Szintillationskristall angekoppelt ist. Die Größe der Bildelemente liegt in der Größenordnung von

10 µm. Die Firma ScyScan gibt ein Auflösungsvermögen von etwa 10 µm an (A. Sassov 1999,2002). Da es sich bei Micro-CTs um Kegelstrahlröntgensysteme handelt, sind dreidimensionale Rekonstruktionsverfahren erforderlich, um die Bilder zu berechnen (G. Wang, S. Zhao and P.-C. Cheng 1998, Th. M. Buzug 2004).

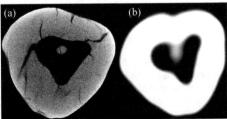

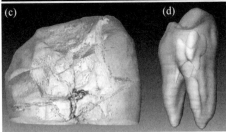

Abb. 6: Vergleichende Untersuchung des Auflösungsvermögens am Beispiel eines archäologischen Zahnfundes. Ein einige 10000 Jahre alter Neandertalerzahn wurde mit einem Micro-CT (a/c) und einem normalen klinischen CT (b/d) untersucht. Die Unterschiede sind sofort zu erkennen. Die feinen Risse, die sich im Laufe der Jahrtausende im Zahn gebildet haben, sind mit dem klinischen CT nicht abzubilden. In der dreidimensionalen Rekonstruktion der Micro-CT-Aufnahme (c) können die Risse als Rissflächen identifiziert werden. Im Vergleich dazu ist in (d) der gesamte Zahn dreidimensional rekonstruiert. Angesichts der Abmessungen eines Zahnes ist auch diese Rekonstruktion schon beindruckend, zeigt aber nicht die Mikrostrukturen der Micro-CT-Aufnahme.

6 Ergebnisse

Abbildung 6 zeigt die Unterschiede im Auflösungsvermögen zwischen dem Micro-CT und einem klinischen CT am Beispiel des Neandertalerzahnes. In einer rekonstruierten Schicht sind im Ergebnis des Micro-CTs sehr feine Risse zu erkennen (a). Die Rekonstruktion mit einem klinischen CT (b) vermag diese Details nicht aufzulösen. Im Vergleich der dreidimensionalen Rekonstruktionen und Visualisierungen (c/d) ist der Unterschied ebenfalls sehr deutlich. Nur in der Micro-CT-Aufnahme sind Rissflächen sichtbar, die durch den gesamten Zahn verlaufen.

Insgesamt sind die Zähne also nur noch fragmentarisch erhalten, wobei der Zahnschmelz durch thermische Einwirkung größtenteils abgeplatzt ist. Auch ist die ursprüngliche Größe durch die Hitzentwicklung deutlich reduziert und die Zahnwurzel dabei leicht verformt. Bei den Zähnen könnte es sich um einen 3 Molar und einen Prämolar einer hominiden Zahnfolge handeln. Genauere anthropologische Untersuchungen stehen derzeit allerdings noch aus. Durch das Alter der Fundschicht mit etwa 200.000 Jahre stellen die Funde vom „Schweinskopf" Hominidenreste dar, die unmittelbar am Beginn der Neandertalerentwicklung stehen, die zu diesem Zeitpunkt in Mitteleuropa begann. Bei den menschlichen Zahnfragmenten aus der Kraterfüllung handelt es sich um die bislang frühesten Menschenreste des Osteifel Vulkanfeldes am Mittelrhein.

7 Literatur

M. Baales 2002, Vulkanismus und Archäologie des Eiszeitalters am Mittelrhein. Die Forschungsergebnisse der letzten 30 Jahre. Jahrb. Röm.- Germ.- Zentralmuseum 49, 1, 2002, 41 – 83.

A.von Berg, H. H. Wegner 2001, Jäger – Bauern – Keltenfürsten. 50 Jahre Archäologie an Mittelrhein und Mosel. Archäologie an Mittelrhein und Mosel 13 (Koblenz).

A. von Berg, S. Condemi, M. Frechen 2000, Die Schädelkalotte des Neandertalers von Ochtendung / Osteifel. Archäologie, Paläoanthropologie und Geologie. Eiszeitalter und Gegenwart 50, 56 – 68.

P. v. d. Bogaard, H. U. Schmincke 1990, Die Entwicklungsgeschichte des Mittelrheinraumes und die Eruptionsgeschichte des Osteifel – Vulkanfeldes. In: W. Schirmer (Hrsg.), Rheingeschichte zwischen Mosel und Maas. Deuqua – Führer 1 (Düsseldorf), 166 – 190

G. Bosinski 1992, Eiszeitjäger im Neuwieder - Becken (3. Auflage). Archäologie an Mittelrhein und Mosel 1 (Koblenz).

G. Bosinski et al 1986, Altsteinzeitliche Siedlungsplätze auf den Osteifel Vulkanen. Jahrb. Röm. Germ. Zentralmuseum 33, 2, 97 – 130

Th. M. Buzug 2004, Einführung in die Computertomographie, Springer-Verlag (Heidelberg).

N. M. De Clerck, D. van Dyck and A. A. Postnov, 2003, Non-Invasive High-Resolution µCT of the Inner Structure of Living Animals, Microscopy and Analysis **1**, 13.

N. J. Conard 2001, River Terraces, Volcanic Craters and Midlle Paleolithic Settlement in the Rhineland. In: N. J. Conard (Hrsg.), Settlement Dynamics of the Middle Paleolithic and Middle Stone Age (Tübingen), 221 – 250.

W. Meyer 1986, Geologie der Eifel (Bonn), 365

A. Sassov 1999, Desktop X-ray Micro-CT, in: Proc. of the DGZiP BB67-CD, Computerized Tomography for Industrial Applications and Image Processing in Radiology (Berlin) 165.

A. Sassov 2002, Desktop X-ray Micro-CT Instruments, Proc. of SPIE **4503**, 282.

A. Sassov 2002, Comparison of Fan-Beam, Cone-Beam and Spiral Scan Reconstruction in X-Ray Micro-CT, Proc. of SPIE **4503**, 124.

J. Schäfer 1990, Der altsteinzeitliche Fundplatz auf dem Vulkan Schweinskopf – Karmelenberg (Dissertation Köln).

J. Schäfer 1987, Der altsteinzeitliche Fundplatz Schweinskopf, Gem. Bassenheim, Kreis Mayen – Koblenz. Arch Korrbl. 17, 1 - 12

H. U. Schmincke 1988, Vulkane im Laacher See - Gebiet. Ihre Entstehung und heutige Bedeutung (Haltern).

G. Wang, S. Zhao and P.-C. Cheng 1998, Exact and Approximate Cone-Beam X-Ray Microscopy, in: P. C. Cheng, P. P. Huang, J. L. Wu, G. Wang and H. G. Kim (Eds.) Modern Microscopies (I) – Instrumentation and Image Processing; World Scientific, Singapore.

Development of a C-Arm-Based Meso-CT for NDT and Educational Purposes

S. Schneider, B. Bruckschen and T. M. Buzug
Department of Mathematics and Technology, RheinAhrCampus Remagen, Suedallee 2, D-53424 Remagen

Abstract

Today's standard computed tomography (CT) scanners are suitable for objects in the range of decimeters with a resolution not better than a millimeter. For very small objects – on the other hand – micro CTs are commercially available showing a resolution in the range of a few microns for an object diameter of typically one or two centimeters. In this paper the development of a mid-scale "meso-CT" based on a conventional C-arm X-ray device is presented. The key benefits of the prototype are that it takes up less room and it is much cheaper than a normal CT. During the development the main tasks were design and realization of the workplace and the data acquisition set-up. Caused by an image intensifier there are some geometrical distortions that must be corrected prior to image reconstruction. To get rid of these distortions a polynomial-landmark based transformation is used as one step in the data pre-processing chain.

1 Introduction

Over the last years none destructive testing (NDT) has grown into a large industry. The areas of application are vast and include many branches like automation, aviation, aerospace and health care. The radiographic testing (RT) is an important part of NDT. It allows inspecting the entire inner structure of an object and not only its surface. The resolution of modern CT scanners is high enough to detect small cracks or other defects in solid materials.

Since a few years devices like micro CTs have improved the resolution properties substantially. However, between the resolution of a normal and a micro-CT exists a gap and the prototype presented here is an approach to bridge it (see fig. 1). Due to relative small investments this prototype can also be used for educational purposes in programs like medical engineering.

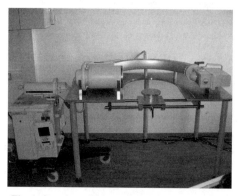

Fig. 1: Overview of the experimental meso-CT set-up.

2 Competing Systems

In a first project step a survey on competing systems has been carried out. The results of this literature and internet study and the trends of actual C-arm based developments are summarized in the following subsections.

2.1 Experimental CT for Education and Research

A system developed at the TU Ilmenau is based on the classical third generation CT scanner with a linear detector array. The object-of-interest is fixed on a turn-tilt plate, which is mounted on a planar desk. For the rotation of the plate a stepping motor is used. To allow a zoom the focus-centre distance (FCD) is adjustable. For that purpose the turn-tilt place can be linearly shifted between the X-ray source and the detector. The linear X-ray sensitive detector array is of 314 mm length and shows a resolution of 5 lp / mm. The line detector-array is basically adapted from industrial applications.

The reconstruction yields grey-level images with a depth of 12 bits. The total acquisition and reconstruction time is about five minutes for a 512 x 512 image. With this set up a resolution of objects below 1 mm is possible. The Ilmenau design is the standard for most of today's CT scanners [1,2].

2.2 First Generation Scanners

Some other universities (University of Gießen and RWTH Aachen) developed CT prototypes for educational purposes that are based on first generation geometry. As an alternative to the X-ray based imagers some systems are based on laser scanning.

Those systems have a laser source and a photo diode as detector. The objects under investigation are synthetic grids. However, as for the X-ray based systems the absorption of light is the basis of object reconstruction.

The geometrical set-up is the same as for the X-ray based systems. The detector is rotated around the object and for every predefined angle step it was linearly translated along the object. The light source on the other side of the object was simultaneously rotated and moved, respectively.

Such a geometrical set-up was used in the beginning of computed tomography. However, the principles of this technique can easily be demonstrated, and, therefore, today it is often used for educational purposes.

2.3 SIREMOBIL Iso C^{3D}

For intra-operative navigation updated data sets of the region-of-change are needed to provide the highest surgical accuracy. Usually, the clinical CT is the only way to create appropriate data sets for surgical planning and navigation. However, the demand of 3D visualization during the surgical intervention is growing. A recent system development providing the possibilities to create on-line data sets during surgery is the Siemens SIREMOBIL Iso C^{3D}. The basic set-up of this system is a conventional C-arm equipped with motor devices. The C-arm performs a single 190° orbital rotation around the patient and acquires 100 digital images. The acquisition time for one rotation is about 120 s. The field of view (FOV) is about 119 mm by a resolution of 256 pixels in all three directions. So the size of the reconstructed voxel is 0.46 mm. The lateral resolution of the system is 9 lp / mm [3-7].

The image quality of the Iso C^{3D} is inferior to CT, but is high enough to provide a good navigation and the speed of the image acquisition makes this system to a valuable tool in navigated surgery. An improved version of the Iso C^{3D} is currently under development. Instead of the X-ray image intensifier (XRII) a new flat panel detector will be used.

2.4 Conclusions from Survey

The results of the brief survey showed that most of the systems are of classical geometries. The third generation geometry is the standard for actual CT scanners. However, the latest trend in development goes to large area flat detectors. Therefore, the cone-beam geometry was chosen as set-up for the system presented in this paper.

3. Hardware Design

The system development started with a demounted and boxed Philips BV 25 T C-arm X-Ray device. It is a system that provides geometrical flexibility as illustrated in Fig. 1.

3.1 Requirements

The Philips BV 25 T is a relatively small dismountable C-arm constructed for military field application. To get an optimal fixation, the C-arm was mounted separately on a plane table. This yields an appropriate geometrical stability preventing of elastic C-arm vibrations during object movements.

However, there are some additional points that potentially affect the imaging quality of the system. The XRII (X-Ray Image Intensifier) detector is known to be sensitive to magnetic fields.

Therefore, ferromagnetic materials and electromagnetic fields of other devices in the lab may cause problems because the intensifier principle is based on magnetic fields as well, so these objects and devices producing spurious fields could disturb the magnetic lens of the XRII leading to unwanted distortions in the images.

3.2. Electro-Mechanical Realization

The realization consists of the following steps:

o Measurement of the C-arm geometry of the Philips BV 25 T,
o construction the object rotation device,
o construction of a guide way consisting of two 20 mm steel rods,
o construction of the C-arm fixation (between XRII and X-ray tube the table has been cut-out and the guide way and the turn-tilt plate were mounted),
o mounting of the turn tilt plate and the stepping motor via four linear roll bearings on the guide way
    (in that way the positioning of the plate is very easy and precise; additionally, movements to the side direction is not permitted),
o connection of the turn tilt plate to the stepping motor.

To provide high angle accuracy a gear with the ratio 1/50 has been installed between the motor and the plate. One step of the motor is 0.036° on the table. The angle resolution is about 10.000 steps on 360 degrees. A principle geometrical limitation is due to the usage of the Philips BV 25 T. The field of view is about 160 mm by a resolution of 547 pixels.

4. Controlling and Timing

Beside the mechanical design, the PC based controlling and timing of the rotation plate and the image acquisition is a key working package of the project presented here.

4.1. Requirements

To establish an accurate object rotation the turn-tilt plate, a control panel and a connection scheme had to be developed. To minimize noise on the signal cables two separate power supplies were constructed supplying the stepping motor and the controlling electronics. To control the speed of the stepping-motor a variable clock signal was installed. To realize automatic image acquisition using the PC, a new trigger card controlling the BV 25 T must be constructed. This board also watches the security door switch, so if the door is open, it is impossible to activate the X-ray. If the door is opened during a running measurement, the X-ray generation will be stopped immediately.

Fig. 2: Stepping motor with rotation encoder attached.

4.2. Stepping-Motor Control

In the project presented here the LabView™ development software has been used to control the voltage for the VCO (Voltage Controlled Oscillator) and to stop the rotation of the plate by controlling the respective driver board. Due to the adjustments made on the VCO-card the clock signal varies between 20 Hz and 400 Hz depending on the controlling voltage (5.2 V – 5.8 V). To prevent the clock signal from becoming unstable a buffer was placed behind the VCO chip. The generated signal is sent to the step-driver board.

On this board the clock signal is transformed to the step signal by the L297 integrated circuit. This signal is required to control the motor coils in the appropriate sequence [8].
Additionally, a driver chip is integrated in this board and the stepping signal can directly be transferred to the stepping motor. This design allows controlling of motor speed and, in that way, variation of the whole scan process time. The motor power is high enough to rotate objects up to 8 kg.

4.3. Position Measurement

To receive a precise angular position from the table the executed motor steps must be acquired. As described above one step of the motor is 0.036° on the table. As shown in Fig. 2 a rotation encoder is attached to the motor. The resolution of the rotation is 2500 steps per 360° of the motor. This angular clock signal is transferred to the PC.

4.4. Image Acquisition

The incoming step signals are counted and on predefined intervals the rotation is stopped, the X-ray is activated and the image is grabbed from the frame buffer.
In this prototype set-up a National Instruments 32 bit RGB frame grabber card is used, and, for testing, an 8 bit grey-level version from the same manufacturer has also been used. The images are stored in the lossless bitmap format.
However, one major problem is that the frame buffer from the BV 25 T has an 8 bit grey-level resolution only. Therefore, the usage of a 32 bit RGB frame grabber board will not improve the image. At this project status, the restriction in grey-level depth will be a serious problem for low contrast objects.

5. Image Pre-processing

The projection images grabbed from the C-arm need some pre-processing steps before using them in a cone-beam reconstruction. Especially, the usage of a XRII produces some characteristic artefacts of the images that must be compensated prior to reconstruction.

5.1. Intensity Homogenization

The first artefact that must be compensated is the inhomogeneity of the XRII detector. This inherent artefact is caused by the design of the device.

As a first step to cope with this problem about 20 so-called empty images are taken (images without an object in the FOV). With this stack of images an average empty image is calculated. The resulting mean image represents the static sensitivity signature of the XRII detector.

As a second step the average empty image is inverted so that the highest grey-level produces the smallest correction value.

The images from the objects and the correction image are multiplied for artefact reduction leading to a flat sensitivity profile (compare Fig. 3).

5.2. Geometric Distortion

The geometric distortion is a typical artefact of XRII detectors inducing problems to use these uncorrected images directly in a 3D reconstruction algorithm. A raw reconstruction would result in 3D images with unacceptable artefacts. A brief explanation of the nature of these distortions is given below.

5.2.1. Working Principle of XRII and Inherent Artefacts

The simplest description of a XRII starts with a first detector layer that is a luminescent screen close to a photo cathode. The incoming X-rays produce a luminescence, which is not very intensive. However, the luminescence photons produce photoelectrons from the cathode due to the photo effect.

These electrons are accelerated to the backside of the XRII where a second luminescent screen is collecting the electrons. The trajectory of the electrons is controlled by an electron optics. The electrons from the cathode are focused on this second screen. The number of electrons is proportional to the intensity. At the backside of the second screen a CCD camera is placed that directly produce digital images of the luminescence.

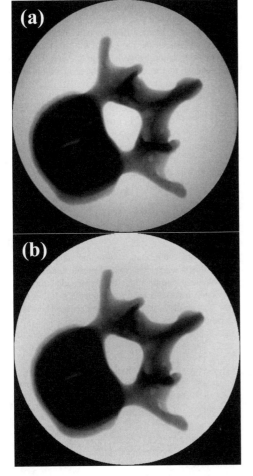

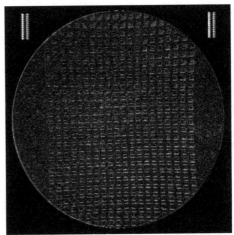

Fig. 4: Image differences between two acquisitions of a grid phantom.

Fig. 3: (a) Raw projection image and (b) intensity corrected projection of a vertebra phantom.

A major problem of this construction is that the fields created in the electron optics are typically imperfect. As a consequence image defects like pincushion or barrel distortion could appear. These inherent artefacts are due to the nature of the XRII and cannot be avoided.

Another problem is the interaction of the electron optics with the magnetic field of the earth. So, when the XRII is rotated in the magnetic field the distortions vary. This is not a major problem because in the project described here the C-arm is mounted on the workbench.

However, the distortions measured in the pictures show a time variance as demonstrated in Fig. 4 that is induced by spurious fields of other electric devices in the laboratory. This problem must be fixed by an appropriate shielding in a later step of the project.

Additionally, the afterglow effect of the luminescent detector material produces "ghost images" of objects, which are not longer in the FOV. In the first project prototype this problem is fixed by an X-ray pause between acquisitions of subsequent projections.

5.2.2. Marker-Based Image Correction

Generally, the geometrical image distortion can be compensated by a fit of an appropriate model f based on a set of n homologous markers or landmarks in the measured image (that must be corrected – see Fig. 5) and in an ideal mathematical, synthetic version of the phantom, respectively.

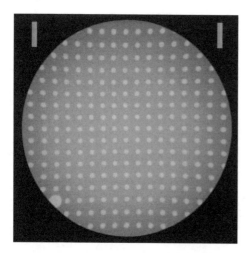

Fig. 5: Al-grid phantom, U=50 KV, I=0.3 mA, t=0.1 s.

The set of n homologous markers is directly used to estimate the parameters of the transformation f relating the points of the measured image to the corresponding points in the ideal distortion-free image

$$(x,y) \rightarrow (x',y') = \left(f_x(x,y), f_y(x,y)\right), \quad (5.1)$$

where (x,y) and (x',y') are the coordinates of homologous points in both images.

5.2.3. Automated Marker Extraction

The automatic marker extraction is currently under development. To find all the markers in the measured image (see Fig. 5), the grey-value homogenization described above has to be applied. This is necessary because of the use of a cross-correlation to compute the centre of all markers.

The cross-correlation is carried out in the frequency domain using the Wiener-Khintchine theorem [9]: The entire image and a small template taken from the image have to be Fourier transformed using the FFT. The inverse FFT of the product of the two spectra is the cross-correlation between the image and the template leading to the centres of the grid points that must be extracted in a final step.

5.2.4. Polynomial Transformation

After the markers are found in the image the coefficients of the transformation eq. (5.1) can easily be computed. The actual transformation or model f is a fifth order polynomial

$$\begin{aligned} x' &= a_{01} + a_{11}x + a_{21}y + a_{31}xy + \ldots + a_{k1}x^5y^5 \\ y' &= a_{02} + a_{12}x + a_{22}y + a_{32}xy + \ldots + a_{k2}x^5y^5 \end{aligned} \quad (5.2)$$

As there are more than k homologous landmarks $(n > k)$, this is an overdetermined problem, and a singular value decomposition (SVD) of calculation of the pseudo inverse is used to produce a solution that is the best result in the least-squares sense. The overdetermined system is given by

$$\begin{pmatrix} x'_1 & y'_1 \\ x'_2 & y'_2 \\ \vdots & \vdots \\ x'_n & y'_n \end{pmatrix} = A \begin{pmatrix} a_{01} & a_{02} \\ a_{11} & a_{12} \\ a_{21} & a_{22} \\ a_{31} & a_{32} \\ a_{41} & a_{42} \\ a_{51} & a_{52} \\ \vdots & \vdots \\ a_{k1} & a_{k2} \end{pmatrix} = \begin{pmatrix} 1 & x_1 & y_1 & x_1y_1 & x_1^2 & y_1^2 & \ldots & x_1^5y_1^5 \\ 1 & x_2 & y_2 & x_2y_2 & x_2^2 & y_2^2 & \ldots & x_2^5y_2^5 \\ \vdots & \vdots & \vdots & \vdots & \vdots & \vdots & \ddots & \vdots \\ 1 & x_n & y_n & x_ny_n & x_n^2 & y_n^2 & \ldots & x_n^5y_n^5 \end{pmatrix} \begin{pmatrix} a_{01} & a_{02} \\ a_{11} & a_{12} \\ a_{21} & a_{22} \\ a_{31} & a_{32} \\ a_{41} & a_{42} \\ a_{51} & a_{52} \\ \vdots & \vdots \\ a_{k1} & a_{k2} \end{pmatrix} \quad (5.3)$$

To solve this linear system of equation (5.3), the pseudo inverse

$$A^+ = A^T (A \cdot A^T)^{-1} \quad (5.4)$$

of the design matrix A has to be computed.

The coefficients

$$\begin{pmatrix} a_{01} & a_{11} & a_{21} & \cdots & a_{k1} \\ a_{02} & a_{12} & a_{22} & \cdots & a_{k2} \end{pmatrix}^T = A^+ \begin{pmatrix} x'_1 & y'_1 \\ x'_2 & y'_2 \\ \vdots & \vdots \\ x'_n & y'_n \end{pmatrix} \quad (5.5)$$

will be used in the polynomial transformation 5.2. First results of this algorithm are promising. But some residual defects remain in the image. Therefore, a higher order method – the thin-plate spline transformation – solving this problem is under investigation.

5.2.5. Thin-Plate Spline Transformation

Another approach for distortion correction is the so-called TSP (thin-plate spline) transformation. Unlike the polynomial transformation, the TSP is an elastic warping which can be considered as an interpolation. The set of homologous landmarks is the same as for the polynomial transformation described above. For the thin-plate spline approach the function f in eq. (5.1) can be written as

$$f_\bullet(x_j, y_j) = a_{0\bullet} + a_{1\bullet} x_j + a_{2\bullet} y_j + \ldots \\ \ldots + \sum_{i=1}^n b_{i\bullet} U(\|(x_j, y_j) - (x_i, y_i)\|) \quad (5.6)$$

The point denotes the index for the x- or y-coordinate. The $U(r)$ is a radial basis function

$$U(r) = r^2 \cdot \ln(r) \quad (5.7)$$

that provides the elastic part of the transformation. The sum gives information about the forces that a pixel is subjected from the surrounding landmarks. The corresponding design matrix can be found in

$$\begin{pmatrix} x'_1 & y'_1 \\ \vdots & \vdots \\ \vdots & \vdots \\ x'_n & y'_n \\ 0 & 0 \\ 0 & 0 \\ 0 & 0 \end{pmatrix} = \begin{pmatrix} 0 & U(r_{12}) & \cdots & U(r_{1n}) & 1 & x_1 & y_1 \\ U(r_{21}) & 0 & \cdots & U(r_{2n}) & 1 & x_2 & y_2 \\ \vdots & \vdots & \ddots & \vdots & \vdots & \vdots & \vdots \\ U(r_{n1}) & U(r_{n2}) & \cdots & 0 & 1 & x_n & y_n \\ 1 & 1 & \cdots & 1 & 0 & 0 & 0 \\ x_1 & x_2 & \cdots & x_n & 0 & 0 & 0 \\ y_1 & y_2 & \cdots & y_n & 0 & 0 & 0 \end{pmatrix} \cdot \begin{pmatrix} b_{11} & b_{12} \\ \vdots & \vdots \\ \vdots & \vdots \\ b_{n1} & b_{n2} \\ a_{01} & a_{02} \\ a_{11} & a_{12} \\ a_{21} & a_{22} \end{pmatrix} \quad (5.8)$$

This approach gives quite good results but in consideration of the huge number of 165 landmarks in the phantom the computation of the corrected image will take extremely long. Another problem of the elastic image correction is that it will produce holes in the image, because it can only be used for the forward transformation.

6. Future Work

In this paper the construction of a C-arm based meso-CT is shown from the beginning. After implementation of all pre-processing steps the a first goal of this project could be achieved.
In the next steps a cone-beam reconstruction must be implemented to reconstruct a 3D volume of the examined objects. Two basic methods are possible. The first method is the Feldkamp algorithm (FDK). It is the three dimensional version of the filtered back-projection developed in 1984 by Feldkamp, Davis and Kress [10]. The alternative that will be followed in the project is the implementation of a Maximum Likelihood method on consumer graphics cards. This method is very popular in the nuclear medicine where the signal noise ratio (SNR) is not as high as in X-ray CT.

References

[1] A. Keller, *Experimenteller Computertomograph für Ausbildung und Forschung Teil 1*, mt medizintechnik **3**, 2002, p. 103 – 105.

[2] S. Gross, A. Keller, U. Krüger, *Experimenteller Computertomograph für Ausbildung und Forschung Teil 2*, mt medizintechnik **4**, 2002, p. 135 – 140.

[3] C. Rock, U. Linsenmaier, R. Brandl, D. Kotosianos, S. Wirth, R. Kaltschmidt, E. Euler, W. Mutschler, K.J. Pfeifer, *Vorstellung eines neuen mobilen C-Bogen- / CT-Kombinationsgeräts (ISO-C-3D)*, Unfallchirurg **104**, 2001, p. 827 – 833.

[4] F. Gebhard, M. Kraus, E. Schneider, M. Arand, L. Kinzl, A. Hebecker, L. Bätz, *Strahlendosis im OP – ein Vergleich computerassistierter Verfahren*, Unfallchirurg **106**, 2003, p. 492 – 497.

[5] D. Kendoff, J. Geerling, L. Mahlke, M. Citak, M. Kfuri, T. Hüfner, C. Krettek, *Navigierte ISO-C-3D-basierte Anbohrung einer osteochondralen Läsion des Talus*, Unfallchirurg **106**, 2003, p. 963 – 967.

[6] E. Euler, S. Heining, S. Wirth, D. Kotsianos, W. Mutschler, *3D-Bildwandler – Erfahrungen zu Aufwand und Benefit*, Trauma und Berufskrankheit **6**, 2004, p. 185 – 190.

[7] D. Kotsianos, S. Wirth, T. Fischer, E. Euler, C. Rock, U. Linsenmaier, K.J. Pfeifer, M. Reiser, *3D imaging with an isocentric mobile C-arm Comparison of image quality with spiral CT*, European Radiology **14**, 2004, p. 1590 – 1595.

[8] F. Prautzsch, *Schrittmotor-Antriebe: präzise positionieren in computergeregelter Technik*, Franzis, Poing, 1991.

[9] T. M. Buzug, *Einführung in die Computertomographie*, Springer-Verlag, Heidelberg, 2004.

[10] L. A. Feldkamp, L. C. Davis and J. W. Kress, *Practical Cone-Beam Algorithm*, J. Opt. Soc. Am. **A6** (1984) pp. 612.

ND
X-Ray Based NDT of Accumulator Membranes

Dirk Thomsen[1], Thorsten Kurz [2], Richard Kaesler[2] and Thorsten M. Buzug[1]
[1] Department of Mathematics and Technology, RheinAhrCampus Remagen, Suedallee 2, D-53424 Remagen
[2] Integral Accumulator KG, Sinziger Str. 47, D-53424 Remagen

Abstract

Accumulators are used in lots of applications. So they are used as a pressure accumulator for quick energy supply to hold a constant pressure (equalization of leakage), for pulsation damping and as an element of suspension. We can differ between 3 types of accumulators: 1. Diaphragm Accumulator, 2. Piston Accumulator 3. Metal-Bellow Accumulator. At applications with high requirements to permeation the diaphragm accumulators are equipped with so called multi-layer diaphragms. During test series some of these accumulators failed because of big gas losses. At a first look, no external damage could be seen, so that the damage was assumed in the internal layer. In this contribution the results of a detailed X-ray based non-destructive testing (NDT) of accumulator membranes are reported. This includes digital X-ray fluoroscopy using a Philips Computed Radiography (PCR) system as well as conventional and micro CT investigations.

1 Introduction

Accumulators were originally developed as energy accumulators. Modern hydro-accumulators are also used for damping, to balance out leakage and volume and for pulsation damping. There are diaphragm and piston accumulators and valve units used in conjunction with diaphragm accumulators.

In extreme situations a membrane leakage has been observed in a testing environment. A first optical inspection did not show any damage of the issue. Thus, due to the multi-layer structure of the membrane an inner-structure damage is suspected to be the cause of the leakage.

The internal structure of the multi-layer diaphragms is shown in figure 1.

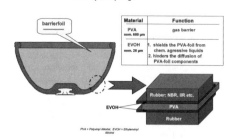

Figure 1: Overview of multi-layer diaphragm assembly.

The diaphragm used in the accumulator consists mainly of nitrile rubber (NBR) and contains an internal layer of polyvinyl alcohol (PVA) which works as a gas barrier. This layer is covered by two thin layers of ethylene vinyl alcohol copolymer (EVOH) which have to shield the PVA-foil from chemically aggressive liquids and prevent the PVA-foil components from diffusion.

2 X-ray Fluoroscopy

As a first low-cost easy to use method in the investigation of gas losses, we used a digital X-ray fluoroscopy system. In Figure 2 the setup and the data of the PCR system are shown. The data do not reveal any significant difference between a new diaphragm (upper part) and the damaged diaphragm from the test series (lower part). It shows no details about the internal structures of the component parts and especially no indication about the cause of the gas loss. The reason for this is the low contrast of X-ray fluoroscopy systems compared to CT X-ray systems.

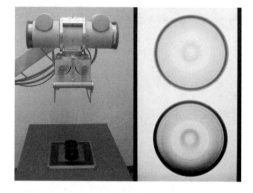

Figure 2: Setup of the X-ray fluoroscopy system with a digital cassette (resolution 2364 x 2964 pixel) and data of reference membrane (top) and membrane after test series (bottom).

3 Conventional CT

To increase contrast we used a Philips Secura CT, a conventional medical spiral CT-System. The 3-D reconstruction of the CT-data in figure 3 shows different areas of reduced absorption which can be divided into two categories:

- large detachments of the inner rubber layer mainly in the center of the diaphragm (see also figure 4 - section through the center of the membrane)

- multiple isolated lineaments located in the concentric strengthened middle part of the diaphragm. These lineaments are orientated on meridians.

With their high contrast the CT-data reveal detailed information on structural irregularities in the membrane. However, for further investigations of the lineaments presented in the dataset a higher spatial resolution is necessary.

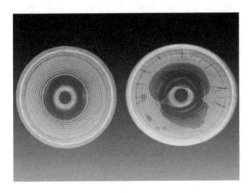

Figure 3: 3-D reconstruction of the CT data of reference membrane (left) and membrane after test series (right).

Figure 4: Section from the center of the membrane.

4 Micro CT

4.1 Spatial Resolution

To combine high contrast and high spatial resolution a micro CT was used.
The relevant components of such CT systems are micro focus X-ray tube and a high resolution 2D-CCD detector array. The spatial resolution is limited by the size of the focus (see figure 5) and the size of the detector elements.

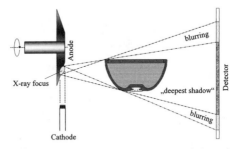

Figure 5: The inclination of the anode controls the optical size of the X-Ray focus. A larger X-ray focus increases the blurring of the image.

The so called modulation-transfer function *MTF* gives the resolution for spectral frequencies q in line pairs per millimetre [1]:

$$MTF_{system}(q) = MTF_{focus}(q)MTF_{detector}(q) = \left|\frac{\sin(\pi\, b_F\, q)}{\pi\, b_F\, q}\right|\left|\frac{\sin(\pi\, b_D\, q)}{\pi\, b_D\, q}\right|$$

Typical focus sizes are $b_f = 1$ mm for conventional medical CTs and $b_f < 10$ μm for micro CTs.
Figure 6 shows the characteristics of the *MTF* of a conventional medical CT and a micro CT, respectively.

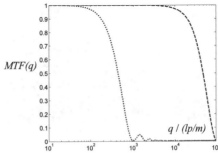

Figure 6: Frequency dependent modulation-transfer-function of conventional CT (doted curve) and micro-CT (dashed curve).

As a consequence of the small focus size of micro CTs the anode current is limited to $I < 100$ µA. This affects the intensity of the X-ray spectrum and limits the range of possible probe materials.

The micro CT by the Belgian company SkyScan shown in figure 7 uses a 12 Bit CCD-Chip with a pixel matrix of 1024 x 1024 and a pixel size of about 10 µm which is coupled via fibre optics to a scintillation crystal. In conventional CTs the detector dimension b_d is typically found in the order of 1 mm.

SkyScan gives a specification of spatial resolution of 10 µm for the 1072 scanner used in the experiments shown here [2-4].

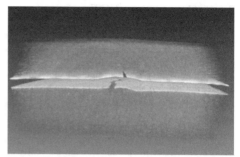

Figure 9: 3-D reconstruction of the membrane rupture.

Figure 7: Setup of micro-CT.

4.2 Imaging Results

Figure 7 displays the setup of the micro CT which can be used to investigate objects up to a size of 2 cm³.

The 3-D reconstruction of the micro CT data of one of the lineaments shown in figure 9 reveals that there are ruptures especially in the covering EVOH layers.

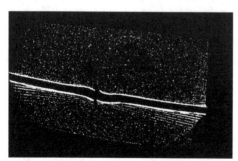

Figure 8: Single tomographic section of the membrane rupture.

5 Microscopic Evaluation

The microscope pictures in figure 10 show a part of the inner EVOH-layer in different enlargements. The higher magnification in figure 10b clearly indicates that the material becomes brittle and fragile along the fold structures.

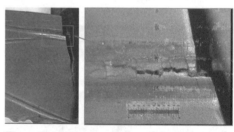

Figure 10: Microscope picture of the fold structure (width approx. 1 mm).

6 Resume

The production of the membranes apparently leads to folds in the internal EVOH/PVA-layer. The cause of gas losses are ruptures in these folds.

References

[1] T. M. Buzug, *Einführung in die Computertomographie*, Springer-Verlag, Heidelberg, 2004.
[2] A. Sassov 1999, *Desktop X-ray Micro-CT*, in: Proc. of the DGZiP BB67-CD, Computerized Tomography for Industrial Applications and Image Processing in Radiology (Berlin) 165.
[3] A. Sassov 2002, *Desktop X-ray Micro-CT Instruments*, Proc. of SPIE **4503**, 282.
[4] A. Sassov 2002, *Comparison of Fan-Beam, Cone-Beam and Spiral Scan Reconstruction in X-Ray Micro-CT*, Proc. of SPIE **4503**, 124.

Lasermesstechnik

Laser-Measurement Technology

Bestimmung der optischen Eigenschaften und Farbwirkung von zahnfarbenen Füllungsmaterialien

Kirsten Weniger[1], Gerhard Müller[2]

[1] Laser- und Medizin-Technologie GmbH, Berlin (LMTB), Fabeckstr. 60-62, 14195 Berlin
[2] Institut für Medizinische Physik und Lasermedizin, Campus Benjamin Franklin, Charité-Universitätsmedizin Berlin, Fabeckstr. 60-62, 14195 Berlin

Kurzfassung

Damit sich eine Füllung nicht sichtbar vom Zahn unterscheidet, sollten deren optische Eigenschaften möglichst gut angeglichen sein. Anhand der Untersuchung von unterschiedlichen dentalen Kompositfüllungsmaterialien konnte gezeigt werden, dass deren optische Eigenschaften durch die Berechnung der optischen Parameter Absorptionskoeffizient μ_a, Streukoeffizient μ_s und Anisotropiefaktor g mittels inverser Monte-Carlo-Simulation aus im Spektrometer gemessenen Transmission- und Remissionswerten präziser und vollständiger als bisher bestimmt und verglichen werden können. Dies ermöglicht mit der Erstellung eines Datenpools die Entwicklung und Anwendung mathematischer Formalismen und weiterer Berechnungsmodelle zur Optimierung von Herstellungsprozessen und Materialeigenschaften. Als Anwendungsbeispiel wurde ein Verfahren zur Vorhersage der Remission mit nachfolgender Berechnung der Farbwirkung nach CIELAB für wählbare Kompositschichtdicken und Mehrschichtmodelle entwickelt und validiert [1].

1 Einleitung

Der optische Eindruck und die Farbwirkung einer Füllung sind bestimmt von den optischen Eigenschaften des Materials wie Absorption, Streuung und Reflexion des Lichts. Die mathematische Beschreibung der Lichtausbreitung in trüben, also nicht transparenten Medien erfolgt mit den zugehörigen physikalischen Parametern, dem Absorptionskoeffizient μ_a, dem Streukoeffizient μ_s und dem Anisotropiefaktor g gemäß der Strahlungstransportgleichung:

$$\frac{dL(\mathbf{r},\mathbf{s})}{ds} = -(\mu_a + \mu_s)L(\mathbf{r},\mathbf{s}) + \frac{\mu_s}{4\pi}\int_{4\pi} p(\mathbf{s}\,\mathbf{s'})L(\mathbf{r},\mathbf{s})d\Omega' + S(\mathbf{r},\mathbf{s})$$

Diese basiert auf dem Teilchencharakter von Photonen und beschreibt die Änderung der Strahlungsdichte $L(r,s)$ [Wcm^{-2}sr^{-1}] am Ort r in Richtung s über den Absorptionskoeffizient μ_a [mm^{-1}] und den Streukoeffizient μ_s [mm^{-1}] als Maß dafür, wie viel gestreut oder absorbiert wird, sowie über eine Streuphasenfunktion p(s,s') als Wahrscheinlichkeit dafür, dass ein Photon aus der Richtung s in die Richtung s' gestreut wird. Aus der Streuphasenfunktion abgeleitet ist der Anisotropiefaktor g, der die Winkelverteilung der Streuereignisse charakterisiert. Er umfasst den Wertebereich von -1 für die totale Rückstreuung bis +1 für die totale Vorwärtsstreuung. Für g = 0 liegt isotrope Streuung vor, alle Streuwinkel sind im gleichen Maß wahrscheinlich [2]. Für die Betrachtung optisch dicker Schichten kann eine Zusammenfassung von Streukoeffizient und Anisotropiefaktor zum reduzierten Streukoeffizienten μ_s' gemäß $\mu_s' = (1 - g)\mu_s$ erfolgen, was die Beschreibung der Streuprozesse auf eine isotrope Phasenfunktion reduziert und nur fern von Quellen und Grenzflächen gültig ist.

Die von der Schichtdicke unabhängigen, optischen Parameter μ_a, μ_s und g können nicht direkt gemessen werden, sondern sind aus an Materialproben in einem Ulbrichtkugel-Spektrometer gemessenen Remissions- und Transmissionswerten zu berechnen. Für diese Berechnung hat sich die inverse Monte-Carlo-Simulation bereits am Beispiel biologischer Gewebe als das genaueste Modell zur Lösung der Strahlungstransportgleichung herausgestellt [3, 4, 5]. Zur Überprüfung der Anwendbarkeit des Verfahrens bei Dentalkomposits wurden die optischen Parameter für drei verschiedene Kompositmaterialien in unterschiedlichen Farbvarianten im sichtbaren Bereich der Wellenlängen von 400 nm bis 700 nm bestimmt und die Differenzierbarkeit bezüglich Material- und Farbvarianten überprüft. Ziel war es, das physikalische Verständnis für die optischen Eigenschaften von zahnfarbenen Komposits zu erweitern und mit der Erprobung des Verfahrens die Voraussetzungen zur Erstellung eines Datenpools zu schaffen.

Die Bestimmung der Farbwirkung von Komposits anhand von Proben einer festgelegten Schichtdicke wird in der Dentalindustrie zur Qualitätssicherung eingesetzt, um einen Vergleich mit Farbstandards derselben Schichtdicke durchzuführen. Dabei werden nach Messung der Remissionswerte in einem Farb spektrometer durch anschließende geräteintegrierte Umrechnung die L*,a*,b*-Werte der Proben nach dem CIELAB-Farbsystem bestimmt. Die Ermittlung von Remission und CIELAB-Werten kann jedoch auch auf Basis von μ_a, μ_s und g für wählbare Schichtdicken erfolgen. Dies wurde am hierzu entwickelten Berechnungsverfahren gezeigt und hinsichtlich der Simulationsgenauigkeit am Beispiel der Schichtdicke 1 mm evaluiert.

2 Verwendete Materialien

Dentalkomposits sind lichthärtende, zahnfarbene Füllungswerkstoffe, die aus einer Matrix mit darin eingebetteten Füllkörpern unterschiedlicher Art und Größe, sowie Zusatzstoffen wie Farbpigmenten bestehen. Untersucht wurden die beiden sich in den Füllkörpern unterscheidenden Hybridkomposits Spectrum und Esthet-X, sowie das zusätzlich mit anderer Matrix, bzw. Gesamtaufbau konzipierte Ormocer Definite der Firma Dentsply DeTrey, Konstanz in unterschiedlichen VITA-Farben (**Tabelle 1**). Bei der Codierung der VITA-Farben stehen die Buchstaben A, B und C für die Farbgruppe und zunehmende Zahlenwerte für eine Abnahme der Helligkeitsstufe.

Tab. 1: Untersuchte Komposits und VITA-Farben

Komposit	VITA-Farben
Spectrum	A1, A2, A3, A3.5, B1, B2, B3, C1, C2, C3
Esthet-X	A1, A2, A3, A3.5, B1, B2, B3, C2, C3
Definite	A1, A2, A3, A3.5, B2, C2

Da für die Messungen im Spektrometer blasenfreie Proben in Dicken kleiner 500 µm mit einem Durchmesser von 22 mm und glatter Oberfläche erforderlich waren, wurde ein spezielles Präparationsverfahren auf Ultraschallbasis entwickelt. Proben mit Inhomogenitäten wie Lufteinschlüssen wurden verworfen.

3 Bestimmung der optischen Eigenschaften

3.1 Messung im Spektrometer

Als Remissions- und Transmissionseigenschaften wurden die makroskopischen optischen Parameter diffuse Remission R_d, totale Transmission T_t, sowie diffuse Transmission T_d von 400 bis 700 nm in 5-nm-Schritten in einem Ulbrichtkugel-Spektrometer (UV/VIS/NIR-Spektrometer Lambda 900, Perkin-Elmer Corporation, Norwalk USA) gemessen. Durch die Verwendung der Ulbrichtkugel kann ein kompletter Datensatz von drei makroskopischen optischen Parametern in einem Gerät unter gleichen Bedingungen gemessen werden.

3.2 Inverse Monte-Carlo-Simulation

Die Monte-Carlo-Simulation dient als statistische Methode zur Berechnung der Lichtausbreitung in trüben Medien, indem die Trajektorien einer Vielzahl von Photonen berechnet und als Ergebnis die Remissions- und Transmissionseigenschaften einer Probe mit bekannten mikroskopischen optischen Parametern ausgegeben werden. Um gerade umgekehrt die unbekannten mikroskopischen optischen Parameter aus den gemessenen Remissions- und Transmissionseigenschaften zu bestimmen, ist das Modell der Monte-Carlo-Methode zu invertieren [3]. Wie im Ablaufdiagramm dieser inversen Monte-Carlo-Simulation (iMCS) in **Bild 1** schematisch dargestellt, werden dazu, von einem mittels der Kubelka-Munk-Theorie geschätzten Startparametersatz für μ_a, μ_s und g ausgehend, die Messwerte simuliert und mit den wahren, gemessenen Werten verglichen. Sind die Differenzen größer als die als Fehlergrenze vorgeschriebene Abweichung, wird eine Gradientenmatrix berechnet, aus der sich neue Werte für einen Parametersatz von μ_a, μ_s und g ableiten lassen. Dieses Vorgehen wird solange wiederholt, bis die gemessenen Werte innerhalb der vorgeschrieben Fehlergrenze wiedergegeben werden. Damit kann dieser Wertesatz der mikroskopischen optischen Parameter akzeptiert werden.

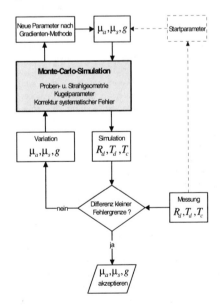

Bild 1 Ablaufdiagramm der iMCS [1]

Eingebettet in die Software Winfit 32 berücksichtigt diese inverse Monte-Carlo-Simulation die geometrischen und optischen Verhältnisse bei der Messung im Ulbrichtkugel-Spektrometer wie Probengeometrie, Blendendurchmesser, Kugelparameter, Strahldivergenz, Durchmesser des Lichtspots auf der Probe, Brechungsindexänderungen und seitliche Verluste von Photonen an Probengrenzflächen, so dass systematische Fehler vermieden werden.

In dieser Arbeit wurde mit 10^5 Photonen gerechnet und eine Fehlergrenze von 0,15 % vorgegeben. Als Streuphasenfunktionen wurden nach Probesimulationen die Henyey-Greenstein-Phasenfunktion [3] für die beiden Hybridkomposits Spectrum und Esthet-X und die Reynolds-McCormick-Phasenfunktion [6] für das Ormocer Definite gewählt. Als weitere Inputgröße für die Simulation wurde der Brechungsindex für die drei Kompositmaterialien jeweils wellenlängenabhängig mittels Abbe-Refraktometrie und nachfolgender Sellmeier-Approximation bestimmt.

3.4 Ergebnisse und Diskussion

Die Abnahme von Absorption und Streuung mit zunehmender Wellenlänge, sowie die zum gelblichen Farbeindruck beitragende, bei niedrigen Wellenlängen hohe Absorption (siehe **Bild 3**) wird auch von anderen Autoren in früheren Untersuchungen zu optischen Eigenschaften von Dentalkomposits festgestellt [7, 8]. Weiterhin wird postuliert, dass die Absorption an den beigemischten Farbpigmenten oder der Matrix [8] und die Streuung hauptsächlich an den Füllpartikeln erfolgt [7], was die Ergebnisse dieser Arbeit ebenfalls bestätigen. Der Absorptionskoeffizient μ_a weist eine wellenlängenspezifische Abhängigkeit von den Farbvarianten auf (siehe **Bilder 2 und 4**). Innerhalb der Materialien ist die Abweichung der μ_a-Werte zwischen den sich bezüglich der Matrix unterscheidenden Ormocer und Hybridkomposits stärker ausgeprägt, als innerhalb der matrixgleichen Hybridkomposits (siehe **Bild 3**). Der Streukoeffizient μ_s, der Anisotropiefaktor g und der reduzierte Streukoeffizient μ_s' zeigen dagegen eine geringere Abhängigkeit von den Farbvarianten (siehe **Bild 4**), unterscheiden sich aber wesentlich innerhalb der drei verschiedene Füllpartikel enthaltenden Materialien (siehe **Bild 3**). Die Streuung wird also von den unterschiedlichen Arten der Farbpigmente geringer beeinflusst als von den Füllpartikeln, bei der Absorption ist es umgekehrt. Bei den Streueigenschaften hebt sich das Ormocer Definite von den beiden Hybridkomposits ab, indem es den höchsten Streukoeffizienten μ_s, sowie mit dem höchsten Anisotropiefaktor g die stärkste Vorwärtsstreuung aufweist. Esthet-X weist für den Streukoeffizienten μ_s im Bereich kurzer Wellenlängen,

also im Blauen, größere Werte auf als Spectrum und Definite (siehe **Bild 3**). Daraus ist zu schließen, dass die Verteilung der Partikelgrößen der Füllstoffe für Esthet-X einen höheren Anteil kleiner Partikel mit Abmessungen in der Größenordnung dieser Wellenlängen aufweist, als die anderen beiden Komposits.

Die Farbe eines Komposits wird durch Zumischung von Farbpigmenten wie Metalloxiden eingestellt. Innerhalb der VITA-Farbgruppen A bis C wird über die Konzentration der Pigmente die Helligkeitsstufe 1 bis 4 vorgegeben. Dies bestätigend ist für die Helligkeitsstufen eine Korrelation mit der Höhe der Absorption festzustellen (siehe **Bild 2**).

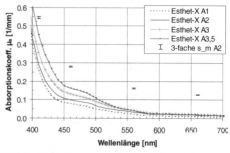

Bild 2 Wellenlängenspezifische Abhängigkeit des Absorptionskoeffizienten μ_a von den Helligkeitsstufen innerhalb der Farbgruppe A bei Esthet-X mit dreifacher Standardabweichung s_m

Nach der statistischen Auswertung zeigten die Ergebnisse des beispielhaft bei 450 nm für niedrige und bei 660 nm für hohe Wellenlängen durchgeführten H-Test nach Kruskal und Wallis, dass sich die untersuchten Farbvarianten bei allen vier Parametern, mit Ausnahme von μ_s bei 450 nm bei Definite, signifikant unterscheiden.

Mit der inversen Monte-Carlo-Simulation kann die getrennte Einzelbestimmung der Anteile von Streuung und Absorption erfolgen. Mit der in bisherigen Untersuchungen von dentalen Komposits verwendeten Kubelka-Munk-Methode [7, 8, 9] ist durch die Berechnung des Absorptionskoeffizienten A_{KM} und des Streukoeffizienten S_{KM} keine getrennte Bestimmung von Streu- und Absorptionsanteilen möglich. S_{KM} entspricht einer Mischung aus Absorption und Streuung in Form von μ_a, μ_s und g [3]. Somit ist auch keine separate Ermittlung der reinen Streueigenschaften getrennt von der Richtung der Streuung möglich, wie es die Bestimmung von μ_s und g erlaubt. Die in der Software WinFit 32 verwirklichte inverse Monte-Carlo-Simulation weist aufgrund der Berücksichtigung von Messgeometrien, Grenzflächen, wellenlängenabhängigen Brechungsindizes und der Korrektur systematischer Fehler gegenüber der Kubelka-Munk-Methode weiterhin eine wesentlich höhere Genauigkeit und Flexibilität auf.

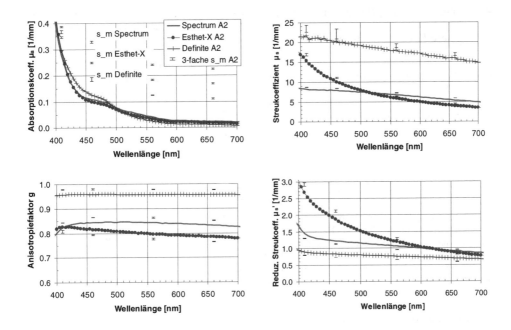

Bild 3 Abhängigkeit der mikroskopischen optischen Parameter von der Wellenlänge für die drei untersuchten Kompositmaterialien am Beispiel der Farbvariante A2 mit dreifacher Standardabweichung des Mittelwerts s_m

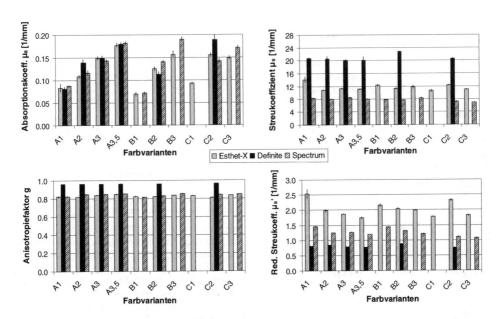

Bild 4 Abhängigkeit der mikroskopischen optischen Parameter von den Farbvarianten für die drei untersuchten Kompositmaterialien am Beispiel der Wellenlänge 450 nm mit dreifacher Standardabweichung des Mittelwerts s_m

4 Bestimmung der Farbwirkung

4.1 Berechnungsverfahren

Aus μ_a, μ_s und g wurden mit Hilfe einer Vorwärts-Monte-Carlo-Simulation (Software MCS-LAYER, LMTB, Berlin) materialabhängig die Werte der diffusen Remission R_d für eine Schichtdicke von 1 mm bei Farbvariante B2 ermittelt. Dabei konnte die Simulationsgenauigkeit durch eine Korrektur des Absorptionskoeffizienten μ_a zu μ_{ak} mit Hilfe von Messwerten einer zusätzlich hergestellten, optisch dichten Kompositprobe der Dicke 5 mm weiter erhöht werden. Anschließend wurden gemäß DIN-Norm 5033 [10] aus den materialabhängig simulierten Werten für die diffuse Remission R_d die X, Y, Z-Normfarbwerte in Bezug auf die Normlichtart D65 und den 2°-Normalbeobachter in 10 nm-Schritten von 400 bis 700 nm berechnet und diese in die L*,a*,b*-Werte nach CIE transformiert. Zur Evaluation des Berechnungsverfahrens wurden Proben von etwa 1 mm Dicke der Farbvariante B2 hergestellt, deren Remissionswerte R_d im Spektrometer gemessen und mit den für die jeweilige Schichtdicke simulierten Daten verglichen. Nachfolgend wurde gemäß DIN-Norm 6174 [11] der Farbabstand ΔE^*_{ab} jeweils aus den simulierten Werten als CIELAB-Probenwert und aus den gemessenen Werten als CIELAB-Bezugswert berechnet.

4.2 Ergebnisse und Evaluierung

Die bei der Farbvariante B2 im Einzelvergleich zwischen den drei Kompositmaterialien bei einer Schichtdicke von 1 mm ermittelten Werte für den Farbabstand ΔE^*_{ab} liegen nur beim Vergleich von Spectrum und Definite im mit ΔE^*_{ab}-Werten < 2 als klinisch akzeptabel definierten Bereich, während zwischen Esthet-X und Spectrum sowie Definite und Esthet-X deutlich sichtbare Farbabstände auftreten (siehe **Tabelle 2**).

Tab. 2: $\Delta L^*, \Delta a^*, \Delta b^*$-Werte, Farbeindruck und -abstand ΔE^*_{ab} für B2 bei 1 mm Schichtdicke 2°-Normalbeobachter und Normlichtart D65

	Esthet-X - Spectrum	Esthet-X - Definite	Definite - Spectrum
ΔL^*-Wert	1,15	-0,25	1,40
Farbeindruck	heller	dunkler	heller
Δa^*-Wert	-1,64	-0,64	-1,00
Farbeindruck	grüner	grüner	grüner
Δb^*-Wert	-5,46	-5,46	0,00
Farbeindruck	blauer	blauer	gleich
ΔE^*_{ab}	5,82	5,51	1,72

Die vergleichende Darstellung der gemessenen Werte für die diffuse Remission R_d und der mit korrigiertem μ_{ak} simulierten Werte für R_d zeigt für die untersuchten Hybridkomposits eine sehr gute Übereinstimmung. Dies gilt ebenso für Definite, bei dem sich nur im kurzen Wellenlängenbereich geringe Abweichungen ergaben (siehe **Bild 5**).

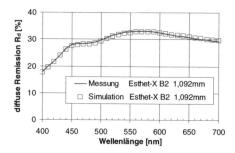

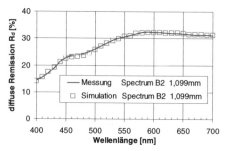

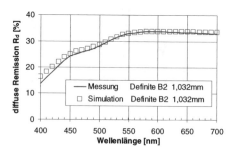

Bild 5 Vergleich der gemessenen und simulierten Werte für die diffuse Remission R_d

Der Farbabstand nach CIELAB zwischen Messung und Simulation liegt bei den beiden Hybridkomposits mit Werten von $\Delta E^*_{ab} = 0{,}42$ für Spectrum und $\Delta E^*_{ab} = 0{,}45$ für Esthet-X im von 50 % der Beobachter nicht wahrnehmbaren Bereich. Bei Definite liegt der ermittelte Wert von $\Delta E^*_{ab} = 1{,}71$ knapp unterhalb der Grenze zum visuell mäßig wahrnehmbaren Bereich von 1,72 bis 4,58, sowie im als klinisch akzeptabel definierten Bereich von $\Delta E^*_{ab} < 2$. Mit der Kubelka-Munk-Methode wurde von GRAJOWER et al. [7] eine wesentlich geringere

Simulationsgenauigkeit mit Werten für den Farbabstand ΔE^*_{FMC} von 3 bis 8 erreicht, die dem mäßig wahrnehmbaren ΔE^*_{ab}-Bereich von 1,72 bis 4,58 entsprechen und somit auch außerhalb der klinischen Akzeptanz liegen.

5 Zusammenfassung und Ausblick

Die Ergebnisse erweitern das physikalische Verständnis für die optischen Eigenschaften von zahnfarbenen Kompositfüllungsmaterialien, da eine umfassende optische Charakterisierung mit Aufschlüsselung der Streueigenschaften in reine Streuung und Richtung der Streuung möglich ist. Die Differenzierung von Materialien und Farbvarianten untereinander war eindeutig möglich. Die vorgestellten Berechnungsverfahren erreichen dabei eine deutlich höhere Simulationsgenauigkeit als die bisher genutzte Kubelka-Munk-Methode und eignen sich zur Erstellung eines Datenpools. Für die Entwicklung von neuartigen Kompositmaterialien kann die Ermittlung von makroskopischen und mikroskopischen Parametern für Experimental- und Referenzmaterialien genutzt werden, um Rückschlüsse aus dem Absorptions- und Streuverhalten ziehen zu können und grundlegende Unterschiede oder Gemeinsamkeiten mit anderen Materialien feststellen zu können. Für den Vergleich der optischen Eigenschaften von Kompositfüllungsmaterialien mit Zahnhartsubstanzen sind weitere Untersuchungen an Schmelz- und Dentinproben nach der vorgestellten Methode erforderlich. Die Berechnung von Remission und Farbwirkung für unterschiedliche Kompositschichtdicken liefert zusätzliche Informationen über das Farbverhalten. Mit der verwendeten Software MCS-LAYER ist dies auch für Mehrschichtmodelle mit bis zur vier Schichten möglich. Dies kann besonders für kombinierte Schichten aus verschiedenen Opazitäten für den Schmelz- und Dentinersatz aufschlussreich sein. Auch Berechnungen der Farbwirkung nach anderen Farbsystemen als dem CIELAB-System sind möglich.

Neben der Berechnung der Farbwirkung kann auf Basis von μ_a, μ_s und g auch eine Berechnung der Eindringtiefen des Lichts mittels Diffusionstheorie erfolgen. Ein Vergleich der Eindringtiefen für verschiedene Kompositmaterialien bei der Wellenlänge, bei der das Absorptionsmaximum des Photoinitiators Kampferchinon liegt, kann beispielsweise Hinweise für den Vergleich zu erwartender Polymerisationseigenschaften geben.

Letztendlich können die vorgestellten Untersuchungen durch genauere Kenntnis von Absorptions- und Streueigenschaften hinsichtlich der Lasermaterialbearbeitung von Kunststoffen, auch außerhalb des Dentalbereichs, von Nutzen sein.

6 Literatur

[1] Weniger, K.: Optische Eigenschaften von dentalen Kompositfüllungsmaterialien. Dissertation, Charité - Universitätsmedizin Berlin, in Müller, G. (Hrsg.): Forschungsberichte 4 / Laser- und Medizin-Technologie Berlin. Berlin: dissertation.de-Verlag im Internet, 2004

[2] Ertl, T.; Roggan, A.; Zgoda, F.: Optische Eigenschaften von Gewebe, in Müller, G.; Ertl, T. (Hrsg.): Angewandte Laserzahnheilkunde, Kap. II-3.1.1, Landsberg: ecomed, 1995

[3] Roggan, A.: Dosimetrie thermischer Laseranwendungen in der Medizin. Untersuchung der optischen Gewebeeigenschaften und physikalisch-mathematische Modellentwicklung, Dissertation, TU Berlin, in Müller, G.; Berlien, H.P. (Hrsg.): Fortschritte in der Lasermedizin 16, Landsberg: ecomed, 1997

[4] Roggan, A.; Albrecht, H.; Dörschel, K.; Minet, O.; Müller, G.: Experimental set-up and Monte-Carlo model for the determination of optical properties in the wavelength range 330-1100 nm, Proc. SPIE 2323 (1995), pp. 21-36

[5] Hammer, M.; Roggan, A.; Schweitzer, D.; Müller, G.: Optical properties of ocular fundus tissue - an in vitro study using double integrating sphere technique. Phys. Med. Biol. 40 (1995), pp. 963-978

[6] Roggan, A.; Friebel, M.; Dörschel, K.; Hahn, A.; Müller, G.: Optical Properties of circulating human blood in the wavelength range 400-2500 nm. J. Biomed. Opt. 4 (1999), pp. 36-46

[7] Grajower, R.; Wozniak, W.T.; Lindsay, J.M.: Optical properties of composite resin. J. Oral. Rehab. 9 (1982), pp. 389-399

[8] Yeh, C.L.; Miyagawa, Y.; Powers, J.M.: Optical properties of composites of selected shades. J. Dent. Res. 61 (1982), pp. 797-80

[9] Taira, M.; Okazaki, M.; Takahashi, J.: Studies on optical properties of two commercial visible-light-cured composite resins by diffuse reflectance measurements. J. Oral. Rehab. 26 (1999), pp. 329-337

[10] Deutsches Institut für Normung e.V.: DIN-Norm 5033. Farbmessung Teil 1 bis 9, Berlin: Beuth-Verlag, 1979-1992

[11] Deutsches Institut für Normung e.V.: DIN-Norm 6174. Farbmetrische Bestimmung von Farbabständen bei Körperfarben nach der CIELAB-Formel, Berlin: Beuth-Verlag, 1979

Die Arbeiten wurden im Rahmen einer Dissertation durchgeführt, unterstützt durch die Laser- und Medizin-Technologie GmbH, Berlin sowie die Firma Dentsply DeTrey durch die Bereitstellung von Geräten und Materialien.

Lasermesstechnik   Laser-Measurement Technology

Ultraschallübertragung über Quarzglasfasern: Endoskopische Gewebeentfernung in der Neurochirurgie

Karsten Liebold, Laser- und Medizin-Technologie GmbH, Berlin, Deutschland

Kurzfassung

In der Neurochirurgie kommen bei der Resektion von Hirntumoren oft Ultraschall-Dissektoren zum Einsatz, mit denen sehr feinfühlig Gewebe abgetragen werden kann. Die derzeit am Markt befindlichen Systeme sind durch den im Handstück befindlichen Schallwandler in Größe und Gewicht nicht für die endoskopische Anwendung geeignet. Eine dämpfungsarme Ultraschallübertragung über Quarzglasfasern erlaubt es, den Wandler räumlich getrennt unterzubringen und mit der dünnen, flexiblen Faser als Applikator endoskopisch zu arbeiten. Gleichzeitig kann über diese Faser Laserlicht, z.B. für die koagulative Blutstillung, übertragen werden. Im Rahmen der Forschungsarbeiten entstand ein Funktionsmuster, das in vitro erfolgreich getestet wurde.

1 Einleitung

In der Neurochirurgie kommen bei der Resektion von Hirntumoren oft Ultraschall-Dissektoren zum Einsatz, mit denen feinfühlig Gewebe abgetragen werden kann, ohne hierbei Kräfte auf die umliegenden Bereiche auszuüben. Dabei erlaubt die selektive Wirkung des Ultraschalls die zügige Abtragung von Parenchymgewebe unter Schonung der festeren Gefäße und Nerven.
Die derzeit am Markt befindlichen Systeme bestehen aus einem Generator und einem Handstück, in dem sich der Ultraschallwandler befindet. Über eine starre Sonotrode wird der Ultraschall appliziert, wobei eine integrierte Spülung und Absaugung die Entfernung des fragmentierten Gewebes erlaubt. Üblicherweise finden piezoelektrische Verbundwandler Verwendung, die mit der Sonotrode in Resonanz betrieben werden, um die notwendigen Amplituden zu erreichen. Aufgrund der dadurch vorgegebenen Baulängen und dem Gewicht der Wandler eigenen sich diese Handstücke nicht für die endoskopische Anwendung in der Neurochirurgie.
Eine dämpfungsarme Ultraschallübertragung über Quarzglasfasern erlaubt es hier, den Wandler räumlich vom Anwendungsort zu trennen und lediglich mit der dünnen, flexiblen Faser als Applikator endoskopisch zu arbeiten. Gleichzeitig kann über diese Faser Laserlicht, z.B. für die koagulative Blutstillung, übertragen werden. Dies ist von großem Interesse, da der Zeitaufwand für die Blutstillung mit der üblichen bipolaren Pinzette einen großen Teil der Zeiteinsparung aufwiegt, die sich ansonsten aus der zügigen Gewebeabtragung mit dem Ultraschall-Dissektor ergibt.

2 Material und Methode

2.1 Ultraschall

Für die Übertragung von Ultraschall der notwendigen Leistung über flexible Quarzglasfasern eignen sich vor allem quasilongitudinale Dehnwellen, denn Biege- und Torsionswellen belasten den Wellenleiter mechanisch zu stark.
Zur Erzeugung dient ein piezoelektrischer Verbundwandler mit einem Stufenhorn als Amplitudentransformator. Angetrieben wird dieser durch einen Ultraschallgenerator, der über das Signal einer passiven Piezoscheibe im Wandler die Frequenz auf Resonanz und die Amplitude auf den gewählten Wert regelt. Dies erlaubt den Ausgleich der durch die betriebsbedingte Erwärmung auftretenden Resonanzänderung und gewährleistet eine gleichmäßige, von der Belastung unabhängige Leistungsabgabe.
Die Biegung des Wellenleiters führt zu einer Resonanzverschiebung, die mit steigender Frequenz an Einfluss verliert. Mit steigender Frequenz wird jedoch auch die Erzeugung größerer Amplituden schwierig. Aus diesem Grund wurde die optimale Ultraschallfrequenz mit Hilfe einer Simulation ermittelt. Bei den gewählten 45 kHz liegt die durch die Resonanzverschiebung verursachte Amplitudenänderung unterhalb von vertretbaren 5 %.

2.2 Übertragung

Auf den Amplitudentransformator des Wandlers kann ein kleines Koppelstück aufgeschraubt werden, in dessen Durchgangsbohrung der opto-akustische Wellenleiter, eine zirka 90 cm lange Quarz/Quarz-Faser (∅= 600/660 µm) mit Hilfe eines speziellen Epoxydharzes eingeklebt ist.

Bei dem im Vergleich zur Wellenlänge kleinen Durchmesser des Wellenleiters kommt es praktisch immer zur Anregung von Biegewellen, die aufgrund der genannten hohen mechanischen Belastung unerwünscht sind. Verhindern lässt sich dies durch eine Führung der Faser in einem dünnen Lumen mit Flüssigkeitsspülung, die im wesentlichen nur die transversalen Anteile dämpft. Gleichzeitig führt die Spülung die durch die Reibung entstehende Wärme ab. Hierbei erwiesen sich Spülspalte im Bereich von 0,1 bis 0,3 mm als optimal. Unterhalb steigt die Dämpfung durch die Reibung mit der Wandung, während oberhalb vermehrt Biegewellen auftreten. Hinsichtlich der Spülung ergeben Raten ab ca. 10 ml/min (0,9 % NaCl) ausreichend niedrige Austrittstemperaturen um die Spülung gleichzeitig zur Unterstützung der Gewebeabsaugung verwenden zu können.

2.3 Laser

Der Schallwandler verfügt über eine zentrale Bohrung, durch die eine vom Laser kommende, schwingungsfrei gelagerte Faser ($\varnothing$ < 400 µm) bis vor die proximale Stirnfläche des Wellenleiters geführt werden kann. Über eine Luftstrecke akustisch entkoppelt, kann hier die Laserstrahlung direkt eingekoppelt werden.

2.4 Applikator

Der Wellenleiter ist im Spülkanal eines koaxialen zweilumigen Schleusensystems geführt, durch das die Spülung und Absaugung während der Gewebeabtragung erfolgt. Das Schleusensystem ist in einem Halter am Gehäuse des Wandlers fixiert, so dass der Wellenleiter im Bereich der Ultraschalleinkopplung nicht gebogen werden kann. Da die Abtragsleistung des Ultraschalls maßgeblich von der aktiven Fläche abhängt, befindet sich am distalen Faserende eine zusätzliche Applikationsspitze (**Bild 1**).

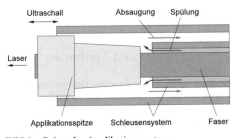

Bild 1 Spitze des Applikationssystems

Das Applikationssystem ist mit einem Außendurchmesser von 2,1 mm für die Verwendung im Arbeitskanal eines Ventrikuloskops ausgelegt.

3 Ergebnis

Entsprechend der dargestellten Konzeption wurde ein Funktionsmuster aufgebaut, das bei einer Frequenz von 45 kHz eine Amplitude von 60 µm erreicht (**Bild 2**). Die Tests erfolgten in In-vitro-Versuchen an Kalbshirn. In Marklager und Cortex war die Abtragsrate dabei von der Geschwindigkeit der Applikatorbewegung begrenzt, in den festeren Bereichen lag sie bei 0,25 cm^3/min.

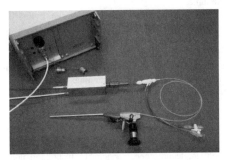

Bild 2 Funktionsmuster mit Generator, Schallwandler und Applikationssystem im Ventrikuloskop

4 Fazit

In ersten In-vitro-Versuchen zeigte das aufgebaute Funktionsmuster die Möglichkeit auf, die Vorteile von Ultraschall-Dissektions-Systemen auch für die endoskopassistierte Neurochirurgie ausnutzen zu können.
Im nächsten Schritt erfolgt nun die klinische Evaluierung des bereits entsprechend DIN EN 60601-1-1 ausgelegten Funktionsmusters.

Das Projekt wurde im Rahmen des Verbundes MINOP II vom BMBF (FKZ 16SV1443) sowie der AESCULAP AG & CO. KG gefördert

5 Literatur

[1] K. Desinger, K. Liebold, J.Helfmann, T. Stein, G. Müller: *A new system for combined laser and ultrasound application in neurosurgery*, Neurological Resarch, Vol. 21, pp 84-88, 1999

[2] K. Desinger: *Untersuchungen zur Übertragung und Wirkung kombinierter Ultraschall-Laserstrahlung für die Chirurgie*, Landsberg: ecomed, 1999

[3] S. N. Makarov, M. Ochmann, K. Desinger: *The longitudinal vibration response of a curved fiber used for laser ultrasound surgical therapy*, J. Acoust. Soc. Am., Vol. 102, pp 1191-1197, Aug. 1997

Konzept eines neuartigen Miniatur-Scanners für kleinste Hohlräume

Karl Stock, Michael Müller, Raimund Hibst
Institut für Lasertechnologien in der Medizin und Meßtechnik
an der Universität Ulm

Kurzfassung

Ziel der vorliegenden Studie ist die Entwicklung des optischen Designs von miniaturisierten, scannenden Laserstrahlführungssystemen auf der Basis von einzelnen Lichtleitern zur Bearbeitung und Untersuchung kleinster Hohlräume.
Für erste grundlegende Untersuchungen an zylindrischen Lichtleitern wurde ein experimenteller Aufbau mit winkelvariabler Laserstrahleinkopplung bei feststehender Lichtquelle und Faser sowie simultaner Detektion des distal abgestrahlten Laserlichts erstellt. Vergleichend wurden Simulationen mit dem optischen Raytracing-Programm Zemax durchgeführt.
Die Ergebnisse demonstrieren bei sehr guter Übereinstimmung von Messung und Simulation die prinzipielle Funktionsweise eines Winkelscannings auf der Basis von Einzelfasern, wobei die Scaneigenschaften in einem weiten Bereich durch die Faser- und Strahlparameter beeinflusst werden können.

1 Einleitung

Sowohl im technischen Bereich zur Inspektion und Bearbeitung schwer zugänglicher Hohlräume als auch im medizinischen Bereich zur möglichst schonenden und minimal invasiven Behandlung stehen heute Endoskope verschiedener Bauart zur Verfügung. Mit zunehmender Miniaturisierung der Bauteile wird man jedoch in Zukunft immer häufiger an die Grenzen der heutigen Endoskope stoßen.
Im Rahmen der vorliegenden Studie soll das Konzept eines völlig neuartigen Laserstrahl-Scanners auf der Basis einzelner Lichtleiter entwickelt werden. Für ein solches System lassen sich verschiedenste Anwendungen in der Medizin und Technik finden, bei denen nicht die Abbildungsqualität heutiger Endoskope, sondern eine filigrane Konstruktion zur automatischen Bearbeitung von sehr kleinen Hohlräumen im Vordergrund steht. Die Verwendung einzelner Lichtleiter soll kleinste Hohlräume mit Durchmesser im 100 µm Bereich zugänglich machen. Als typische Anwendungen sind eine formgebende Ausgestaltung von Hohlräumen kleinster Bauteile (z. B. Mikropumpen) oder eine, den intakten Zahnschmelz weitgehend konservierende, Entfernung einer „unterminierenden" Karies denkbar.
Es ist bekannt, dass der Winkel, unter dem in einen Lichtleiter eingekoppelt wird, auch über größere Übertragungsstrecken zum Teil erhalten bleibt [1]. Diese Tatsache wird meist als nachteilig angesehen, weil so eine schlechte, z.B. schiefe Einkopplung zu einer unsymmetrischen Abstrahlcharakteristik der Faser führt. Ansatz unseres neuartigen Laserstrahl-Scanners ist die gezielte Variation des Einkoppel-Winkels zur Variation der Abstrahlcharakteristik. In **Bild 1** ist das Schema eines solchen Laserstrahlführungssystems mit seinen einzelnen Komponenten dargestellt.

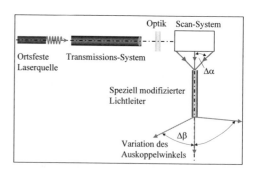

Bild 1 Schema des Miniaturscanners mit seinen einzelnen Komponenten.

2 Material und Methoden

Zunächst wurden grundlegende Untersuchungen an kurzen Lichtleitern durchgeführt. Zur Minimierung des experimentellen Aufwands wurden die Fragestellungen zunächst anhand von optischen Simulationen bearbeitet und anschließend einzelne Lichtleiter hergestellt und untersucht.
Hierzu wurde ein experimenteller Aufbau erstellt, mit dem bei feststehender Lichtquelle und Faser unter variablem Winkel in Lichtleiter eingekoppelt werden kann (siehe **Bild 2**). Hierbei fokussiert eine Linse

(f = 80 mm) das Licht eines HeNe-Lasers über eine Scannereinheit in den Brennpunkt eines Hohlspiegels (f = 25 mm), wodurch der Laserstrahl nach Reflexion am Hohlspiegel weitgehend kollimiert in die Einkoppelebene der Faser fällt. Der Scanspiegel und die Faser sind so angeordnet, dass der Auftreffpunkt des Lasers in der Scannerachse über den Hohlspiegel auf die Einkoppelfläche der Faser 1:1 abgebildet wird. Auf diese Weise kann durch eine Spiegelbewegung der Einstrahlwinkel in die Faser horizontal von -15° bis +20° variiert werden, ohne dass sich der Fleck ($1/e^2$-$\varnothing = 266$ µm) im Einkoppelort bewegt (gemessene max. laterale Verschiebung: 20 µm). Die Drehung des Scanspiegels erfolgte dabei manuell, wobei die Winkelauflösung mithilfe einer speziellen Ableseeinheit ca. 0,2° beträgt. Zur Justage kann der Lichtleiter über eine X-Y-Z-Verschiebeeinheit positioniert werden. Die Detektion des Laserlichts nach der Faser erfolgte über eine CCD-Kamera, die axial verschiebbar und um den Auskoppelort drehbar gelagert ist. Die Bilder wurden über einen Framegrabber (MV-Delta, Matrix Vision GmbH) aufgezeichnet und mithilfe eigens erstellter Programme in Matlab (MathWorks Inc., Version 12) ausgewertet.

Als Lichtleiter wurden konventionelle Quarz- bzw. Saphirfasern mit unterschiedlichen Durchmessern verwendet, die auf die entsprechenden Längen und Geometrien geschliffen und poliert wurden.

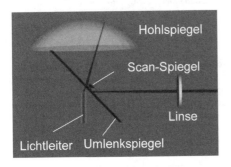

Bild 2 3D-Layout des simulierten Versuchaufbaus.

Zu Beginn der optischen Simulationen wurde die Anzahl der Fasermoden bestimmt, die bei den betrachteten Faserdurchmessern zu erwarten sind [2]. Da diese im Bereich von ca. 70000 (Faserdurchmesser 200 µm) bis ca. 1,7Mio (Durchmesser 1 mm) liegen, ist eine rein geometrisch optische Betrachtung des Strahlverlaufs im Lichtleiter ausreichend. Um eine möglichst hohe Übereinstimmung zwischen Experiment und Simulation zu erhalten, wurde der bestehende experimentelle Aufbau mit allen optischen Komponenten simuliert (Zemax, Version EE). Die Intensitätsverteilung nach dem Lichtleiter wurde zur weiteren Auswertung in Matlab bzw. zum anschließenden Vergleich mit den Messungen als ASCII-Datei abgespeichert.

Zunächst wurden an kurzen zylindrischen Fasern der Einfluss verschiedener Lichtleiter-Parameter, wie Länge, Durchmesser und Brechungsindex, auf die Übertragungseigenschaften und Scanmöglichkeiten untersucht. Weiterhin wurden Lichtleiter mit rechteckigem Querschnitt, getaperte Lichtleiter und Lichtleiter mit unterschiedlich modifizierten Ein- bzw. Auskoppelenden simuliert bzw. hergestellt und getestet.

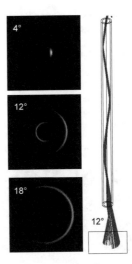

Bild 3 Simulationen an einem kurzen zylindrischen Lichtleiter (Quarz, Durchmesser 1000 µm, Länge 26 mm) bei verschiedenen Einkoppelwinkeln; links: Strahlprofil in 3 mm Entfernung zum distalen Lichtleiterende (Detektorgröße 5 mm x 5 mm); rechts: Strahlverlauf im Lichtleiter bei 12°.

3 Ergebnisse

3.1 Grundlegende Beobachtungen an zylindrischen Lichtleitern

3.1.1 Funktionsprinzip

In **Bild 3** sind die simulierten Strahlprofile dargestellt, die man bei verschiedenen Einkoppelwinkeln im Abstand von 3 mm nach einem zylindrischen Lichtleiter (Länge 26 mm, Durchmesser 1 mm, Quarz-Kern und -Cladding) erhält. Wie zu sehen ist, führen die Reflexionen an der zylindrischen Mantelfläche des Lichtleiters zu einer zunehmend kegelförmigen Abstrahlcharakteristik. Das punktförmigen Strahlprofil auf dem Detektor erweitert sich mit zunehmender Anzahl der Reflexionen zu einer Sichel bis hin zu einem geschlossenen Ring. Wenn sich die Anzahl der Reflexionen um Eins erhöht, springt der Punkt bzw. die Sichel auf die andere Seite. Abhängig vom Durchmesser des eingekoppelten Laserstrahls existiert ein bestimmter Winkelbereich, in dem ein Teil des Laserstrahls direkt austritt, während ein Teil bereits nochmals reflektiert wird. Dies führt zu der Entstehung von zwei Punkten bzw. Sicheln (s. Bild 3, mittleres Detektorbild bei 12°).

3.1.2 Einfluss verschiedener Parameter auf die Scan-Eigenschaften

Einfluss der Lichtleiterlänge

Aus einfachen geometrischen Überlegungen ergibt sich eine proportionale Zunahme der Anzahl der Reflexionen im Lichtleiter mit dessen Länge. Wie sowohl die Messungen als auch die Simulationen zeigen, führt dadurch erwartungsgemäß eine größere Länge zu einer stärkeren Ausbildung der Ringstruktur bei gleichem Einstrahlwinkel (siehe **Bild 4**).

Bild 4 Simulierte Strahlprofile in 20 mm Abstand zum Faserende bei verschiedenen Längen des Lichtleiters; Einkoppelwinkel 10°, Detektorgröße 8 mm x 8 mm.

Einfluss des Lichtleiterdurchmessers

Mit zunehmendem Durchmesser des Lichtleiters nimmt zum einen die Anzahl der Reflexionen ab und die Krümmung der zylindrischen Mantelfläche ist geringer. Wie **Bild 5** zeigt, führen beide Faktoren erwartungsgemäß zu einer geringeren Ausprägung der Ringstruktur.

Einfluss des Brechungsindex

Die Untersuchungen zum Einfluss des Brechungsindex auf die Übertragungseigenschaften der Lichtleiter zeigen eine geringere Ausprägung der Ringstruktur bei größerem Brechungsindex. Dies lässt sich durch die höhere Brechung des eingekoppelten Lichtstrahls zur Lichtleiter-Achse hin erklären. Der Strahlverlauf in dem Lichtleiter ist damit vergleichsweise flacher und damit die Anzahl der Reflexionen bei gleichem Einstrahlwinkel geringer.

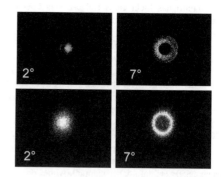

Bild 5 Einfluss des Faserdurchmessers: Oben: Simulation bei Faserdurchmesser 500 µm; unten: Messung bei Durchmesser 200 µm; Faserlänge 40 mm, Detektor 6,4 mm x 4,8 mm.

Einfluss des Laserstrahldurchmessers

Mit zunehmendem Durchmesser des Laserstrahls sind die beschriebenen Sprungstellen mit zwei gleichzeitig auftretenden Abstrahlrichtungen ausgeprägter. Im Extremfall entspricht der Strahldurchmesser in etwa dem Faserdurchmesser. Dann erhält man bereits bei kleinen Winkeln einen geschlossenen Ring.

3.2 Lichtleiter mit rechteckigem Querschnitt

Durch die plane Mantelfläche des Lichtleiters mit rechteckigem Querschnitt entfällt die Ausbildung der Ringstruktur, d.h. das punktförmige Strahlprofil des eingekoppelten Laserstrahls bleibt erhalten. Die Richtungsänderung bei Erhöhung der Reflexionsanzahl sowie die Strahlaufteilung im Bereich der Sprungstellen sind erwartungsgemäß auch hier zu beobachten.

3.3 Getaperte Lichtleiter

Bei Verwendung eines getaperten Lichtleiters (konisch zulaufende Form, siehe **Bild 6**) wird der Winkel des eingekoppelte Laserstrahls bei jeder Reflexion an der geneigten Mantelfläche vergrößert. Dies führt letztendlich zu einer Winkelübersetzung, abhängig von der Geometrie des getaperten Lichtleiters (Länge, Durchmesser und Taperwinkel). Bei Einkopplung in das dünnere Ende des Lichtleiters erhält man entsprechend eine Untersetzung.
Die Geometrie des getaperten Lichtleiters lässt sich auch so wählen, dass es zu einer einkoppelwinkelabhängigen Verletzung der Totalreflexion und damit zu einem seitlichen Austreten des Laserstrahls an der Mantelfläche kommt. Die Lage der Austrittsstelle ent-

lang des Lichtleiters variiert mit dem Einkoppelwinkel, d.h. ein axialer Scan ist möglich.

Bild 6 Simulierter Strahlverlauf zur Winkelübersetzung bei einem getapertem Lichtleiter; Länge 35 mm, Taperwinkel 0,8°, Einkoppelwinkel 18°, Auskoppelwinkel 69°.

3.4 Lichtleiter mit modifiziertem Ein- bzw. Auskoppelende

Durch Modifikation der Lichtleiterenden lassen sich ebenfalls die Übertragungseigenschaften der Lichtleiter variieren. So kann beispielsweise durch Anschrägen der Einkoppelfläche bereits bei koaxialer Einkopplung (Einkoppelwinkel 0°) ein schräger Strahlverlauf in dem Lichtleiter und damit eine ringförmige Abstrahlcharakteristik erzeugt werden. Ein Anschrägen des distalen Lichtleiterendes führt zu einem Verkippen der Abstrahlrichtung. Bei Auftreten von Totalreflexion an der schrägen Endfläche bzw. wenn die Auskoppelfläche verspiegelt wird, tritt der Laserstrahl seitlich aus. Je nach Winkel der Anschrägung kann dadurch auch eine retrograde Abstrahlung realisiert werden (siehe **Bild 7**).

4 Diskussion

Als ein wesentliches Ergebnis konnte in den vorliegenden Untersuchungen gezeigt werden, dass der Winkel, unter dem in einen Lichtleiter eingekoppelt wird, auch über längere Strecken erhalten bleibt. Dies ist die entscheidende Voraussetzung für ein mögliches Scannen unter Verwendung einzelner Lichtleitern. Weiterhin wurde gezeigt, welchen Einfluss die verschiedenen Parameter des Lichtleiters bzw. des Laserstrahls auf die Übertragungseigenschaften haben. Das Verständnis erlaubt eine systematische Vorgehensweise beim Design der Lichtleiter, angepasst an die jeweilige Anforderung.

Mit einfachen zylindrischen Lichtleitern ist ein rotationssymmetrischer Winkelscan möglich. Mögliche Anwendungen sind beispielsweise variable Beleuchtungssysteme, auch in Verbindung mit Endoskopen. Denkbar ist auch die Spektroskopie in Hohlräumen, z.B. um den zentralen Bereich und den Randbereich selektiv anzuregen.

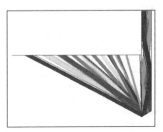

Bild 7 Lichtleiter mit rechteckigem Querschnitt und verspiegeltem angeschrägtem distalen Ende; farblich kodierte Darstellung des Strahlverlaufs bei verschiedenen Einkoppelwinkeln.

Wie gezeigt wurde, lässt sich ein punktförmiger Scan durch die Verwendung von Lichtleitern mit rechteckigem Querschnitt realisieren.

Die wenigen gezeigten Beispiele der Modifikation der Lichtleiterform und Lichtleiterenden sollen die mannigfaltige Variationsmöglichkeiten demonstrieren. Die daraus resultierende Breite der Übertragungseigenschaften lässt auf ein weites, zukünftiges Anwendungsfeld schließen.

Als weitere Schritte werden die Möglichkeiten eines zweidimensionalen Winkelscans untersucht werden und erste Demonstrationsmuster entwickelt und aufgebaut werden. Ein wesentliches Ziel ist jedoch eine möglichst zügige Umsetzung in erste Anwendungen.

Die hier vorgestellten grundlegenden Untersuchungen sind Teil eines Forschungsprojektes, welches von der Landesstiftung Baden Württemberg finanziert wird.

5 Literatur

[1] Schröder, G., Treiber, H.: Technische Optik. 9. erw. Aufl., Würzburg: Vogel, 2002
[2] Young, M.: Optik, Laser, Wellenleiter. 4. Aufl., Berlin Heidelberg: Springer 1997

No-Motion OCT-Verfahren

Edmund Koch[1], Alexander Popp[1], Peter Koch[2], Dennis Boller[2]
[1]Klinisches Sensoring und Monitoring, Medizinische Fakultät der TU-Dresden, Deutschland
[2]Medizinisches Laserzentrum Lübeck, Deutschland

Kurzfassung

Die Optische Kohärenz-Tomographie (OCT) ist ein innovatives bildgebendes Verfahren, das es erlaubt, auch in stark streuendem Gewebe, wie der Haut, Details mit hoher Auflösung zu erkennen. In der klassischen OCT ist für jeden A-Scan eine mechanische Verlängerung des Referenzarms um die Messtiefe erforderlich. In diesem Beitrag werden unterschiedliche Verfahren präsentiert, die es erlauben ohne mechanisch bewegte Teile einen A-Scan auszuführen. Die notwendigen Informationen werden in allen Fällen von einem Zeilensensor geliefert. Ein erstes Verfahren beruht auf dem Young-Interferometer und erreicht mit einem üblichen Zeilensensor (1024 Elemente) nur eine Messtiefe von 150 µm. Eine Abtastung der Interferenzen mittels Amplitudengitter erlaubt eine Erhöhung der Fringe-Frequenz um ca. eine Zehnerpotenz. Damit verbunden ist eine Erhöhung der Messtiefe auf ca. 1 mm. Als Drittes werden Ergebnisse der Fourier-Domain-OCT (FDOCT) vorgestellt.

1 Einleitung

Die Optische Kohärenz Tomographie (OCT) [1] ist ein bildgebendes Verfahren, das es auch in stark streuendem Gewebe wie der Haut erlaubt, Bilder mit hoher Auflösung zu gewinnen. Insbesondere in der Ophthalmologie [2,3] und der Dermatologie [4,5] haben sich eine Vielzahl von Anwendungen ergeben. Ein Interferometer verbunden mit einer kurzkohärenten Lichtquelle wird zur Bestimmung der Tiefeninformation benutzt. In der Zeit-Domänen OCT (TDOCT) (Siehe **Bild 1**) wird die Modulation des Ausgangssignals eines Interferometers bei Veränderung der Länge des Referenzarms aufgezeichnet und in ein Tiefenprofil umgesetzt. Die Änderung der Länge des Referenzarms entspricht damit der Messtiefe in Luft. In Gewebe reduziert sich die Messtiefe um den Faktor des Brechungsindexes. Um die Interferenzen möglichst rauscharm zu detektieren, werden Bandpassfilter eingesetzt. Voraussetzung für die optimale Dimensionierung der Filter ist eine möglichst lineare Längenänderung des Referenzarms. Da für jeden Tiefenscan, im weiteren in Analogie zur Ultraschalltechnik als A-Scan bezeichnet, eine komplette Änderung der Länge des Referenzarms notwendig ist, soll, um Artefakte durch Bewegung des Objekts zu vermeiden, die Änderung der Länge des Referenzweges möglichst schnell erfolgen. Deshalb wurden dafür viele Möglichkeiten erprobt. Zunächst wurde die Position des Spiegels im Referenzarm elektrodynamisch oder piezoelektrisch verfahren. Um den benötigten Hub zu verkleinern, wurden Anordnungen mit Vielfachreflexion zwischen parallel bzw. keilförmig angeordneten Spiegeln erprobt. Weiterhin wurde die elastische Dehnung von Fasern, die um Zylinder aufgewickelt wurden, getestet. Versuche, eine kontinuierliche Rotationsbewegung zu verwenden, erzielten entweder keine ausreichende Linearität oder

der duty cycle war sehr klein. Vergleichsweise schnelle lineare Änderungen der Referenzstrecke können durch eine Anordnung aus Gitter und Drehspiegel [6] erzielt werden. Letztendlich wird die Geschwindigkeit dieser TDOCT-Systeme limitiert durch die Geschwindigkeit der mechanischen Verschiebung des Referenzarms. Eine gewisse Steigerung lässt sich noch durch Parallelisierung erreichen, in dem zeitgleich mehrere Objektzeilen erfasst werden, doch erhöht dies auch die Kosten der Systeme erheblich. Weitere Nachteile sind, dass es bei hohen Geschwindigkeiten zu unerwünschten Vibrationen des Gesamtsystems kommen kann und dass die Kosten des Referenzscanners nicht unerheblich zum Gesamtpreis beitragen.

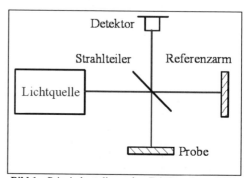

Bild 1 Prinzipdarstellung der Zeit-Domänen OCT. Licht einer kurzkohärenten Lichtquelle wird in einem Michelson-Interferometer auf die Probe und einen Spiegel im Referenzarm geschickt.

Es besteht deshalb ein Bedarf nach einem OCT-Verfahren, welches ohne Referenzscanner auskommt. Solche Verfahren werden im Weiteren als „No-

Motion-OCT" bezeichnet, obwohl zur Bewegung des Lichtstrahls über die Probe weiterhin mechanische Strahlablenker zum Einsatz kommen.

In dieser Arbeit werden drei No-Motion-OCT Verfahren beschrieben. Neben dem Michelson-Interferometer sind andere Interferometer bekannt, die die Längendifferenz von Proben- und Referenzarm auf eine Ortskoordinate projizieren. Da die Interferenzen linear in den Ortsraum transformiert werden, wird diese Form Lineare OCT (LOCT) genannt [7,8]. Dieses prinzipiell sehr einfache Konzept erreicht mit üblichen Zeilensensoren jedoch nur eine geringe Messtiefe. Deswegen wird als zweites ein Konzept beschrieben, dass die Messtiefe dieses Systems um ca. eine Zehnerpotenz vergrößert (Lineare OCT mit Maske). Zuletzt wird ein System erläutert, das es erlaubt, die Ortsinformation aus einem Michelson-Interferometer ohne Veränderung der Länge des Referenzarms zu gewinnen [2,9]. Dieses als Fourier-Domain OCT (FDOCT) bezeichnete Verfahren wertet die Interferenzen eines nicht abgeglichenen Interferometers spektroskopisch aus. Eine Fouriertransformation des Spektrums führt dann auf die Tiefeninformation. Da das Spektrum mittels Zeilen- oder Array-Sensor schnell ausgelesen werden kann, lassen sich sehr hohe Geschwindigkeiten erreichen.

2 No-Motion OCT

2.1 Lineare OCT

2.1.1 Theorie

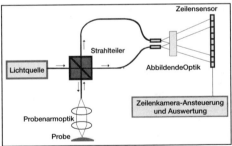

Bild 2 Prinzipaufbau der Linearen OCT. Der Zeilensensor ist stark vergrößert dargestellt.

Schickt man das Licht von Proben- und Referenzarm als ebene Welle aus unterschiedlichem Winkel auf einen Zeilensensor (Siehe **Bild 2**), so beobachtet man die Interferenzen als räumliche Intensitätsschwankungen auf dem Zeilensensor. Die Intensität I_F in der Detektorebene ist:

$$I_F(x) = I_P + I_R + 2\sqrt{I_P I_R} \cos(\kappa_I x) \cdot \gamma(x) \quad (1)$$

wobei I_P die Intensität im Probenarm, I_R die Referenzintensität und $\gamma(x)$ der Modulationsgrad ist, der von der Kohärenz zwischen Proben- und Referenzlicht abhängt. κ_I ist die Trägerkreisfrequenz der Modulation und vom Winkel α zwischen den interferierenden Strahlen und der Wellenlänge λ abhängig (Siehe **Bild 3**)

$$\kappa_I = 2\pi \alpha / \lambda. \quad (2)$$

Für eine kurzkohärente Lichtquelle entstehen die Interferenzen nur in einer kleinen Umgebung um die Position, für die Proben- und Referenzarm gleich lang sind. Unterschiedliche Streuer innerhalb der Probe führen deshalb zur Modulation des Signals an verschiedenen Positionen des Zeilensensors.

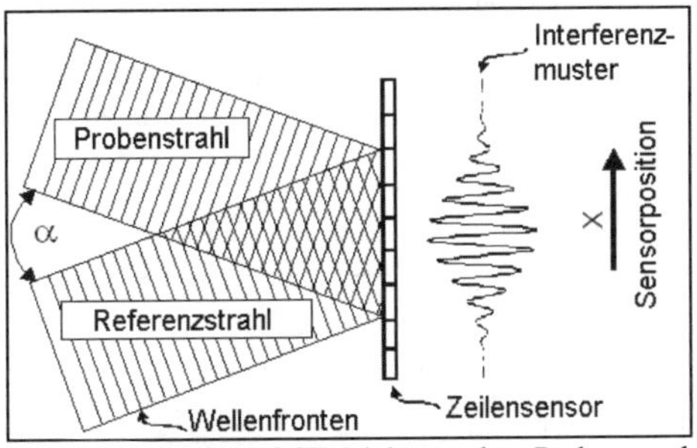

Bild 3 Prinzip der LOCT. Licht aus dem Proben- und Referenzarm interferiert auf dem Zeilensensor

Die Empfindlichkeit eines Zeilensensors für verschiedene Ortsfrequenzen wird durch die Modulations-Transfer-Funktion (MTF) beschrieben. Für eine eindeutige Bestimmung aller Frequenzen müssen nach dem Nyquist-Theorem mindestens zwei Pixel des Zeilensensors pro Interferenzperiode vorhanden sein. Um eine ausreichende Modulation zu gewährleisten sollten besser drei Pixel pro Interferenzstreifen genutzt werden. Der Messbereich für ein LOCT-System mit einem Zeilensensor mit N-Pixeln beträgt:

$$\Delta L_{max} = \frac{N\lambda}{2P} \quad (3)$$

wobei λ die Zentralwellenlänge und P die Anzahl der Pixel pro Interferenzperiode auf dem Zeilensensor ist. Für einen Zeilensensor mit N = 1024 Pixeln und einer Wellenlänge von λ = 830 nm führt dies mit P = 3 auf einen Messtiefe ΔL_{max} von ca. 150 µm.

2.1.2 Experimenteller Aufbau

Bild 2 zeigt den verwendeten Aufbau des LOCT-Systems. Licht einer Superlumineszenzdiode (SLD) wird über einen Strahlteiler (Faserkoppler) zum Teil auf die Probe gelenkt, der andere Teil wird direkt auf einen Eingang des Young-Interferometers geführt. Nicht eingezeichnet sind Komponenten zum Abgleich der Länge von Proben und Referenzstrahl und Komponenten zur Drehung der Polarisation. Da die Messtiefe dieses LOCT-Systems mit ca. 150 µm für Messungen an der Haut zu klein ist, wurde zur Demonstration der Leistungsfähigkeit des Systems die Oberfläche eines mikromechanischen Beschleunigungssensors zweidimensional abgetastet. Dazu wurde der Schwingungssensor mittels zweier Linearversteller vor der Probenoptik bewegt.

2.1.3 Ergebnisse

Aus jedem einzelnen A-Scan wurde die Position der stärksten Interferenzen ermittelt und in eine Höhe umgerechnet. Diese Höhe wurde in **Bild 4** als Funktion der Position der beiden Linearversteller aufgetragen. Der aufgetragenen Höhendifferenzen betragen nur ca. 15 µm. Die Kantenlänge des aufgenommenen Bereiches beträgt 1,4 mm. Man erkennt sehr gut, dass die Bestimmung der Position der maximalen Interferenzen sehr viel genauer möglich ist, als die Kohärenzlänge der Lichtquelle, die ca. 10 µm betrug.

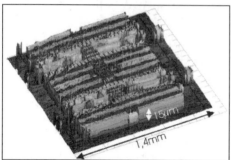

Bild 4 Oberfläche eines mikromechanischen Beschleunigungssensors mittels LOCT erfasst.

2.1.4 Diskussion

Die Genauigkeit der Bestimmung der Position des Maximums der Interferenzen kann sehr viel genauer als die Wellenlänge des Lichts erfolgen. Die Genauigkeit des gesamten Sensorsystems hängt jedoch stark von dem Verhalten der Glasfasern ab, da Referenz- und Probenstrahl über unterschiedliche Fasern geführt werden. Temperatureffekte und Bewegung der Fasern führen zu Veränderungen der optischen Länge der Fasern, die die Messgenauigkeit des Systems limitieren.

Die Messtiefe des LOCT-Systems lässt sich durch Wahl einer Sensorzeile mit mehr Pixeln vergrößern. Es sind heute CCD-Sensorzeilen mit mehr als 10 000 Pixeln verfügbar, so dass bei entsprechender Auslegung der Messbereich auch für die üblichen Anwendungen in der OCT ausreicht, doch wird zum einen ein solches System vergleichsweise groß, zum anderen muss das Licht von Proben- und Referenzstrahl auf der gesamten Länge des Zeilensensors mit einer Höhe von wenigen µm zusammengeführt werden. Auch wird die Geschwindigkeit des Systems dann durch die Auslesegeschwindigkeit und AD-Wandlung für die 10 000 Elemente der CCD-Zeile limitiert. Es wurde deshalb nach einem System gesucht, das eine Vergrößerung der Messtiefe erlaubt, ohne eine entsprechend Anzahl von Pixeln zu benötigen.

2.2 Lineare OCT mit Maske

2.2.1 Theorie

Für eine einfachere Abtastung der Interferenzen sollte die Frequenz des Signals reduziert werden. Aus der Signalverarbeitung ist dafür die Methode der Frequenzmischung bekannt. Um eine (Orts-)Frequenzmischung auf dem Zeilensensor zu erzielen, wird das Licht unmittelbar vor dem Detektor durch eine periodische Maske geschickt. Die Transmission durch die Maske kann durch:

$$T(x) = \tfrac{1}{2}(1 + \cos(\kappa_M x)) \qquad (4)$$

beschrieben werden. Das Signal hinter der Maske I_M ergibt sich dann als Produkt aus I_F und T. Durch die Mischung treten nun mehrere Frequenzen auf dem Zeilensensor auf. Bei geeigneter Wahl von κ_I und κ_M liegt davon nur $|\kappa_I - \kappa_M|$ zwischen 0 und der Niquist-Frequenz des Zeilensensors. Das Eingangssignal nach Gleichung (1) ist amplitudenmoduliert, enthält dementsprechend weitere Frequenzen um κ_I. Damit keine Verzerrungen entstehen, sollten alle diese Frequenzen $\kappa_I - \kappa_M$ nach der Mischung positiv oder negativ und kleiner als die Niquist-Frequenz sein.

2.2.2 Experimenteller Aufbau

Der verwendete Aufbau ist in **Bild 5** gezeigt. Das Licht einer SLD wird über einen Strahlteiler auf die Probe geführt. Die plane Faserendfläche dient als Referenzfläche. Proben- und Referenzlicht werden auf einer Faser zum Young-Interferometer geführt. Ein Strahlteiler spaltet den Strahl in zwei Teilstrahlen auf. Eine Zylinderlinse sammelt die Strahlen auf den Zeilensensor. Die Maske wurde nach Entfernen des Fensters direkt auf den Sensorchip geklebt.

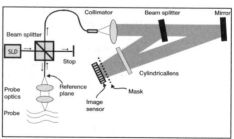

Bild 5 Optischer Aufbau für die LOCT mit Maske

Weil Proben- und Referenzlicht über die gleiche Faser geschickt werden, ist der Aufbau unempfindlich gegen Temperaturänderungen und Bewegungen der Faser. Dies hat darüber hinaus den Vorteil, dass auf die sonst benötigten Polarisationssteller zur Kompensation der von den Glasfasern verursachten Doppelbrechung verzichtet werden kann.

2.2.3 Ergebnisse

Um das LOCT-System zu testen, wurde ein Deckglas als Probe verwendet. Das resultierende Sensorsignal ist in **Bild 6** aufgetragen. Deutlich sind zwei Bereiche mit Interferenzen zu erkennen, die auf den Reflex von Vorder- und Rückseite zurückzuführen sind. Einer dieser Bereiche ist nochmals vergrößert dargestellt. Obwohl die Lichtquelle eine Kohärenzlänge von ca. 10 µm hat, sind aufgrund der Frequenzmischung nur wenige Oszillationen zu erkennen.

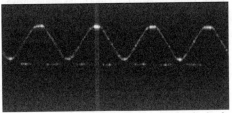

Bild 7 B-Scan einer Schraube. Die Stärke der hochfrequenten Oszillationen ist in ein Intensitätssignal umgesetzt. Der dargestellte Bereich hat eine Größe von 940µm • 2000 µm

2.2.4 Diskussion

Durch Verwendung einer feinen Maske vor dem Zeilensensor konnte der Messbereich eines LOCT-Systems um eine Größenordnung gesteigert werden. Eine solche Maske könnte im Herstellungsprozess des Zeilensensors leicht aufgebracht werden, wodurch sich die Probleme bei dem nachträglichen Aufbringen der Maske vermeiden ließen.

Durch die Maske wird die Lichtintensität um 50% reduziert. Weiterhin nimmt der Kontrast der Interferenzen um 50% ab. Weil in diesem Aufbau Proben- und Referenzlicht über die gleiche Faser geführt werden, wird der Einfluss von Temperaturänderungen oder Bewegung der Faser ausgeschlossen. Weiterer Vorteil dieses Aufbaus ist das Entfallen von Polarisationsstellern. Man bezahlt dies mit einer weiteren Reduktion der Stärke der Interferenzen, da in beiden Armen des Young-Interferometers Proben- und Referenzlicht vorhanden ist. Das Verfahren bietet sich dort an, wo nicht das maximale Signal-Rausch-Verhältnis notwendig ist, aber eine schnelle, preiswerte und einfache Messtechnik benötigt wird.

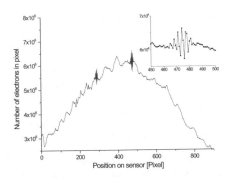

Bild 6 Lichtintensität auf dem Zeilensensor mit einem Deckglas als Probe. Rechts oben ist ein Bereich mit Interferenzen vergrößert dargestellt. Aufgrund der Mischung erscheinen nur wenige Interferenzstreifen.

Die niederfrequente Modulation auf dem Signal ist vermutlich auf Interferenzen durch einen kleinen Abstand zwischen Sensorchip und Maske zurückzuführen. Durch die verwendete Maske ist der Messbereich um ca. eine Größenordnung gestiegen. Das Wechselsignal am Detektor ist durch das Verfahren auf 25% gefallen. 50% der Intensität werden von der Maske absorbiert und durch die Frequenzmischung nimmt die Stärke der Modulation um 50% ab. Für den Fall eines Shot-Noise begrenzten Systems entspricht dies einer Abnahme des S/N um 9 dB. Um zu zeigen, dass das Verfahren auch für 2-dimensionale Bilder (B-Scans) geeignet ist, wurde der Probenstrahl auf eine Schraube gerichtet. Die Kontur der Schraube ist in **Bild 7** zu erkennen. Zur Bildgebung wurde die Stärke des hochfrequenten Wechselsignals auf dem Detektor in ein Intensitätssignal umgesetzt.

2.3 Fourier Domain OCT

Aufgrund der eingeschränkten Dynamik ist das LOCT-System mit Maske für Messungen an biologischen Proben mit ihrem kleinen Streuamplituden und hoher Dämpfung nur sehr eingeschränkt geeignet. Gesucht ist deshalb ein Konzept, das ohne bewegte Teile eine mindestens gleich hohe Dynamik wie TDOCT-Systeme bei möglichst großer Messgeschwindigkeit erreicht. Schaut man noch mal auf das Prinzip der TDOCT (Siehe Bild 1) so erkennt man, dass die Information über das Probenlicht auch bei fester Länge des Referenzarms im interferierenden Licht vorhanden ist. Abhängig vom Gangunterschied zwischen Proben- und Referenzlicht kommt es bei der Überlagerung zu Interferenzen, die sich im Spektrum des Lichts auswirken. Da die Frequenz der Interferenzen als Funktion der reziproken Wellenlänge ein Maß für die Differenz zwischen Proben- und Referenzweg ist, benötigt man eine Fouriertransformation zur Gewinnung der Tiefeninformation. Das Verfahren wird deshalb als Fourier-Domain-OCT (FDOCT) bezeichnet (Siehe **Bild 8**).

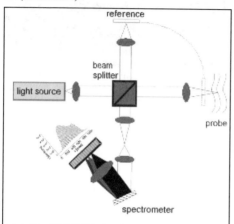

Bild 8 Prinzip der FDOCT. Licht aus einem nicht abgeglichenen Michelson-Interferometer wird spektral zerlegt. Eine Fouriertransformation des Spektrums führt dann auf die Tiefeninformation

2.3.1 Theorie

Durch Interferenz des Lichtes von Proben- und Referenzarm erhält man hinter einem Michelson-Interferometer als Funktion der Wellenlänge Interferenzen:

$$I_F(\lambda) = I_P + I_R - 2\sqrt{I_P I_R} \cos\left(\frac{2\pi n d}{\lambda}\right) \quad (5)$$

wobei I_P und I_R die Intensität aus dem Proben- und Referenzarm sind, d der Unterschied zwischen dem Proben- und Referenzweg ist, n der Brechungsindex in diesem Bereich. Bezeichnen wir mit $L = nd/2$ den unterschiedlichen optischen Abstand vom Strahlteiler und gehen von der Wellenlänge zur Wellenzahl $v = 1/\lambda$ über:

$$I_F(v) = I_P + I_R - 2\sqrt{I_P I_R} \cos(4\pi L v) \quad (6)$$

Bestimmt man die Intensität I_F als Funktion der Wellenzahl v, so kann man durch Fouriertransformation die Länge L aus der Frequenz und I_P aus der Intensität der Interferenzen bestimmen. Durch den Spektralbereich der Lichtquelle bzw. des Spektrometers eingeschränkt werden die Interferenzen nur in einem Bereich von Wellenzahlen $\Delta v = v_1 - v_2$ mit einem Abtastintervall δv bestimmt. Nach Gleichung 6 entsprechen hohe Frequenzen der Interferenzen auch großen Abständen L. Die höchste Frequenz der Interferenzen, die bestimmt werden kann, ergibt sich deshalb aus dem Nyquist-Theorem zu dem halben Wert von $1/\delta v$:

$$L_{max} = \frac{1}{4\delta v} = \frac{\lambda^2}{4\delta\lambda} \quad (7)$$

Eine Oszillation der Interferenzen im Wellenzahlbereich Δv entspricht einem Abstand L_{Min} von:

$$L_{Min} = \frac{1}{2\Delta v} = \frac{\lambda^2}{2\Delta\lambda} \quad (8)$$

Um zwei Probenpunkte in unterschiedlicher Tiefe sicher trennen zu können, muss der Abstand mindestens doppelt so groß sein, wie der minimale Abstand L_{Min}. Demnach ist die Auflösung ΔL des FDOCT-Systems:

$$\Delta L = \frac{1}{\Delta v} = \frac{\lambda^2}{\Delta\lambda} \quad (9)$$

Für eine genauere Darstellung der Theorie sei auf [9,10] verwiesen.

2.3.2 Experimenteller Aufbau

Aufgrund der Anforderungen an Messbereich und Auflösung des FDOCT-Systems wurde das Spektrometer selbst entwickelt. Angepasst an die gewünschte Messtiefe wurde eine Superlumineszenzdiode ausgewählt und ein entsprechendes hochauflösendes Spektrometer konzipiert. Hierbei wurden die Eigenschaften des CMOS Zeilensensors berücksichtigt. Dieser Zeilensensor mit 1024 Elementen kann mit Frequenzen bis zu 20 MHz ausgelesen werden. Die verfügbare AD-Wandlerkarte limitiert die Ausleserate momentan auf 5 MHz. Der gesamte optische Aufbau wurde in ein staubdichtes Gehäuse eingebaut. Das Spektrometer, sowie alle elektrischen Komponenten wurden in ein Gehäuse mit den Außenmaßen (32•27•10 cm) integriert (Siehe **Bild 9**).

Bild 9 Foto des OCT-Gerätes. Zentral ist das Spektrometer, im Hintergrund die Elektronik zu erkennen. Rückseitig ist die Verbindung zum Computer, auf der Vorderseite sind der optische und elektrische Anschluss für den Applikator zu erkennen.

Die Ansteuerung, Auswertung und Darstellung der Daten erfolgt auf einem Standard-PC.

2.3.3 Ergebnisse

Mit dem entwickelten FDOCT-System lassen sich hochauflösende Schnittbilder der oberen Gewebeschichten mit einer Bildwiederholrate von bis zu 10 Bildern pro Sekunde erzeugen. Bei hohen Messraten kommt es jedoch zu einer Abnahme der Dynamik, so dass meist mit 3-4 Bildern pro Sekunde gearbeitet wird. Experimentell konnten die aus Gleichung 7 und 9 berechneten Werte für die Messtiefe des Systems von ca. 2 mm, und die Auflösung von etwa 8 µm bestätigt werden. Die Auswertung und Darstellung der Daten kann auf üblichen PC's in Echtzeit erfolgen. **Bild 10** zeigt ein OCT-Bild der Haut am Unterarm.

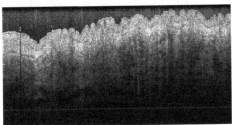

Bild 10 OCT Bild der Haut am Unterarm. Neben der Furchung der Haut ist als oberste sehr dünne Schicht das Stratum Corneum zu erkennen. Darunter folgt die Epidermis gefolgt von der Dermis. Bei den echoarmen Zonen nahe dem unteren Bildrand handelt es sich vermutlich um Blutgefäße. Das Bild zeigt einen Bereich von 6 mm in der Breite und 1,5 mm in der Tiefe.

2.3.4 Diskussion

Mittels FDOCT unter Verwendung von SLD's können sehr kompakte und preiswerte OCT-Geräte hergestellt werden, die eine hohe Dynamik und hohes Auflösungsvermögen besitzen. Die erzielbare Messtiefe ist für Anwendungen an der Haut völlig ausreichend. Die Verwendung des Gerätes setzt nur wenige Fachkenntnisse voraus. Zu erwartende Verbesserungen an den SLD's und den Zeilensensoren können das Auflösungsvermögen bei gleich bleibender Messtiefe weiter steigern. Durch Verwendung eines größeren Spektralbereichs ist auch eine funktionelle OCT möglich, die es erlaubt verschiedene Zelltypen zu unterscheiden [11].

3 Zusammenfassung

Es wurden drei verschiedene OCT-Verfahren beschrieben, die ohne mechanisch bewegte Teile einen Tiefenscan ausführen können. Während die LOCT mit üblichen Zeilensensoren in der Messtiefe sehr limitiert ist, erlaubt die LOCT mit Maske eine deutlich größere Messtiefe, erreicht aber eine geringere Dynamik. Etwas komplexer, aber sehr viel leistungsfähiger ist das Verfahren der FDOCT. Hier lassen sich sehr hohe Messgeschwindigkeiten, wie auch eine sehr hohe Dynamik erzielen. Entsprechende Geräte können sehr kompakt und robust aufgebaut werden.

4 Literatur

[1] D. Huang, E.A. Swanson, C.P. Lin, J.S. Schuman, W.G. Stinson, W. Chang, M.R. Hee, T. Flotte, K. Gregory, C.A. Puliafito and J.G. Fujimoto: Optical Coherence Tomography, Science 254, 1178-1181 (1991).

[2] A. F. Fercher, C. K. Hitzenberger, G. Kamp, S. Y. El-Zaiat, "Measurements of intraocular distances by backscattering spectral interferometry" Opt. Comm 117, 43-48 (1995)

[3] Hrynchak, P. and T. Simpson. "Optical coherence tomography: an introduction to the technique and its use." Optom.Vis.Sci. 77.7 (2000): 347-56

[4] J. Welzel: Optical coherence tomography in dermatology: a review, Skin Res Technol (2001) Feb; 7(1):1-9

[5] G. Petrova, et. Al. Optical Coherence tomography using tissue clearing for skin disease diagnostics, SPIE Proceedings 5140 168-177 (2002)

[6] Andrew M. Rollins, Manish D. Kulkarni, Siavash Yazdanfar, Rujchai Ung-arunyawee and Joseph A. Izatt; Optics Express Vol. 3, No. 6 (1998) 219-229

[7] C. Hauger, L. Wang, M. Wörz, and T. Hellmuth, „Theoretical and experimental characterisation of a stationary low coherence interferometer for optical coherence tomography"in Optical Coherence Tomography and Coherence Techniques Proc. SPIE 5140, 60-68, (2003)

[8] C. Hauger, M. Wörz, and T. Hellmuth, „Interferometer for optical coherence tomography " Applied Optics, 42, No.19, pp. 3896-3902, (2003)

[9] G. Häusler, M.W. Lindner: "Coherence Radar" and "Spectral Radar" – New Tools for Dermatological Diagnosis, Journal of Biomedical Optics (3) (1998) 21-31

[10] M.A. Choma, M.V.Sarunic, C.Jang, J.A. Isatt: Sensitivity advantage of swept source and Fourier domain optical coherence tomography, Opt. Express 11, 2183-2189 (2003)

[11] B. Hermann, et.al. Precision of quantitative absorption measurement with spectroscopic optical coherence tomography, SPIE Proceedings 5140 84-86 (2002)

Videofluoreszenz-Mikroskopie für die Beobachtung von Oberflächenkoronargefäßen am isoliert schlagenden Herzen

Alexander Popp, Sebastian Stehr, Andreas Deußen, Edmund Koch, Medizinische Fakultät der Technischen Universität Dresden, 01307 Dresden, Deutschland

Kurzfassung

Wir präsentieren die Entwicklung eines Mikroskopaufbaus, mit dem die oberflächennahen Koronargefäße eines *schlagenden* isolierten Rattenherzen visualisiert werden. Anhand der Bildsequenzen sollen die Gefäßweitenregulation und die Adhäsion von Leukozyten am Endothel der Oberflächenkoronargefäße unter Gabe von volatilen Anästhetika analysiert werden. Zur Fokussierung des Objektivs auf die Myokardoberfläche in der diastolischen Phase der Herzbewegung wird der Abstand zum Objektiv mit einer Lichtschranke bestimmt und mit einem piezoelektrischen Stellelement entsprechend fokussiert. Diese Autofokussierung gelingt bisher nur bei verlangsamter Herztätigkeit. Zur Kontrastierung der Blutgefäße und der Leukozyten wird je ein verschiedener Fluoreszenzfarbstoff verwendet. Zur Anregung der Fluoreszenz wurden verschiedenartige Lichtquellen getestet.

1 Einführung

1.1 Medizinische Fragestellung

Mit Verschiebung der Alterspyramide und zunehmender Inzidenz von Koronarstenosen in der Bevölkerung steigt der Anteil an Hochrisikopatienten, die sich großen chirurgischen Operationen unterziehen müssen [1]. Daher kommt der Prävention von kardialen Ereignissen während der Anästhesie eine große Bedeutung zu. Neben kurzzeitigen ischämischen Episoden kann es auch zu kritischen Ischämien bis hin zur akuten Infarzierung des Myokards mit letalem Ausgang kommen.

Begünstigend für die Entstehung kritischer Ischämien sind neben lokalen Faktoren wie endothelialer Dysfunktion, atherosklerotischem Wandumbau und vorbestehender Stenosierung generelle Faktoren wie Hypoxie/Ischämie durch akute Blutung oder Kreislaufdepression und Veränderung der Blutviskosität. Für das Outcome der Patienten, welche intra- oder postoperativ eine myokardiale Ischämie erleiden, spielt das Ausmaß des Reperfusionsschadens nach Ischämie eine entscheidende Rolle.

In diesem Zusammenhang wird die Aktivierung polymorphkerniger neutrophiler Granulozyten (PMN) mit konsekutiver Freisetzung von reaktiven Sauerstoffspezies und hydrolytischen Enzymen bei gleichzeitig verminderter antioxidativer Kapazität [2] als einer der wesentlichen pathogenetischen Faktoren angenommen. Dabei konnte an Ischämie/Reperfusionsmodellen gezeigt werden, dass eine Leukozytendepletion oder die Inhibition der Neutrophilenadhäsion und Sauerstoffradikalfreisetzung den Gewebeschaden reduziert und einen Outcome-Vorteil bietet [3-5].

In verschiedenen Untersuchungen wurden kardioprotektive Effekte von volatilen Anästhetika experimentell [6,7] und auch klinisch [8,9] demonstriert. Dabei konnte am isolierten Meerschweinchenherzen gezeigt werden, dass der kardioprotektive Effekt volatiler Anästhetika wahrscheinlich zumindest teilweise über eine Verminderung der postischämischen Adhäsion von Neutrophilen vermittelt wird [10]. Die bisher publizierten Studien lassen jedoch weder auf die Lokalisation der Granulozyten (intra- oder extravasal) noch auf die Abhängigkeit vom regionalen Blutfluss schließen. Auch die Auswirkung der volatilen Anästhetika auf die Mikrozirkulation konnte bisher nicht untersucht werden.

1.2 Fluoreszenzmikroskopie

Mikroskopische Verfahren zur Visualisierung der Mikrozirkulation haben in den letzten Jahrzehnten zur Erweiterung des Verständnisses von pathophysiologischen Zusammenhängen in verschiedensten Organen beigetragen [11-13]. Aus technischen Gründen haben sich jedoch Modelle für die Betrachtung des schlagenden Herzens als schwierig erwiesen. Im Einzelnen liegen die Herausforderungen in den großen Bewegungsamplituden der Herzoberfläche während der Systole, der Variation der Ruhelage in der Diastole und zudem in der hohen Frequenz der Herztätigkeit von etwa 300 Schlägen pro Minute (Ratte).

Bei bisherigen Ansätzen zur Visualisierung der Mikrozirkulation am Herzen wurde daher entweder die Herztätigkeit mechanisch durch Platzierung von Stahlnadeln eingeschränkt [14-18], oder es wurden medikamentös induzierte asystole Herzen beobachtet [19-21]. Andere Gruppen betrachteten die Herzoberfläche mit Objektivanordnungen unter gleich bleiben-dem Auflagedruck auf die Oberfläche, um so den Abstand zum Objektiv konstant zu halten [19]. Die Beeinflussung von wichtigen Messgrößen, wie des Arteriolendurchmessers (flächiger Kontakt) bzw. der Zu- oder Abflüsse aus der Umgebung (Ringauflage), sind bei diesen Methoden nicht unbedingt auszuschließen. Ein wirklich berührungsfreies Autofokussystem wurde für die Untersuchungen von schlagenden Herzen noch nicht eingesetzt. Autofokusalgorithmen, die auf der Echtzeitauswertung der Bildschärfe beruhen, erreichen nicht die nötige Geschwindigkeit [22].

Unser Ansatz zur Visualisierung der Mikrozirkulation am isolierten, schlagenden Herzen strebt eine berührungsfreie und während der Diastole scharfe, fluoreszenzkontrastierte Bildaufnahme an.

2 Material und Methoden

2.1 Modell des isolierten schlagenden Herzens

Das isolierte, isovolumetrisch arbeitende (schlagende) Herz ist ein etabliertes Versuchsmodell, an dem kardiale Fragestellungen ohne systemische Einflüsse (z.B. neuroge Faktoren, systemischer Gefäßwiderstand, Volumenstatus), aber dennoch eng an physiologischen Bedingungen untersucht werden können. Zur Organentnahme erfolgt zunächst eine Anästhesie der Ratte mit Sevofluran unter der Glasglocke, danach wird Thiopental i.p. als Narkose und Heparin i.p. zur Antikoagulation appliziert. Nach Eröffnung des Zwerchfells und des Thorax wird das Herz mit 4°C Krebs-Henseleit unterkühlt, entnommen, präpariert und im Organbad mit 37°C warmer Krebs-Henseleit-Lösung flusskonstant (10ml/min) über die Aorta ascendens perfundiert. Das isolierte Herz schlägt spontan mit etwa 300 Schlägen pro Minute. Im linken Ventrikel wird ein mit destilliertem Wasser gefüllter Ballonkatheter zur Messung des linksventrikulären Druckes und zur Erzeugung einer gewissen Vordehnung des Ventrikels (Preload) platziert. Zur Erfassung des Koronarflusses wird ein Flow-Sensor (Transonic Flowprobe, Fa. Transonic Systems Inc., New York, USA) zwischen Perfusatpumpe und Aortenkanüle eingebracht.

2.2 Videomikroskopie

Das Auflichtfluoreszenzmikroskop (DMR, Fa. Leica) wird für die Beobachtung des schlagenden Herzens modifiziert. Der Objekttisch wird entfernt und durch einen motorisierten x-y-z Tisch ersetzt, der die genaue Positionierung des Herzens erlaubt und genug Arbeitsabstand für die Unterbringung des Organbads bietet. Die Abbildung der Herzoberfläche geschieht durch ein Wasserimmersionsobjektiv (HCX APO L 10x/0.30 W) mit einer Numerischen Apertur von 0.3 ins Unendliche und dann mit einer Tubuslinse auf den Chip einer CMOS Kamera. Im Tubus sitzt ein Filtersatz (B/G/R) für die Trennung von Fluoreszenz und Beleuchtung. Die Tiefenschärfe von ca. 3 µm erfordert eine entsprechend genaue Fokussierung des Objektivs, die auf der Basis einer berührungsfreien Abstandsmessung erfolgt und im folgenden näher beschrieben wird (siehe **Bild 1**).

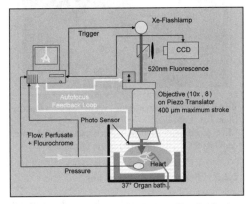

Bild 1 Funktionsschema der Videomikroskopie des isoliert schlagenden Herzens mit fokussierbarem Objektiv und Regelschleife

2.2.1 Abstandssensor

Eine Lichtschranke, bestehend aus einer Infrarot-Leuchtdiode mit engem Abstrahlwinkel (Siemens SFH 480-2) und einer doppelten Silizium Photodiode (Siemens KOM 2125), wird an dem 10x Objektiv so angebracht, dass die Oberfläche des Herzens wenige Mikrometer in die Lichtschranke hineinragt, wenn sie in der Fokusebene des Objektivs liegt (siehe **Bild 2**). Das resultierende Signal der Photodioden wird verstärkt und per Analog/Digitalwandlung in den PC eingelesen. Mit Hilfe einer vorher aufgenommenen und als Polynom angenäherten Kalibrierkurve wird die Abweichung von der Fokusposition aus dem Lichtschrankensignal berechnet.

2.2.2 Piezoobjektivversteller

Zur Verstellung der Objektivposition bei der Fokussierung ist zwischen Mikroskoptubus und Objektiv ein Piezoverstellsystem (MIPOS 5, Fa. Piezosysteme Jena) eingebaut (siehe Bild 2). Das System verfügt über eine geregelte Positionierung mit eingebautem Dehnungsmessstreifen. Die eingebaute Regelung ist für die Ausregelung des Abstands mit der Herzfrequenz zu langsam. Deshalb wird die eingebaute Regelung überbrückt und die Piezoverstelleinheit wird allein mit Hilfe der Lichtschranke positioniert.

Bild 2 Mit Piezotranslator verstellbares Objektiv (hier 10x Luftobjektiv) mit Lichtschranke (links IR-LED, rechts photoempfindlicher Detektor) und unter dem Objektiv das isolierte Herz im Organbad.

2.2.3 Fluoreszenzmarkierung

Zur Kontrastierung der Blutgefäße wird dem Perfusat der Fluoreszenzfarbstoff FITC (gebunden an Dextran) zugesetzt. FITC besitzt bei 490 nm das Anregungsmaximum und fluoresziert bei 520 nm. Bolusgaben des Farbstoffs dienen zur Unterscheidung von Arteriolen und Venolen über die Anflutungsrichtung. Ein entsprechendes Filter ist in den Strahlengang des Mikroskops eingebaut.
Für die Markierung der Leukozyten wird ein zweiter Fluoreszenzfarbstoff verwendet (z.B. Cell Tracker Red, Fa. Molecular Probes).

2.2.4 Lichtquelle

Zur Anregung der Fluoreszenz wurden mehrere Lichtquellen getestet. Es gilt, eine möglichst hohe Lichtleistung im Spektralbereich der Fluorochromanregung zu balancieren gegenüber einer Belastung der Physiologie und des Gewebes durch die starke Lichtbestrahlung. Getestet wurden eine 50 W Hg-Dampflampe und ein 488 nm Halbleiterlaser (Sapphire, Fa. Coherent) mit 20 mW Leistung (siehe Abschnitt 3.3).
Da beide Lichtquellen Nachteile besitzen, wird in Zukunft eine speziell entwickelte Xenon-Doppelblitzlampe (Fa. Rapp Optoelectronics) eingesetzt, die genügend Licht in 100 µs kurzen Pulsen zur Verfügung stellt und das Gewebe in der restlichen Zeit schont. Der doppelte Blitz kann dazu verwendet werden, die Geschwindigkeiten von markierten Leukozyten zu bestimmen.

2.2.5 Kamera

Die Bilderfassung erfolgt mit einer 640x480 Pixel CMOS Schwarzweiß-Kamera mit maximal 100 Bildern pro Sekunde mit *Progressive-Scan* Zeilenabtastung (A602f, Fa. Basler). Die Verschlusszeit ist einstellbar und die Auslösung kann mit einem externen Triggersignal erfolgen.

2.2.6 Datenerfassung und Regelung

Die impedanzgewandelten Photodiodenströme und die Regelspannung für den Piezo werden von einer Multi I/O Karte (Fa. National Instruments) digitalisiert bzw. ausgegeben. Der Regelalgorithmus und die Bilderübertragung über IEEE 1394 Schnittstelle sind in LabVIEW® (National Instruments) implementiert.
Es wurden verschiedene Regelalgorithmen und Steuerungen für die Objektivfokussierung getestet.

3 Ergebnisse

3.1 Bewegungsamplitude des Herzens

Das isolierte Herz kontrahiert in der Systole, so dass die seitliche Oberfläche nach außen gedrückt wird. In der Diastole geht die seitliche Oberfläche wieder annähernd in die Ausgangsstellung zurück. Da das Myokard in der Diastole perfundiert wird, sind nur scharfe Bilder von dieser Ausgangstellung von Interesse. Zwei Experimente bringen Aufschluss über die Bewegungsamplituden und insbesondere über die Abweichung der diastolischen Position.

3.1.1 Messung mit telezentrischem Objektiv

Ein isoliertes, schlagendes Herz wird von der Seite mit einem telezentrischen Objektiv aufgenommen. Das Abbildungsverhältnis des Objektivs von 1:1 bleibt innerhalb des Telezentriebereichs ohne perspektivische Verzerrung konstant. In **Bild 3** wird eine Aufnahme vom diastolischen Ausgangspunkt mit einer Aufnahme aus der Systole verglichen. Die Gesamtamplitude der Herzoberfläche ist ca. 1,1 mm. Das untere Bild ist durch die Bewegung des Herzens zwischen den beiden Teilbildern (interlaced Zeilenabtastung der Video Norm) durch Streifenartefakte unscharf. Eine Analyse der gesamten Videosequenz ergab, dass die diastolische Ausgangsposition in einem Bereich von +/- 180 µm variiert. Bei einer Numerischen Apertur des Mikroskopobjektivs von 0,3 bestimmt die Schärfentiefe von 3 µm die Genauigkeit, mit der das Objektiv auf die Oberfläche nachfokussiert werden muss. Der montierte Piezoaktuator mit 400 µm Gesamthub reicht also aus, um die Variation der diastolischen Position auszugleichen. Voraussetzung ist die hinreichend genaue Messung des Abstands mit der Lichtschranke.

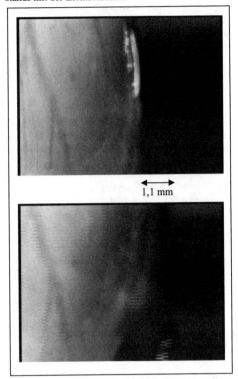

Bild 3 Seitliche Aufnahmen eines isoliert schlagenden Herzens mit einem telezentrischen Messobjektiv, oben in der Diastole unten während der Systolenbewegung. Trotz der Streifenartefakte lässt sich ein Gesamthub von 1.1 mm abschätzen

3.1.2 Messung mit Lichtschranke

Die Bewegung des schlagenden Herzens wird mit der Lichtschranke aufgenommen. In **Bild 4** erkennt man einen annähernd periodischen Verlauf mit 270 Schlägen pro Minute. Die diastolischen Phasen sind die oberen Umkehrpunkte (bei etwa 45 µm), deren Lage über den gesamten Zeitraum der Messung von 10 Sekunden mit +/- 1,4 µm wenig mehr als die Messgenauigkeit von +/- 0,6 µm abweicht. Die Umkehrpunkte sind etwa 5 ms lang innerhalb der von der Schärfentiefe vorgegebenen Toleranz. Dies würde nur eine maximale Belichtungszeit von 5 ms erlauben, um ein scharfes Bild pro Diastole zu erhalten. Das Rauschäquivalent der Lichtschrankenmessung bleibt mit 0,6 µm innerhalb der Toleranz, die durch die Schärfentiefe vorgegeben ist.

3.2 PID Regelung des Abstands

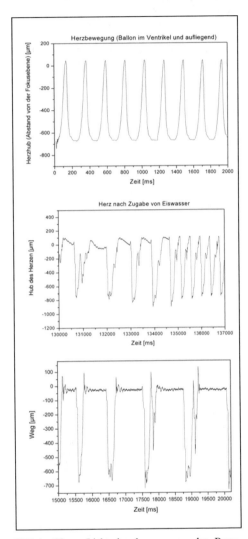

Bild 4 Oben: Lichtschrankenmessung der Bewegung des frei schlagenden Herzens; Mitte: Verlängerung der diastolische Ruhephase durch Kaltwassergaben in den Ventrikel. Unten: Bei Kaltwassergabe ist eine Fokusregelung des Objektivs wegen der verlängerten Ruhephase in der Diastole möglich.

Um das Objektiv zu fokussieren und die Zeitspanne zur Aufnahme scharfer Bilder zu verlängern, wird ein PID Algorithmus implementiert, der unterhalb einer Schwellenabweichung vom Sollabstand zwischen Objektiv und Herzoberfläche einsetzt und diese Abweichung ausregeln soll. Bei normaler Herzfrequenz erweist sich die Regelung als zu langsam. Erst wenn die

Diastole durch Kaltwassergaben in den Ventrikel des Herzens verlängert wird, ist eine Regelung möglich (siehe Bild 4 unten). Die Begrenzung der Regelbandbreite ist durch den Tiefpass bestimmt, der eine Mitkopplung des schwingungsfähigen Systems aus Piezoversteller und Objektivmasse verhindert. Dieses Frequenzverhalten wird darum näher untersucht.

3.2.1 Frequenzgang der Piezoverstellung

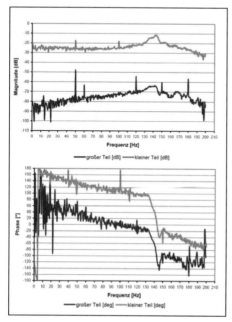

Bild 5 Amplituden (oben) und Phasengang (unten) des Piezoverstellers mit Objektiv. Abgesehen von Peaks bei Vielfachen von 50 Hz sind erste starke mechanische Resonanzen bei ca. 140 Hz erkennbar

Der Frequenzgang (siehe **Bild 5**) des Piezoverstellers mit Objektivmasse zeigt erste starke Resonanzen bei ca. 140 Hz, mit einer Phasendrehung, die bei 130 Hz beginnt. In das System wird deswegen ein Tiefpass mit einer -3dB Grenze von 70 Hz eingebaut. Es konnte in der Praxis kein PID Parametersatz gefunden werden, der eine Regelung des Objektivs auf die schnellen Abstandsänderungen des uneingeschränkt schlagenden Herzens erlaubt, ohne dass das System aufschwingt oder zu träge wird. Als Alternative bleibt, das Objektiv auf den diastolischen Umkehrpunkt des vorhergehenden Herzschlags einzustellen und scharfe Bilder durch entsprechend kurze Belichtungszeiten zu gewährleisten.

3.3 Lichtquelle und Belichtungszeiten

Zur Belichtung bzw. Anregung der Fluoreszenz des FITC Farbstoffs wurden verschiedene Lichtquellen getestet und die minimal erreichbaren Belichtungszeiten abgeschätzt.

3.3.1 Quecksilber Dampflampe

Eine Hg-Hochdruckdampflampe mit 50 W elektrischer Leistung besitzt innerhalb des Anregungsspektrums von FITC keine starke Emissionslinie und liefert nur etwa 14 mW Lichtleistung hinter dem Anregungsfilter. Dies ist für eine minimale Belichtungszeit von etwa 40 ms ausreichend. Es können also keine artefaktfreien Aufnahmen des schlagenden Herzens erstellt werden, sondern nur Aufnahmen des asystolen Herzens (siehe **Bild 6**).

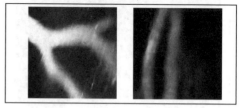

Bild 6 Aufnahmen von Arteriolen des asystolen Herzens nach Bolusgaben von FITC-Dextran, links mit 80 ms, rechts mit 40 ms Belichtungszeit.

3.3.2 488 nm Halbleiterlaser

Ein Halbleiterlaser (Sapphire, Fa. Coherent) liefert 20 mW Dauerstrichleistung bei 488 nm nah dem Anregungsmaximum von FITC. Der Laserstrahl wurde mit einem Kollimator aufgeweitet und in den Auflichtstrahlengang des Mikroskops eingekoppelt. Mit dem Laser sind Belichtungszeiten unter 4 ms möglich. Es gibt allerdings zwei entscheidende Nachteile, nämlich zum einen die mangelnde Flexibilität der Anregungswellenlänge und zum anderen eine erhebliche Belastung durch die 20 mW Dauerbestrahlung für das Herzmuskelgewebe mit Beeinflussung der Physiologie.

4 Ausblick

Die scharfe Aufnahme der Mikrozirkulation des uneingeschränkt schlagenden Herzens erweist sich als ein technisch schwieriges Problem. Wenn man sich nur auf ein 5 ms kurzes Fenster um den Umkehrpunkt in der Diastole beschränkt, auf den das Objektiv aktiv fokussiert wird, lässt sich mit kurzen Belichtungszeiten (<100 µs) eine scharfe Aufnahme gewährleisten. Die dazu nötige triggerbare Xe-Blitzlampe wird als nächstes implementiert. Als Doppelblitz ausgelegt ermöglicht sie auch die Bestimmung von Geschwindigkeiten der mit einem zweiten Fluoreszenzfarbstoff markierten Leukozyten. Wir hoffen so, in Kürze Messungen mit Ergebnissen zu den medizinischen Fragestellungen präsentieren zu können.

5 Literatur

1. Priebe, H. J. The aged cardiovascular risk patient. Br.J Anaesth. 85: 763-778, 2000.
2. Metnitz, P. G., C. Bartens, M. Fischer, P. Fridrich, H. Steltzer, and W. Druml. Antioxidant status in patients with acute respiratory distress syndrome. Intensive Care Med. 25: 180-185, 1999.
3. Heindl, B., F. M. Reichle, S. Zahler, P. F. Conzen, and B. F. Becker. Sevoflurane and isoflurane protect the reperfused guinea pig heart by reducing postischemic adhesion of polymorphonuclear neutrophils. Anesthesiology 91: 521-530, 1999.
4. Hernandez, L. A., M. B. Grisham, B. Twohig, K. E. Arfors, J. M. Harlan, and D. N. Granger. Role of neutrophils in ischemia-reperfusion-induced microvascular injury. Am.J Physiol 253: H699-H703, 1987.
5. Suzuki, M., W. Inauen, P. R. Kvietys, M. B. Grisham, C. Meininger, M. E. Schelling, H. J. Granger, and D. N. Granger. Superoxide mediates reperfusion-induced leukocyte-endothelial cell interactions. Am.J Physiol 257: H1740-H1745, 1989.
6. Obal, D., H. Scharbatke, H. Barthel, B. Preckel, J. Mullenheim, and W. Schlack. Cardioprotection against reperfusion injury is maximal with only two minutes of sevoflurane administration in rats. Can.J Anaesth. 50: 940-945, 2003.
7. De Hert, S. G., S. Cromheecke, P. W. ten Broecke, E. Mertens, I. G. De Blier, B. A. Stockman, I. E. Rodrigus, and P. J. Van der Linden. Effects of propofol, desflurane, and sevoflurane on recovery of myocardial function after coronary surgery in elderly high-risk patients. Anesthesiology 99: 314-323, 2003.
8. Martini, N., B. Preckel, V. Thamer, and W. Schlack. Can isoflurane mimic ischaemic preconditioning in isolated rat heart? Br.J Anaesth. 86: 269-271, 2001.
9. Conzen, P. F., S. Fischer, C. Detter, and K. Peter. Sevoflurane provides greater protection of the myocardium than propofol in patients undergoing off-pump coronary artery bypass surgery. Anesthesiology 99: 826-833, 2003.
10. Kowalski, C., S. Zahler, B. F. Becker, A. Flaucher, P. F. Conzen, E. Gerlach, and K. Peter. Halothane, isoflurane, and sevoflurane reduce postischemic adhesion of neutrophils in the coronary system. Anesthesiology 86: 188-195, 1997.
11. Hudetz, A. G. Blood flow in the cerebral capillary network: a review emphasizing observations with intravital microscopy. Microcirculation. 4: 233-252, 1997.
12. Menger, M. D., S. Richter, J. I. Yamauchi, and B. Vollmar. Intravital microscopy for the study of the microcirculation in various disease states. Ann.Acad.Med.Singapore 28: 542-556, 1999.
13. Menger, M. D., M. W. Laschke, and B. Vollmar. Viewing the microcirculation through the window: some twenty years experience with the hamster dorsal skinfold chamber. Eur Surg.Res. 34: 83-91, 2002.
14. Tillmanns, H., S. Ikeda, H. Hansen, J. S. Sarma, J. M. Fauvel, and R. J. Bing. Microcirculation in the ventricle of the dog and turtle. Circ.Res. 34: 561-569, 1974.
15. Chilian, W. M., C. L. Eastham, and M. L. Marcus. Microvascular distribution of coronary vascular resistance in beating left ventricle. Am.J Physiol 251: H779-H788, 1986.
16. Habazettl, H., P. F. Conzen, B. Vollmar, H. Baier, M. Christ, A. E. Goetz, K. Peter, and W. Brendel. Dilation of coronary microvessels by adenosine induced hypotension in dogs. Int.J Microcirc.Clin.Exp. 11: 51-65, 1992.
17. Habazettl, H., B. Vollmar, M. Christ, H. Baier, P. F. Conzen, and K. Peter. Heterogeneous microvascular coronary vasodilation by adenosine and nitroglycerin in dogs. J Appl.Physiol 76: 1951-1960, 1994.
18. Ashikawa, K., H. Kanatsuka, T. Suzuki, and T. Takishima. A new microscopic system for the continuous observation of the coronary microcirculation in the beating canine left ventricle. Microvasc.Res. 28: 387-394, 1984.
19. Kaneko, T., H. Tanaka, M. Oyamada, S. Kawata, and T. Takamatsu. Three distinct types of Ca(2+) waves in Langendorff-perfused rat heart revealed by real-time confocal microscopy. Circ.Res. 86: 1093-1099, 2000.
20. Knisley, S. B., R. K. Justice, W. Kong, and P. L. Johnson. Ratiometry of transmembrane voltage-sensitive fluorescent dye emission in hearts. Am.J Physiol Heart Circ.Physiol 279: H1421-H1433, 2000.
21. Laurita, K. R. and A. Singal. Mapping action potentials and calcium transients simultaneously from the intact heart. Am.J Physiol Heart Circ.Physiol 280: H2053-H2060, 2001.
22. Bocker, W., W. Rolf, W. U. Muller, and C. Streffer. A fast autofocus unit for fluorescence microscopy. Phys.Med.Biol. 42: 1981-1992, 1997.

Ultraschnelle holografische Gesichtsprofilvermessung mit vollautomatischer Hologrammentwicklung

Natalie Ladrière [1], Susanne Frey [1], Andrea Thelen [1], Sven Hirsch [1], Jens Bongartz [3], Dominik Giel [1], Peter Hering [1,2]

1 Forschungszentrum caesar, Ludwig-Erhard-Allee 2, D-53175 Bonn
2 Institut für Lasermedizin, Universität Düsseldorf, Universitätsstrasse 1, D-40225 Düsseldorf
3 RheinAhrCampus Remagen, Südallee 2, 53424 Remagen

Zusammenfassung

Die ultraschnelle holografische Gesichtsprofilvermessung ist ein neues Verfahren zu dreidimensionalen, hochauflösenden Erfassung des Gesichts und zur Erzeugung realistischer, texturierter Computermodelle. Diese werden in der Dokumentation und Planung von Operationen im Bereich der Mund-, Kiefer-, und Geschichtschirurgie sinnvoll eingesetzt. Als Speichermedium für das mit Hilfe eines kurzgepulsten Lasersystems erzeugten Hologramms wurde ein Silberhalogenidmaterial verwendet. Der Einsatz einer Röntgenfilmentwicklungsmaschine zur vollautomatischen Maschinenentwicklung von Amplitudenhologrammen machte eine Evaluierung geeigneter Entwickler notwendig.

1 Holografische Gesichtsprofilvermessung

Die Gestalt des Gesichts wird dadurch bestimmt, wie sich das Weichgewebe auf den Schädel anordnet. Bei chirurgischen Eingriffen zur Behebung von Defekten, wie zum Beispiel Frakturen des Schädels oder angeborene Fehlbildungen, ist neben der funktionellen Wiederherstellung auch auf ein optimales ästhetisches Resultat zu achten.

Das holografischen Prinzip kann zur Vermessung der Gesichtsoberfläche genutzt werden. Mit einer sehr kurzen Aufnahmedauer (35ns) wird ein Hologramm aufgezeichnet, welches anschließend zeitunkritisch rekonstruiert, digitalisiert und zu einem dreidimensionalen Modell verarbeitet werden kann. Verwackelungen, wie sie bei anderen Oberflächenvermessungsmethoden auftreten können, sind auf Grund der extrem kurzen Belichtungszeit ausgeschlossen. Diese Daten bilden in Kombination mit der Information über die knöcherne Struktur aus der Computertomographie eine ideale Grundlage zur Planung von opertiven Eingriffen.

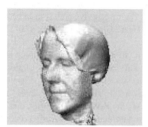

Bild 1 CT-Daten mit überlagerten holografischen Daten.

2 Holografisches Prinzip

Bei einer Fotografie wird die Intensitätsverteilung des vom Objekt gestreuten Lichtes in der fotografischen Emulsion gespeichert. Die Information der relativen Phase der einzelnen Elementarwellen, die von unterschiedlichen Punkten des Gesichts ausgehen, geht verloren. So wird von einem dreidimensionalen Objekt nur ein zweidimensionales Bild erstellt.

Bei der Holografie wird das Objekt mit kohärentem Licht, d.h. Licht mit einer festen Phasenbeziehung, beleuchtet [1]. Das vom Gesicht gestreute Licht wird auch hier in einem fotografischen Material gespeichert. Im Unterschied zur Fotografie wird diese Objektwelle jedoch auf der Fotoplatte noch mit einer ebenfalls kohärenten Referenzwelle überlagert. Beide Wellen interferieren miteinander und bilden in der Emulsion ein Interferenzmuster. Hierdurch werden sowohl die Amplitude als auch die Phase der Elementarwellen gespeichert [2]. Nach der chemischen Verarbeitung kann das so entstandenen Hologramm mit der phasenkonjugierten Referenzwelle beleuchtet werden und es entsteht ein dreidimensionales, reelles Bild des Objektes an dessen ursprünglichen Position.

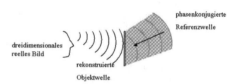

Bild 2 Rekonstruktion des reellen Bildes mit der phasenkonjugierten Referenzwelle.

2.1 Holografische Kamera

Im Forschungszentrum caesar (center of advanced european studies and reasearch) wird für die Hologrammaufnahme das Kamerasystem GP-2J der Firma Geola verwendet. Es handelt sich dabei um einen frequenzverdoppelten Nd:YLF Festkörperlaser mit einer Pulsdauer von 35ns und einer Wellenlänge von 526,5nm. Der Objektstrahl wird vor dem Verlassen der Kamera durch zwei Streuscheiben diffus gestreut, so dass einerseits eine gleichmäßige Objektbeleuchtung gewährleistet ist und andererseits die Fokussierbarkeit des kohärenten Lichts stark herabgesetzt wird. Somit ist eine Aufnahme mit geöffneten Augen möglich. Die starke Modenselektion mit anschließender Verstärkung des Laserstrahls durch einen zusätzlichen Kristall ermöglicht eine große Kohärenzlänge (> 3m) bei gleichzeitig hoher Energiedichte.

3 Amplituden- und Phasenhologramme

Nach der Hologrammaufnahme werden die Hologramme einer chemischen Verarbeitung unterzogen. Hierbei unterscheiden sich Phasenhologramme erheblich von Amplitudenhologrammen. Erstere werden nach der Entwicklung in ein Bleichbad gebracht, wobei das reduzierte Silber aus der Emulsion gelöst und in einer anschließenden Wässerung ausgewaschen wird [3]. Das Silberhalogenid bleibt in der Emulsion zurück. Amplitudenhologramme werden bereits in der Entwicklung zu einer geringeren optischen Dichte ($\approx 0,6$) entwickelt und anschließend fixiert. Durch diesen Vorgang wird das unbelichtete Silberhalogenid in lösliche Salze umgewandelt und anschließend ausgewaschen. Zurück bleibt lediglich das reduzierte, schwarz erscheinende Silber.

Bei der Rekonstruktion erfährt die Referenzwelle bei dem Durchlaufen des Phasenhologramm eine Phasenmodulation auf Grund unterschiedlicher Brechungsindizes an belichteten und unbelichteten Stellen. Bei einem Amplitudenhologramm wird die Amplitude der Referenzwelle durch Lichtabsorption an belichteten und unbelichteten Stellen variiert. Durch die Lichtabsorption besitzen Amplitudenhologramme in der Regel eine geringere Beugungseffizienz und somit ein lichtschwächeres Bild als Phasenhologramme. Durch die Körnigkeit der fotografischen Emulsion kommt es bei der Rekonstruktion zur Lichtstreuung des Rekonstruktionslichtes. Die Lichtstreuung verursacht ein Rauschen, das die eigentliche Bildinformation überlagert. Da Phasenhologramme nach der Entwicklung noch viele Silberhalogenid-Körner enthalten, tritt bei ihnen eine erhöhte Lichtstreuung und somit ein vermehrtes Rauschen auf. Im Gegensatz dazu ist die Lichtstreuung und somit auch das Rauschen bei Amplitudenhologrammen auf Grund einiger weniger Silberkörner in der Emulsion geringer. [4].

Ein weiterer Vorteil der Amplitudenhologramme ist die Verarbeitungsmöglichkeit mit kommerziellen Fixieren. Für Phasenhologramme gibt es kaum kommerzielle Schwarz/Weiß- Bleichbäder mit einer längeren Haltbarkeit die für die Verarbeitung der Hologramme in einer Entwicklungsmaschine geeignet sind.

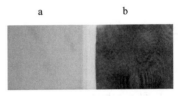

Bild 3 (a) ist ein Phasenholgramm, das wie üblich fast vollständig transparent ist, (b) ein Amplitudenhologramm mit einer optischen Dichte von etwa 0,6.

4 Automatische Entwicklung

Um die chemische Verarbeitung möglichst einfach zu gestalten, wurde im Forschungszentrum caesar eine Röntgenfilmentwicklungsmaschine für die automatische Hologrammentwicklung angeschafft. Dabei sollten insbesondere neue Entwickler getestet und optimale Verarbeitungsbedingungen für Amplitudenhologramme in der Maschine bestimmt werden. Im Vorfeld konnte durch Tests gezeigt werden, dass die Amplitudenhologramme für eine optimale Beugungseffizienz eine mittlere optische Dichte von etwa 0,6 aufweisen sollten. Für die Verarbeitung in der Maschine kann kein Glasträger verwendet werden, sie wurden in Aufnahme- und Rekonstruktionssystem durch flexible Filmträgern ersetzt. Hierfür wurde eine Vakuumansaugplatte benutzt, wie sie auch in der FH Köln bei Herrn Prof. Dr. J. Gutjahr verwendet wird. Diese ermöglicht eine schnelle und plane Fixierung des Film, welcher anschließend in der Maschine verarbeitet werden kann. Ein weiterer Vorteil des Filmträgers ist zudem der wesentlich geringere Preis.

Entwickler

Getestet wurden vier kommerzielle und ein selbst angesetzter Entwickler. Bei den kommerziellen Entwicklern handelte es sich um die Feinkornentwickler Atomal und Refinal (Agfa), den Papierent-

wickler Neutol Plus (Agfa) und den Dokumentenfilmentwickler Dokumol (Tetenal).
Die Feinkornentwickler sind speziell für die Anwendung in der Maschine entwickelt worden. Die beiden anderen kommerziellen Entwickler wurden auf Grund ihrer hohen bei der Entwicklung erreichten Gradation ausgewählt. Für die Verarbeitung in der Maschine sind sie vom Hersteller nicht vorgesehen. Bei dem Selbstansatz handelt es sich um einen Ascorbinsäureentwickler, der auch bereits in der Fachhochschule Köln für die Holografie benutzt wird.

4.2 Ergebnisse der Maschinenentwicklung

Die mit den Feinkornentwicklern erstellten Hologramme erzeugten alle ein reelles Bild, das zu kontrastarm für eine Digitalisierung war. Für die hochauflösende Holografie, die immer auch mit einer geringen Filmempfindlichkeit gekoppelt ist, sind Feinkornentwickler auf Grund besonderer Substanzen, die ein feines Korn bewirken sollen, nicht geeignet. Durch die speziellen Zusätze soll ein Zusammenballen der Körner vermieden werden. Hierzu werden die Silberhalogenid-Körner in kleinere Gebilde zerlegt, was zu einer Empfindlichkeitseinbuße führt. Silberhalogenide, die nach der Entwicklung einen Latentkeim besaßen und entwicklungsfähig waren, werden zu nicht entwickelbaren Keimen (sogenannte Subbildkörner). Es tritt dabei ein Informationsverlust auf, der zu einem schlechten Ergebnis führt.
Durch die hohe Gradation der beiden anderen kommerziellen Entwickler soll das holografische Gitter möglichst eine große Modulation aufweisen, wodurch das reelle Bild positiv beeinflusst wird. Die Ergebnisse waren für das bei caesar angewendete Auswertesystem jedoch im reellen Bild zu kontrastarm. Der selbstangesetzte Ascorbinsäure-Entwickler zeigte hinsichtlich der ausgewerteten Hologramme das beste Ergebnis und ist zudem über die Zugabe von Ascorbinsäure regenerierbar. Eine automatische Entwicklung von Folienhologrammen konnte somit erreicht werden [5].

Mobiles System

Für die klinische Anwendung wurde in Kooperation mit Prof. Dr. H. F. Zeilhofer von der Universitätsklinik für wiederherstellende Chirurgie im Universitätsspital Basel ein mobiles holografisches System installiert. Das von der Firma Geola in Zusammenarbeit mit caesar konzipierte System lässt sich innerhalb von 20 Minuten überall aufbauen und kann durch einen speziellen Verschluss-Belichtungsmechanismus bei Tageslicht betrieben werden.
Das Gerät besitzt, wie das bei caesar verwendete System, eine Pulsdauer von 35ns und eine Wellenlänge von 532nm. Auch hier beträgt die Kohärenzlänge mehr als drei Meter. Maximal kann eine Energie von 1,5 J pro Puls abgegeben werden.

Bild 4 Das komplette mobile Holografiekamera-System.

Literatur

[1] Lexikon der Physik, Band 3, Spektrum Akademischer Verlag, 1999, S. 255-257
[2] L. Bergmann, C. Schaefer: Lehrbuch der Experimental-Physik, 9. Auflage, Band 3, Walter de Gruyter, 1993, S. 428-447
[3] P. Hariharan: Optical Holography Principles, techniques, and applications, 2 Auflage, Cambridge University Press, 1996, S. 102-103
[4] J. Eichler, G. Ackermann: Holografie, Springer Verlag, 1993 S. 74 –77, 187-189
[5] N. Ladrière: „Optische und chemische Aspekte der hochauflösenden, vollautomatischen Hologrammentwicklung" Diplomarbeit, Fachhochschule Köln, Studiengang Photoingenieurwesen, Bonn, 2004, S. 58-61

Laserosteotomie mit gepulsten CO_2-Lasern

Martin Werner[1], Said Afilal[1], Mikhail Ivanenko[1], Manfred Klasing[1] und Peter Hering [1,2]
[1] Forschungszentrum caesar, Ludwig Erhard Allee 2, 53175 Bonn, www.caesar.de
[2] Institut für Lasermedizin, Universität Düsseldorf, 40001 Düsseldorf, www.ilm.uni-duesseldorf.de

Kurzfassung

Die Laserosteotomie - das Schneiden von Knochen mit Hilfe eines Lasers - ist ein geeignetes Werkzeug, um die oftmals schwer traumatisierende Bearbeitung von Knochen mit mechanischen Werkzeugen wie Sägen, Bohrern und Meißeln während medizinischer Eingriffe zu ersetzen. Die entwickelte Schneidetechnik mit kurz gepulsten CO_2-Lasern in Kombination mit einem feinen Luftwasserspray hat sich als sehr gut biologisch verträglich erwiesen und erreicht in geeigneten Anwendungen Schnittgeschwindigkeiten, die mit denen mechanischer Werkzeuge vergleichbar sind. Das physikalische Prozessmodell der Knochenablation basiert auf der schnellen Verdampfung des knocheneigenen bzw. des von außen dem Knochen mit dem Luftwasserspray zugeführten Wassers. Der so entstehende hohe Dampfdruck zerreißt den Knochen förmlich an der Schnittstelle. In der vorliegenden Arbeit wird der wasserbasierte Ablationsprozess genauer untersucht, indem Rinderknochen mit verschiedenem Wassergehalt sowie reines synthetisch hergestelltes Hydroxylapatit bei identischen Bestrahlungsparametern mit einem kurzgepulsten TEA-CO_2-Laser bestrahlt werden. Aus den gewonnenen Daten zur Ablationseffizienz sowie aus elektronenmikroskopischen Bildern der bestrahlten Proben und Röntgenfluoreszenzspektren kann das angenommene thermo-mechanische Ablationsmodell bestätigt werden.

1 Einleitung

Der Einsatz der Laserosteotomie in diversen chirurgischen Feldern bietet viele Vorteile gegenüber herkömmlichen mechanischen Schneidewerkzeugen. Durch ein kontaktloses Arbeiten wird die Traumatisierung des Gewebes reduziert. Ferner ist die Abwesenheit von metallischem Abrieb ein wichtiger Vorteil für die NMR Diagnostik gerade im neurochirurgischen Bereich. Der Einsatz des Laser erlaubt es zudem sehr schmale Schnitte von einer Breite unter 200 µm mit einer Tiefe von bis zu 9 mm in kompakten Knochen zu erzeugen. Eine besondere Bestrahlungstechnik, bei der der Schnitt künstlich auf ca. 1 mm aufgeweitet wird, ermöglicht Schnitttiefen bis zu 15 mm in kompaktem, spongiösem und mehrschichtigem Knochengewebe [1]. Der wohl größte Vorteil liegt jedoch in der Kopplung des Lasers mit einem Strahlscanner, der es ermöglicht beliebige Schnittgeometrien auszuführen [2]. Durch die Computersteuerung der Laserstrahlposition und der Schnittgeometrie ist so eine optimale Anbindung an Operationsplanungs- sowie intraoperative Navigationssysteme möglich. Die gute biologische Verträglichkeit von Laserosteotomien mit gepulsten CO_2-Lasern in Verbindung mit der Applikation eines feinen Luftwassersprays wurden ausgiebig in histologischen Studien untersucht und belegt [3,4,5]. Der in unsrer Gruppe, basierend auf einem zuverlässigen Industrie CO_2-Laser, entwickelte Prototyp osteoLAS wurde inzwischen bereits erfolgreich in vier Serien von Tierversuchen eingesetzt [6].

1.1 Knochenaufbau

Das Trockengewicht ausdifferenzierten Knochens besteht in etwa aus 30 – 40 % organischen Bestandteilen, wobei es sich hierbei im wesentlichen um Kollagen handelt, sowie aus entsprechend 60 – 70 % anorganischen Bestandteilen, wobei hierbei das mineralische Hydroxylapatit $(Ca_{10}(PO_4)_6(OH)_2)$ dominiert. Die Hydroxylapatitkristalle haben eine Länge von 20 - 100 nm und eine Breite von 1,5 – 3 nm. Die Kristallnadeln sind in mehreren Schichten entlang ihrer Längsachsen den Kollagenfasern angelagert. Im nativen Knochen ist in dieser Extrazellulärmatrix Wasser eingelagert. Das Wasser nimmt in etwa 20% des gesamten Knochenvolumens ein [7].

1.2 Thermo-mechanisches Ablationsmodell

Auf Grund der sehr hohen Absorption von Strahlung im Wellenlängenbereich zwischen 9 und 11 µm in Knochengewebe [8] eignet sich der CO_2-Laser, dessen wichtigste Emissionslinien bei 9,6 bzw. 10,6 µm liegen, hervorragend zur Knochenablation. Die hohe Gesamtabsorption im Knochengewebe kommt im wesentlichen durch die starke Absorption im Hydroxylapatit des Knochens zustande. So wird die einfallende Laserstrahlung in einer nur einige Mikrometer dicken Schicht an der Oberfläche des Knochens absorbiert und die Energie des Laserpulses dort deponiert. Die deponierte Energie wird schnell an das Wasser im Knochen übertragen. Durch den schnellen Temperaturanstieg kommt es zu einer explosionsartigen Verdampfung des im Knochen befindlichen Wassers, welches den Knochen dann lokal in Mikroexplosionen förmlich zerreißt [9,10]. Man spricht folglich von einem thermo-mechanischen Ablations-prozess. Die Temperaturen liegen hierbei deutlich unterhalb der Schmelztemperatur des mineralischen Knochenbestandteils dem Hydroxylapatit, die 1280°C beträgt [11]. Für den beschriebenen Ablationsprozess ergibt sich eine Energiedichteschwelle (Jcm^{-2}) sowie eine Intensitätsschwelle ($Jcm^{-2}s^{-1}$) die abhängig ist von der Dauer des Laserpulses [12, 13]. Daraus ergibt sich als Bedingung für eine effektive thermo-mechanische Ablation eine schnelle lokalisierte Energiedeposition, sowie ein ausreichender Wassergehalt des Knochengewebes. In dieser Arbeit soll der Unterschied im Ablauf der Ablation bei verschiedenem Wassergehalt des Knochengewebes untersucht werden, um die Richtigkeit des thermo-mecha-nischen Ablationsmodell für den Abtrag von Knochengewebe mit gepulsten CO_2-Lasern zu bestätigen.

2 Methoden

2.1 Probenmaterial

Für die Experimente wurden Stücke einer Größe von ca. 1 x 2 x 0,5 cm (Breite x Höhe x Tiefe) kompakten Knochens aus dem Rinderfemur verwendet. Es wurden insgesamt vier verschiedene Knochenpräparationen für die Experimente verwendet. Ein Teil der Knochenproben wurde über einen Zeitraum von mehren Wochen luftgetrocknet. Danach wurde eine Gruppe dieser Knochen in Ethanol über insgesamt 10 Tagen lang entwässert. Diese Gruppe wird im folgenden als Ethanolpräparation bezeichnet. Eine weitere Teil der luftgetrockneten Knochen wurde über einen Zeitraum von 4 Tagen in Aceton entwässert. Diese Gruppe wird im folgenden als Acetonpräparation bezeichnet. Beide Probengruppen wurden vor der Laserbestrahlung 24 h in einem Excikator mit Trockenmittel aufbewahrt. Die dritte Probengruppe wurde nach der Lufttrocknung nicht weiter behandelt und wird dementsprechend als Lufttrocknung bezeichnet. Als Kontrollgruppe wurde frischer Knochen verwendet, der nach der Zerteilung in Probestücke in Wasser gelagert wurde, um den natürlichen Wassergehalt des Knochens zu erhalten. Diese Gruppe wird als frischer Knochen bezeichnet. Zusätzlich wurde zum Vergleich noch eine synthetisch hergestellte Hydroxylapatit-Keramik als Probematerial für die Laserbestrahlung verwendet. Diese besitzt eine Reinheit von über 99% und enthält dementsprechend kein eingeschlossenes Wasser.

2.2 Ablationsexperimente

Für die Ablationsexperimente wurde ein kurzgepulster TEA (transversely excited at atmospheric pressure) CO_2-Laser (Edinburgh Instruments MTL3-GT) verwendet. Der Laserpuls besteht aus einem Peak mit steilen Flanken und einer Halbwertsbreite (FWHM) von ca. 50 ns sowie einem Pulsschwanz von etwa 900 µs. Die Gesamtpulsdauer liegt also im sub-µs Bereich. Die Pulsenergie betrug im Experiment. E_{puls} ca. 30 mJ und es wurde eine Pulswiederholfrequenz von 35 Hz gewählt. Die verwendete Wellenlänge lag bei 9,57 µm (Linie 9P22). Der Laserstrahl wurde mit Hilfe eines sphärischen Vorfokussierspiegels (R = 2300 mm) und einer ZnSe-Linse der Brennweite F = 127 mm auf einen Fokusdurchmesser von $2w_0$ = 260 µm fokussiert. Daraus ergibt sich eine Energiedichte im Fokus von ca. 55 J/cm^2. Die Knochen- bzw. Hydroxylapatitproben wurden jeweils mit Hilfe eines motorisierten Präzisionsverschiebetisch (Owis DC 30) transversal zum fokussierten Laserstrahl mit einer Geschwindigkeit von v = 4mm/s verschoben. Der Strahlfokus lag dabei auf der Knochenoberfläche. So wird durch die Laserablation ein Schnitt in der Probe erzeugt. Es wird ein Pulsüberlapp n definiert, das ist ein Maß dafür wie stark sich die einzelnen aufeinanderfolgenden Laserpulse, bei der Bewegung des Knochens durch den Laserstrahl auf die Probe fallen, lokal überschneiden. Es gilt $n = f w_0 / v = 1,14$. Die äquivalente Pulsanzahl N_{eq} gibt an viele Pulse effektiv an ein und dieselbe Stelle fallen bei N Durchgängen der Probe durch den Laserstrahl. Es ergibt sich N_{eq} = n N. Die Proben wurden dann jeweils N = 32, 64, 128 und teilweise 256 mal durch den Laserstrahl gefahren. Während der Bestrahlung der fünf verschieden Proben (Knochen nach Ethanol- bzw. Acetonpräparation, nach Lufttrocknung, frische Knochen sowie reines Hydroxylapatit) wurden verschiedene Einstellung des Sprays verwendet, das auf die Ablationsstelle gerichtet wurde. Bei der Bestrahlung der frischen Knochen als Kontrollgruppe wurde das herkömmliche Luftwasserspray verwendet, welches das bei der Laserbestrahlung verdampfte Wasser kontinuierlich ersetzt. Bei allen übrigen Proben wurde ein Stickstoffstrom auf die Probe gerichtet (Reinheit 5.0 aus einer Druckflasche), um zu vermeiden, dass die entwässerten Knochen während der Bestrahlung wieder Wasser aus der Atmosphäre aufnehmen. Die entwässerten Knochen aus der Ethanol- bzw. Acetonpräparation wurden erst unmittelbar vor der Bestrahlung aus dem Excikator entnommen und direkt in den Stickstoffstrom positioniert.

2.3 Optische Auswertung

Nach der Laserbestrahlung der Proben wird das - aufgrund der gaußförmigen Intensitätsverteilung annähernd keilförmige - Schnittprofil der entstanden Laserschnitte mit Hilfe eines optischen Mikroskop (Olympus SZX 12) digital aufgenommen und anschließend mit Hilfe eines Grafikprogramms vermessen. Es werden die Werte für die Maximaltiefe des Schnittes D, der Breite an der Probenoberfläche w_{top} und der Querschnittsfläche A des Schnittprofils gemessen. Aus diesen Daten kann dann unter Einbeziehung der Pulsenergie E_{puls} und der Pulswiederholfrequenz f die effektive Ablationsenergie W_{eff} (in J/mm^3) berechnet werden. Die effektive Ablationsenergie gibt an wie viel zugeführte Laserenergie inklusive aller Verlustmechanismen wie Wärmeleitung, Absorption usw. notwendig ist, um ein bestimmtes Volumen des Probenmaterials abzutragen. Die gewonnenen Daten wurden in **Abbildung 1** und **Abbildung 2** im Abschnitt 3.1 dargestellt.

2.4 Rasterelektronenmikroskopische Auswertung

Zusätzlich zur optischen Auswertung wurden die bestrahlten Proben noch unter dem Rasterelektronenmikroskop (LEO Supra 55) untersucht. Hierbei wurden Elektronenbilder des Schnittbereichs der bestrahlten Knochenproben mit Ethanol- und Acetonpräparation, der bestrahlten frischen Knochenproben, sowie des bestrahlten Hydroxylapatits erstellt. Ferner wurde ein Vergleich der Röntgenfluoreszenzspektren bestrahlten Knochens aus der Ethanolpräparation bzw. frischen Knochens im Bereich der Laserbestrahlung, sowie im Bereich der nicht bestrahlt wurde, angestellt. Die Resultate dieser Untersuchungen sind in Abschnitt 3.2 bzw. 3.3 dargestellt.

3 Ergebnisse

3.1 Physikalische Ergebnisse

Aus den Ergebnissen der optischen Auswertung kann man die beiden unten aufgeführten Graphen, die Abtragstiefe D gegen die äquivalente Pulsanzahl N_{eq} (**Abbildung 1**) sowie die effektive Ablationsenergie W_{eff} gegen die Abtragstiefe D (**Abbildung 2**), für die fünf Probematerialien auftragen.

Man kann erkennen, dass der Abtrag von frischem Knochen mit Luftwasserspray signifikant effektiver ist als der Abtrag von entwässerten Knochen. In **Abbildung 1** ist ersichtlich, dass die Abtragstiefe für entwässerte Knochen bei identischer äquivalenter Pulsanzahl niedriger liegt als für den frischen Knochen. Aus **Abbildung 2** erkennt man, dass die effektive Ablationsenergie W_{eff} bei gleicher Abtragstiefe D für entwässerte Knochen größer ist als für frische Knochen. In beiden Abbildungen steigen die jeweiligen Differenzen zwischen frischen und entwässerten Knochen mit steigender äquivalenter Pulsanzahl N_{eq} bzw. Abtragstiefe D an. Ein Grund hierfür könnte sein, dass die äußerste Knochenschicht schnell Feuchtigkeit aus der Luft aufnimmt und so Wasserreserven für den Ablationsprozess in der obersten Schicht zur Verfügung stehen. Zwischen den einzelnen Entwässerungsmodalitäten sind keine nennenswerten Unterschiede in der Ablationseffizienz sichtbar, weder in **Abbildung 1** Tiefe D gegen äquivalente Pulsanzahl N_{eq} noch in **Abbildung 2** effektive Ablationsenergie W_{eff} gegen Tiefe D. Der Abtrag von reinem Hydroxylapatit ist in etwa um einen Faktor 10 weniger effektiv als der Abtrag von frischem Knochen. Dies ist aus den beiden Graphen (**Abbildung 1 und 2**) unten ersichtlich.

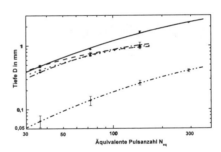

Abbildung 1: Tiefe D gegen äquivalenten Pulsanzahl N_{eq} : (———) frischer Knochen; (- - -) Knochen aus Ethanolpräparation; (· · ·) Knochen aus Acetonpräparation; (- · - · -) luftgetrockneter Knochen; (- · · - · · -) reines Hydroxylapatit;

Abbildung 2: Effektive Ablationsenergie W_{eff} gegen Tiefe D: Symbole wie in Abbildung 1;

3.2 Rasterelektronenmikroskopische Untersuchung

In **Bild 1** ist eine rasterelektronischenmikroskopische (REM) Aufnahme der Schnittfläche eines Laserschnitts (N = 32) in frischen Knochen zu sehen. Wäh-

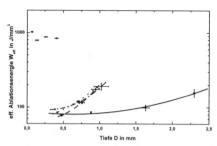

rend der Bestrahlung wurde ein Luftwasserspray benutzt. Vergleicht man diese Aufnahme mit **Bild 2** einer entsprechenden REM Aufnahme eines laserbestrahlten Knochens (N = 32) aus der Ethanolpräparation, bei dem ein Stickstoffstrom während der Bestrahlung verwendet wurde, so ist der auffälligste Unterschied das Auftreten von Schmelztropfen im Schnittbereich bei dem entwässerten Knochen.

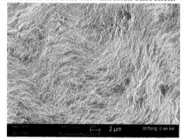

Bild 1: Laserbestrahlter frischer Knochen (N=32)

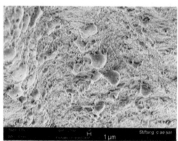

Bild 2: Laserbestrahlter entwässerter Knochen (N=32) aus Ethanolpräparation mit Schmelztropfen

Deutlich sichtbar sind erstarrte Tropfen von geschmolzenem Material. Bei dem frischen Knochen treten keine Schmelzspuren auf. Auch bei dem laserbestrahlten Hydroxylapatit (N=128) (**Bild 3**) treten deutliche Spuren von Schmelzprozessen auf. Der gesamte Schnittbereich ist von einer wieder erstarrten Schmelzschicht bedeckt.

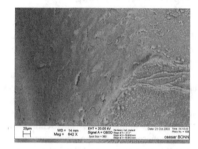

Bild 3: Laserbestrahltes Hydroxylapatit (N=128) mit Schmelzspuren

3.3 Röntgenfluoreszenzspektren

Des weiteren wurden mit Hilfe eines EDX-Detektors (Energy Dispersive X-ray Spectroscopy) (Fa. Oxford), der eine Energieauflösung von 133 eV besitzt, Röntgenfluoreszenzspektren von der unbestrahlten sowie der laserbestrahlten Oberfläche eines Knochens aus der Ethanolpräparation und eines frischen Knochens aufgenommen und miteinander verglichen. Dieses Verfahren liefert nur relative Aussagen. Aus dem Vergleich der Spektren kann jedoch auf den abgelaufenen Prozess bei der Laserbestrahlung geschlossen werden. In den Röntgenfluoreszenzspektren der Proben tauchen die charakteristischen Röntgenlinien der Elemente Kohlenstoff (C), Kalzium (Ca), Sauerstoff (O) und Phosphor (P) auf. Hydroxylapatit enthält Ca, P und O, Kollagen enthält C und O. P und Ca sind also als ein Indikator für Hydroxylapatit zu betrachten. Die beschleunigten Elektronen (20keV) dringen nur wenige µm in die Probe ein, daher erhält man durch die Röntgenfluoreszenzspektren nur Informationen über eine dünne Schicht an der Oberfläche der untersuchten Probe. Betrachtet man nun die Veränderung in den Spektren, die durch die Laserbestrahlung hervorgerufen wurde im Fall von frischen Knochen (Abb. 3 - unbestrahlte Knochenoberfläche, Abb. 4 - Laserschnittfläche), so ist eher eine Verlagerung in Richtung der niederenergetischen C bzw. Ca und O Peaks zu sehen. Das bedeutet eine Abnahme des relativen Hydroxylapatitanteils bzw. eine Zunahme des Kollagenanteils durch die Lasereinwirkung.

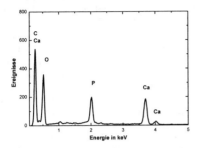

Abb. 3: Unbestrahlte Oberfläche von frischem Knochen

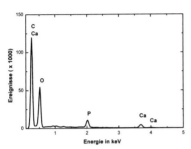

Abb. 4: Laserschnittfläche in frischem Knochen

Ganz anders ist die Veränderung des Röntgenfluoreszenzspektrums durch die Laserbestrahlung im Falle des entwässerten Knochens aus der Ethanolpräparation. (Abb. 5 - unbestrahlte Oberfläche, Abb. 6 - Laserschnittfläche; beides Knochen aus Ethanolpräparation). Hier ist eine deutliche Verschiebung hin zu den energiereichen P und Ca Peaks zu beobachten, dies weist auf einen Anstieg des Hydroxylapatitanteils bzw. eine Abnahme des Kollagenanteils im Schnittbereich hin bei einer Laserbestrahlung ohne Wasser.

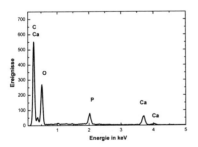

Abb. 5 unbestrahlte Oberfläche von Knochen aus Ethanolpräparation

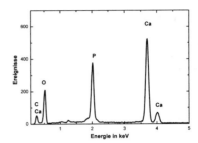

Abb. 6: Laserschnittfläche in Knochen aus Ethanolpräparation

4 Diskussion

Aus den im Abschnitt 3.1 aufgeführten Ergebnissen ist ersichtlich, dass eine Ablation von Knochenmaterial auch ohne einen gewissen Wassergehalt möglich ist, aber deutlich weniger effektiv ist. Für reines Hydroxylapatit ist die Effizienz nochmals deutlich geringer. Dieses Verhalten unterstützt, das auf explosionsartiger Wasserverdampfung beruhende, angenommene thermo-mechanische Ablationsmodell, welches in Abschnitt 1.2 vorgestellt wurde für den Knochenabtrag mit gepulsten CO_2-Lasern. Die Tatsache, dass auch ein Abtrag von entwässerten Knochen ohne Wasserzufuhr und selbst von reinem Hydroxylapatit möglich ist, lässt auf einen weniger effektiven konkurrierenden Abtragsprozess schließen, der auf dem Schmelzen und schließlich dem Verdampfen des Materials beruht und bei wesentlich höheren Temperaturen abläuft. Diese Vorstellung unterstützen die elektronenmikroskopischen Bilder die in Abschnitt 3.2 gezeigt sind. Bei der Laserbestrahlung von entwässertem Knochen und reinem Hydroxylapatit treten deutliche Schmelzspuren auf, d.h. es wird Material geschmolzen und dieses erstarrt dann wieder und ist in den REM Bildern sichtbar. Diese Schmelzspuren treten im Laserschnittbereich bei frischem Knochen mit ausreichendem Wassergehalt nicht auf. Die in Abschnitt 3.3 vorgestellten Röntgenfluoreszenz-spektren lassen darauf schließen, dass im Falle der Laserbestrahlung von entwässertem Knochen geschmolzenes Hydroxylapatit sich an der Oberfläche des Schnittspaltes nach der Ablation wieder niederschlägt und erstarrt und so den relativen Anstieg der Häufigkeit von Phosphor und Kalzium im Schnittbereich des Knochen aus der Ethanolpräparation erklärt. Dieses Verhalten erfordert das Erreichen von Temperaturen über 1280°C, um das Hydroxylapatit zu schmelzen. Bei der Laserknochenbearbeitung mit ausreichendem Wassergehalt werden dies Temperaturen bei weitem nicht erreicht.

Danksagung

Vielen Dank an Frau Dr. Barbara Wehner von der REM-Gruppe des Forschungszentrum caesar für die freundliche Unterstützung bei den rasterelektronenmikroskopischen Untersuchungen der Proben.

Literatur

[1] Afilal S., Ivanenko M., Werner M., Hering P.: Osteotomie mit 80 µs-Laserpulsen, Fortschritt-Berichte VDI,17. Biotechnik/Medizintechnik **231** (2003), S. 164-169
[2] Hering P., Ivanenko M., Hartmann M. : 3-D Beam Scanning Head Improves Bone Surgery, Europhotonics **7** (2002), S.32-33
[3] Ivanenko M., EyrichG., Bruder E., Hering P. : In vitro incision of bone tissue with a Q-switch CO_2 laser. Histologicalexamination, Lasers in Life Sciences, **9** (2000), S 171-179
[4] Ivanenko M., Fahimi-Weber S., Mitra T., Wierich W., Hering P. :Bone Tissue Ablation with sub-µs Pulses of a Q-switch CO_2 Laser: Histological Examination of Thermal Side-Effects, Lasers Med Sci **17** (2002), S. 258-264
[5] Frentzen M., Götz W., Ivanenko M., Afilal S., Werner M., Hering P. : Osteotomy with 80-µs CO_2 laser pulses – histological results, Lasers Med Sci **18** (2003), S. 119-124
[6] Hänisch: Osteogenese nach Laserosteotomie mit einem CO2-Laser im vergleich zur Osteotomie mit konventioneller Sägetechnik – eine tierexperimentelle Studie -, Dissertation Tierärztliche Fakultät, Ludwig- Maximilians-Universität München, (2004)
[7] Tillmann B., Töndury G.: Anatomie des Menschen, Bd. I Bewegungsapparat, Kapitel 2 (Rauber/Kopsch), Georg Thieme Verlag, Stuttgart, 1998
[8] Kar H., Ringelhan H.: Grundlagen und Technik der Photoablation, Band 6 Advances in Laser Medicine, Hrsg: Müller, Berlien, Ecomed Landsberg/Lech, 1992
[9] Forrer M., Frenz M., Romano V., Altermatt H.J., Weber H.P., Silenok A., Istomyn M., Konov V.I. : Bone-Ablation Mechanism Using CO_2 Lasers of Different Pulse Duration and Wavelength, Appl. Phys. B **56** (1993), S.104-112
[10] Hibst R. : Technik, Wirkungsweise und medizinische Anwendung von Holmium- und erbium-Lasern, Kapitel 3, Fortschritte in der Lasemedizin, Hrsg: Müller, Berlien, Ecomed Landsberg/Lech, 1996
[11] Corcia J.T., Moody W.E.: Thermal Analyisis of Human Dental Enamel, Journal of Dental Research **53** (1974), S. 571-580
[12] Mitra T.: Ablation biologischen Hartgewebes mit gepulsten IR-Lasern, Dissertation, Math.-Nat. Fakultät, Heinrich-Heine-Universität Düsseldorf, 2002, (http://diss.ub.uni-duesseldorf.de/ebib/diss/show?dissid=220)
[13] Afilal S.: Ablationsmechanismen von biologischem Hartgewebe bei Bestrahlung mit kurzgepulsten CO_2-Lasern, Dissertation, Math.-Nat. Fakultät, Heinrich-Heine-Universität Düsseldorf, 2004, (http://diss.ub.uni-duesseldorf.de/ebib/diss/show?dissid=905)

Acoustic monitoring of bone ablation using pulsed CO_2 lasers

Antje Rätzer-Scheibe[1], Manfred Klasing[1], Martin Werner[1], Mikhail Ivanenko[1], Peter Hering[1,2]

[1]*caesar (center of advanced european studies and research), Ludwig-Erhard-Allee 2, 53175 Bonn, Germany*
[2]*Institute of Laser Medicine, University of Düsseldorf, Universitätsstrasse 1, 40225 Düsseldorf, Germany*

Abstract

Laser osteotomy offers remarkable advantages, e.g. free cut geometry, over the conventional mechanical saw. Unlike the saw, however, the laser lacks haptic feedback during cutting. Based on ablation noise analysis, we are developing an acoustic-feedback system for laser osteotomy to obtain in situ information on the ablation within the tissue. We used a pulsed TEA CO_2 laser (wavelength 9.57µm, pulse length 1µs/50ns FWHM, pulse energy 25 mJ) and piezoelectric transducers for sound detection. Various bone specimens as well as reference materials were studied. For the determination of the ablation crater depth, we analyzed the time delay between the laser-induced acoustic signal from the surface of the specimen and the bottom of the ablation crater. The possibility of controlling the cut depth in material with known acoustical properties with high precision is demonstrated. For acoustic tissue differentiation, we analyzed acoustic spectra initiated by laser pulses in different materials. The spectra show specific material-based features. This will prompt surgeons with information about the transition from compact bone to other materials, e.g. soft tissue.

1 Introduction

Osteotomie, which is the transection of bone, nowadays still is performed with drills, oscillation saws and chisels. This comprises the danger of vibrational trauma and deposition of metal impurities within the bone tissue. The implementation of a non-contact laser beam treatment circumvents this risk and, due to the small spot size [1], has the additional advantage of not beeing limited in incision shape. Initial investigations using continuous wave lasers featured thermal damage of bone tissue near the incision area [2]. By the application of short-pulse lasers this downside was overcome [3,4,5]. An adequate selection of laser parameters aligned to the properties of bone tissue constituents lead to a high efficiency of the cutting process [6,7]. CO_2-lasers are very promising for this application due to their suitable wavelength and beam shape [8]. The wavelength of the used laser mode (9.57 µm) is absorbed strongly by the mineral hydroxyapatite, which is one of the main constituents of bone, and water, which is the driving force of the ablation process [9]. In this environment the ablation depth is about 10 µm, resulting in little material removal per pulse and thus good control over the osteotomie procedure.

The ablation process is accompanied by audible and ultrasonic noise generation. The sound signal corresponds to a pressure wave which is generated by thermoelastic stresses and ejection of matter (**Fig 1**) [10,11].

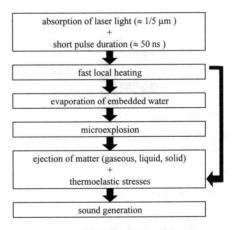

Figure 1 **Principle of sound generation by laser ablation**

To achieve sufficient accuracy in the surgery, a good anatomical knowledge from pre- and intraoperative data is needed. As the IR laser lacks haptic and optic feedback, the sound signal offers a potential control mechanism that is independent of conventional imagination-based data [11,12,13,14]. This way damage of surrounding and enveloped tissue can be prevented even if preoperative data exhibits inaccuracies.

In this paper we investigate the ablation noise of a CO_2 laser-based system that is currently developed at the caesar institute for osteotomy and other applications [15]. A comparative analysis of ablation features and features of the associated air- and structure-borne sound signals is carried out. In particular, the sound propagation delay is used for acoustic depth measurement whereas the spectral analysis of acoustic signals induced in different materials serves for tissue discrimination

2 Experimental

The schematic diagram of the experimental setup is depicted in **Fig. 2**. A TEA-CO_2 laser (λ=9.57 µm) (MTL-3 Edinburgh Instruments Ltd) was used to deliver an energy of about 30 mJ at a pulse duration of 50 ns (FWHM). The beam was guided through a set of mirrors onto a focusing unit, consisting of a focusing mirror (f=11500 mm) and lens (f=127 mm). The resulting spot size on the surface of the specimen had a diameter of about 300 µm. An energy density of 42 J/cm² and/or peak intensity of 840 MW/cm² was thus realized.

Various bone specimens as well as reference materials were studied.

For analysis of structure-borne sound, a piezoelectric transducer manufactured of a PVDF foil was mounted on the backplane of the sample with ultrasound gel. For analysis of air borne sound, a commercial transducer based on a piezoelectric ceramic (PCB-M132A31) was used. It was positioned in front of the specimen at a distance of 40 mm and an inclination angle of 20° with respect to the incident direction of the laser beam.

The recorded audio signals were stored in a digital sampling oscilloscope (Tektronix 3052) and transferred to a PC for Fourier transformation. In this manner, two signals were obtained: the time domain signal and the frequency domain signal.

The study was divided into two parts. In the first part, the relation between the sound propagation delay and the depth of ablation holes was studied. In the second part the relation of the spectra of the acoustic signal with the ablated material was investigated.

Evaluation of the ablation crater geometry was achieved with polymathylacrylate (PMMA) samples in conjunction with optical microscopy.

The depth D and the geometry of the PMMA craters were analyzed using an imaging software on the optical image captured by a CCD camera that is coupled to the microscope (Olympus SZX12).

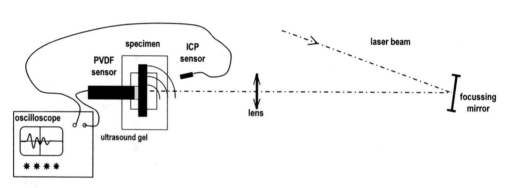

Figure 2 **Experimental setup**

3 Results and Discussion

3.1 Incision Depth Control

A typical oscillogram of an acoustic signal captured by the structure borne transducer is represented in **Fig 3**. An electro-magnetic interference (EMI) signal followed by the actual sound signal is seen. The EMI signal occurs in the moment of pulse emission. It originates from the pulse generator of the laser.

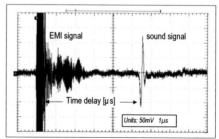

Figure 3 **Oscillogram of structure-borne sound signal**

As speed of light is negligible compared to speed of sound, the moment of pulse emission can be assumed to coincide with the beginning of the ablation process. Therefore, the time delay between the EMI and the sound signal is a measure of the distance traversed by the sound waves from the site of generation to the site of detection. Regarding the structure-borne transducer mounted on a material of known speed of sound, the distance to the ablation site can be determined. Conversely, with knowledge of ablation site position relative to the transducer, the speed of sound of the traversed material can be derived. We used the latter approach in order to evaluate the accuracy of the propagation delay method.

Initial measurements were implemented on reference samples prepared by drilling a series of holes with varying depth. A laser pulse was transmitted on the bottom of each hole. Special attention was paid to an accurate alignment of the sensor with the laser beam for each hole. As the sample thickness behind each hole was known from the drilling process, the speed of sound was derived. By this means the speed of sound for polycarbonate (PC) was determined to be 2039±44 m/sec which is close to the reference data of 2220 m/sec from the manufacturer (**Fig. 4**). For a similarly prepared sample of cattle corticalis bone we obtained 2853±92 m/sec which lies within the range typically found in literature.

speed of sound (cattle bone):

2853 ± 92   m/sec (Exp.)

ca. 2600-2900   m/sec (Lit.)

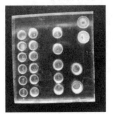

speed of sound (PC=Polycarbonate) :

2039 ± 44   m/sec (Exp.)

2220   m/sec (Lit.)

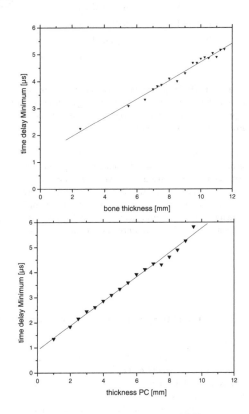

Figure 4 **Left: Image of the specimen. Right: Time delay versus thickness for cattle bone and PC.**

We also utilized the ablation process itself for the preparation of holes in order to confirm that the method can operate in situ during the cutting process. Here we used transparent Polymethylacrylate (PMMA) for simplified crater depth analysis. With varying number of laser pulses, craters of different depth were produced. For each crater we recorded the acoustic signal induced by the first and the last laser pulse of the drilling process. The difference between their time delays is a measure of the crater depth. Again the speed of sound was verified. The range of the obtained speed of sound data 2351±125 m/sec corresponds well with the value of 2240 m/sec quoted by the manufacturer (**Fig. 5**).

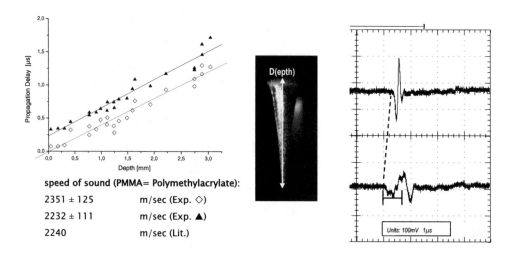

Figure 5 Left: Propagation delay versus crater depth. Upper line, difference between last and first propagation delay time. Lower line, propagation delay time evaluated from a single pulse, see text. Center: Image of an ablation crater. Right: Acoustic signal generated by the first and the last pulse in the upper and lower curve, respectively.

Thorough analysis revealed that the acoustic signal generated at the bottom of a crater is superimposed with contributions from the total crater surface and also the front plane of the sample. Thus information from the surface as well as from the crater bottom is contained in the overall acoustic signal. This is reflected by the fact that only the initial fall of the acoustic signal is moving towards shorter propagation delay time while the maximum remains fixed in time. This enables us to derive the sound propagation between crater bottom and surface from a single ablation pulse-induced signal. The speed of sound obtained that way is 2232±111 m/sec (Fig. 5).

3.2 Material-Specific Spectral Features of Ablation Noise

Spectral analysis of the ablation noise from various materials was performed with an airborne ultrasonic sensor, see **figure 6**. There is only little spectral content at frequencies above 50 kHz for the non-biological specimens aluminum (Al) and PMMA compared to the biological specimens cattle bone and flesh. It is noted that only the biological specimens contain significant amounts of water.

It could thus be argued, that the microexplosion resulting from the evaporation of water accounts for the high frequencies in the biological specimens. Moreover, flesh and bone are easily distinguished by their characteristic spectral features in the frequency range up to 200 kHz. We assume that this is due to the larger relative content of water and the lack of hydroxyapatite in flesh compared to bone.

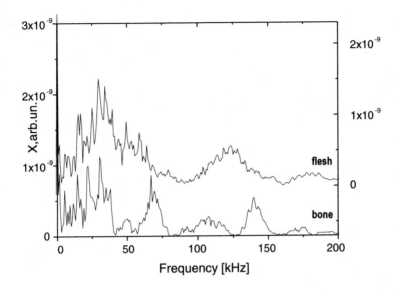

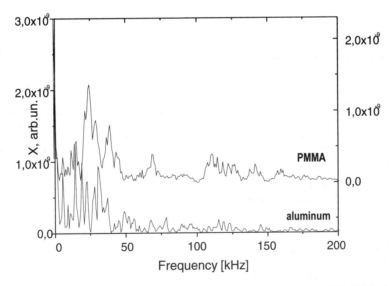

Figure 6 **Fourier spectra of acoustic signals generated during material ablation. Left: Biological tissues, cattle flesh upper, cattle bone lower spectrum. Right: Non-biological material, PMMA upper, Al lower spectrum.**

4 Conclusion

It was shown that propagation delay measurements enable the in situ control of cutting depth of a pulse laser system in osteotomy applications. In addition it was demonstrated that characteristic spectral features of the ablation noise allow accurate discrimination of flesh and bone. However this approach will acquire reliable real time software-based analysis of the spectrum as for example demonstrated by Boesnach et al. [12]. An acoustic feedback system will eventually provide online control over the depth of the incision and ancillary recognition of the tissue exposed to the laser.

5 Literature

[1] Afilal S., Ivanenko M., Werner M., Hering P.: Osteotomie mit 80 µs-Laserpulsen. Fortschritt-Berichte VDI,17. Biotechnik/Medizintechnik **231** (2003), S. 164-169

[2] Horch, H. H.: Laser-Osteotomie und Anwendungsmöglichkeiten des Lasers in der oralen Weichteilchirurgie. Westdeutsche Kieferklinik der Universität Düsseldorf, GSF-Bericht, Kap. 6, S. 78 - 194, 1978

[3] Ivanenko M., EyrichG., Bruder E., Hering P.: In vitro incision of bone tissue with a Q-switch CO_2 laser: Histological examination. Lasers in Life Sciences, **9** (2000), pp. 171-179

[4] Ivanenko M., Fahimi-Weber S., Mitra T., Wierich W., Hering P. :Bone Tissue Ablation with sub-µs Pulses of a Q-switch CO_2 Laser: Histological Examination of Thermal Side-Effects. Lasers Med Sci **17** (2002), pp. 258-264

[5] Frentzen M., Götz W., Ivanenko M., Afilal S., Werner M., Hering P.: Osteotomy with 80-µs CO_2 laser pulses – histological results. Lasers Med Sci **18** (2003), pp. 119-124

[6] Mitra T.: Ablation biologischen Hartgewebes mit gepulsten IR-Lasern. PhD thesis, Math.-Nat. Fakultät, Heinrich-Heine-Universität Düsseldorf, 2002, (http://diss.ub.uni-duesseldorf.de /ebib/diss/liste_jahr?fak=2&year=2002)

[7] Afilal S.: Ablationsmechanismen von biologischem Hartgewebe bei Bestrahlung mit kurzgepulsten CO_2-Lasern. PhD thesis, Math.-Nat. Fakultät, Heinrich-Heine-Universität Düsseldorf, 2004, (http://diss.ub.uni-duesseldorf.de /ebib/diss/liste_jahr?fak=2&year=2004)

[8] Kar H., Ringelhan H.: Grundlagen und Technik der Photoablation, Fortschritte in der Lasermedizin. Hrsg.: G.J. Müller, H. P. Berlien, ecomed Bd. 6, Seite: 42 - 164, 1992

[9] Gutknecht N.: Lasertherapie in der zahnärztlichen Praxis, Die Anwendung der unterschiedlichen Lasertypen in ihren jeweiligen Spezialgebieten. Quintessenz Verlags-GmbH, 1999

[10] Altshuler G. B., Belikov A. V., Boiko K. N., Erofeev A. V. and Vitiaz I.V.: Acoustic response of hard dental tissues to pulsed laser action. In G. B. Altshuler und R. Hibst (Hrsg.), Dental Applications of Lasers, Vol. 2080 from Proc. SPIE, 199

[11] Nahen K.: Akustische Online-Kontrolle der Infrarot-Photoablation biologischer Gewebe. Dissertation, Technisch-Naturwissenschaftliche Fakultät, Medizinische Universität zu Lübeck, 2001

[12] Boesnach I., Hahn M., Muldenhauer J., Beth T., Spetzger U.: Analysis of Drill Sound in Spine Surgery, Medical Robotics, Navigation and Visualization (MNRV), Remagen, 11.-12. März 2004.

[13] Tangermann K. and Uller J.: Application of an Er:YAG laser in oral and craniomaxillofacila surgery. LaserOpto, 33(1):40-45, 2001

[14] Specht H., Bende T., Jean B. and Fruehauf W.: Noncontact photoacoustic spectroscopy (NCPAS) for photoablation control: data acquisition and analysis using cluster analysis. In P. O. Rol, K. M. Joos, F. Manns, B. E. Stuck und M. Belkin (Hrsg.), Ophthalmic Technologies IX, Vol. 3591 aus Proc. SPIE

Lasertechnik
Laser Technology

Prozessüberwachung beim Laserreinigen und Laserschichtabtrag durch Anwendung der Plasmaanalyse mittels "low resolution" Spektrometer

Marco Lentjes, Laserzentrum FH Münster, Steinfurt, Deutschland
Klaus Dickmann, Laserzentrum FH Münster, Steinfurt, Deutschland
Johan Meijer, Universität Twente, Enschede, Niederlande

Kurzfassung

Bei der Laserreinigung und dem Laserabtrag an Mehrschichtsystemen ist es von großer Bedeutung, eine Beschädigung der zu erhaltenden Schicht zu vermeiden. Ein Verfahren zur Überwachung des Laserabtragsprozesses ist die "Laser Induzierte Plasma Spektroskopie" (LIPS) [1,2,3,4]. Unterschiede im Schichtaufbau bewirken Veränderungen im Spektrum und der Intensität des Plasmas. Ein Nachteil gebräuchlicher LIPS-Systeme sind die hohen Investitionskosten und die komplexe Handhabung. Neueste Entwicklungen in der Spektrometer-Technologie bieten inzwischen sehr kompakte Systeme mit einfacher Handhabung und niedrigen Investitionskosten [5,6]. Im Laserzentrum der FH Münster wurde ein "low cost"-LIPS-System zur Laserreinigung/-abtrag in einem Excimer-Laser und Nd:YAG-Laser (mit Frequenzvervielfachung) integriert, um den Abtragprozess zu überwachen und ggf. zu regeln. Die Schichterkennung funktioniert über lineare Korrelation mit vorher aufgenommenem Referenzspektrum [7,8,9]. Wenn der Korrelationskoeffizient sich einem vorher festgelegten Wert nähert, wird der Abtrags-/Reinigungsprozess gestoppt. Erste Tests haben positive Ergebnisse gezeigt und werden zusammen mit dem Verfahren in diesem Artikel beschrieben.

1 Einleitung

Im Allgemeinen findet das Abtragen von Schichten mit gepulster Laserstrahlung bei hohen Intensitäten statt; hieraus resultiert ein Plasma über der bestrahlten Fläche. Die Eigenschaften des Plasmas sind vornehmlich abhängig von der Laserwellenlänge, Laserintensität und chemischen Zusammensetzung der abgetragenen Materialien. Die Plasmastrahlung wird charakterisiert durch Dauer, Spektrum und Intensität.
Spektroskopische Untersuchungen von laserinduzierten Plasmen werden in vielen verschiedenen Bereichen angewendet. Die bekannteste ist die quantitative und qualitative Elementerkennung zur Bestimmung der chemischen Zusammensetzung [10]. Die besonderen Vorteile von LIPS liegen in seiner Schnelligkeit und der Möglichkeit, Materialien bei allen Aggregatzuständen analysieren zu können.
LIPS als Prozesskontrolle bei der Laserreinigung von Kulturgütern ist schon in vielen Applikationen erfolgreich getestet worden. Die Schichtidentifikation erfolgt hier mittels Elementerkennung mit konventionellem, hochauflösendem Spektrometer, basierend auf "intensified CCD-arrays". Diese besitzen ein großes Signal-/ Rausch-Verhältnis und eine hohe Empfindlichkeit. Nachteilig sind die hohen Investitionskosten, Komplexität und die aufwendige Handhabung des Systems.
In dieser Studie wird die Plasmastrahlung verwendet zur Online-Prozessüberwachung bei der Laserreinigung und dem Laserabtrag mittels eines "low resolution" Spektrometer "mit non-intensified CCD-array". Dieses Spektrometer ist gekennzeichnet durch seine große spektrale Bandbreite, niedrigere Auflösung und Empfindlichkeit. Die Investitionskosten und die Komplexität sind deutlich niedriger als bei gebräuchlichen LIPS-Systemen.
Durch diese Eigenschaften eignet sich dieses System nicht zur Bestimmung einzelner Elemente. Es hat sich jedoch gezeigt, dass für die Schichterkennung dieses nicht zwingend nötig ist [11,12]. Jede Schicht hat ihren eigenen spektralen "fingerprint" und unterscheidet sich hiermit von den anderen Schichten in einem Mehrschichtsystem. Mittels linearer Korrelation ist es trotz Spektrometer mit geringerer Auflösung möglich, einzelne Schichten anhand von Referenzspektren zu detektieren.

2 Laserinduziertes Plasma

Für das Induzieren von Plasmen bei LIPS-Experimenten werden gepulste Laser angewendet. Eine Vielzahl unterschiedlicher Prozesse findet statt, abhängig von verschiedenen Parametern wie: Laserwellenlänge, Pulsdauer, Pulsenergie, Spotgröße und elementare Zusammensetzung der Probe. Das physikalische Prinzip kann unterteilt werden in drei Zeitabschnitte (**Bild 1**).
Im ersten Zeitabschnitt (a) erwärmt und verdampft der Laser eine dünne Schicht der Probe durch linearer oder nicht-linearer Absorption der Laserstrahlung.

Die einzelnen freien Elektronen in dem Dampf absorbieren weiterhin Photonen und erwärmen sich durch inverse Bremsstrahlung. Es bildet sich eine Elektronenwolke, die in einem Lawineneffekt ionisiert wird. So entsteht Plasma, ein Mix von Atomen, Ionen und Elektronen.

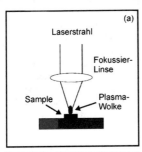

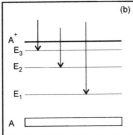

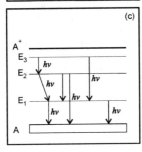

Bild 1 Zeitlicher Verlauf von laserinduzierten Plasmen (a) Plasma-Erzeugung (b) Breitband-Weiß-Licht-Emission (c) Atom-Linien-Emission [13]

Der zweite Zeitabschnitt (b: Dauer von einigen Hundert Nanosekunden) ist gekennzeichnet durch eine kontinuierliche, breitbandige Weißlicht-Strahlung, die verursacht wird durch die Bremsstrahlung der freien Elektronen und Elektron-Ion-Rekombination. Einige Linien ragen aus dem Strahlungsspektrum hervor und können als elementare Linien identifiziert werden.

Der letzte Zeitabschnitt (c: Dauer von einigen Mikrosekunden) wird dominiert durch diskrete Spektrallinien, verursacht durch Rückfall im Energie-Niveau (z. B. von $E_3 \rightarrow E_2$ unter Ausstrahlung von $h \cdot v$), korrespondierend zu den Elementen in der Probe. Die Intensität der Linien ist dabei proportional zu der Konzentration der Elemente in der Probe. Dieser Zeitabschnitt wird verwendet für konventionale LIPS-Messungen.

Dass Material für die LIPS-Analyse verdampft werden muss, ist in diesem Fall kein Nachteil, da lediglich die Plasma-Strahlung dieses Prozesses genutzt wird. Durch die Anwendung von optimalen Laserparametern beim Abtragen und Reinigen kann die Laserenergie nicht beliebig erhöht werden, um eine helle Plasma-Strahlung zu erzeugen. Im Fall von geringer Plasma-Strahlung wird das Detektieren mit einfachem Spektrometer kritisch oder evt. unmöglich. Dadurch, dass die Schichterkennung bei einfachem Spektrometer nicht via Elementerkennung geschieht, ist man nicht beschränkt auf den Zeitabschnitt c, sondern kann auch Zeitabschnitt b nutzen. Der Vorteil von Abschnitt b ist die höhere Plasma-Intensität als in Abschnitt c. Der Nachteil ist die geringere Differenz zwischen Spektren von unterschiedlichen Schichten durch die starke, kontinuierliche Weißlicht-Strahlung.

3 Integration des Spektrometers in zwei Lasersysteme (IR,UV)

Zum Testen der Möglichkeiten des Miniatur-Spektrometers als Überwachungssensor während der Laserreinigung und des Abtrags ist das Spektrometer-System implementiert in zwei Lasersyteme, ein KrF-Excimer-Laser und ein Nd:YAG-Reinigungssystem.

Das Spektrometer ist ein kundenspezifisch konfiguriertes „Ocean Optics HR2000 UV-VIS" Miniatur Faser Optik Spektrometer. Zur Einkopplung der Plasma-Strahlung in das Spektrometer wird eine 2 m lange 600 µm Glasfaser benutzt. Die Plasma-Strahlung wird mittels eines Kollimators in die Glasfaser fokussiert. Die Liniendichte des Gitters (300 Linien/mm) und der Eintrittsspalt ergeben eine Spektralbandbreite von 200 bis 1100 nm mit einer Auflösung von 2 nm. Das Spektrometer ist außerdem ausgestattet mit einer LIPS-Erweiterung, die es ermöglicht, eine Trigger-Verzögerungszeit von -125 µs bis +125 µs einzustellen. Ein Computerprogramm steuert Laser und Spektrometer. Der elektronische „Shutter" hat eine fixe Belichtungszeit von 2 ms.

Der erste experimentelle Aufbau hat als Laserstrahlquelle einen KrF-Excimer-Laser (Lambda Physics LPX 305i, 248 nm, 1 – 50 Hz, 20 – 40 ns). Der optische Aufbau ist ein Standard-Aufbau, der beim spezifischen Excimer-Masken-Abbildungsverfahren eingesetzt wird [14] (**Bild 2**). Die Kollimatorlinse und Glasfaser sind hinter dem letzten dielektrischen Spiegel angebracht, was sich in der Praxis als vorteilhaft erweist.

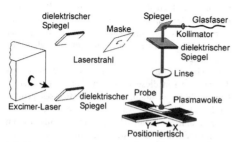

Bild 2 Glasfaser und Kollimator implementiert im 248 nm Excimer-Masken-Abbildungsverfahren-Aufbau

In dem zweiten Aufbau sind Kollimator und Glasfaser integriert in das Handstück des Nd:YAG-Reinigungslasers mit beweglichem Spiegelarm (**Bild 3**). Der Reinigungslaser Typ SAGA 220/10 von Thales ist ein blitzlampengepumpter, gütegeschalteter Nd:YAG-Laser mit Möglichkeit zur Frequenz-Verdopplung, -Verdreifachung und Vervierfachung (1064 nm, 532 nm, 355 nm, 266 nm, 1 - 10 Hz, 7 ns), basierend auf dem Oszillator-/Verstärker Prinzip. Kollimator und Glasfaser sind in dem letzten Spiegelhalter hinter dem 45°-Spiegel angeordnet.

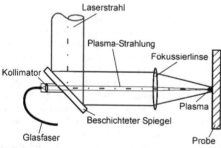

Bild 3 Strahlengang im Handstück des beweglichen Spiegelarms (Nd:YAG-Reinigungslaser); Kollimator und Glasfaser sind im letzten Spiegelhalter vor der Fokussierlinse integriert

4 Korrelationskoeffizient zum Erkennen von Schichten

Der Unterschied eines konventionellen hochauflösenden Spektrometers zu einem Miniatur-Spektrometer mit einfachem (non-intensified) CCD-array (niedrige Auflösung) liegt in der Information, die aus dem Plasmaspektrum herausgezogen werden kann. Beim hochauflösenden Spektrometer ist die Schichterkennung via Elementerkennung möglich. Da diese Methode aufgrund der niedrigen Auflösung beim Miniatur-Spektrometer sehr schwierig ist, muss hier eine andere Technik angewendet werden, um Schichten via Plasma-Spektroskopie voneinander zu unterscheiden. Spektren verschiedener Schichten sind einzigartige "fingerprints", die sich in der spektralen Intensität und totalen Intensität unterscheiden (s. **Bild 8a bis d**). Es gibt mehrere Möglichkeiten diese "fingerprints" durch Vergleich mit Referenzspektren einer Schicht zuzuweisen.

In dieser Studie wurde das Erkennen von Schichten via lineare Korrelation mit vorher aufgenommenen Referenzspektren durchgeführt.

Der lineare Korrelationskoeffizient ist ein dimensionsloser Index mit dem Wertebereich $-1 \leq r \geq 1$ und ein Maß dafür, inwieweit zwischen zwei Datensätzen eine lineare Abhängigkeit besteht. Der lineare Korrelationskoeffizient r wird wie folgt ausgedrückt:

$$r = \frac{\sum_{i=1}^{n}(x_i - \bar{x})(y_i - \bar{y})}{\sqrt{\sum_{i=1}^{n}(x_i - \bar{x})^2}\sqrt{\sum_{i=1}^{n}(y_i - \bar{y})^2}} \quad (1)$$

Hierin ist $\bar{x}$ der Mittelwert von allen x_i-Werten und $\bar{y}$ der Mittelwert von allen y_i-Werten. Ein Datensatz besteht maximal aus 2048 x- und y-Werten, wenn zwei Spektren miteinander korreliert werden (das angewendete Spektrometer verfügt über einen CCD-array mit 2048 Pixel). Die Intensität der Pixel des Referenzspektrums formen den x-Datensatz, und die Intensität der Pixel von dem neu gemessenen Spektrums formen den y-Datensatz. Jedes Pixel resultiert hierdurch in einen x_i-Wert (Referenzspektrum) und in einen y_i-Wert (neu gemessenes Spektrum), der in eine Punkt-Grafik übereinander aufgetragen werden kann (**Bild 4**).

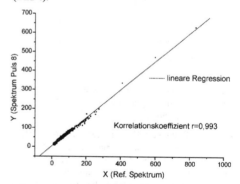

Bild 4 Korrelation vom Spektrum des 8. Abtragspulses mit einem vorher gespeicherten Referenzspektrum. Der Wert des Korrelations-Koeffizienten von $r = 0,993$ zeigt für diesen Fall einen nahezu idealen linearen Zusammenhang des x_i, y_i-Werte

Die Grafik in **Bild 4**, die die Korrelation zwischen einem Abtragspuls und einem vorher gespeicherten Referenzspektrum darstellt, hat einen Korrelationskoeffizient, der dem Wert 1 sehr nahe kommt. Es ist erkennbar, dass beide Spektren fast identisch sind, woraus in den meisten Fällen die Schlussfolgerung gezogen werden kann, dass beide Spektren zur gleichen Schicht (Material) gehören.

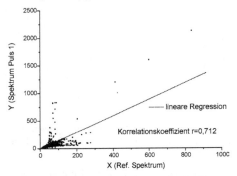

Bild 5 Korrelation vom Spektrum des 1. Abtragspulses mit einem vorher gespeicherten Referenzspektrum. Geringer linearer Zusammenhang zwischen den unterschiedlichen (x_i, y_i) Werten. Korrelationskoeffizient $r = 0{,}712$

Bild 5 stellt eine Grafik dar mit einer geringeren linearen Abhängigkeit der einzelnen Punkte als in Bild 4. Das Spektrum des ersten Laserpulses wird korreliert mit dem Referenzspektrum einer abweichenden Schicht. Diese beiden Spektren haben eine geringe Korrelation, die in der Grafik durch eine "chaotische Punkteverteilung" erkennbar ist und einen niedrigen Korrelationskoeffizient, $r = 0{,}712$ ergibt.

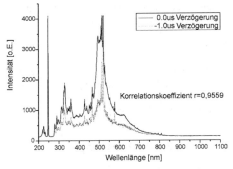

Bild 6 LIPS-Spektren von reinem Kupfer, gemessen mit unterschiedlicher Zeitverzögerung. (plasmainduziert mittels Excimer-Laser, 248 nm, 0,13 GW/cm², Spotgröße 2,6 mm²)

Ein großer Vorteil dieser Korrelationsmethode ist das Herausfiltern von Plasma-Intensitäts-Schwankungen, verursacht durch z.B. Variation der Laserpulsparameter und Variationen in der Verzögerungszeit. **Bild 6** zeigt zwei Kupfer-Spektren gemessen mit unterschiedlicher Zeitverzögerung. Trotz unterschiedlicher Intensitäten sind sich beide Spektren ähnlich im Linienverlauf, woraus sich ein Korrelationskoeffizient von $r = 0{,}9559$ ergibt.

5 Experimentelle Resultate

Zum Testen des Korrelationskoeffizienten als Entscheidungsfaktor für kontrollierte Laserreinigung oder -Abtrag, wurden Versuche an einem definierten Mehrschichtaufbau durchgeführt. Das hierfür angewendete Mehrschichtsystem ist eine „3M Laserbeschriftbare Folie" Typennummer 7848 (Schichtaufbau dargestellt in **Bild 7**).

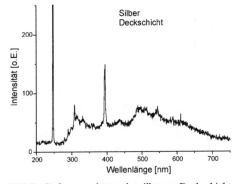

Bild 7 Schichtaufbau "3M Laserbeschriftbare Folie"

Diese Probe wurde verwendet, da sie aus bekannten homogenen Schichten besteht. Diese können bei gleichem Laserparameter mit wiederholbaren Pulszahlen abgetragen werden, was die Kontrolle des Miniatur-Spekrometers als Prozessüberwachungssystem vereinfacht.

Zuerst wurde von jeder Schicht ein Referenzspektrum aufgenommen (**Bild 8a bis 8d**).

Bild 8a Referenzspektrum der silbernen Deckschicht. Aufgenommen beim 3. Laserabtragspuls, Zeitverzögerung 0,5 µs, Excimer-Laser, 248 nm, 0,05 GW/cm², Spotgröße 1 mm²

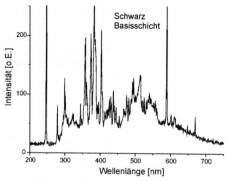

Bild 8b Referenzspektrum schwarze Basisschicht. Aufgenommen beim 8. Laserabtragspuls, Zeitverzögerung 0,5 µs, Excimer-Laser, 248 nm, 0,05 GW/cm², Spotgröße 1 mm²

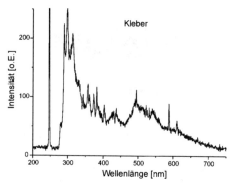

Bild 8c Referenzspektrum Kleberschicht. Aufgenommen beim 45. Laserabtragspuls, Zeitverzögerung 0,5 µs, Excimer-Laser, 248 nm, 0,05 GW/cm², Spotgröße 1 mm²

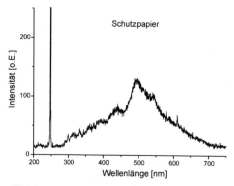

Bild 8d Referenzspektrum Schutzpapierschicht. Aufgenommen beim 58. Laserabtragspuls, Zeitverzögerung 0,5 µs, Excimer-Laser, 248 nm, 0,05 GW/cm², Spotgröße 1 mm²

Die in allen Spektren stark ausgeprägte 248 nm-Linie kommt durch Sättigung der Pixel aufgrund der starken Laserstrahlung zustande.

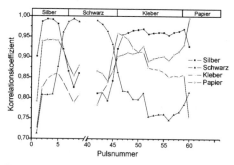

Bild 9 Korrelationskoeffizienten aufgetragen über die Anzahl der Laserabtragspulse für die vier Referenzspektren

Die Korrelationskoeffizienten wurden beim Laserabtrag nach jedem Laserpuls zwischen dem aufgenommenen Spektrum und den vier Referenzspektren berechnet. In **Bild 9** sind die Korrelationskoeffizienten über die Pulsnummern für die vier Referenzspektren aufgetragen. Der Balken oben in der Grafik zeigt, innerhalb welcher Schicht abgetragen wird. Betrachtet man die Linie, die zur Korrelation dem schwarzen Referenzspektrum zugewiesen ist, ist zu erkennen, dass sich beim Eintreffen in die schwarze Schicht der Korrelationskoeffizient dem Wert Eins nähert. Der höchste Korrelationswert wird erreicht bei gleicher Pulszahl (8), bei dem auch das Referenzspektrum aufgenommen wurde. Die Korrelation mit den weiteren drei Referenzspektren ergibt das gleiche Ergebnis; der höchste Wert wird erreicht bei fast gleicher Pulszahl wie das Referenzspektrum aufgenommen wurde. Geringe Änderungen der Pulszahl werden verursacht durch Laserpulsvariationen.

Somit konnte gezeigt werden, dass die Anwendung des Korrelationskoeffizienten geeignet ist, um definiert den Reinigungs- bzw. Abtragsprozess in einer vorgegebene Schicht zu stoppen.

6 Zusammenfassung

Erste Tests an einer Mehrschicht-Testfolie haben gezeigt, dass das Miniatur-Spektrometer mittels linearer Korrelation ein geeignetes Sensor-System für kontrollierte Laserreinigung und Laserschichtabtrag ist. Die Untersuchungen haben gezeigt, dass es möglich ist, den Prozess des Laserabtrags in einer vorher festgelegten Schicht anzuhalten.

In einem nächsten Schritt soll ein Computerprogramm den Prozess in Echtzeit automatisch ausführen, so dass der PC entweder den Abtragsprozess stoppt oder ein Signal emittiert, wenn der Laser die zu erhaltende Schicht erreicht hat.

Das Korrelieren von Spektren mit Referenzspektren und der Korrelationskoeffizient als Entscheidungsfaktor in diesem Prozess bringen den großen Vorteil, dass Laserpuls-Variationen keinen bedeutenden Einfluss haben auf einen positiven Verlauf des Prozesses. Geringere Laserpuls-Variationen Könnten sonst bereits große Änderungen in einem LIPS-Spektrum bewirken. Dadurch, dass beim Korrelieren von Spektren nicht einzelne Linien miteinander verglichen wurden, sondern der lineare Zusammenhang von zwei vollständigen Spektren miteinander verglichen wird, werden diesen Änderungen herausgefiltert.

In Fällen von fast gleichen LIPS-Spektren für zwei unterschiedliche Schichten wird das Unterscheiden beider Schichten via lineare Korrelation sehr schwierig oder unmöglich.

7 Danksagung

Dieses Projekt wird gefördert von der EUREGIO (INTERREG III Program) innerhalb des Vorhabens "EUREGIO CENTER für Art Restauration Technology" (ECEACT).

8 Literatur

[1] R. Teule, H. Scholten, O.F. vd Brink, R.M.A. Heeren, V. Zafiropulus, R. Hesterman, M. Castillejo, M. Martin, U. Ullenius, I. Larsson, F. Guerra-Librero, A. Silva, H. Gouveia, M.B. Albuquerque, Controlled UV laser cleaning of painted artworks: a systematic effect study on egg tempera paint samples, Journal of Cultural Heritage 4, 2003, pp. 209-215

[2] J.H. Scholten, J.M. Teule, V. Zafiropulos, R.M.A. Heeren, Controlled laser cleaning of painted artworks using accurate beam manipulation and on-line LIBS-detection, Journal of Cultural Heritage 1, 2000, pp. 215-220

[3] S. Klein, J. Hildenhagen, K. Dickmann, T. Stratoudaki, V. Zafiropulos, LIBS spectroscopy for monitoring and control of the laser cleaning process of stone and medieval glass, Journal of Cultural Heritage 1, 2000, pp. 287-292

[4] I. Gobernado-Mitre, A.C. Prieto, V. Zafiropulos, Y. Spetsidou, C. Fotakis, On-Line Monitoring of Laser Cleaning of Limestone by Laser-Induced Breakdown Spectroscopy and Laser-Induced Fluorescence, Appl. Spectr. Vol. 51, No. 8, 1997, pp. 1125-1129

[5] M. Sabsabi, R. Heon, V. Detalle, L. St-Onge, A. Hamel, Comparison between intensified CCD and non-intensified gated CCD detectors for LIPS analysis of solid samples, Laser Induced Plasma Spectroscopy and Applications, Vol. 81 of OSA Trends in Optics and Photonics Series, OSA, Washington D.C., 2002, pp. 128-130

[6] J.E. Carranza, E. Gibb, B.W. Smith, D.W. Hahn, J.D. Winefordner, Comparison of nonintensified and intensified CCD detectors for laser-induced breakdown spectroscopy, Appl. Optics, Vol. 42, No. 30, 2003, pp. 6016-6021

[7] I.B. Gomushkin, B.W. Smith, H. Nasajpour, J.D. Winefordner, Identification of Solid Materials by Correlation Analysis Using a Microscopic Laser-Induced Plasma Spectrometer, Anal. Chem. Vol. 71, No. 22, 1999, pp. 5157-5164

[8] I.B. Gomushkin, A. Ruiz-Medina, J.M. Anzano, B.W. Smith, J.D. Winefordner, Identification of particulate materials by correlation analysis using a microscopic laser induced breakdown spectrometer, Journal of analytical atomic spectrometry, Vol. 15, 2000, pp. 581-886

[9] J.M. Anzano, I.B. Gomushkin, B.W. Smith, J.D. Winefordner, Laser-Induced Plasma Spectroscopy for Plastic Identification, Polymer engineering & science, Vol. 40, 2000, pp. 2423-2429

[10] F. Colao, R. Fantoni, V. Lazic, A. Morone, A. Santagata, A. Giardini, LIBS used as a diagnostic tool during the laser cleaning of ancient marble from Mediterrean areas, Appl. Phys.79, 2004, pp. 213-219

[11] M. Lentjes, D. Klomp, K. Dickmann, Sensor concept for controlled laser cleaning via photodiode, abstract book LACONA V, 5th International Conference on Lasers in the Conservation of Artworks, Osnabrück, Germany, September 15-18, 2003, pp. 193-195

[12] M. Lentjes, K. Dickmann, Sensor-System für Echtzeitkontrolle bei der Laserreinigung, Journal of the University of Applied Sciences Mittweida, No. 1, 2003, pp. 68-71

[13] M. Kompitsas, F. Roubani-Kalantzopoulou, I. Bassiotis, A. Diamantopoulou, A. Giannoudakos, Laser induced plasma spectroscopy (LIPS) as an efficient method for elemental analysis of environmental samples, EARSel eProc., Vol 1, No. 1, Paris 2001, pp. 130-138

[14] K.-H. Gerlach, J. Jersch, K. Dickmann, L.J. Hildenhagen, Design and performance of an excimer-laser based optical system for high precision microstructuring, Optics and Laser Technology, Vol. 29, No. 8, 1997, pp. 439-447

Untersuchung und Manipulierbarkeit der Photophysik einzelner Fluorophore mit Hilfe der Zwei-Farben Einzelmolekülspektroskopie

Robert Kasper, Mike Heilemann, Philip Tinnefeld, Markus Sauer

Angewandte Laserphysik und Laserspektroskopie, Universität Bielefeld
Universitätsstr. 25, 33615 Bielefeld, e-mail: rkasper@physik.uni-bielefeld.de

Kurzfassung

Die Entwicklung der Fluoreszenzspektroskopie an einzelnen Molekülen eröffnet die Möglichkeit, ein einzelnes System und dessen Eigenschaften zu beobachten, während die Ensemblespektroskopie durch die Mittelung über viele Systeme diese Information nicht enthält. Eine Reihe von verschiedenen Fluoreszenzfarbstoffen für unterschiedliche spektrale Bereiche dienen hierbei als Sonden, z.B. gekoppelt an Biomolekülen zur Untersuchung biologischer Prozesse oder in Polymeren zur Untersuchung nanoskopischer Eigenschaften.

Hier soll das Prinzip der konfokalen Mikroskopie an einzelnen durch Laserlicht angeregten Chromophoren erläutert werden. Am Beispiel eines Cyaninfarbstoffes wird gezeigt, wie ein einzelnes Molekül mittels zweier verschiedener Anregungswellenlängen zwischen unterschiedlichen elektronischen Zuständen gezielt und reversibel geschaltet werden kann. Ein auf diese Weise optisch schaltbares Molekül könnte die Basis für einen digitalen Datenspeicher auf einzelmolekularer Ebene darstellen.

1 Experimenteller Aufbau

Um die Beobachtung einzelner Moleküle in Lösung oder auf Oberflächen zu ermöglichen, müssen die erhaltenen Signale vom Hintergrund separiert werden, d.h. das Signal-zu-Rausch (S/R)-Verhältnis verbessert werden. Anfang der 90er Jahre wurde dazu die konfokale Fluoreszenzmikroskopie für die Einzelmoleküldetektion etabliert [1].

1.1 Konfokale Mikroskopie

Im Gegensatz zur herkömmlichen Weitfeldmikroskopie, bei welcher das Bild in der Zwischenbildebene erzeugt und vergrößert wird, wird bei der konfokalen Mikroskopie das Licht außerhalb der Brennebene des Objektivs ausgeblendet. Dies wird zum einen durch ein punktförmiges Anregungsvolumen, dem Laserfokus, erreicht, zum anderen durch das Ausblenden störenden Hintergrunds durch ein Lochblende im Detektionsstrahlengang. Das typische Beobachtungsvolumen bei einem konfokalen Mikroskop liegt damit im Bereich von einigen Femtolitern. Der Vorteil dieser Methode ist somit die punktgenaue Abrasterung einer dünnen Schicht eines Objektes, ohne störende Hintergrundsignale außerhalb der Bildebene [2].

Im hier verwendeten Aufbau wird ein Argon-Ionen-Laser (488 nm) über einen dichroitischen Anregungsstrahlteiler durch das Objektiv (Zeiss, 100x, NA = 1.45) in ein inverses Mikroskop (Zeiss Axiovert 200M) eingekoppelt und in die Probe fokussiert. Das Fluoreszenzlicht wird mit dem gleichen Objektiv gesammelt und durch den Strahlteiler und eine Tubuslinse auf eine Lochblende(100 µm) abgebildet. Ein Langpassfilter (500 ALP) blendet gestreutes Anregungslicht vom Fluoreszenzsignal aus. Mittels einer zweiten Linse im Detektionsstrahlengang wird das Fluoreszenzsignal auf die als Detektoren verwendeten Avalanche-Photodioden (APD) abgebildet. Die spektrale Trennung des detektierten Fluoreszenzlichtes erfolgt dabei mit Hilfe dichroitischer Strahlteiler und Emissionsfilter, welche vor die jeweiligen APDs positioniert werden.

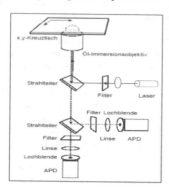

Bild 1: Aufbau eines konfokalen Mikroskops.

Im hier verwendeten Aufbau wurde neben dem Argon-Ionen-Laser ein HeNe-Laser (632.8 nm) eingekoppelt. Das gesammelte Fluoreszenzlicht wurde mit einem dichroitischen Strahlteiler (680DCLP, Chroma) auf zwei APDs abgebildet, vor den Detektoren wurden Emissionsfilter (700DF75 bzw. 665DF50) platziert. Für Oberflächenscans wird ein x,y-Piezotisch eingesetzt.

1.2 Einzelmolekülspektroskopie

Durch das Reduzieren des Anregungs- und Detektionsvolumens mittels der konfokalen Mikroskopie ist es nun prinzipiell möglich, einzelne Moleküle in Lösung oder auf Oberflächen zu beobachten. Dabei sind die bei der auf Fluoreszenzemission basierenden Detektion einzelner Moleküle beobachteten Farbstoffmoleküle nur wenige nm groß und können somit als Punktlichtquelle betrachtet werden. Die Größe der Punktabbildungsfunktion hängt jedoch von der optischen Auflösungsgrenze des Mikroskops ab, die je nach Anregungswellenlänge bei etwa 250-400 nm liegt. Um Farbstoffmoleküle einzeln betrachten und voneinander trennen zu können, ist es notwendig, diese in einem Abstand voneinander zu positionieren, der größer ist als der auflösungsbegrenzte Durchmesser einer Punktabbildungsfunktion (**Bild 2**). Dies wird, sowohl in Lösung als auch auf Oberflächen, durch das Arbeiten mit stark verdünnten Lösungen erreicht.

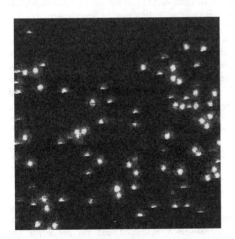

Bild 2: 20 x 20 µm² Rasterscan in Lösung immobilisierter Farbstoffe (Cy5), an DNA gebunden. Anregung mit 633 nm, 14 kWcm^{-2}.

Ein Kriterium für das Beobachten einzelner Moleküle ist das Vorkommen von Fluoreszenzauszeiten, welche durch Übergänge des Moleküls in nichtfluoreszierende Zustände auftreten. Dabei unterscheidet man zwischen „Blinken", also reversiblen Auszuständen (meist handelt es sich dabei um Triplettzustände), und „Bleichen", der Photozerstörung eines einzelnen Farbstoffmoleküls, beispielsweise durch Sauerstoff.

2 Materialien und Methoden

Die photophysikalischen Eigenschaften eines Farbstoffmoleküls hängen maßgeblich von der näheren Umgebung des Moleküls ab. So zeigen beispielsweise auf Glasoberflächen abgelegte Chromophore sehr heterogene Verteilungen ihrer Fluoreszenzeigenschaften, wie beispielsweise Emissionswellenlänge oder Fluoreszenzlebensdauer. Eine Möglichkeit, diese Eigenschaften besser zu kontrollieren, ist das Beobachten von Molekülen in wässriger Umgebung. Dafür müssen diese mittels einer Ankermethode auf der Oberfläche fixiert werden.

Eine etablierte und gut kontrollierbare Methode dafür ist die in diesem Fall benutzte Biotin-Streptavidin-Wechselwirkung.

2.1 Immobilisierung der Farbstoffe

In den hier vorgestellten Experimenten wird eine Glasoberfläche zunächst mit dem Protein Albumin (bovine serum albumine, BSA) belegt. Etwa 10% dieser Proteinmoleküle tragen einen Biotin-Anker, welcher von einem zweiten Protein, Streptavidin, spezifisch gebunden wird. Streptavidin bietet des weiteren die Möglichkeit, weitere Biotin-Anker zu binden, und so werden die Farbstoffmoleküle (Cy5, Amersham Biosciences) ebenfalls über einen Biotin-Anker an die Oberfläche fixiert. Als Abstandshalter zwischen der Proteinoberfläche und dem Farbstoff wird eine etwa 20 nm lange DNA benutzt, damit unspezifische Wechselwirkungen einzelner Farbstoffmoleküle mit den Proteinen vermieden werden.

Die hier beschriebene Methode zur Immobilisierung von Farbstoffmolekülen unter wässrigen Bedingungen bietet den Vorteil, dass die chemische Umgebung der Moleküle beeinflusst werden kann. Dabei ist von besonderem Interesse, die Sauerstoffkonzentration zu verringern, da dieser eine entscheidende Rolle bei der Zerstörung der Chromophore durch irreversible Photooxidation bildet [3].

2.1.1 Oxygen-Scavenging

Die Fluoreszenzeigenschaften einzelner Farbstoffmoleküle wird, wie bereits erwähnt, durch die Gegenwart von Sauerstoff wesentlich beeinflusst. Dabei kommt es sowohl zur Depopulierung nicht-fluoreszierender Triplettzustände als auch zur endgültigen Photozerstörung eines Farbstoffes, dem „Bleichen". Besonders das Ausbleichen von Farbstoffen durch Photooxidation ist eine Konkurrenzreaktion zu den reversiblen

Prozessen dieser Moleküle, welche untersucht werden sollten, und entzieht die Moleküle der weiteren spektroskopischen Beobachtung [4]. Dies gilt besonders für die hier verwendeten Cyaninfarbstoffe Cy5 und Alexa647. Um die Nebenreaktion der Photooxidation und anschliessendes Ausbleichen der Chromophore zu vermeiden, wurde der Sauerstoff enzymatisch aus der wässrigen Umgebung entzogen [5,6]. Dazu wird eine Mischung von Enzym und Substraten zugegeben, die den Sauerstoff in mehreren Schritten chemisch umwandeln und dem System wirkungsvoll entziehen, ohne jedoch zu störenden Hintergrundsignalen zu führen.

Das Resultat einer so präparierten Oberfläche zeigt **Bild 3**. Die vorher runden Fluoreszenzspots sind nun stark zerklüftet, eine Folge nun deutlich längerer Aufenthaltsdauern der Chromophore in Triplettzuständen. Gleichzeitig zeigt sich jedoch, dass die Stabilität der Farbstoffe zugenommen hat, da das Ausbleichen durch den Entzug von Sauerstoff reduziert wurde.

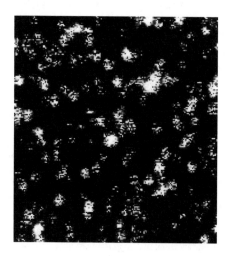

Bild 3: $10 \times 10 \mu m^2$ Scanbild von Cy5-Molekülen in wässriger, sauerstofffreier Umgebung; die stark zerklüftete Struktur der „Spots" resultiert aus deutlich längeren Triplettaufenthaltsdauern der Chromophore.

2.1.2 Triplett-Quenching

Befindet sich ein Chromophor in einem Triplettzustand, so können keine Fluoreszenzphotonen detektiert werden. Es ist somit wünschenswert, unter Beibehaltung der Stabilität der Moleküle die Länge der Triplettzeiten deutlich zu reduzieren.
Eine Möglichkeit zur Verkürzung der Triplettzeiten liegt in der Zugabe von „Triplett-Quenchern", also Molekülen, die gezielt den Triplett-Zustand des Chromophors über Stoßprozesse entvölkern. Zu diesen Substanzen gehören Schwefelverbindungen wie Mercaptoethylamin (MEA) und β-mercaptoethanol (BME) in millimolarer Konzentration.
Neben der bloßen Verkürzung der Triplettzustände kann so auch unterschieden werden zwischen triplettabhängigen und –unabhängigen Folgereaktionen der Chromophore, indem diese entsprechend zugelassen oder verhindert werden [7].

3 Ergebnisse

Cyaninfarbstoffe sind seit langem die mit am besten untersuchten Farbstoffe für verschiedene Anwendungsbereiche. Besonders auf Einzelmolekülebene gibt es eine Anzahl an Publikationen über die photophysikalischen Eigenschaften der Chromophore [8]. Dabei ist insbesondere gezeigt worden, dass es verschiedene intermediäre Zustände auf dem Weg zur Photozerstörung gibt [9]. Da nicht alle dieser Reaktionspfade ausschließlich von Sauerstoff abhängig sind, sollte eine Gleichgewichtslage entstehen, wenn Sauerstoff aus dem System entfernt wird. Im Interesse der hier vorgestellten Arbeit liegt es, diese intermediären Zustände und die Abhängigkeit zur Umgebung genauer zu untersuchen und mit Hilfe eines zweiten Laser mit kürzerer Wellenlänge zu beeinflussen.

Im folgenden Abschnitt sind die Resultate dieser Versuche zusammengefasst. Dabei wurden die Cyaninfarbstoffe Cy5 und Alexa647 unter den beschriebenen wässrigen Bedingungen untersucht. Konkurrenzreaktionen, wie die Photooxidation oder Übergänge in Triplettzustände, wurden selektiv ausgeschaltet und die photophysikalischen Prozesse bei Verwendung von alternierender Anregung mittels zweier Laserwellenlängen, 488 nm und 632.8 nm, untersucht.

3.2 Photophysik von Cyaninfarbstoffen in wässriger Umgebung mit Zweifarben-Laseranregung

In den folgenden Experimenten wurde der Cyaninfarbstoff Cy5 (Struktur **Bild 4A**) an eine 60-Basenpaar-DNA gekoppelt und auf die zuvor beschriebene Art und Weise auf der Oberfläche immobilisiert. Bei Entfernung von Sauerstoff und Zugabe eines Triplettquenchers mit einer Konzentration von etwa 100 mM zeigt sich, dass bei einer Anregung mit 14 kWcm^{-2} bei 632.8 nm die Mehrheit der Moleküle in einen nicht-fluoreszierenden Auszustand geht. Wird nun die gleiche Oberfläche mit etwa 4 kWcm^{-2} und 488 nm nochmals abgerastert, so zeigt sich im darauf folgenden Scan mit 632.8 nm, dass die meisten Moleküle wieder im fluoreszierenden Zustand sind (**Bild 4B-D**). Der gleiche Vorgang kann für die gleiche

Stelle auf der Oberfläche und die gleichen Moleküle mehrfach wiederholt werden, d.h. es handelt sich um einen stark reversiblen Zustand, bei dem die Fluoreszenzeigenschaften der Moleküle zu einem großen Anteil nicht zerstört werden.

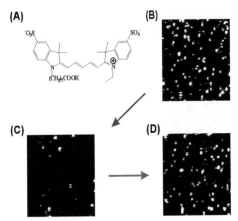

bei wird zunächst mit der Anregung bei 632.8 nm der Auszustand generiert, anschließend das Molekül für etwa 2 s mit 488 nm bestrahlt und wiederum mit 632.8 nm angeregt. Ein Beispiel einer solchen Wiederholungssequenz zeigt **Bild 5B**, das entsprechende Molekül konnte dabei von 21 Zyklen insgesamt 20 mal geschaltet werden, das entspricht einer Verlässlichkeit von 95%.

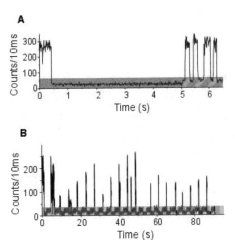

Bild 4: (A) Struktur des Cyaninfarbstoffs Cy5; (B) Scan einer 5 x 5 µm² Oberfläche von immobilisierten Cy5-Molekülen mit 8 kWcm^{-2} und 632.8 nm (C) wiederholter Scan mit 632.8 nm (D) nach Abrasterung der gleichen Oberfläche mit 488 nm wird die Oberfläche erneut mit 632.8 nm gescannt; die meisten Moleküle sind zurückgekehrt in einen fluoreszierenden Zustand.

Um den gezeigten Mechanismus genauer zu charakterisieren, wurden Zweifarben-Laseranregungsexperimente an einzelnen Molekülen durchgeführt. Dadurch ist es prinzipiell möglich, eine Aussage über die Leistungsabhängigkeit des Schaltprozesses als auch die benötigte „Schaltdauer" zu treffen. Des weitern ist es von Interesse, wie groß die Verlässlichkeit dieses Schalters auf Einzelmolekülebene ist. Eine hohe Verlässlichkeit ist eine Grundvorrausetzung, um dieses System für Datenspeicherprozesse in Betracht zu ziehen.

Um die Zeitdauer zu bestimmen, für welche der Farbstoff einer Anregung mit 488 nm unterzogen werden muss, wird das Chromophor zunächst mit 632.8 nm in den nicht-fluoreszierenden, jedoch reversiblen Zustand gebracht. Nachdem für einige Sekunden keine Fluoreszenz detektiert wird, wird der zweite Laser zusätzlich angeschaltet. Beobachtet wird nun eine Fluktuation zwischen dem fluoreszierenden und dem nicht-fluoreszierenden Zustand, wobei die Auszeiten des Farbstoffs bei einigen hundert ms liegt (**Bild 5 A**). Um die Verlässlichkeit des Systems genauer zu überprüfen, wurde ein einzelnes Chromophor möglichst häufig zwischen dem fluoreszierenden und dem nicht fluoreszierenden Zustand hin- und hergeschaltet. Da-

Bild 5: zeitlicher Verlauf des Fluoreszenzsignals eines Cyaninmoleküls; (A) Präparieren des Auszustandes mit 632.8 nm und anschließend gleichzeitige Anregung mit 632.8 und 488 nm (B) alternierende Anregung eines Moleküls mit 632.8 nm und 488 nm.

4 Zusammenfassung

Bis heute war das Fusionsprotein GFP (green fluorecent protein) das einzige unmodifizierte Farbstoffmolekül, das sich photoinduziert an- und ausschalten ließ [10]. Mit den vorliegenden Messdaten konnten wir zeigen, dass sich einfache und kommerziell erhältliche Cyaninfarbstoffe wie Cy5 und Alexa647 im Bereich von einigen hundert ms reversibel Schalten lassen. Die Verlässlichkeit eines solchen Moleküls als optischer Schalter lag dabei bei bis zu 95%. Eine genauere mechanistische Charakterisierung befindet sich noch in Untersuchung, wobei verschiedenste Techniken aus dem Bereich der Einzelmolekülspektroskopie zum Einsatz kommen.

Das hier präsentierte System ist von besonderem Interesse als möglicher Datenspeicher auf Einzelmolekülebene.

Literatur

[1] Rigler, R.; Mets, U.; Widengren, J.; Kasuk, P.: *Biophysical Journal,* **1993**, 22, 169-175

[2] Pawley, James, B.: *Handbook of biological confocal microscopy,* Plenum Press, New York, **1995**

[3] Piestert, O.; Barsch, H.; Buschmann, V.; Heinlein, T.; Knemeyer, J.P.; Weston, K.D.; Sauer, M.: *Nano Letters.* **2003**, 3, 979

[4] Turro, Nicholas J.: *Modern Molecular Photochemistry*, University science books, Sausalito, **1991**

[5] Zhuang, X.; Kim, H.; Pereira, M.J.B.; Babcock, H.P.; Walter, N.G.; Chu, S.: *Science,* **2002**, 296, 1473

[6] Ha, T.; Rasnik, I.; Cheng, W.; Babcock, H.P.; Gauss, G.H.; Lohman, T.M.; Chu, S.: *Nature,* **2002**, 419, 638

[7] Song, L.; Varma, C.A.G.O.; Verhoeven, J.W.; Tanke, H.J: *Biophysical Journal,* **1996**, 70, 2959

[8] Widengren, J; Schwille, P.: *J. Phys. Chem A,* **2000**, 104, 6416

[9] Ha, T.; Xu, J.: *Phys. Rev. Lett.,* **2003**, 90, 223002

[10] Dickson, R.M.; Cubitt, A.B.; Tsien, R.Y.; Moerner, W.E.: *Nature,* **1997**, 388, 355.

Druckwellen in Gewebe bei Ablation mit ultrakurzen Laserpulsen – Modellmessungen in Wasser

J. Eichler, Kim Beop-Min, L. Dünkel, C. Schneeweiss, L. Cibik
Technische Fachhochschule Berlin, University of Applied Sciences

Zusammenfassung

Beim Einsatz ultrakurzer Laserpulse zum Abtragen von biologischem Gewebe entstehen Druckwellen. Zur Untersuchung dieses Effektes wurde Wasser als Modellsubstanz verwendet. Die Dauer der Laserpulse betrug 140 fs bis 10 ps. Interferometrische Messungen ergaben Drücke im Bereich 0,1- bis 1 kbar bei Tiefen bis zu 300 μm. Von der eingestrahlten Laserenergie wird etwa 1 % in Druckwellen umgewandelt.

1. Einleitung

Ultrakurze Laserpulse ($10^{-13} - 10^{-11}$ s) sind für den Einsatz in der Medizin (z.B. Hornhautchirurgie, Zahnmedizin) aus folgenden Gründen interessant:
- hohe Pulsleistung bei kleiner Energie
- hohe Präzision bei der Anwendung am Gewebe
- geringe thermische Schäden des angrenzenden Gewebes

Beim Abtragen von Gewebe entstehen durch den Rückstoß Druckwellen, die in dieser Arbeit untersucht werden.

2. Experimenteller Aufbau

Es wurde ein Ti:Saphir-Laser mit Verstärker benutzt (800 nm, maximal 1 mJ, 0,12-10 ps). Als Modellsubstanz für Gewebe wurde Wasser gewählt. Der Druck führt zu einer Änderung der Brechzahl im Wasser, was mit einem Mach-Zehender-Interferometer untersucht wurde. Die Interferogramme wurden zeitverzögert mit einem Farbstofflaser aufgenommen. Typische Interferogramme zeigt Bild 1.

 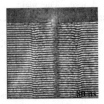

Bild 1: Typische Interferogramme der Druckwellen (Bildausschnitt 0,6x0,6 mm²). Der Laserstrahl fällt von oben ein. a) Halbkugelförmige Druckwelle (0,14 ps, 75 ns nach dem Laserpuls, 240 μJ). b) Zylindrische Druckwelle (10 ps, 80 ns nach dem Laserpuls, 450 μJ).

3. Ergebnisse

Bild 2 zeigt die Schwelle der Energiedichte beim Abtragen von Wasser. Der maximale Druck der halbkugelförmigen Druckwellen in Abhängigkeit von der Tiefe r im Wasser kann durch die Gleichung (1) berechnet werden und ist in Bild 3 dargestellt:

$$p = c \cdot \frac{F}{F_{th}} \cdot \frac{1}{r} \quad (\text{mit } c \approx 2{,}5 \cdot 10^{-2} \text{ bar} \cdot \text{m}) \quad (1)$$

Bei kleineren Leitungsdichten entstehen zylinderförmige Druckwellen.

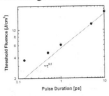

Bild 2: Schwelle der Energiedichte beim Abtragen von Wasser F_{th}. Die Schwelle ist durch das Auftreten einer Druckwelle definiert.

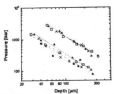

Bild 3: Spitzendruck als Funktion der Tiefe unter der Oberfläche bei $F/F_{th} = 3$ und 1 (▲ = 140 fs, □ = 500 fs, × = 1 ps, • = 5 ps).

4. Diskussion

Der Druck beim Abtragen von Wasser (und Gewebe) liegt bei Tiefen im 100 μm-Bereich um 1000 bar. Es werden 1 % der Laserenergie in mechanische Energie (Druck) umgewandelt. Eine Schädigung beim Abtragen von Gewebe bei dem Druck wird in der Literatur nicht beobachtet.

Literatur

B.-M. Kim, S. Reidt, J. Eichler, *J. Appl. Phys.* **94**, 709-715, 2003

New Spatial Resolving Detector for Infrared Laser Radiation

P. Kohns[1], M. Oehler[1], Th. M. Buzug[1], G. Kokodi[2] and V. Kuzmihov[3]
1) Department of Mathematics and Technology, RheinAhrCampus Remagen, Suedallee 2, 53424 Remagen
2) National University of Pharmacy, Ukraine, 61002, Charkov, Pushkinskaya Str. 35
3) Karazin Charkov National University, Ukraine, 61077, Charkov, Swobody sqw. 4

Abstract

We propose a new kind of spatial resolving laser detectors consisting of grids from thin metal wires. The main idea is that the wires will change their electric conductivity if they are heated by laser radiation. By placing an array of wires in a laser beam the measurement of the change of conductivity of each wire delivers information about the spatial distribution of intensity. The resolution and the total absorption depend on the spatial density of the wires. Two or more detectors can be used simultaneous at several z-positions of the beam. This will allow a very convenient and fast adjustment of the laser.

1 Introduction

Lasers are widely used in science and industry. An important field of application is material processing, where lasers are used since 20 years. In all laser machines it is necessary to have devices for measuring laser radiation parameters like the power of continuous lasers or the pulse energy of pulsed lasers, intensity distribution in the cross-section of the beam, beam diameter, or the beam divergence.

These data must be measured for beam diameters from several 10 micrometers (e.g. near the focus of a laser cutting machine) to several decimeters (e.g. in the free beam of a laser machine). The measurement of the parameters of the unfocused beam with a diameter about 10 cm is very important because all multi-kilowatt laser machines generate such a wide beam directly at the laser station, and the measurement of the beam diameter and beam divergence directly at the laser source will show the quality of laser operation and will allow a prediction of the minimum focus diameter and therefore the quality of the laser beam. If the beam quality at the laser source is bad there is no optics which can improve the beam quality at the work piece.

In the visible and near infrared range (wavelength range from 0.4 to 1.1 µm) good photoelectric detectors of radiation exist both as single detectors and arrays of 1000 x 1000 elements and more. Precise devices for measuring power of continuous radiation and pulse energy are developed. The diameter of radiation beams is usually less than 2 cm.

Unfortunately, the most powerful laser machines (i.e. carbon dioxide lasers) do not work within the operating spectral range of these devices, and the laser beams usually have larger diameters. In the spectral regions which are of highest interest for laser machining the choice of suited laser detectors is rather limited. In principle a user has to choose between thermal receivers like pyroelectric detectors or thermopiles on the one hand or photoelectric low temperature receivers on the other hand [1].

Photoelectric receivers must be cooled down well below 0°C or even to the temperature of liquid nitrogen. They are inconvenient in work and are very expensive. In addition, their active area is too small to be useful; typical sizes are less than 10 mm. Pyroelectric detectors and thermopiles are much cheaper and easier to use. But to measure a powerful beam it is necessary to use attenuators of laser radiation.

We propose a solution that is derived from the reconstruction methods of computed tomography.

2 Proposed Set-Up

Grids from thin metal wires are proposed to solve this problem. They may be used to measure the intensity distribution in the beam of a continuous wave or pulsed laser. The main idea is that the wires will change their electric conductivity if they are heated by laser radiation. By placing an array of wires in a laser beam the measurement of the change of conductivity of each wire delivers information about the spatial distribution of intensity.

The resolution and the total absorption depend on the spatial density of the wires, where a higher density allows a higher resolution, and a lower density allows placing the wire grid directly in the laser beam with very small losses of power and distortion of the shape of the beam cross-section. Two or more detectors can be used simultaneously at several z-positions of the beam, provided that only a few wires (i.e. at reduced resolution) are used. This will allow a very convenient online-control of the long term stability of the laser beam (compare figure 1).

In order to reduce the absorption and the diffraction at the grids thin wires will be used. A typical Pt wire with a diameter of 25 µm has an electrical resistance of a few Ohms per cm length. In order to get a lot of information we use several equidistant wires and perform measurements at several rotation angles of the wires where the axis of rotation is parallel to the light beam (z-axis). In the following text the term "projection" is used for the measurement of the change of resistance of one wire at one rotation angle. The number of projections is the number of wires times the number of angles.

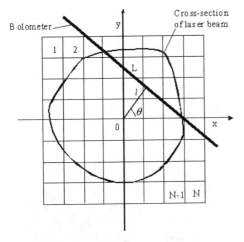

Fig. 1: Geometry of the problem: Cross section of laser intensity distribution and one wire with coordinates. The pixels of the illuminated area are numbered from 1 to N.

3   Mathematical Model

The mathematical problem of the determination of the laser radiation distribution is similar to the inverse problem of computed tomography. The change of the electric conductivity of the wire and, consequently, the signal from wire with distance l from the center of the observation area (see fig. 1) is proportional to

$$r(l,\theta) = \int_{wire} I(x,y)ds \quad (1)$$

Integration is along the length of the wire. The position of the wire is determined by the parameters l and angle of rotation θ. $I(x,y)$ is the intensity distribution of the laser radiation. The task is to find this function using the observable signals $r(l,\theta)$ which can be measured at various values of parameters l and θ.

For solving of this problem the inverse Radon transformation can be used [2,3]. The main difference to computed tomography is the lower number of wires (projection directions) and, therefore, the lower resolution. A typically system will consist on 8 wires and we will measure at 8, 10 or 12 angles. Therefore, we get a spatial resolution of 8 x 8 pixels. This is sufficient to control the stability of a laser system used in material processing.

Due to the low number of measurements, i.e. projections, and pixels we can use the exact algebraic method for image reconstruction, i.e. the reconstruction of the intensity distribution. Following the notation given in [3] we introduce:

1. The "pixel vector" f is a one-column-vector where the number of elements equals the number of pixels N. The value of the nth element equals the average intensity impinging on the nth pixel.

2. The "projection vector" p is a one-column-vector where the number of elements equals the number of projections M (Number of wires multiplied with the number of angles). The value of the mth element equals the change of resistance measured at the parameters of the mth projection.

3. The system matrix A contains M rows and N columns. The element a_{ij} is given by the spatial overlap of pixel j and projection i. The matrix A can be calculated if the number of pixels, the number of projections and the angle step are known.

If the system matrix A is known – It can be calculated by simple geometrical considerations – the projection vector p can be defined as

$$p = A \cdot f. \quad (2)$$

In practice we measure p and have to calculate f. Due to the fact that A is a square matrix only in the case that the number of projections equals the number of the pixels a more general approach to solve equation (2) for f must be chosen [3,4].
In general, that means in case of M≥N, we obtain f by

$$f = (A^T A)^{-1} \cdot A^T \cdot p. \quad (3)$$

Therefore, the problem of algebraic reconstruction consists of two steps:

1. Calculation of the system matrix A
2. Calculation of the so-called pseudo inverse $(A^T A)^{-1} A^T$

The problem to calculate the element a_{ij} of the system matrix A is very similar to the calculation of the length of intersection of a ray (=projection) with a square box (=pixel) in optical ray tracing [5]. In order to achieve an algorithm for a ray-box intersection we define the box by two horizontal $(y_0 < y < y_1)$ and two vertical lines $(x_0 < x < x_1)$. If the ray is described by a starting point (x_e, y_e) and a normalized direction (x_d, y_d), then the geometry of the ray is given by the parametric equation

$$(x,y)(t) = (x_e, y_e) + t \cdot (x_d, y_d). \qquad (4)$$

First, we compute the ray parameter t_{xmin}, where the ray hits the vertical line $x=x_0$

$$t_{xmin} = (x_0 - x_e)/x_d. \qquad (5)$$

We then carry out similar computations for t_{xmax}, t_{ymin} and t_{ymax}. The ray hits the box, if and only if the intervals $[t_{xmin}, t_{xmax}]$ and $[t_{ymin}, t_{ymax}]$ overlap, i.e. if their intersection is nonempty. The intersection length is given by the overlap of the two intervals, if the direction vector has unit length (compare figure 2).
Of course the algorithm requires some additional assumptions in order to avoid division by zero. In addition, to the described calculation of the exact system matrix we used a simpler algorithm to calculate a second type of system matrix, the approximated system matrix, where a_{ij} equals 1 if and only if pixel i and projection j overlap. If this is not the case, a_{ij} vanishes. This approximated matrix is often used in medical tomography where the system matrix contains 10^9 elements or more.

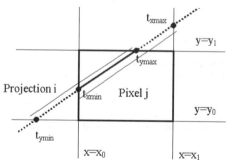

Fig. 2: Calculation of overlap of pixel j (square with bold lines) and projection i (dashed / solid line). In this example, the overlap of pixel j and projection i is given by the solid line and $a_{ij}=t_{ymax}-t_{xmin}$.

Several system matrices A with different numbers of pixels and projections were calculated. First we examined square system matrices where the angle step equals 180° divided by the number of angles. We found that all of these matrices are singular. Exact algebraic reconstruction is not possible in these cases. Therefore, we calculated non-square matrices where the number of projection is larger than the number of pixels. With Mathlab^{TM} only a few seconds were needed to calculate the pseudo inverse of A even if A consist of 100 000 elements, provided that the pseudo inverse exists. However, such a large number of elements will not be achieved in our experiments.

4 Simulation

In the following calculations we used a realistic case of 8 x 8 pixels, 8 wires, and 12 angles with angle step 15°. In this case the pseudo inverse can be computed. For the simulations we generated a cross-like "phantom" illumination of 64 pixel elements as shown in figure 3. A graphical representation of the calculated system matrix is shown in figure 4. The system matrix consists of 96 x 64 elements. The projections calculated by equation (2) are shown in the so called sinogram (raw projection data) in figure 5.

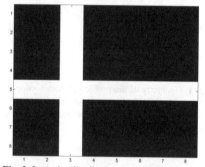

Fig. 3: Intensity distribution used for simulation

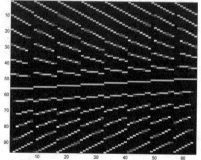

Fig. 4: The system matrix of the simulation. Dark points represent vanishing matrix elements.

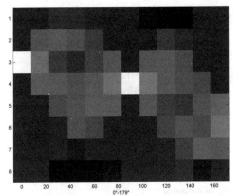

Fig. 5: Calculated sinogram of the cross-like irradiation distribution.

In order to show that even with the low number of pixels and projections a reconstruction of the illumination is possible we calculated the pixel vector using eq. (3). The result is shown in figure 6, where the exact system matrix A is used. Obviously, an exact reconstruction is possible.

the series of pixel vectors converges to the pixel vector f. We used the following iteration [3]:

1. initial value: $f^{(0)} = (0,0,0,..0)$
2. $p^{(n)} = A f^{(n)}$
3. correction: $f^{(n)} = f^{(n-1)} - (a_i f^{(n-1)} - p_i) a_i^T / |a_i|^2$
   The variable i is randomly distributed.
4. repeat step 2. If the change of the value of the elements of the pixel vector fall below a threshold the procedure is stopped.

It is straight forward to write a Mathlab™ script where the iteration step is written in one single line of code:

f=f-((a(i,:)*f-p(i)/(a(i,:)*a(i,:)')*a(i,:)');

We performed 70 and 600 steps of iteration. In figure 7 it is shown that the reconstruction converges to the starting pixel field very fast, if the exact system matrix is used for reconstruction.

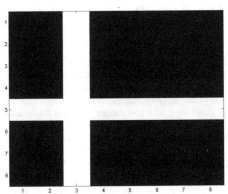

Fig. 6: Reconstruction on the basis of the raw projection data (sinogram of fig. 5) by equation (3).

In addition, we used the established algebraic reconstruction technique (ART) [6]. Usually ART is used if the design matrix has many (10^9 or more) elements. In this case it is not possible to calculate the pseudo inverse of the system matrix directly.

Instead, an iterative approach is used. ART starts with an initial pixel vector $f^{(0)}$ and calculates a series of pixel vectors $f^{(0)}, f^{(1)}, ..f^{(n)},...$ .Using the design matrix A the nth projection vector is calculated as $p^{(n)} = A f^{(n)}$. The difference between $p^{(n)}$ and the measured projection vector p is used to calculate $f^{(n+1)}$. In most cases

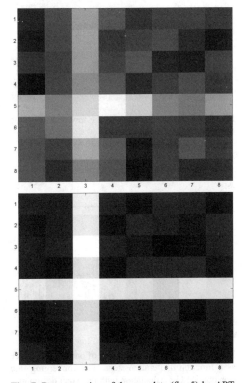

Fig. 7: Reconstruction of the raw data (fig. 5) by ART. The exact system matrix was used for reconstruction. We performed 70 (upper figure) and 600 iteration steps.

After 10 000 steps of iteration there is no difference between the original and the reconstructed image. For comparison we show in figure 8 the reconstruction of the intensity image if the approximated system matrix is used. Even after 10000 steps of iteration the quality of reconstruction is very bad and noisy.

Finally, we examined the convergence of ART for one pixel. The result is shown in figure 9. In this figure we show the value of the 5th pixel in the 1st row for an ART reconstruction with the exact system matrix (curve converging to a fixed value) and a reconstruction where the approximated design matrix is used. Obviously, the exact matrix must be used for reconstruction. However, due to the low number of pixels this is not a problem.

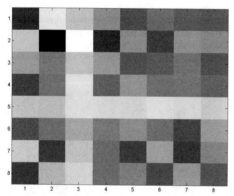

Fig. 8: ART-Reconstruction of the raw data by using the approximated system matrix.

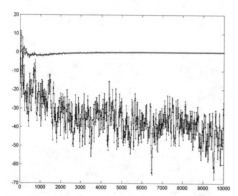

Fig. 9: Reconstructed irradiation of one pixel by ART as a function of iteration steps. The converging line was obtained by calculation with the exact design matrix. The diverging noisy line was obtained by using the approximated design matrix.

5 Experimental results

A wire grid with 6 wires was used for the first measurements. The distance of the wires was 6 mm. The wires were strips from Pt-Ag alloy with cross-section sizes of 100 x 10 μm. The resistance of the wires was a few Ohms. As a light source we used a halogen tungsten lamp the electrical input of which could be tuned from zero to 200 Watt. The lamp was placed in a housing with elliptical reflector. We measured the change of electrical resistance of the wires at 9 angles (0°, 15°, .., 120°). Therefore we have 6 x 6 pixels and 6 x 9 projections. In figure 10 we show the calculated irradiation of four selected pixels as a function of the signal measured by a power meter placed at the location of the pixels. The measurements were performed for three different electrical powers of the lamp.

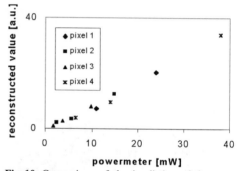

Fig. 10: Comparison of the irradiation of four selected pixels. The irradiation obtained by reconstruction of the measured sinogram is shown as a function of the reading of a powermeter placed at the location of the pixel. The measurements were performed for three power levels of the lamps. Therefore the figure contains e.g. three triangles representing the data of pixel 3 at the three power levels of the lamp.

6 Conclusion and Discussion

The measurement of the intensity distribution of radiation is possible by the use of a simple wire bolometer. This allows for example the online-control of laser material processing machines. In order to get automatic measurements it is necessary to rotate the wires automatically and to perform the data acquisition automatically.

Due to these processes can be performed very slowly in the time frame of seconds they can be controlled by a single chip controller, e.g. a PIC or an ATMEL, which in addition may be equipped with an analog/digital-converter. For further data processing, especially the reconstruction of the irradiation, the measured data may be transferred to a PC. On the other hand, because the most time-consuming calculation is that of the pseudo-inverse, which does not change if the number of wires and the measuring angles do not change, the pseudo-inverse of the measuring situation can be stored in the PIC. In this case the user has a stand-alone device.

In the near future the following base research will be performed:

a) It is very likely that the change of resistance does not depend linearly on the input power. Therefore we have to introduce a step (which is not performed in medical tomography) to eliminate the nonlinearity of the dependence of resistance on laser heating. This calculation can be performed for each pixel separately.

b) A second source of nonlinearity is given by the thermal conductivity of the wires. If the diameter of the wires is small compared to the size of a pixel (i.e. the distance of the wires) this problem should be neglectible.

c) We used Pt wire in order to avoid oxidation of the wires. Of course this material is very expensive. We note that in most cases it is not possible to place the wires in protecting gas because there are only a few expensive window materials available which withstand the high power of a carbon dioxide laser. A possible solution may be the use of Tantalum wires, because this metal builds an oxide layer. In addition it has a very high melting point. We will perform experiments with a scanning electron microscope equipped with EDX to observe the oxidation of this material due to the laser radiation.

After the first experiments the proposed detector seems to be a promising tool to measure the parameters of a laser.

7 References

[1] M. Bass (ed.), Handbook of Optics, Wiley, NY, 1995.

[2] G. T. Herman, Image reconstruction from projections, The fundamentals of computerized tomography, N.Y., Toronto, 1980.

[3] Th. M. Buzug, Einführung in die Computertomographie, Springer, Berlin, 2004.

[4] F. Natterer, The Mathematics of Computerized Tomography, Classics in Applied Mathematics, SIAM Monographs on Mathematical Modelling and Computing, Philadelphia, 2001.

[5] P. Shirley, Realistic Ray Tracing, A.K.Peters Ltd, Canada, 2003.

[6] A. C. Kak and M. Slaney, Principles of Computerized Imaging, Classics in Applied Mathemathics 33, IEEE Press, NY, 1988.

Realisierung eines monochromatischen Laser-Doppler Profilsensors unter Verwendung von Frequenzmultiplexing

P. Pfeiffer, T. Pfister, L. Büttner, K. Shirai, J. Czarske
Laser Zentrum Hannover e.V. (LZH), Dep. Laser Metrology, Hollerithallee 8, 30419 Hannover
e-mail: tp@lzh.de

Abstract

In diesem Beitrag wird ein Laser-Doppler Profilsensor für die Vermessung von Strömungsprofilen vorgestellt der sich durch eine Ortsauflösung von $\approx 10~\mu m$ innerhalb eines Messvolumens von 1mm von auszeichnet. Durch die Verwendung von Frequenzmultiplexing (FDM) können Geschwindigkeiten im Bereich von 3cm/s bis ca. 40 m/s mit der selben Signalverarbeitungstechnik gemessen werden. Die praktische Anwendbarkeit des Sensors wird durch Windkanalmessungen demonstriert.

1 Einleitung

Seit der Entwicklung des ersten Laser-Doppler-Anemometers 1964 durch Yeh und Cummins [1] haben sich LDA-Techniken in der Strömungsmesstechnik als Geschwindigkeitsmessverfahren etabliert. Die Laser-Doppler-Anemometrie zeichnet sich besonders, als nicht-invasives Verfahren durch hohe Zeitauflösung und geringe Messfehler aus. Kommerziell verfügbare LDA-Systeme weisen eine Ortsauflösung von 50-200 μm auf. Dies ist für Strömungssysteme, wie sie in der Mikrofluidik und Biomedizin auftreten häufig nicht ausreichend. Neue technische Entwicklungen z.B. im Bereich der Mikrofluidik verlangen nach Systemen die einen geringen Messfehler und eine hohe Zeit- und Ortsauflösung aufweisen. Als Anwendungsbeispiele aus der Mikrofluidik sollen hier Mikroreaktoren genannt werden. In diesem Beitrag wird auf ein am LZH (Laser-Zentrum-Hannover) entwickeltes Messverfahren [2] [3] eingegangen, das eine ortsaufgelöste Geschwindigkeitsmessung innerhalb des Messvolumens ermöglicht. Um dem Anwender viele Freiheiten zu bieten, ist das System modular konzipiert. Ein Anwendungsgebiet des Sensors liegt in der Messung von Grenzschichtströmungen, wie sie z.B. an Flugzeugtragflächen auftreten.

2 Laser-Doppler-Profilsensor

2.1 Prinzip

In kommerziell verfügbaren LDA-Systemen ist die Ortsauflösung durch die Ausdehnung des Messvolumens begrenzt. Zwar kann diese durch eine Verkleinerung des Detektionsvolumens oder durch eine stärkere Fokussierung der Sendestrahlen gesteigert werden, allerdings wird in beiden Fällen der Messfehler der Geschwindigkeit erhöht.

Bedingt durch die Wellenfrontkrümmung des Gaußschen Strahl und der daraus resultierenden Streifenabstandvariation ist die gemessene Frequenz von der Position des Streuteilchens innerhalb des Messvolumens abhängig, ohne das diese ermittelt werden kann. Es wird eine systematische Messabweichung der Messgröße Geschwindigkeit in das System induziert. Dieser Fehler kann durch ein zweites Interferenzstreifensystem korrigiert werden, welches vom ersten unterscheidbar sein muss. Die Position und Geschwindigkeit des Streuteilchens kann dann über die Auswertung der beiden Doppler-Frequenzen erfolgen. Das eigentliche Grundprinzip des Laser-Doppler-Profilsensors beruht auf der Erzeugung eines divergierenden und eines konvergierenden Interferenzstreifensystems mit den Streifenabständen $d_1(z)$ und $d_2(z)$. Die Kanaltrennung erfolgt mittels Frequenzmultiplexing. Aus dem Quotienten der beiden Dopplerfrequenzen $f_1(z)$ und $f_2(z)$ kann die Kalibrationsfunktion $q(z)$ gebildet werden [2] [3]:

$$q(z) = \frac{f_1(z)}{f_2(z)} = \frac{v/d_2(z)}{v/d_1(z)} = \frac{d_1(z)}{d_2(z)} \quad (1)$$

Zu beachten ist, dass die Kalibrationsfunktion nicht mehr von der Messgröße Geschwindigkeit abhängig ist. Durch die Bestimmung der Position aus der Kalibrationsfunktion kann auf die lokalen Streifenabstände geschlossen und daraus die Geschwindigkeit berechnet werden:

$$v(z) = f_1(z) \cdot d_1(z) = f_2(z) \cdot d_2(z) \quad (2)$$

2.2 Aufbau

Der Messaufbau des Sensor ist in **Abb. 1** schematisch dargestellt. Als Strahlquelle wird ein frequenzverdoppelter Nd-YAG Laser (λ=532nm) verwendet der bei einer optischen Ausgangsleistung von 70mW im transversalen Grundmode TEM00 betrieben werden kann. Mit einem Strahlteilerwürfel wird der Strahl zunächst in 2 Teilstrahlen aufgeteilt. Die weitere Aufteilung in

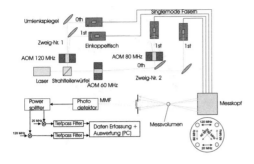

Abbildung 1: Schematischer Aufbau

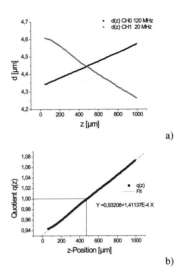

Abbildung 2: a) Verlauf der Interferenzstreifenabstände über das Messvolumen. Ch0:Δf 120 MHz, Ch1:Δf 20 MHz b) Daraus resultierende Kalibrationsfunktion q(z). Die Länge des Messvolumens ergibt sich zu 1mm.

4 LDA-Teilstrahlen erfolgt mittels akustooptischen Modulatoren (AOMs), durch Nutzung der 0. und +1. Beugungsordnungen. Das hindurch laufende Licht (0.Beugungsordnung) behält seine Frequenz f_{Laser} bei, für die +1 Beugungsordnungen ergibt sich $f_{BGO1.} = f_{Laser} + f_{AOM}$. Auf diese Weise werden 4 Sendestrahlen erzeugt, denen unterschiedliche Trägerfrequenzen (0, 60, 80, 120 MHz) aufgeprägt sind. Daraus werden 2 Sendestrahlpaare gebildet, deren Differenzfrequenzen Δf 120 und 20 MHz betragen. Die Differenzfrequenzen sind so gewählt, dass sie nur zwischen den entsprechenden Teilstrahlen auftreten. Die Teilstrahlen werden in Singlemode Fasern eingekoppelt und zu einem optischen Messkopf geführt, der die Abbildungsoptik und optional die Empfangsoptik des Systems bildet. Der Messkopf ist für einen Arbeitsabstand von 30cm konzipiert. Durch einen Austausch oder die Wahl einer anderen Justage des Messkopfes kann das System einfach an verschiedene Anwendungen angepasst werden. Nach der Detektion wird das Signal zunächst mit einem Power-Splitter aufgeteilt und mit den entsprechenden Referenzsignale der AOMs ins Basisband runtergemischt. Durch das Runtermischen der Trägerfrequenzen ins Basisband und die anschließende Tiefpassfilterung erfolgt die Kanaltrennung mittels Frequenzdemultiplexing. Dieses Vorgehen bietet den Vorteil, dass so auch der DC-Gleichanteil des Messsignals beseitigt wird, was besonders für die Messung niedriger Geschwindigkeiten von Bedeutung ist. Die eigentliche Signalauswertung erfolgt dann nach der Erfassung mit einem Digital/Analog Umsetzer im PC. Die beiden Doppler-Frequenzen werden mit einer FFT-Signalverabbeitungstechnik bestimmt.

2.3 Kalibration

Für die Kalibration wird ein in der Frequenz stabilisierter Chopper verwendet, an den ein 4μm Draht angebracht wurde. Aus dieser Anordnung ergibt sich ein Geschwindigkeitsnormal. Der Chopper wurde mit einem motorisierten μm-Tisch entlang der optischen Achse durch das Messvolumen verfahren. An jeder Messposition wurden 20 Doppler-Signale aufgenommen, um durch die Mittelung den statistischen Fehler zu reduzieren. Aus der bekannten Frequenz des Choppers $f_{Chopper}$ und dem Radius r, auf dem sich der Draht bewegt, kann die Geschwindigkeit $v = 2\pi r \cdot f_{Chopper}$ berechnet und damit der lokale Streifenabstand $d_i(z) = \frac{v}{f_i(z)}$ der beiden Interferenzstreifensysteme aus der jeweiligen gemessenen Doppler-Frequenz $f_i(z)$ bestimmt werden. Durch eine geeignete Justage der Strahltaillen der Sendestrahlpaare relativ zum Kreuzungspunkt werden ein konvergierendes und ein divergierendes Streifensystem erzeugt und damit eine monoton steigende Kalibrationsfunktion q(z) erreicht. Der Verlauf der Streifenabstände und die sich daraus ergebende Quotientenfunktion q(z) mit einer Steigung von 0,14/mm sind in **Abb. 2** dargestellt. Die absolute Meßunsicherheit der Geschwindigkeit konnte zu $\frac{\Delta v}{v} = 2,6 \cdot 10^{-3}$ bestimmt werden. Die Standardabweichung bzgl. der Position (Ortsauflösung) ist kleiner 10 μm und beträg im Mittel $\approx 6,6\mu m$ (siehe **Abb. 3b**). Die longitudinale Ausdehnung des Messvolumens ist $\approx 1mm$.

3 Windkanalmessung

Der Sensor wurden am Lehstuhl für Strömungsmechanik (LSTM) in Erlangen für Windkanalmessung eingesetzt. Für die Messungen stand ein Windkanal Göttin-

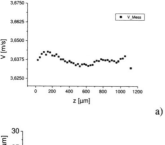

a)

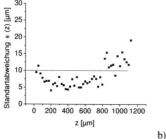

b)

Abbildung 3: a) Verlauf der gemessenen Geschwindigkeit über das Messvolumen. b) Genauigkeit der Ortsauflösung innerhalb des Messvolumens

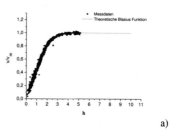

a)

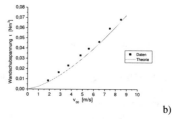

b)

Abbildung 4: a) Normiertes Grenzschichtprofil und theoretisches Blasius-Profil. Messdaten und Theorie stimmen gut überein. b) Aus den Messdaten berechnete Wandschubspannung bei verschiedenen Anströmgeschwindigkeiten im Vergleich zur Theorie

ger Bauart zur Verfügung. Der Windkanal zeichnet sich durch seine geschlossene Bauart mit offener Messstrecke aus. Um Grenzschichtprofile untersuchen zu können wurde eine Glasplatte parallel zur Strömungsrichtung im Windkanal angebracht. Die Detektion erfolgte in Vorwärtssteurichtung. DEHS (Diethylhexalsebacate)-Tropfen mit einem mittleren Durchmesser von 2,5 μm wurden als Tracerpartikel eingesetzt. Es wurden laminare Grenzschichtprofile bei Reynoldszahlen Re kleiner $1 \cdot 10^6$ aufgenommen. Die Reynoldszahl $Re(x) = \frac{xv_\infty}{\nu}$, die den Strömungszustand charakterisiert, ergibt sich aus der kinematischen Viskosität von Luft $\nu \approx 1,5 \cdot 10^{-5} \frac{m^2}{s}$, der Messposition x relativ zur Vorderkante der Glasplatte und der Anströmgeschwindigkeit v_∞. Die erfassten Strömungsprofile können durch Maßstabsfaktor für Ort und Geschwindigkeit, unabhängig von der Messposition x auf der Platte und der Anströmgeschwindigkeit v_∞ inneinander überführt werden, d.h. sie sind selbstähnlich. Als Ergebnis ergibt sich das sogenannte Blasius-Profil.

$$v \rightarrow \frac{v}{v_\infty} \quad (3)$$

$$z \rightarrow \eta = z\sqrt{\frac{v_\infty}{2\nu x}} \quad (4)$$

Da eine numerische Lösung zur Beschreibung eines Blasius-Profils [4] [3] existieren können die Messergebnisse leicht mit der Theorie überprüft werden. **Abb. 4a** zeigt die gute Übereinstimmung zwischen Messdaten und Theorie. Die niedrigste erfasste Strömungsgeschwindigkeit war dabei 3cm/s. Aus der Steigung des Profils nahe der Glasplatte und der Anströmgeschwindigkeit kann auf die Wandschubspannung $\tau(x, v_\infty)$ geschlossen werden. Die Wandschubspannung für laminare Strömungen ergibt sich mit der dynamischen Viskosität μ nach Blasius zu:

$$\tau(x, v_\infty) = 0,332 \cdot \mu\sqrt{\frac{v_\infty^3}{2\nu x}} \quad (5)$$

Die Wandschubspannung ist ein Maß für die auf den unströmten Körper an dieser Stelle wirkende Scherkraft. Für die strömungsmechanische Optimierung von Körpern ist sie von Interesse, da durch aufintegrieren der Wandschubspannung über die Oberfläche des umströmten Körper der Reibungswiderstand W bestimmt werden kann. Der Sensor kann zur Messtechnischen Erfassung der Wandschubspannung eingesetzt werden. Dies geht aus der in **Abb. 4 b** vorgestellten Messreihe und deren Vergleich mit dem theoretischen Verlauf der Wandschubspanng hervor.

4 Zusammenfassung

Ein faseroptischer Laser-Doppler-Profilsensor wurde vorgestellt. Das System besteht aus einem Messkopf und einem Bereich in dem die Sendestrahlen generiert und in Fasern eingekoppelt werden und ist somit modular konzipiert. Das System weißt eine Ortsauflösung kleiner 10 μm, bei einem relativen Messfehler der Geschwindigkeit von 0.26 Prozent, innerhalb eines Messvolumens von ungefähr 1mm auf. Die Verwendung von Frequenzmultiplexing ermöglicht den Einsatz nur einer monochromatischen Laserquelle und die Messung niedriger Strömungsgeschwindigkeiten, wie sie an umströmten Körpern in Wandnähe auftreten. Die kleinste gemessene Geschwindigkeit konnte zu 3cm/s bestimmt werden. Die Anwendbarkeit des Sensor wurde durch Windkanalmessungen in der Praxis verifiziert. Die gewonnenen Messdaten zeigen eine gute Übereinstimmung mit der Theorie. Anwendungsfelder des Sensors liegen z.B in der Turbulenzforschung, der Biomedizin zur Untersuchung von Strömungen in Blutpumpen oder der Produktionstechnik zur Formvermessung rotierender Objekte bei gleichzeitiger Abstands- und Geschwindigkeitsmessung.

5 Danksagung

Die Autoren danken S. Becker und H. Lienhart vom LSTM-Erlangen für die Möglichkeit Windkanalmessungen durchführen zu können. Besonderer Dank gilt der Deutsche Forschungs Gemeinschaft DFG für die Projektförderung (CZ55/9-2).

Literatur

[1] YEH, Y. und H.Z. CUMMINS: *Localized fluid measurement with He-Ne laser spectrometer.* Appl. Phys. Letter, 4(10):176 – 178, 1964.

[2] J. CZARSKE: *Laser-Doppler velocity profile sensor using a chromatic coding.* Meas. Sci. Technol., 12:52–57, 2001.

[3] J. CZARSKE, L. BÜTTNER, T. RAZIK, H. MÜLLER: *Boundary layer velocity measurement by a laser Doppler profile Sensor with micometer spatial resolution.* Meas.Sci.Technol., 13:1979 – 1989, 2002.

[4] H. SCHLICHTING, K. GERSTEN: *Grenzschicht-Theorie.* Springer-Verlag, Berlin/ Heidelberg/ New York, 1997.

Autorenverzeichnis

Afilal 277
Altenburger 62
Altmeyer 40
Andre 195

Barniko 187
Baumert 172
Belitz 74, 92
Beop-Min 301
Bildhauer 161
Blume 114
Boller 262
Bongartz 122, 167, 274
Bourauel 108, 152
Breeuwer 104
Bruckschen 240
Büll 74
Buth 104
Büttner 308
Buzug 18, 24, 30, 62, 108, 122, 181, 208, 214,
220, 234, 240, 246, 302

Castellanos 74, 98
Cibik 301
Czarske 308

de Putter 104
Deußen 268
Dickmann 290
Dillinger 68
Dillmann 86
Dirnagl 200
Dünkel 301

Eichler 301
Ellegast 158
Ermert 40
Evbatyrov 133

Freund 187
Frey 274

Giel 274
Grünendick 40
Grünwald 62
Gubaidullin 133, 139, 146
Gürtler 195, 200

Haller 62
Halling 4, 6, 10, 68
Hartmann 161, 165, 167
Hau 56
Hauptmann 187
Heiden 45, 56

Heilemann 296
Hering 122, 274, 277, 283
Hermes 187
Herpers 45, 56
Herzog 36, 74
Hetmann 56
Hibst 258
Hirsch 274
Hoffmann 40
Holz 24, 114
Hoogeveen 104
Hoppin 4, 6, 10

Ivanenko 122, 277, 283

Jäger 108, 152

Kaesler 246
Kaiser 74
Kasper 296
Keilig 108, 152
Khodaverdi 36
Klasing 277, 283
Knütel 40
Koch 262, 268
Kohl-Bareis 24, 195, 200
Kohns 302
Kokodi 302
Kose 104
Krücker 36
Kunkel 133
Kurz 246
Kuzmihov 302

Lackas 4, 6, 10
Ladrière 274
Laffargue 104
Laschinski 167
Leinen 68
Leithner 200
Lentjes 290
Liebert 189
Liebold 256
Lindauer 200

Macdonald 189
Mang 214, 220
Marklewitz 30
Meijer 290
Möller 189
Morsnowski 177
Müller 92, 208, 214, 220, 250, 258
Müller-Deile 177
Musmann 14, 18

Neeb	80
Obrig	189
Oehler	18, 62, 302
Ostrowitzki	62
Patzak	68
Perez	36
Pfaffmann	30
Pfeiffer	308
Pfister	177, 308
Pietrzyk	36
Popp	262, 268
Preissler	165
Rahimi	108
Ramm	68
Rätzer-Scheibe	283
Reichert	133
Reimann	108, 152
Reinhold	30
Richarz	165
Rohm	195
Rongen	187
Rouet	104
Royl	200
Ruhlmann	30, 62
Sabri	74
Sauer	296
Schäbe	128
Schiek	187
Schmitz	24, 45
Schneeweiss	301
Schneider	240
Schramm	4, 6, 10, 14
Schulyzk	161
Schumann	30
Schwan	68

Seibt	45
Sellien	200
Shah	80
Shirai	308
Silex	187
Solaiyappan	114
Stehr	268
Steinbrink	189
Stock	258
Strothjohann	181
Stuber	114
Sturm	187
Subke	226
Suslov	68
Tass	187
Thelen	274
Thomsen	208, 246
Tinnefeld	296
Tolxdorff	98
v.d. Bosch	104
Visser	104
Vogelbruch	68
Vogt	40
von Berg	234
Voss	172
Wabnitz	189
Wagenknecht	74, 86, 92, 98
Weber	14, 18, 51
Weniger	250
Werner	277, 283
Wildenburg	80
Winkler	68
Wischnewski	74, 86
Wolters	104
Zwoll	68

Faszination Wiss

D/A/CH 2004
Internationale Dreiländertagung
Engpassmanagement und
Intra-Day-Energieaustausch
15.-16.06.2004 in München
Hrsg.: Energietechnische Gesellschaft
im VDE (ETG) Fachbereich V3 Energiewirtschaft
2004, 114 S., DIN A4, kart.
ISBN 3-8007-2835-4
110,– € / 174,– sFr*

mit CD-ROM

EMV 2004
Elektromagnetische Verträglichkeit /
12. Internationale Fachmesse
und Kongress für Elektromagnetische
Verträglichkeit
10.-12. Februar 2004, Düsseldorf
Hrsg.: Feser, K.
2004, 812 S., 17 cm x 24 cm, geb.
ISBN 3-8007-2810-9
110,– € / 174,– sFr*

mit CD-ROM

VDE VERLAG GMBH · Berlin · Offenbach
Bismarckstraße 33 · 10625 Berlin
Telefon: (030) 34 80 01-0 · Fax: (030) 341 70 93
E-Mail: vertrieb@vde-verlag.de · www.vde-verlag.de